Geological Evolution of the Colorado Plateau
of Eastern Utah and Western Colorado

Including the San Juan River, Natural Bridges, Canyonlands, Arches, and the Book Cliffs

Robert Fillmore

THE UNIVERSITY OF UTAH PRESS
Salt Lake City

The Defiance House Man colophon is a registered trademark
of the University of Utah Press. It is based upon a four-foot-tall,
Ancient Puebloan pictograph (late PIII) near Glen Canyon, Utah.

Library of Congress Cataloging-in-Publication Data
Fillmore, Robert, 1957-
Geological evolution of the Colorado Plateau of eastern Utah and western Colorado, including the San Juan River, Natural Bridges, Canyonlands, Arches, and the Book Cliffs / Robert Fillmore.
p. cm.
Includes bibliographical references and index.
ISBN 978-1-60781-004-9 (pbk. : alk. paper)
1. Geology, Structural—Colorado Plateau—History. 2. Geology, Structural—Utah—History. 3. Geology, Structural—Colorado—History. 4. Geology, Stratigraphic. I. Title.
QE169.F545 2010
557.88'1—dc22

2010024930

Contents

Figures

Color Plates

Color figures follow page 336

Preface

I first glimpsed the Colorado Plateau in the rainy dark of a warm spring night in the form of black, towering, square-cut silhouettes. The next morning I woke, damp but warm, to what even today is one of the most fantastic landscapes I have ever witnessed. Puffy clouds veiled the middle ground of the Six-Shooter peaks and the brilliant orange cliffs of Wingate Sandstone across the valley. The summits glowed in the dawn light, against what was surely the bluest sky that had ever existed. Lime-green leaves sprouted from the cottonwoods and the air smelled of sage. There was no trace of human existence, save the dirt track I had driven the night before. I looked down at the cow pie that I had unwittingly pulled beneath my head for a pillow and realized that this was the place that I had been looking for. As a kid fresh from the Oklahoma plains, I had never even imagined anything like this. Shaking my head to dislodge the clingy bits of my makeshift pillow, I shed my sleeping bag and began to move, unaware that I was being drawn into a lifelong obsession.

The year was 1978 and I had just finished a long winter of skiing and washing dishes. I had a month to explore, give or take a few weeks, and I used this time to full advantage. Since then, seeking the vistas and exploring the nooks and crannies of the Plateau country has been a wonderful obsession, one that led me directly to another: geology. Hundreds of questions swirled through my head. How did this place get like this? Why here? And on and on. Luckily the questions have never stopped.

After another five years of living the dirtbag life (I say that with great affection) of a climber, skier, and dishwasher, I enrolled in geology courses at Western State College in Gunnison, Colorado, just down the valley from my home in Crested Butte. I loved those courses, and the teachers who shared my passion for this strange and wild place were a great inspiration. Finding that I couldn't get enough, I went on to graduate school, first to Northern Arizona University in Flagstaff for a master's degree and finally to the University of Kansas for a Ph.D.

Kansas was far from the mountains and canyons that inspired me, but I visited often while working toward my ultimate goal of living on or near the Colorado Plateau—without washing dishes for a living. After various temporary teaching positions, much badgering, perseverance, and extraordinary luck, I landed back at Western State College in Gunnison, where I reside today. It has been (and continues to be) an amazing journey.

But enough about me: what about this book you hold in your hands? It has been seven years in the making and a true labor of love. Why so long? My previous book took about three years to write, but that was when I was younger and more energetic, with no permanent full-time job (none of those joyous faculty meetings to attend) and no kids. I now have two hilarious boys, which is like having two little entertainment centers that can't be turned off.

In any event, here is my book, the goal of which is to convey, as clearly as possible, the most recent and plausible interpretations of the geology of this fantastic corner of the Colorado Plateau. It is exactly the kind of book that I was looking for when I first came here. Although there are many books on the geology of the Colorado Plateau, I find most of them to be too general, giving the region only a broad-brush treatment. Many are old and outdated, and some are simply not accurate. This book, however, contains the details gleaned from current geologic literature, but is accessible and of interest to both curious travelers and visiting geologists, professional and student alike. This is not an easy mixture of audiences for which to write, so I would suggest that if the curious layperson finds some part dense and difficult to understand, skip it; there will be no exams. For the geologist who craves more detail, the geological references are listed by chapter.

The interpretations of geologic history presented here are the culmination of the past 150 years of geological studies—the lifetime work of hundreds of geologists. I have read their published papers and distilled, simplified, and interpreted their results in an effort to make them comprehensible to anybody who is interested—anyone, who like me, is intrigued by a bizarre landscape that is unlike any other. It is a place that would easily take several lifetimes to know well, and that is part of the obsession. You can never learn or explore it all; there will always be mysteries.

Chapter 1

An Introduction to the Science of Geology

Before diving directly into the geologic history of southeast Utah and far western Colorado, a basic background in the science of geology is helpful. For the uninitiated, this means a brief lesson on the basic aspects of geology, but one related to this part of the Colorado Plateau—essentially an abbreviated version of the Geology 101 course that I teach every semester, but without the tests. This includes (in this order) the basic principles on which the science is based, rocks and their origins, folds and faults, and geologic time. The more complex aspects of these and other topics are addressed in the text as they become relevant. That way there is a ready (and pertinent) example, and the reason for defining that term or explaining that feature or process is immediately clear.

Principles of Geology

Uniformitarianism

One of the most universally applied principles in geology is **uniformitarianism**. This principle states that *the physical, chemical, and biological processes that occur today are the same as those that have occurred in the geologic past and thus can be used to interpret features of that past*. In its essence, "the present is the key to the past." We can observe modern processes such as volcanic eruptions, the flow of rivers and transport of sediment, and the growth of coral reefs, for example. We can then study the products of these processes and compare these products to those in the rock record in an effort to interpret ancient processes. In short, the products of these modern processes can be used to infer that similar or identical ancient rocks and their various features were formed by the same processes.

In the formative years of geology, the principle of uniformitarianism was used to determine the origin of the black volcanic rock basalt, which is common in both the rock record and recent volcanic eruptions around the world, notably in the

Hawaiian Islands. In the eighteenth century basalt was interpreted by some geologists as something that precipitated from sea water, a concept known as neptunism (named for Neptune, Roman god of the sea), which was loosely derived from biblical interpretations for the origins of rocks. Eventually the students of Abraham Werner, who was a strong proponent of neptunism, actually observed basalt flowing as erupting lava and collected samples of it to compare to earlier collected samples of basalt that had supposedly precipitated from sea water. They were immediately convinced that all basalt originates as fluid lava and solidifies as it cools. Werner was never convinced, but the science moved forward without him.

In another example, our interpretations of ancient river deposits come directly from observations of modern rivers and their deposits, as well as detailed experiments in giant flumes. The sedimentary structures in sand and gravel deposits of modern rivers are identical to those in sandstone and conglomerate in many ancient deposits, regardless of their age (Fig. 1.1). Moreover, these structures can be reproduced in flumes where variables can be controlled, relating them to specific velocities of flow and allowing scientists to observe their formation through the clear plexiglass sides. This notion can be simplified to a more basic assumption: that water moved and deposited sediment in ancient rivers exactly the same way that it does today. Gravity pulls water downhill and causes sediment (mud, sand, and gravel) to be deposited whenever the flow of that water drops below a certain velocity. This works the same way today as it did billions of years ago.

In reality, uniformitarianism is only the preliminary test in interpreting the geologic past. Upon the initial examination of a group of rocks, this principle is used to speculate on their origins and to form hypotheses on their mode of formation. From there, the various possibilities or hypotheses are tested further as more details become evident. All the characteristics of the rocks—including the chemical and mineral content, textures, structures within the rock, and the overall geometry of the rock body—are used to test the various hypotheses. Some possibilities may be discarded as more details come to light. Eventually a specific mode of formation is *interpreted* based on all the available evidence. This interpretation is then placed into a regional context to determine its fit with surrounding, associated rocks. These interpretations are under constant scrutiny, and with more time and study by other, later geologists, our understanding of the geologic setting becomes further refined. In some cases, an earlier interpretation is changed drastically. These types of changes and refinements are common, as evidenced by examples presented throughout this book.

The key word here is *interpret*—a word that is vital to the science of geology. Many of the rocks and features that we study formed millions of years ago. We can never *prove* their origins. Take, for instance, the statement that the lower part of the Jurassic Morrison Formation is *interpreted* to have been deposited by east-flowing rivers that had a source in eastern Nevada. There are no time machines to

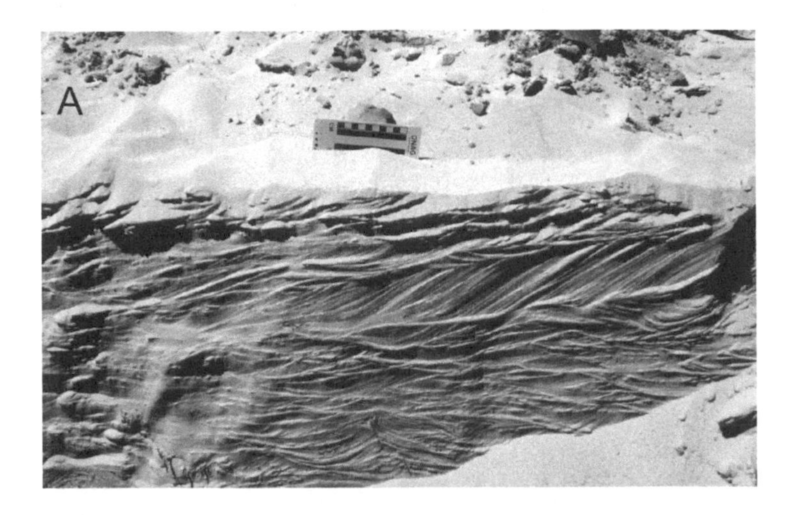

FIG. 1.1. Top, cross-stratification in modern unconsolidated sand in a trench dug in a Colorado River sandbar in the Grand Canyon; bottom, similar cross-stratification in Upper Cretaceous sandstone (~80 million years old) in southwest Utah.

go back 150 million years ago to *prove* the existence of these rivers or mountains. Instead, we state that "all the available evidence suggests that it was deposited by east-flowing rivers."

The mountains that fed the Morrison rivers are gone, long ago reduced by erosion to a featureless plain. They have been buried by the detritus of later uplifts and then uplifted again to form other mountains. The rivers are also gone, but the record of their passage remains. It is from these preserved sediments, long cemented into rock, that we obtain pieces of this ancient puzzle. The mountains long ago were reduced to rubble and planed flat, but it is that rubble, deposited as sand and gravel in the river channels, that provides a glimpse at the makeup of these disintegrated mountains: these sediments *were* the mountains. The east-directed flow of the rivers is revealed by structures in the sedimentary rocks such as cross-stratification, and by textural features such as regional changes in grain size. The sediment particles in the Morrison gradually become finer (smaller) with distance from the mountainous source. This occurs as the particles are reduced in size by abrasion during transport in the river, and as the rivers become less energetic and unable to carry large particles. All of these features can be used to trace rivers in reverse to their highland sources. The sediments become coarser (larger) toward their highland source as the site of the ancient mountain front becomes closer. Eventually, in this direction the Morrison disappears completely. From this the location of the mountains can be inferred. Here no deposition was occurring, only the active removal of rock by erosion, which effectively funneled the detritus to the Morrison rivers. Thus the location of the source mountains is indicated by a void in the rock record—in this case the absence of Jurassic rocks.

Law of Faunal Succession and the Theory of Evolution

Another principle that is essential to Earth history is the **law of faunal succession**. This well-established principle states that *fossil organisms (plants and animals) succeed one another in a definite and recognizable order, with each geologic formation having a different assemblage of fossils from that in the formations above it and below it.* Throughout geologic time, new species have constantly appeared through evolution while others have disappeared due to extinction; because specific fossil organisms existed during restricted time spans, they can be used to determine the age of strata in which they are found. Moreover, a formation of fossil-bearing rocks in one locality may be considered the same age as strata elsewhere that contain those same fossils, whether rocks in the Grand Canyon are being compared to a succession in Utah or Kazakhstan. This demonstration of age equivalence is an important task in the geological sciences and is known as **correlation**. Although there are many methods for correlating rock units, fossils are the simplest and most convenient, and remain the standard method for dating and correlating fossiliferous strata.

The law of faunal succession is validated by the **theory of evolution**, *which states that life on Earth has developed gradually through geologic time, from a few simple organisms to a variety of more complex organisms.* Although this theory was first stated by Charles Darwin in 1859 based on his observations of modern plants and animals, it is the fossil record that confirms its validity. Life has changed dramatically over the past ~600 million years by two different processes. Many organisms changed markedly, evolving into recognizably different organisms. These changes occurred at varying rates; some evolved rapidly, giving rise to new organisms over periods of less than a million years, while changes in others was slow, over perhaps several million years. At the same time, some organisms became extinct, disappearing forever, never to be seen in subsequent strata. These abrupt extinction events are important because they define sharp time lines in the geologic record. The theory of evolution has held true through many vigorous debates over the past 150 years, and no evidence for its invalidation has been discovered despite the thousands of scientists that have constantly tested it. In fact, each discovery of a new fossil has actually strengthened the theory of evolution, refining

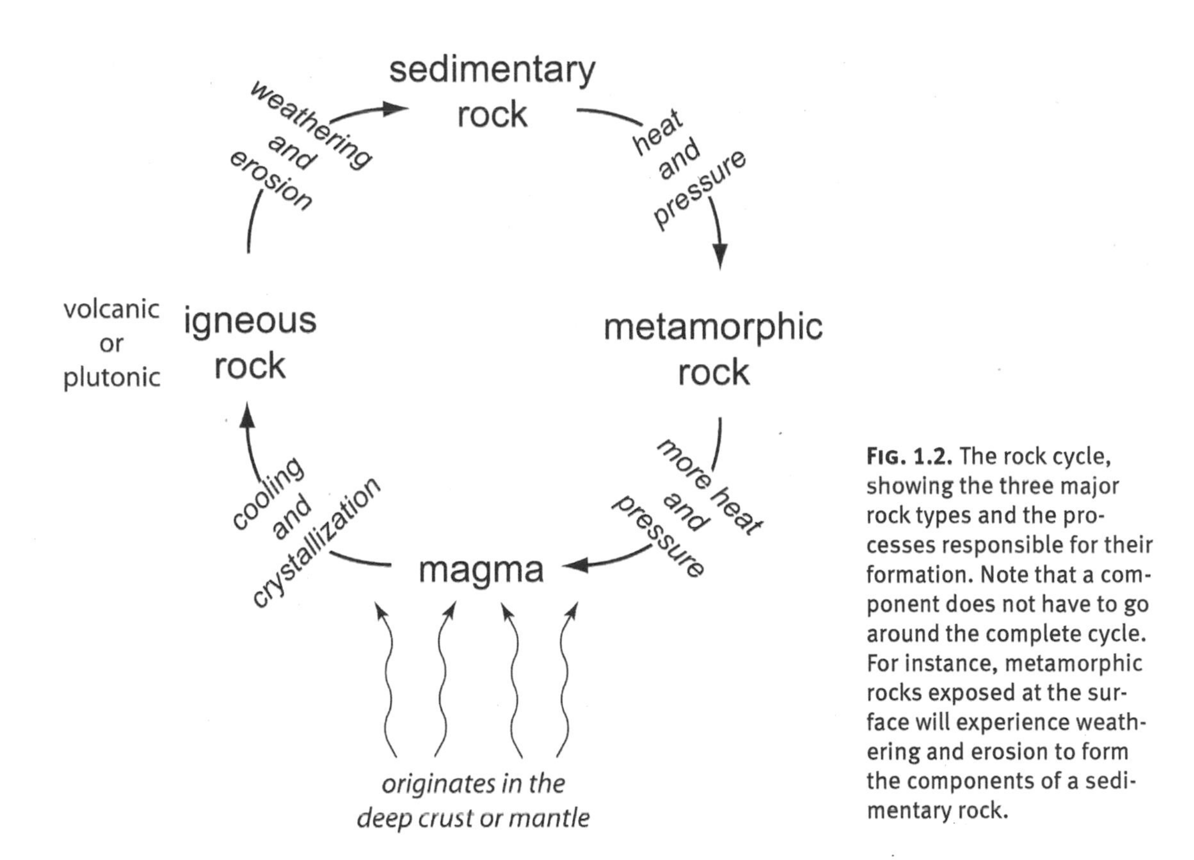

FIG. 1.2. The rock cycle, showing the three major rock types and the processes responsible for their formation. Note that a component does not have to go around the complete cycle. For instance, metamorphic rocks exposed at the surface will experience weathering and erosion to form the components of a sedimentary rock.

the fossil record and "evolving" an increasingly precise tool for determining the age of rocks and correlating different successions.

The Rock Cycle and the Origins of Rocks

The rock cycle, although not a formal principle, provides a simple introduction to the formation of the three basic rock types through a steplike series of processes. The progressive transformation of one rock type to another is shown by these processes (Fig. 1.2), which are occurring constantly at the surface and within the Earth.

The cycle begins (and ends) with **magma**, molten rock that originates in the mantle or lower crust (Fig. 1.2). Magma is less dense than surrounding rock, so it rises slowly toward the Earth's surface: hot, molten rock pushing through cooler, solid rock. Physically this is the same as a hot air balloon rising buoyantly through the cooler atmosphere. An important general aspect of geology is that hot material expands and so is less dense than its cooler counterpart. As magma rises through the crust, it has two possible fates. First, it may stall in the upper crust, where it cools slowly to crystallize in the subsurface and form a **plutonic** (or intrusive) rock, such as granite. Alternatively, the magma may rise into the shallow crust and find a conduit to the surface. In this scenario magma leaks to the surface and erupts as lava, where it cools very rapidly to form a **volcanic** (or extrusive) rock. Plutonic and volcanic rocks collectively are referred to as **igneous** rocks because they originate from magma.

The formation of rock and its subsequent exposure at the surface guides the next process of the rock cycle, weathering and erosion, which leads to the formation of sedimentary rocks. In general, rocks form in the subsurface or under high temperatures, conditions under which they are most stable. Conversely, when exposed at the surface, they become unstable and are broken down through the weathering process, most prominently by the chemical and physical actions of water. It should be noted that all rocks at the surface, regardless of origin, are eventually reduced by weathering. This process results in two distinct products. The first is solid particles in the form of clay, sand, and gravel that eventually are deposited, buried, and cemented to form **clastic sedimentary rocks** such as shale, sandstone, and conglomerate. The second product consists of elements (e.g., Na, Ca, and Si) derived from the chemical weathering of rocks and put into solution, eventually to concentrate in lake and sea water. These elements may bond with others and precipitate from these waters to form **chemical sedimentary rocks** (e.g., halite [NaCl] and gypsum [$CaSO_4 \cdot 2H_2O$]). Another path for these elements in solution is absorption by marine organisms such as corals and molluscs, in which $CaCO_3$ is biochemically precipitated as hard parts. These hard parts accumulate on the sea floor to form limestone. Ultimately the weathered products of all rocks are recycled to generate the various types of sedimentary rocks.

The process that follows the formation of sedimentary rocks is deep burial and exposure to higher temperatures and pressures, converting the original rock to **metamorphic rock**. This occurs while the original rock is in a solid state, and it remains a solid on its path to becoming a different rock. Changes during metamorphism include the development of different minerals and textures. The greater the temperature and pressure, the more dramatic are the changes in the original rock type. Any rock type, including igneous and metamorphic, can be metamorphosed, but because they form at high temperatures and pressures, it takes greater increases in these conditions to affect them. This is especially true when comparing these high-temperature rock types to sedimentary rocks, which form in near-surface conditions. For instance, a temperature of 600°C will metamorphose sandstone but will simply warm granite.

Metamorphic rocks are not exposed in the part of the Colorado Plateau discussed in this volume and are rare elsewhere on the Plateau. Most metamorphic rocks in the region are found along the margins of the Plateau, where they make up the basement, that very old complex of metamorphic and plutonic rocks on which the sedimentary layers are deposited. Across most of the Colorado Plateau this basement complex is covered by thousands of feet of sedimentary rocks that characterize this geologic province. The nearest exposures of metamorphic rocks are to the east, in Westwater Canyon, where they have been exhumed by the Colorado River. Metamorphic rocks are also exposed in the canyons that cut into the Uncompahgre Plateau in far western Colorado. The only other locality for metamorphic rocks on the Colorado Plateau is in the depths of the Grand Canyon, where the Colorado River again cuts through the thick sedimentary veneer. Although the metamorphic rocks on and around the Colorado Plateau are very old, with ages measured in billions rather than millions of years, in many places around the world metamorphic rocks may be much younger. The Colorado Plateau is unique in its relative stability for the past 500 million years. While surrounding regions, particularly to the west, have experienced mountain-building and large-scale tectonic activity, the Plateau region has experienced only mild folding and faulting. There has been no event on the Colorado Plateau in the past 500 million years that would create the high temperature and pressure conditions required for large-scale metamorphism.

The final process in the circular path of the rock cycle is a further increase in temperature, which returns a rock to its ultimate origin. A rise in temperature into the range of ~650°C to 1300°C, above that which produces metamorphism, melts the rock to produce magma. The wide range of temperatures at which melting occurs reflects the different melting temperatures of the different rock types. Upon cooling of the magma, an igneous rock is formed, and the cycle begins again.

Rock Types and Minerals That Make Them

It is important to distinguish between rocks and minerals. Rocks are composed of minerals. Minerals are made up of elements. Take, for example, the common plutonic rock granite. Granite typically contains the minerals quartz, potassium feldspar, and biotite, with minor amounts of other minerals. Each of these minerals has a specific chemical composition. The common mineral quartz has the simple chemical formula of SiO_2—that is, one atom of silicon for every two atoms of oxygen. The mineral composition of a rock is controlled by the chemical content of its parent material, whether that original material was magma or some preexisting rock. This will become evident in this section, which examines in detail the classification and origins of the various rock types.

All rocks are classified and named based on two fundamental characteristics: texture and mineral composition. Both features are readily identifiable, and in addition to their use in classification, they are vital in deciphering a rock's history and origins. *Texture* refers to the size, shape, and arrangement of the grains or crystals that make up a rock. The texture of a rock allows glimpses into the past at such things as the cooling history of an igneous rock, the energy level of flowing water that eventually produced a sedimentary rock, or the pressure and temperature conditions under which a metamorphic rock formed. *Mineral composition* reflects the chemical makeup of a rock and ultimately its origins. It can be used to interpret such things as the source of the magma that cooled to form an igneous rock, or the original composition of a weathered parent rock that was broken down to produce the grains in a sandstone or conglomerate.

Minerals

Rocks are formed of minerals by various processes. To fully understand rocks, their building block minerals must also be understood. A **mineral**, by definition, is *a naturally occurring, inorganic solid substance with a specific chemical composition and crystal structure*. In addition, most minerals exhibit distinctive physical characteristics such as hardness, specific gravity, and color, among other things, that make them easy to identify. While thousands of different minerals have been identified so far, only a few are common at or near Earth's surface. Some of these common rock-forming minerals, as they are known, are used as examples below, where minerals are examined in further detail. *Naturally occurring* and *solid* are easily definable characteristics; man-made substances are not true minerals. A mineral must be solid, so water and magma are not minerals, although they become minerals upon crystallizing into a solid.

A *specific chemical composition* is a key factor in defining each mineral. The simple example of quartz (SiO_2) is noted above, but all minerals are not so simple. The common, black sheetlike mineral biotite has the unwieldy chemical formula $K(Mg,Fe)_3(AlSi_3O_{10})(OH)_2$. A minor variation in chemistry is permissible in this case and is shown in the "$(Mg,Fe)_3$" part of the lengthy chemical formula. This indicates that three atoms in each unit molecule of biotite can be filled by Mg and/or Fe in variable amounts. The exact ratio of these two elements is not specified, only that there will be three of them in total. In contrast, the closely related mineral muscovite has the same crystal structure, but its chemical formula is $KAl_2(AlSi_3)O_{10}(OH)_2$. Muscovite is a silver-white, sheetlike mineral that, except for color, resembles biotite. The Mg and Fe that gives biotite its black color is replaced by K and Al in muscovite, which provide the silver-white color. No expensive chemical analyses are needed to recognize the difference between these two related minerals: it is clearly displayed in the color. However, a word of caution is in order. Although color is significant in this case and in many other minerals, color is unimportant in some minerals as it can vary widely with small amounts of impurities.

The *crystal structure* in a mineral is outwardly reflected by the shape and geometry of the crystal faces and the angles between those crystal faces. Like chemical composition, every mineral type has a unique crystal structure. This ultimately is controlled by the angles in the chemical bonds between the atoms that make up the mineral. These angles, formed on a molecular level, are reflected in the crystal structure. Even if a mineral occurs as a shapeless grain with no obvious crystal faces, as most do, its unseen internal structure is identical to a geometrically perfect crystal. If, for example, quartz occurs as a rounded sand grain or as a perfect 5-cm-long crystal, the silicon and oxygen atoms that bond to form quartz share an identical geometry.

Igneous Rocks

Igneous rocks, as stated earlier, solidify as magma cools, either slowly in the subsurface to form a plutonic rock, or rapidly at the surface to form a volcanic rock. The classification of igneous rocks hinges on two fundamental features: texture and mineral composition. Texture in igneous rock refers to the size and relative amount of crystals, which is governed by the rate of cooling. This, in turn, is determined by *where* they cooled. Texture tells whether a rock is plutonic or volcanic in origin. Because plutonic rocks crystallize slowly beneath the surface, they are completely crystalline, with crystals large enough to see with the naked eye. As magma cools gradually in the warm subsurface, microscopic crystals form from the liquid. These act as a nucleus for further growth. The crystals grow larger with time, adding onto the tiny crystal molecule by molecule. The slower the rate at which magma cools, the larger the crystals are able to grow. Over this prolonged period, all the liquid magma is able to solidify into a completely crystalline plutonic rock.

Alternatively, volcanic rocks form when magma is pushed onto Earth's surface and cools very rapidly. In some instances cooling is so rapid that no minerals are able to crystallize from the **lava**, as magma is called upon reaching the surface. In this case, the various atoms that comprise the melt are not able to bond and organize to form minerals, and the melt solidifies into a noncrystalline glass. Small crystals may be present in volcanic rocks, but they typically cannot be seen without a magnifying glass. In other cases crystals may form in the magma as it rises slowly through the crust and grow relatively large. As the lava erupts at the surface, however, the remaining liquid solidifies rapidly to form glass or microscopic crystals. These volcanic rocks contain a few larger crystals set in a matrix of glass or tiny crystals.

Mineral composition is the other facet of igneous classification and reflects the bulk chemistry and genesis of the magma. In general, the three categories include silicic, intermediate, and mafic (Fig. 1.3). All contain silicon (Si) in variable amounts, but this component decreases appreciably between the silicic (~70 percent Si) and mafic (~50 percent Si) end members. Whether of plutonic or volcanic origin, composition may be reflected in the color. The decrease in Si in mafic rocks is accommodated by an increase in iron (Fe) and magnesium (Mg), which bond with the various other elements to form black minerals and ultimately a black rock. At the other end, silicic rocks contain more Si and relatively little Fe and Mg, so they tend to be light-colored, although the exact color varies. Intermediate rocks fall compositionally between the two end members and typically are shades of gray. Thus the color of igneous rocks as controlled by chemical content can be used as an approximate guide for classification.

Basalt is a dense, black mafic volcanic rock. Its plutonic equivalent is gabbro, also a dense, black rock (Fig. 1.3). These similarities stem from a composition rich in Fe and Mg. The differences between these two rocks lie in grain size and origin. Basalt is dominated by microscopic crystals of the feldspar mineral Ca-plagioclase in a matrix of black glass. It may also contain sparse larger crystals (millimeter-scale) of the minerals olivine and hornblende, which formed as the parent magma rose to the surface. Gabbro contains the same mineral assemblage but is more coarsely crystalline and contains no glass. A single batch of magma rising through the crust potentially could split, with part crystallizing into gabbro deep in the subsurface while the other fraction continues upward to solidify as a basaltic lava flow at the surface. Chemically these two rocks would be identical, but texturally they would be very different.

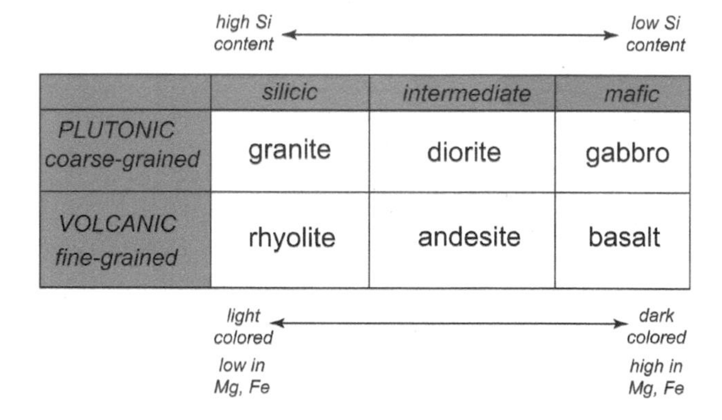

Fig. 1.3. Classification of igneous rocks based on chemical composition and location of crystallization (subsurface vs. surface).

Sedimentary Rocks

Sedimentary rocks of all types symbolize the diverse landscape of the Colorado Plateau. It is the basic characteristics of these rocks—their seemingly infinite horizontal lines (stratification) that stretch to a vanishing point, and a spectrum of colors ranging from crimson red to more somber hues of brown and gray—that make this geologic province so unique. Like the igneous rocks, sedimentary rocks are divided into two broad categories based on composition and origins. Both of these types, the clastic and the chemical/biochemical rocks, are products of weathering and the disintegration of preexisting rocks.

Clastic sedimentary rocks

Clastic rocks are formed of pebbles, sand, silt, and clay—the solid, residual fragments of some earlier formed rock. The sediments travel under the influence of gravity, water, and wind to eventually be deposited and buried. After burial, chemical-rich groundwater infiltrates the loose sediments, slowly precipitating mineral material into the pore spaces between the grains, "cementing" them together into a rock. The cement, as it is known, most commonly is $CaCO_3$ or SiO_2. This process of turning loose sediment into a sedimentary rock is called *lithification*.

Texture in clastic sedimentary rocks is the basis for their classification and primarily describes grain size, which, in turn, reflects the energy of the depositing medium. It is a simple concept: larger particles require greater energy to move. Shale, siltstone, sandstone, and conglomerate are the basic categories of clastic classification (Fig. 1.4). Deposition of the submicroscopic particles that comprise shale depends on a complete absence of current energy in the water through which it is moving. Any current will keep the tiny clay grains suspended in the water column. It is only after a period of quiet that clay settles from this suspended state to the bottom of the sea, lake, or river. Shale typically represents deep marine or lake deposition, or the low-energy parts of a river system, such as an eddy in the channel or an adjacent floodplain. Conglomerate, in contrast, represents a high level of energy to move the clasts, as the gravel particles are called. The size of the clasts in conglomerate is variable and ranges from pebbles (4 to 64 mm), to cobbles (64 to 256 mm), to boulders (more than 256 mm). Even within a conglomerate deposit, boulder conglomerate portrays a higher energy setting than a pebble conglomerate. These relations will be clearly illustrated throughout this book as specific cases arise in discussing the geologic history of the Colorado Plateau.

sedimentary rock name	dominant clast size	size in mm	comments
conglomerate or breccia	boulder		*breccia* is simply conglomerate with angular clasts
		256	
	cobble		
		64	
	pebble		
		4.0	
	granule		
		2.0	
sandstone	coarse		sand grains are further subdivided into fine, medium, and coarse grains
	medium		
	fine		
		0.062	
siltstone	silt		*mudstone* is a mixture of silt, clay, and fine sand
		0.004 mm	
shale	clay		

FIG. 1.4. Classification of clastic sedimentary rocks and the sediments that comprise them.

The composition of clastic rocks is tied directly to the makeup of the source rocks that were originally reduced to sand and pebbles. However, this original source terrane composition will be modified in the resulting sandstones and conglomerates because many of the common rock-forming minerals rapidly break down during the processes of weathering and transport. Because of this, many sandstones are compositionally simple. The most common sand grain types are quartz, chert, and feldspar. There are two reasons for this. First, these are some of the most abundant minerals at the surface, so they are readily available as rocks break into fragments. Second, of all the common minerals, these are the most resistant to chemical and physical weathering, whether they are tumbling in a river, bouncing along the ground in high winds, or simply sitting at the surface breaking down chemically. The other common silicate minerals present in many rocks are not as durable and either disintegrate quickly to unrecognizable clay-sized particles or chemically decay to elemental components such as Fe, Si, and Ca. These typically go into solution, eventually forming chemical sedimentary rocks or precipitating between the sediments as cement.

Quartz makes up ~80 to 100 percent of the sand grains in most sandstone. It is the most abundant mineral in the continental crust of Earth and is much harder than the other common silicate minerals. It is overwhelmingly the most abundant detritus in the breakdown of rocks. Sandstone with ≥ 95 percent quartz is formally classified as *quartz arenite*. In general, however, a quartz-rich sandstone is called a *quartzose sandstone*.

Chert is another common component in sandstone and conglomerate. It forms originally as a secondary mineral in limestone and dolomite. Like quartz, its chemical composition is SiO_2, but its component crystals are microscopic. It is also known as microcrystalline quartz. Chert is physically as hard and resistant as quartz, but its microcrystalline structure makes it slightly less resistant chemically. Detrital chert is generated when sedimentary rocks are uplifted and eroded. The relatively soft carbonate (limestone and dolomite) host rock is destroyed rapidly, leaving chert as the only evidence for its existence. Chert fragments are common in Jurassic and Cretaceous fluvial deposits across the Colorado Plateau. These sediments attest to the uplift and erosion of Paleozoic sedimentary rocks to the west during a prolonged Mesozoic mountain-building event in Nevada and California.

Potassium feldspar, also known as orthoclase ($KAlSi_3O_8$), is an important but sporadic component in sandstone. This pink, blocky mineral comes from the erosion of granite highlands. Compared to quartz and chert, it is less resistant to physical and chemical decay, but it is abundant in granitic rocks. When feldspar is exposed for long periods in low-relief areas, it breaks down chemically to form clay, which then goes on to form the sedimentary rock shale. Because of this, the presence of feldspar in sandstone indicates a high-relief granite source terrane from which the mineral was rapidly eroded, deposited, and buried, thus minimizing its chemical weathering and alteration. Sandstone composed of quartz and an appreciable amount (~25 percent) of feldspar is called *arkose* or an *arkosic sandstone*.

Biochemical sedimentary rocks

Limestone is the dominant biochemical sedimentary rock and forms mostly in marine environments, but it is also common in lake (lacustrine) settings. Limestone is made of the minerals calcite and aragonite, both composed of $CaCO_3$ (calcium carbonate) but with different crystal structures. These minerals form in many ways, but most commonly through water-dwelling organisms that biologically precipitate calcium carbonate in the form of shells and hard parts. Modern organisms include corals, algae, gastropods, and oysters, to name only a few, but ancient limestones contain a multitude of calcareous fossil organisms that have varied considerably over time.

The Ca^{2+} ion that combines with carbonate (CO_3^{2-}) to form $CaCO_3$ comes from the earlier chemical destruction of preexisting rock through the process of weathering. When calcium is liberated from the original rock, it is put into solution as ions that are easily mobilized in water. Calcium may go into groundwater, eventually to precipitate as calcite cement in the formation of sedimentary rocks, or it may be transported on the surface in rivers, eventually to be swept into the sea or lakes, where it concentrates. Upon entering the sea, a variety of organisms absorb calcium and, through metabolic processes, convert it to hard parts of $CaCO_3$. After dying, these shells and skeletal material accumulate on the sea floor, eventually to be buried and cemented into limestone. Like clastic particles, these $CaCO_3$ shells may be broken down by wave activity into smaller fragments to form $CaCO_3$ sand and mud. The components of limestone vary widely, and, in fact, some may precipitate directly from Ca-rich water without the aid of living organisms. All the possible components of limestone are discussed in detail in the following chapter on the Pennsylvanian history of the region, when limestone deposition in shallow marine environments dominated the setting. Regardless of its composition, limestone is a valuable indicator of climatic conditions in the geologic past.

The deposition of large volumes of limestone requires a rigid set of environmental conditions—specifically, clear, clean, shallow water and a warm climate. Today limestone deposition is concentrated in the seas of the lower latitudes, less than 30° north and south of the equator, and well away from areas where large rivers empty into the sea. Calcareous organisms exist around the world, adapting to a variety of settings. However, in order to occur on a large enough scale to produce limestone, these ideal conditions are required. Clear water and an absence of clastic sediment are vital because many of these organisms are filter feeders, receiving nutrients by filtering micro-organisms from circulating water. Clay, silt, and sand clog these filter feeders, killing them. In addition, the abrasive action of hard quartz sand rolling on the sea floor with $CaCO_3$ particles destroys those softer particles. Mud-clouded water inhibits limestone production in other ways. Many calcareous organisms are phototropic, meaning they derive their energy from sunlight. Muddy water inhibits the passage of light. This also explains the requirement of shallow water. Finally, warm water provides optimal conditions for the various organisms to reproduce and thrive. Some of these organisms exist in cooler, higher-latitude seas or in deeper water below the photic zone, but they are sparse and do not occur in great enough numbers to produce limestone.

When looking at marine limestone deposits, the setting that should be envisioned is the Bahaman Islands and other similar tropical oceanic settings: those places with clear, shallow blue water, coral reefs, and white (calcareous) sand beaches. These are the places that limestone is being generated in large quantities today. In fact, most of these islands, as well as the Florida peninsula, are composed completely of limestone accumulated over millions of years and continuing today. Much of our knowledge of limestone deposition comes from the work of the hard-working geologists who have sacrificed so much of their time scuba diving, snorkeling, and walking along these shorelines to study modern processes in these places.

Coal is a black, burnable sedimentary rock composed dominantly of organic carbon. It is the compacted and concentrated product of plant material: leaves, stems, and wood that flourished in swamps millions of years ago. These ancient swamps formed along the low-lying coastal areas where rivers met the sea, in settings analogous to the modern coastal wetlands of the Mississippi River delta. Dank, swamp waters are ideal for coal formation because of the abundance of vegetation and the oxygen-deficient nature of the water. In the presence of oxygen, plant material quickly decomposes to produce CO_2 gas, and no coal is produced. The formation of a one-foot-thick coal bed requires an original 100-foot thickness of plant material. Through a combination of burial, compaction, and time, impurities in the plant material (e.g., oxygen, nitrogen, and hydrogen) are driven out, and carbon becomes increasingly concentrated. The greater the carbon content, the higher the quality of the coal.

Coal deposition was especially important on the Colorado Plateau during Late Cretaceous time (100 to 65 million years before present), when much of the region was submerged beneath a vast seaway. As the lengthy, north-trending shoreline of this sea shifted east and west with oscillations in sea level, the lush coal swamps along its western shores followed. Throughout geologic time seas moved in and out of the region, yet coal deposition was confined to the Cretaceous Period. The main factor was the Late Cretaceous climate, which was particularly warm and humid, spawning many swampy deltas where large east-flowing rivers met the sea. This produced thick and widespread coal deposits in Utah, as well as in Arizona, New Mexico, and Colorado.

Chemical sedimentary rocks

Chemical sedimentary rocks are dominated by the **evaporites**, salts of various chemical compositions that precipitate directly from water as it is reduced by evaporation. These salts form layered crystalline deposits that settle to the lake or sea floor. The type of salt precipitated depends on the chemical content of the water, and the volume of water that has evaporated. Thick evaporite deposits on the Colorado Plateau crystallized from sea water, which produces a predictable succession of gypsum, halite, and potassium salts. The first to precipitate is gypsum ($CaSO_4 \cdot 2H_2O$), which begins to crystallize from sea water after 80 percent of the water has evaporated. It is only after a loss of 90 percent of the water that halite (NaCl) crystallizes. The depletion of Na due to halite precipitation, and a further concentration of the brine, finally triggers the precipitation of potassium salts, including sylvite (KCl) and a host of others. In the complete evaporation of a body of sea water, halite makes up 80 percent of the evaporites that are generated.

The evaporation of large volumes of water that is required to precipitate evaporites occurs in a warm, arid climate. The presence of evaporites, among other diagnostic features, in Pennsylvanian through Jurassic deposits on the Colorado Plateau indicates a long-lived aridity in western North America. Shallow seas left mostly thin gypsum deposits on the Plateau, although the Pennsylvanian Period saw some of the thickest evaporite deposits in the geologic history of North America.

Evaporites—chiefly halite, but also gypsum and potash—play an important role in the evolution of the canyon country of southeast Utah and western Colorado. More than 6000 feet of evaporites were deposited in the Paradox basin during the Pennsylvanian Period. The cyclic fall and rise of sea level periodically isolated the basin, and then reconnected it with the open sea to the west. When sea level dropped, the influx of fresh water was restricted; the resulting drawdown by evaporation in the basin caused gypsum, halite, and sometimes potash to precipitate. The burial by subsequent sediments of this thick evaporite succession caused the less dense, easily mobilized halite to flow viscously upward to escape the pressure of overlying sediments. This eventually resulted in the development of salt anticlines, which have been one of the driving forces in landscape development. Deposition of these evaporites as part of the Paradox Formation and their subsequent role in landscape development are discussed in detail in the following chapter.

Rock Deformation

The deformation of rocks through folding, faulting, and/or fracturing is a reflection of the various stresses imposed on them throughout their existence. The type, geometry, and orientation of the deformation features are used to decipher the types of stresses and their origins. In some regions multiple deformation events have affected the same rocks. By studying these features, geologists can unravel a history of deformation, just as a depositional history can be worked out from the analysis of a succession of sedimentary rocks.

Three types of stress are possible, with each leaving a distinctive mark: *compressional stress*, or squeezing; *tensional stress*, or stretching; and *shearing stress*. Rocks may respond to these stresses by brittle or plastic deformation. *Brittle deformation* occurs when the rocks break to form faults or fractures. *Plastic deformation* takes place when the rock bends under stress, producing folds of various types.

Faults are a common product of stress, and the type of fault accurately portrays the type of stress responsible. Faults are fractures across which the bodies of rock on either side have moved past each other. The fracture that accommodates this movement is the fault plane. Bodies of rock separated by the fault plane my move up and down relative to each other, or they may move laterally. The terms *upthrown* and *downthrown* refer to the relative vertical motion across the fault plane. In reality both blocks do not commonly move along the fault plane. Typically only one side actually moves up or down, hence the term *relative movement*. Faults are classified by the orientation of the fault plane and the relative movement of the blocks on either side.

Normal faults form during tensional stress when the rock is pulled apart. With a normal fault the downthrown or downdropped block slides down the slope of the fault plane, leaving the block on the other side high (Fig. 1.5a). This is the upthrown block. Normal faulting due to tensional stress results in a thinner but wider body of rock.

Reverse faults develop from compressional stress as the rock is squeezed from the sides. Reverse faults form when the upthrown block is pushed up the dipping fault surface (Fig. 1.5b). Squeezing by compression and reverse faulting thickens the rock body to form mountains on a large scale, but causes the region to narrow.

Thrust faults are a type of reverse fault that also form during regional compression. The upthrown block is pushed up the fault plane, but the main difference is the low angle of the thrust fault plane, which slopes at a shallower angle of less than 35° (Fig. 1.5c). Thrust faults are most common in thick sedimentary successions, where older rocks are thrusted up and over younger rocks. However, they may form in any rock type. Thrust-faulted mountains of sedimentary rocks in the Nevadan and Sevier orogenic belts in Nevada and western Utah were important sources for sediment on the Colorado Plateau during the Jurassic and Cretaceous periods.

Large-scale, regional pulses of deformation generate mountain ranges. An *orogeny* is a discrete mountain-building event; each orogeny is unique in possessing a specific style of deformation and taking place over a specific time interval within a discrete geographic region. The term derives from *oros*, which is Greek for "mountain." These events and the resulting mountain ranges have been given names so they may be easily and clearly referred to. For instance, the previously

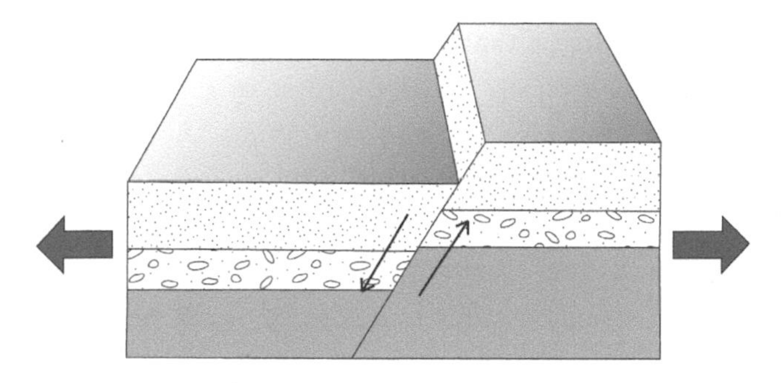

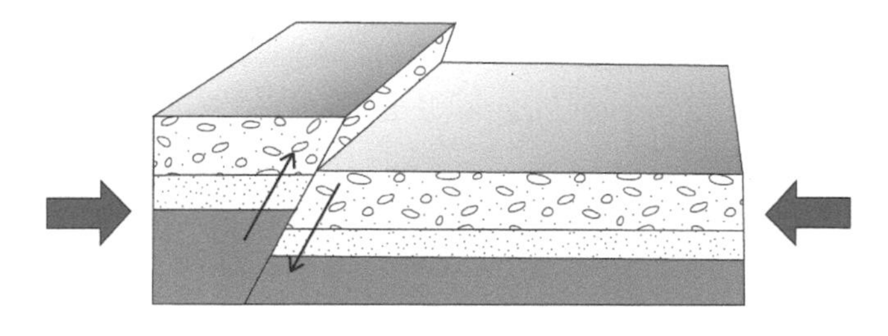

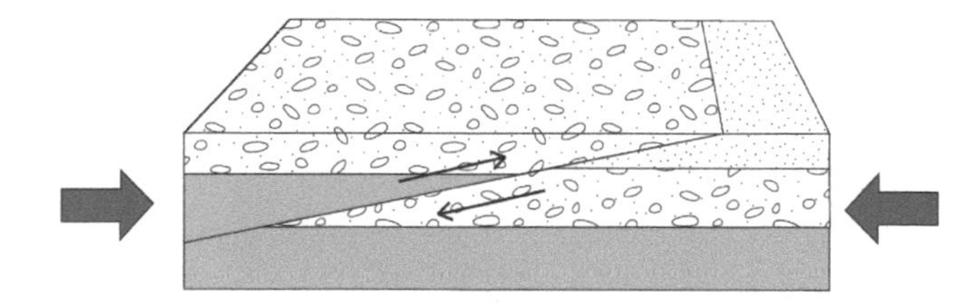

Fig. 1.5. Block diagrams of various fault types: A, a normal fault caused by extensional stress, as shown by the large arrows beside the block. Small arrows within the block diagram show the relative motion of the block on either side of the fault plane. B, a reverse fault caused by compressional stress. C, a thrust fault, which is a low-angle reverse fault also caused by compression. These faults can occur on all scales, from microscopic to the size of mountain ranges.

mentioned Sevier orogenic belt was a linear mountain range that formed during the Cretaceous in western Utah and eastern Nevada from thrust-faulting of sedimentary rocks as they were shoved eastward. The source of this intense compression lay along the west coast of North America during this time.

Folds originate dominantly due to compression, although they can form in any stress regime, and may be associated with faults. The most frequent method of fold formation is by horizontal compression, where lateral pushing causes stratified rocks to buckle into undulations, just as wrinkles develop in a rug on a hardwood floor when it is kicked.

Folds are classified by their geometry and orientation. Uparched folds are **anticlines** (Fig. 1.6a). The crest of an anticline is an imaginary line known as the **fold axis**. The axis splits the fold as evenly as possible into two limbs, both of which slope down and away from the axis. In contrast, downwarped folds are called **synclines** (Fig. 1.6a). The lowest part of the trough-like fold forms the axis that divides the fold into two limbs that slope down toward each other and the axis. Anticlines and synclines typically occur together as directed horizontal compression creates multiple "wrinkles" in the strata. When these folds occur together, they share limbs, and their axes form parallel lines. The orientation of these axes is a significant aspect of folds. Using the wrinkled rug analogy, the fold axes are oriented perpendicular to the direction from which the compressional force originated. Thus folds can provide an important piece of the puzzle that is the geologic history of a region.

Despite the previous general explanation on the genesis of folds, the *salt anticlines* and related synclines that dominate southeast Utah and southwest Colorado are unique in having formed by vertical forces rather than the aforementioned horizontal compression. The salt anticlines that have contributed so much to the regional landscape developed by the horizontal and vertical flow in the subsurface of viscous salt. The evolution of these unique folds is addressed in detail in the following chapter but is briefly recounted here—in part to demonstrate that folds form in a variety of ways, but more important, it is in the interest of understanding some of the Plateau region's most spectacular geologic structures.

The salt anticlines began to form as soon as the great thickness of Pennsylvanian-age salt was loaded from above by a blanket of younger sediments. When pressure is exerted on salt, it flows slowly and viscously away from that pressure. The salt eventually rises, pushing overlying sediments upward as well. In the salt anticline region of Utah and Colorado it is believed that northwest-trending fractures and faults in deep basement rocks shifted up and down, causing salt to flow horizontally away from some areas, only to rise in other areas. This process continued intermittently from the Permian to the present, a period of ~290 million years. Where the salt rose upward to escape the pressure, overlying strata were pushed up into linear northwest-trending anticlines. In the intervening areas between the anticlines, where the salt was withdrawn, overlying strata were downfolded into synclines. Overall the consistent northwest trend of these fold axes resemble those generated by horizontal compression; however, it was the vertical pressure of viscous salt that ultimately generated these folds.

Another type of fold that is common across the Colorado Plateau is the *monocline*, a simple fold that, true to its name, consists of a single sloping limb that connects approximately flat-lying strata on either side of the single limb (Fig. 1.6b). These regional-scale, ramplike folds typically are underlain at deep levels by reverse faults. As displacement occurs on these reverse faults, overlying sedimentary rocks are folded into a monocline. This indicates an origin from horizontal compressional stress. In the area under consideration here, Comb Ridge on the Utah-Arizona border, the San Rafael Swell in east-central Utah, and the fold that

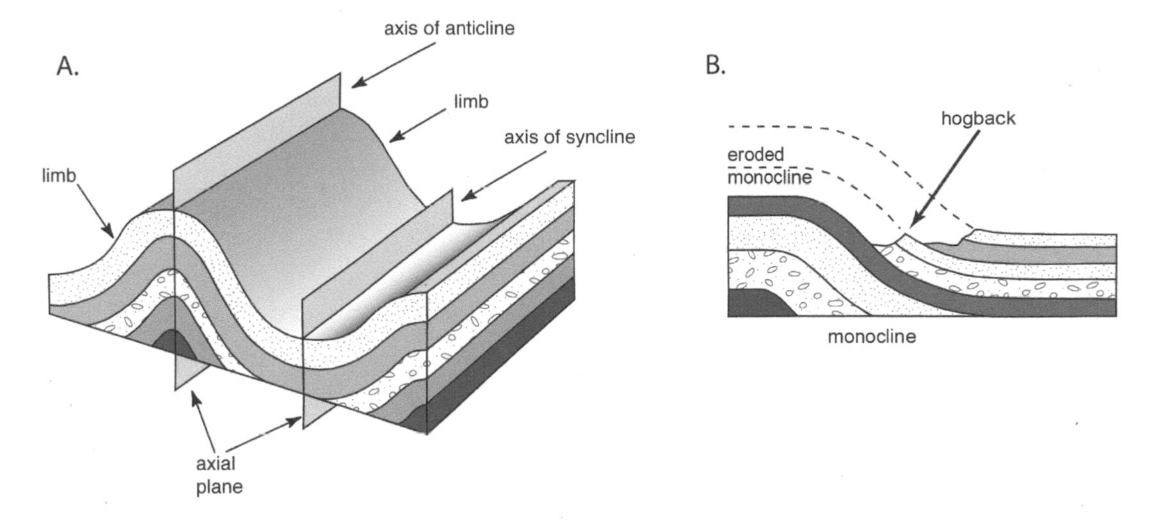

Fig. 1.6. Various types of folds exposed across the Colorado Plateau. A, an anticline/syncline pair. The anticline is the uparched fold on the left with the limbs that dip down away from the axis, or center, of the fold. The syncline is the downwarped fold on the right side in which the limbs on either side of the axis dip down toward the syncline axis. B, a monocline: a single-limbed, ramplike fold in which a lower and an upper area of approximately flat-lying strata are connected by a single limb.

defines Colorado National Monument near Grand Junction provide spectacular, world-class examples of these crustal contortions.

Joints are cracks in rock that form during brittle deformation under extensional stress. They develop as fractures when the rock is pulled apart, but the rock on either side of the fracture does not shift laterally or vertically, as in faults. Joints do, however, create discontinuities and weaknesses in once-solid rock. When rock is exposed at the surface, overland flow of water may be funneled into these minute fractures, gradually widening them. Joints typically occur in systematic sets, forming large areas of parallel joints with a consistent regional trend. Systematic joint sets are common in resistant sandstones across southeast Utah, particularly on the flanks of salt anticlines where folding induced localized extension (Fig. 1.7). In fact, the linear slickrock fins and intervening furrows of Arches National Park, which host the abundant arches, formed from the erosion of joints that opened during deformation of Salt Valley salt anticline. These features are classic and superbly exposed examples of systematic joints. Photos of these sheer, joint-formed ribs appear in most introductory geology textbooks.

Geologic Time

Time is a unique aspect of geology that sets it apart from other sciences. Present data put the age of Earth at 4.57 billion years, an incomprehensible number to most people. Geologic time typically is measured in tens or hundreds of mil-

FIG. 1.7. Aerial photos of parallel fractures on the west limb of the Salt Valley anticline in Arches National Park. Figure 1.7A shows fractures in the Moab Member of the Curtis Formation that have yet to be deeply incised, although they are developed enough that trees have taken root. Figure 1.7B shows the underlying Slick Rock Member of the Entrada Sandstone and the formation of large, parallel fins as the fractures become deeply eroded.

lions of years. These huge numbers and the difficulty of their comprehension are not lost on geologists, who deal with such numbers on a daily basis. But while we understand the depths of geologic time, we do not dwell on it. Commonly, features and events are placed in a simple, more palatable sequential order: an "oldest, old, young, youngest" framework known as **relative time**. Used to decipher geologic history, this relative time scale is based on various observable relations demonstrating that one feature is older (or younger) than another. On the other hand, the ability to place a real number, or **numerical age,** on a feature or event through radiometric dating is a real advance. Take, for instance, a rock that has been dated at 43 million years. All the relative ages associated with this rock now have a numerical age reference. Those older are now known to be more than 43 million years old, and those younger are less than 43 million years in age. Thus these relations begin with relative ages, but with more detailed work and information they evolve toward a much-refined numerical age framework. This is the path of geologic time from a human perspective, as it evolved from a relative time scale with no concept of numerical age to a well-established combination of relative and numerical ages that has become the present geologic time scale. The path is a continuing one that is undergoing constant refinement. These refinements, however, occur in smaller and smaller increments as the geologic time scale becomes more precise.

The Geologic Time Scale and Relative Time

The geologic time scale is the standard reference for the overwhelming span of time used to describe Earth history (Fig. 1.8). The earliest version was constructed in the early 1800s using relative age relations, long before any hint of numerical ages. The geologic time scale has been divided and subdivided into numerous segments of varying lengths, notably into eons, eras, periods, and epochs (Fig. 1.8). For the purposes of this book, and most geologic discussions in general, periods and epochs are used. Subsequent chapters of this book are divided by periods.

The origins of the geologic time scale go back more than 200 years to Europe with the naming of different rock units and establishing their relative ages. It began as an informal exercise to recognize these units and simplify their discussion. The method for establishing relative age was the use of the **Law of Superposition**, which states that in sedimentary and volcanic rocks the unit that lies below a specified unit is older, and that which overlies it is younger. In 1795 the first discrete group of strata was named the Juras, for the Jura Mountains of France and Switzerland. This was followed by an underlying succession called the Trias, due to the three distinct rock types in the group. Overlying the Trias and the Juras were the Cretaceous rocks, named for the white, chalky limestone that comprises the unit. (*Creta* is Latin for "chalk.") The recognition and naming of these units began as a somewhat random task, with the units designated

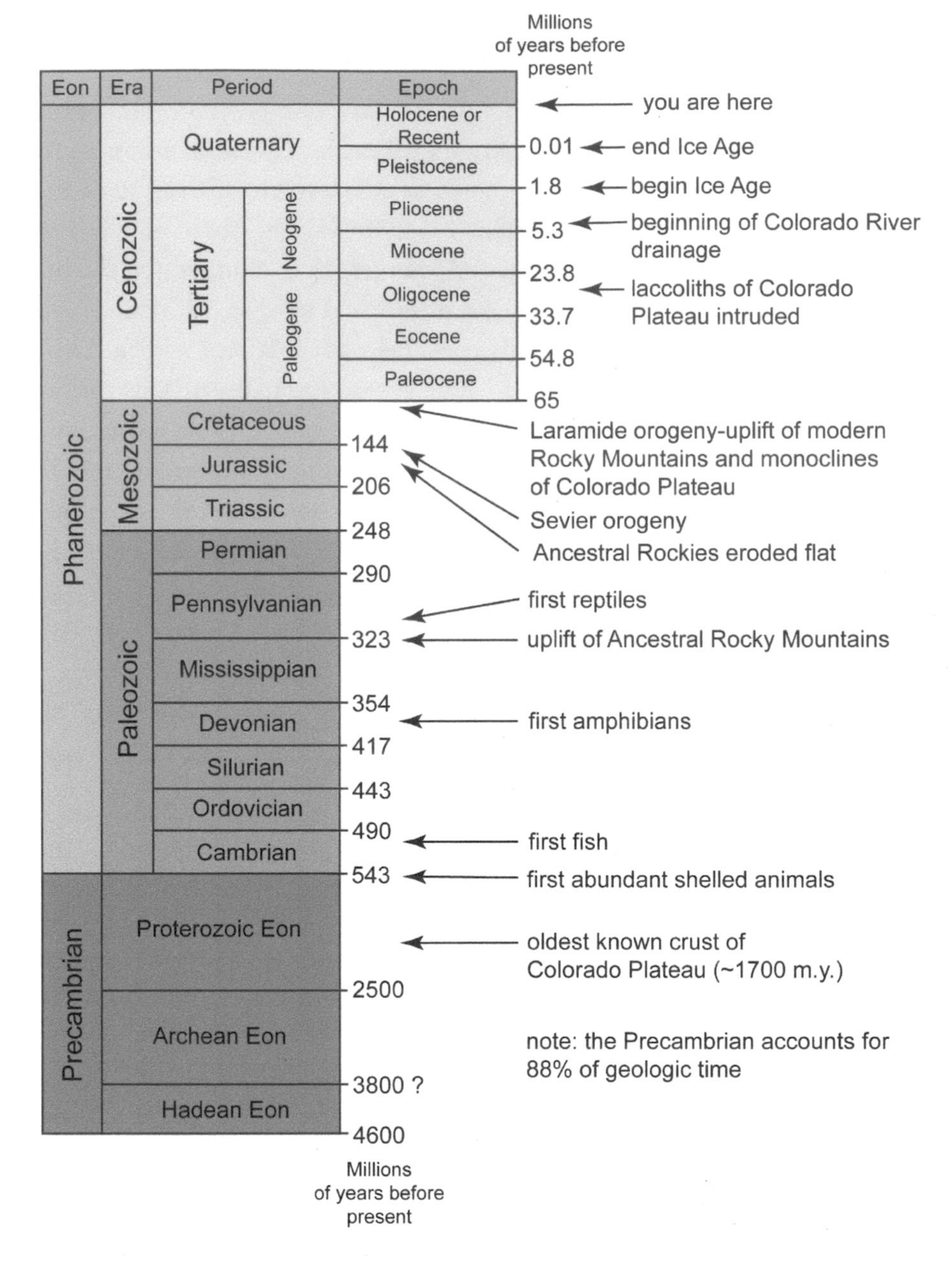

Fig. 1.8. The various breakdowns of geologic time and some of the more important events and milestones in geologic history. Note that the time scale is not linear. Absolute ages are from the Geological Society of America (1999).

as needed. Their relations, however, were accurately documented, and today these age relations remain useful for establishing the time periods of the Mesozoic Era.

The definition of older Paleozoic rock units followed closely as the British geologists Adam Sedgwick and Roderick Murchison set out to systematically establish and name the complete succession of strata across Europe. Cambrian rocks were named in 1835 for Cambria, the ancient Roman name for Wales. Next to be named were the overlying Silurian rocks, after the ancient Welsh tribe the Silures. After naming the rest of the Paleozoic succession, they recognized that the top of Sedgwick's Cambrian succession contained the basal part of Murchison's Silurian rocks. A bitter argument ensued that was not resolved in their lifetimes. In 1879, after their deaths, the problem was solved by establishing the Ordovician succession between the Cambrian and Silurian rocks to include the problematic strata.

It must be emphasized that these names were given to rock units, and relative ages were clearly established in this process. Eventually these early names for rock units became something completely different: namely, designations of the time periods and epochs that are used today (Fig. 1.8). So what originated as tangible, easily observable rock units evolved into abstract units of relative time that are the components of the modern geologic time scale, complete with numerical ages.

During the process of defining the strata of Europe, geologists noted the numerous fossils they contained. It was eventually realized that three long periods of time could be defined based on differences in fossil content. Thus the three eras were defined as the first true time designations of the rock record. The Paleozoic (meaning "ancient life") was defined as encompassing Cambrian to Permian strata. Early in the naming process it was recognized that fossils in these rocks were limited to marine-dwelling invertebrates. The succeeding Mesozoic Era, which includes Triassic through Cretaceous strata, was designated based on the fossils of large vertebrate reptiles, including the dinosaurs. Mesozoic means "middle life." Finally, the Cenozoic Era, which extends to the present, is dominated by mammals and marine invertebrates that are similar to those seen in the modern seas. Cenozoic refers to "modern life."

What of the time before the Paleozoic Era? These older, mostly non-fossiliferous igneous and metamorphic rocks account for more than 88 percent of Earth history (Fig. 1.8). In the initial stage of time scale development, this earliest interval was called "Cryptozoic," meaning "obscure life," or simply "Precambrian," which is what it is commonly called today. Because these ancient igneous and metamorphic rocks are complexly intermixed and so do not conform to the Law of Superposition, further subdivision was not possible. It is only since the advent of radiometric dating and numerical ages that the Precambrian could be divided into the Hadean, Archean, and Proterozoic eons (Fig. 1.8).

Radiometric Dating and Numerical Ages

Radiometric dating of rocks yields the numerical ages that augment the earlier established relative ages and allow for the placement of numbers in millions of years on the modern geologic time scale (Fig. 1.8). The application of radiometric dating was not possible until the discovery of radioactivity. This occurred by accident in 1896 by Henri Becquerel and his assistants Marie and Pierre Curie. Even then, a further understanding of radioactivity was required before it could be applied to dating rocks. This was first attempted in 1907 by Bertram Boltwood, who measured the ratios of lead and uranium in rocks from ten different localities from around the world. His resulting dates ranged from 410 to 2200 million years, surpassing by an order of magnitude what anyone had imagined as the total age of the Earth. Although these ages were later shown to be too old by ~20 percent, they were surprisingly good for a first experiment and opened many scientists' eyes to the possibilities. It has only been in the past 50 years that experimental methods and equipment have improved to the point that radiometric dating and the resulting numerical ages have seen widespread and regular application to geologic problems.

Radiometric dating is limited by several aspects of the method. First, only a few rock types are datable. The rock must contain minerals that incorporate radioactive elements into their crystal structure. Minerals such as potassium feldspar, biotite, muscovite, and zircon are common targets for radiometric dating. Second, only plutonic and volcanic rocks can be reliably dated to determine the age of their origin. When an igneous rock forms from a melt, certain radioactive elements become tightly held as part of the crystal structure in the minerals. Upon crystallization and cooling, these minerals are a closed system, with no elements entering or exiting after the rock is formed. As long as that mineral remains a closed system, the radiometric date obtained from that mineral represents the time at which it crystallized from a melt, as the igneous rock formed.

The corollary to this is that sedimentary rocks cannot easily be dated directly by these methods. Consider, for example, a grain of potassium feldspar in an arkosic sandstone. It is a mineral that is commonly analyzed for radiometric dating, but what would the age of this grain tell? Certainly not the age of the sandstone, because it did not crystallize when the sandstone was deposited. Instead it would yield the age of the original granite source rock that was eroded to supply the sand for this sandstone. Thus it is no help in determining the age of the sandstone. In other situations, the age of the source terrane for clastic sediments may be the objective of radiometric dating, especially if the source terrane is no longer obvious, but it tells nothing of the age of the strata under consideration. Given how expensive and time-consuming radiometric dating is, such investigations must be carefully planned.

If sedimentary rocks cannot be dated directly by radiometric methods, how are numerical ages applied to sedimentary rocks? Essentially, there must be an association with igneous rocks. A common situation is volcanic rock interbedded with sedimentary rocks. The volcanic rock can be dated, providing a numerical age for the surrounding sedimentary rocks. One example that is common in the Mesozoic rocks of the Colorado Plateau is widespread volcanic ash that was erupted high into the atmosphere from volcanoes along the west coast. This ash was distributed eastward by prevailing winds to form thin but extensive volcanic units. Upon deposition that ash layer created a single time line, deposited in a geological instant. Because the ash deposits (or **tuff,** as the rock type is known) typically contain small crystals of datable minerals, a numerical age is readily obtained. This also is how numerical ages are applied to the geologic time scale. Era, period, and epoch boundaries were recognized much earlier by relative age relations and changes in fossil content, and are now recognized from localities around the world. The objective of putting numerical ages on these boundaries is realized by finding and dating volcanic rocks as close as possible to the boundaries. As volcanic and other datable units are discovered in new localities, closer to or on an actual boundary, the numerical age edges closer to the true age of the boundary that was chosen by other criteria. This explains the occasional adjustment of numerical ages on the geologic time scale.

A brief, simplified explanation of the processes involved in radiometric dating is in order. As stated earlier, radiometric dating requires the incorporation of atoms of radioactive elements into the atomic structure of minerals as they crystallize from a melt. Certain radioactive elements fit into a limited number of mineral structures, so only a few minerals are datable. Radioactive elements are inherently unstable and, through time, will turn into different stable elements through the process of radioactive decay. It is during this process that radiation is emitted.

In general, atoms are cored by a dense nucleus composed of protons and neutrons. This nucleus is surrounded by a swarm of electrons buzzing around the nucleus like a cloud of gnats, only better organized. It is the protons and neutrons that dominate the process of radioactive decay. The number of protons in an atom dictates what element it is. Stable, non-radioactive atoms typically contain equal numbers of protons and neutrons. There are also the less common atoms that do not have equal numbers of these components, which are known as **isotopes**. Some of these are the unstable radioactive isotopes—atoms that, through the decay process, eventually become a stable element. It is this path from unstable radioactive atom to a stable element that is the basis for radiometric dating. The unstable isotope becomes stable through a single reaction or a steplike series of them. These reactions, which emit radioactivity, may involve the removal of protons and/or neutrons or their addition to the nucleus. Because the proton controls what element the atom is, the addition or subtraction of proton(s) changes the element.

The common isotope system of K-Ar (potassium-argon) is a useful and relatively simple one to illustrate the radiometric dating process. Potassium is an abundant component in many igneous minerals, so the K-Ar dating method is readily applied to these rocks. In simplest terms, the process begins with the unstable isotope of ^{40}K. The superscripted "40" refers to the isotope's atomic weight, which is the sum of the protons and neutrons in the atom's nucleus.

The element potassium is defined by its 19 protons, so an atom of ^{40}K also has 21 neutrons. ^{40}K decays to the stable isotope ^{40}Ar. In this step the sum of the protons and neutrons remains the same, but a proton is subtracted from the sum as it decreases to 18 protons. How does this happen? First, each proton has a +1 charge, and each electron has a -1 charge; neutrons have no charge and so are neutral. In the conversion from K to Ar a positive-charged proton in the nucleus captures an electron from the surrounding electron cloud to convert the proton into a neutral particle, a neutron. Thus the ^{40}K atom stabilizes by converting a proton to a neutron and becoming ^{40}Ar. The number of protons goes from 19 to 18. The atomic weight remains at 40 because nothing is lost from the nucleus; a proton has simply converted to a neutron.

Dating a mineral requires analysis by a mass spectrometer, which measures the amount of ^{40}K and ^{40}Ar. The ratio of these isotopes is used to calculate the age. The K-Ar system has a **half-life** of 1.31 billion years (Ga), meaning that after 1.31 Ga half of the original ^{40}K atoms in the mineral have decayed to ^{40}Ar. After another 1.31 Ga half of the remaining ^{40}K atoms decay to ^{40}Ar, meaning that an additional quarter of the original ^{40}K that was in the mineral when it crystallized. In general, during a given half-life, half of the original existing unstable atoms decay. Half-life varies greatly depending on the system. Over 2.62 Ga (2 half-lives for K-Ar) the $^{40}K/^{40}Ar$ ratio would be 1:3. Three-quarters of the original ^{40}K atoms would have decayed to ^{40}Ar, and only a quarter of the ^{40}K would remain. Knowing two things—the ratio of ^{40}K to ^{40}Ar, and the half-life of the isotopic system—the age of the mineral can be calculated. In reality the calculations are much more complicated than that, but this is the basic premise. Bertram Boltwood's first radiometric dates obtained in 1907 were too high by 20 percent due to an incomplete understanding of the half-life of the U-Pb isotopic system he was using. Today the half-lives are much refined, the actual analyses are more precise, and many calculations have been developed to reduce errors. Like all aspects of analytical science, however, the entire process will continue to be refined with better technology and new methods.

The most common and precise radiometric dating technique in use today is the $^{40}Ar/^{39}Ar$ system, which is a modified and much refined version of the K-Ar method. Several U-Pb (uranium-lead) isotopic systems are also used for dating, especially igneous intrusive rocks. All these systems have different half-lives and occur in a variety of minerals, substantially expanding the possibilities for radiometric dating. While radiometric dating is a relatively recent development (relative to methods for relative dating developed in the early 1800s), in determining the age of a rock unit or geologic event, it has become a standard tool in the geologist's repertoire, although a fairly expensive one. Still, anywhere that a real number is used to describe the geologic age of something, including the geologic time scale, it comes from a radiometric date.Gendaeribusa vid magnime voluptas

The Pennsylvanian Period

The Rise of the Ancestral Rocky Mountains

The span of geologic time known as the Pennsylvanian Period began 320 Ma (mega-annums, or millions of years) and ended 34 million years later, at ~286 Ma. The Pennsylvanian and the subsequent Permian Period trumpet the end of the Paleozoic Era, that vast segment of time when life on Earth was dominated by marine invertebrates. The shallow seas, which blanketed much of the continental landmass, were teeming with organisms—strange corals, the clamlike brachiopods, crinoids, and many other now-extinct life forms. It was during this time that vertebrate amphibians, only recently evolved from the fishes, began to venture increasing distances from the sea.

The Pennsylvanian landscape onto which these higher life forms crawled was mostly coastal swampland—vast regions covered with shrubs and towering trees. As these jungles of greenery died and were replaced by others, their carbon-rich remains were buried and compressed, eventually becoming important coal resources that are exploited around the world today. Abundant high-grade Pennsylvanian coal attests to humid, tropical conditions in which vegetation thrived. However, Pennsylvanian rocks in western North America point to a far different setting than found elsewhere on Earth, one in which aridity dominated and fresh water was scarce. In fact, blowing sand, a sure sign of an arid climate, dictated the setting of western North America, and especially the Colorado Plateau, for the next 200 million years.

The global tectonic setting also played a vital role in the genesis of the Pennsylvanian landscape. At the end of the previous Mississippian Period, all the continental landmasses began to assemble into the giant supercontinent of Pangea (Greek for "all lands") (Fig. 2.1). This great landmass affected ocean circulation and influenced climatic conditions. Where earlier continental shorelines came

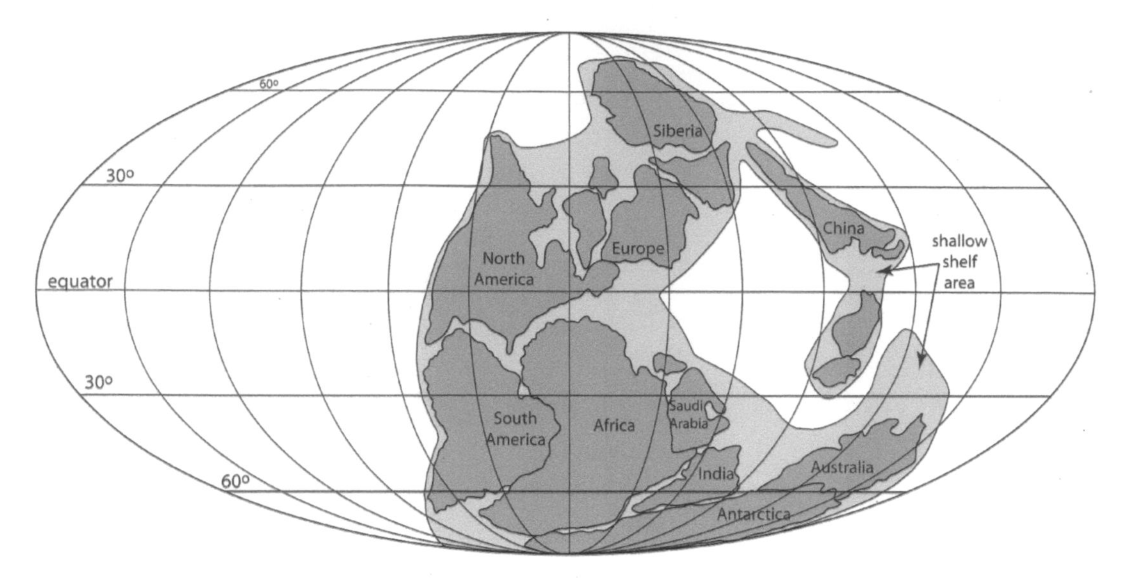

FIG. 2.1. Paleogeographic reconstruction of the supercontinent Pangea during the Pennsylvanian Period (306 Ma). After Scotese and Golonka 1992.

together, squeezing the seas that had separated them, high mountains rose, forced aloft by the intense compression of two lands competing for space with no place to go but up. Thus, these sutures were marked by colossal stone monuments: chains of mountains commemorating vanished oceans.

In southwest North America, compression generated by collision along the continental margins penetrated into the interior regions. Mountain ranges were thrust upward, far from the active borders where such things are normally expected. Thus the Ancestral Rocky Mountains, ancient precursors to our modern ranges, were born from the previously horizontal landscape of what is now New Mexico, Colorado, and Utah. Even as these mountains rose above adjacent basins during Pennsylvanian-Permian time, they were slowly being scraped flat by the forces of erosion. As is the fate of all mountains, their great masses were methodically, grain by grain, transferred into the adjacent lowlands, there to accumulate and be transformed into sedimentary rock. It is this sedimentary record from which we read today, like pages of a history book bound beneath the surface, but now available along the canyons and mesas of the Colorado Plateau and the high mountains of western Colorado.

An important aspect of Pennsylvanian sedimentary rocks worldwide is their cyclic patterns. During this time sea level rose and fell with a stunning regularity that is reflected in strata around the world. A growing body of evidence suggests these cyclic oscillations were caused by the waxing and waning of glaciers in the Southern Hemisphere, where much of the Pangean landmass was centered on the southern polar region (Fig. 2.1). It is generally agreed that climatic oscillations were caused by recurring variations in the planet's orbit and the tilt of its axis. Pennsylvanian cyclothems, as the repetitive rock successions are known, are a current topic of hotly debated research. Undoubtedly more will be known about their genesis in the near future.

Tectonic Setting

During the Pennsylvanian, the Four Corners area was one of tectonic unrest. Uplift of the Ancestral Rockies, particularly the Uncompahgre highlands, had a profound influence on the rock types and structures seen throughout the eastern Colorado Plateau today. The Uncompahgre was the westernmost expression of these ancient mountain ranges (Fig. 2.2). The northwest-trending range occupied much of western Colorado, stretching southeastward from today's Colorado-Utah border across the modern Uncompahgre Plateau and San Juan Mountains into northern New Mexico (Fig. 2.2). Bordering the Uncompahgre to the southwest, and occupying much of southeast Utah, was the Paradox basin (Fig. 2.2); such pairings of mountain belts and sedimentary basins is a recurrent theme in geologic history around the world. This rapidly subsiding trough was a catchment for the sediment washed from the highlands. Early in its history the Paradox basin was inundated by sea. Later, in the Permian, the sea receded westward, and terrestrial conditions replaced the marine environment with sand dunes and rivers whose headwaters lay in the Uncompahgre highlands.

On the east side of the Uncompahgre highlands lay another basin, the Central Colorado trough (CCT) (Fig. 2.2). Smaller and narrower than the vast Paradox basin, it was bounded to the east by the ancestral Front Range highlands, yet another segment of the Ancestral Rockies. Although long since worn flat and covered by a deep sea, this range was re-uplifted in more "recent" times (~65 million years ago) to form the modern Front Range.

During the Pennsylvanian and Permian periods the CCT received sediment from the Uncompahgre highlands to the west and from the Front Range uplift to the east. Like its sister basin to the west, the sea periodically inundated the CCT, especially early in its history.

The exact origin of the Ancestral Rocky Mountains remains enigmatic, as it is mostly circumstantial evidence that provides an explanation. Distinct mountain-building events, or orogenies, typically are tied to compressional forces generated along a continental margin, most often due to collision with another landmass or some other large-scale tectonic encounter. In such settings, uplift normally is limited to the region adjacent to the margin. In rare instances the compressional stress may affect areas far inland to produce seemingly unexplainable interior mountains. Such a scenario has for many years been proposed as the origin of the Ancestral Rockies. Proponents of this idea maintain that a collision between South America and North America in the vicinity of today's Gulf of Mexico gener-

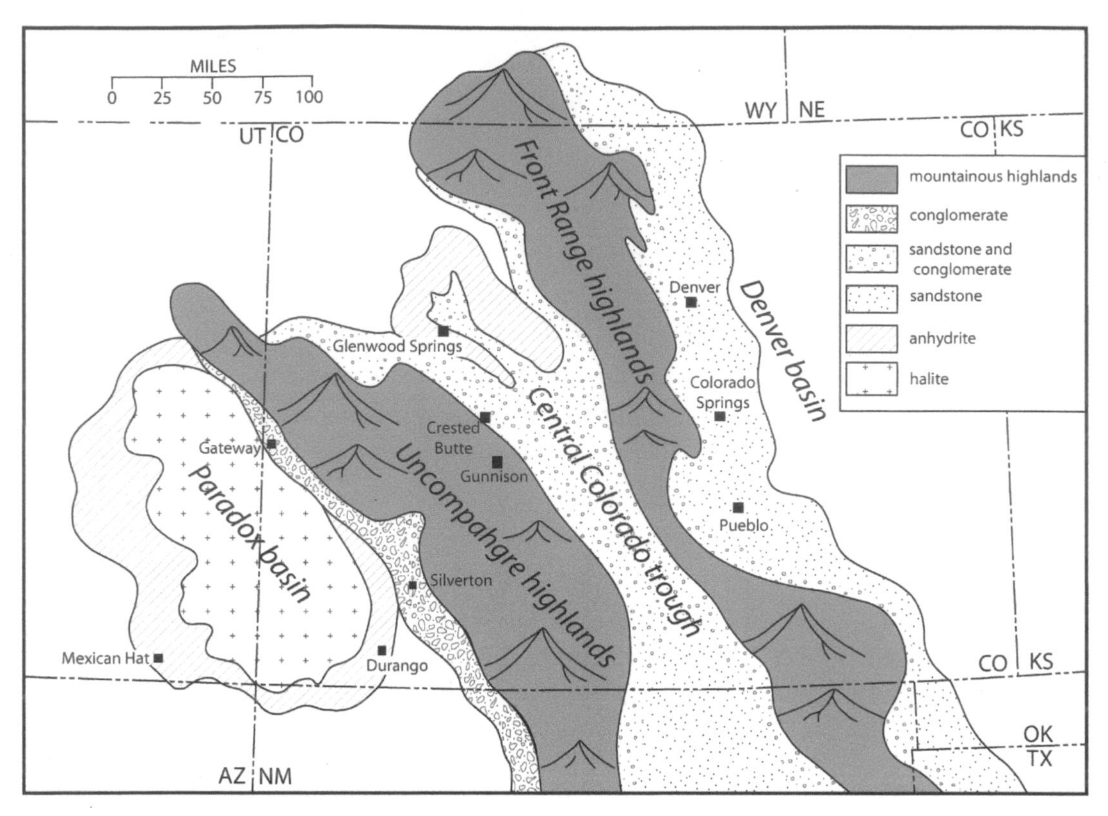

Fig. 2.2. Paleogeography of the Middle Pennsylvanian Ancestral Rocky Mountains, which influenced the geology of the Colorado Plateau. The map shows the locations of the highlands and intervening sedimentary basins, and the general types of deposits. Modified after Mallory 1960.

ated enough compression to raise mountains in Colorado and New Mexico. Recent discoveries to the southwest in Mexico, however, suggest an alternate source of compression from the southwest. Such a relationship fits much better with the orientation of the Ancestral Rockies and their adjacent basins. The following section will discuss evidence for both of the proposed settings. Geologists have yet to reach a consensus on the origin of the Ancestral Rockies.

The Dogmatic View: Compression from the Southeast

During Late Mississippian–Early Pennsylvanian time the supercontinent of Pangea was under construction; that is, large landmasses were colliding and joining as the world's continents converged. The collision of Europe and Africa with the east margin of North America produced the Appalachian Mountains. This was followed closely by the stitching of the north part of South America with North America in the cusp of what is now the Gulf of Mexico (see Fig. 2.1). This convergence resulted in further upheaval in the form of the Marathon-Ouachita orogenic belt, an east-west-trending mountainous region that rippled through present-day south Texas, Oklahoma, and Arkansas to form a continuous mountain chain with the Appalachians. The Marathon-Ouachita belt has since been eroded and deeply buried beneath later sediment.

Until recently, most available evidence linked uplift of the Ancestral Rocky Mountains in Colorado and New Mexico with continental collision along the southern margin of North America. The overlap in the timing of activity for these mountains with those of the Ouachita-Marathon belt to the south is considered by many to be too close for coincidence.

Data for the timing of these mountainous uplifts comes from the sedimentary record, preserved in the deep basins that form alongside orogenic belts. Uplift is always coupled with subsidence in an adjoining area. Telltale signs in the basins for uplift in the adjacent mountain belt include accelerated subsidence rates, told by thick accumulations of sediments over a relatively short period, and/or an influx of coarse sediment driven by an increase in relief in the adjacent source area.

When observed from a chronological standpoint, the east-to-west wave of mountain building for southern North America and possibly its interior becomes evident (Fig. 2.3). The record is as follows:

1. Latest Mississippian (> 320 Ma): rise of the Ouachita uplift in eastern Oklahoma and Arkansas.
2. Earliest Pennsylvanian (~320 Ma): rapid uplift of the Ouachita belt; slow initiation of the Ancestral Rocky Mountains.
3. Middle Pennsylvanian (~305–285 Ma): Ancestral Rockies experienced rapid uplift; activity in the Ouachitas began to wane.
4. Pennsylvanian-Permian (~286 Ma): Arbuckle uplift in Oklahoma and Marathon belt in Texas began uplift and shed abundant sediment into deep basins that bounded them to the north.
5. Early Permian (< 280 Ma): most tectonic activity in the region is over except for the Marathon belt and surrounding areas.

By the Middle Permian, uplift was sporadic and minor throughout the region. The major activity in all areas was the slow filling of the sedimentary basins as the dormant highlands continued to be scoured by erosion.

The geometry of the relationship between a southeastern source for compression and the orientation of the Ancestral Rockies has always been difficult to explain. Mountain belts, and the folds and faults that are their building blocks, typically are elongate perpendicular to the stress that formed them, whether that stress is compressional or extensional. For instance, the Ouachita-Marathon belt is parallel to the collision zone between South America and North America. Likewise, the Appalachian Mountains formed parallel to the suture zone along the east edge of North America, where Africa and Europe docked about the same time

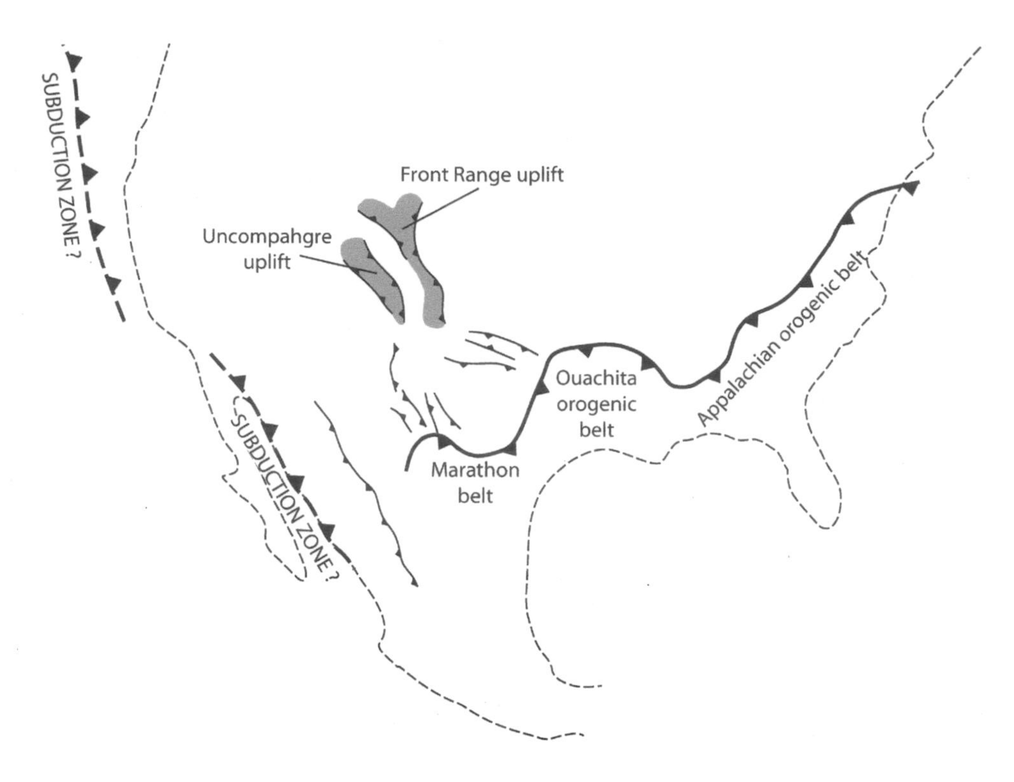

FIG. 2.3. Schematic map showing Pennsylvanian tectonic features associated with the construction of Pangea (Appalachian, Ouachita, and Marathon orogenic belts), the Ancestral Rocky Mountains and associated uplifts immediately to the south, and the hypothetical locations of Pennsylvanian subduction zones to the west (thick dashed lines). Note that the outline of North America is shown only for reference to modern geography. The teeth indicate the uplifted sides of the faults. Modified from Ye and others 1996.

(Fig. 2.3). This is the relationship found in all mountain belts, ancient and modern. It is simple to explain and visualize. The puzzle arises when the Ouachita-Marathon collisional event is called upon to generate the Ancestral Rockies, which trend north-northwest, in the opposite direction that they would be expected to be oriented.

In the past this conflict between stress direction and the orientation of faults was resolved by calling on preexisting weaknesses in the crust of the region. According to a series of papers by geologists Don Baars and Gene Stevenson, deep-seated northwest- and northeast-trending fractures were formed in basement rocks ~1700 million years ago. Northwest-trending fractures were later reactivated any time that compressional or extensional stress rippled through the region. According to this hypothesis these fractures experienced vertical movement in the Pennsylvanian as the Ancestral Rockies were shoved upward and again in the Tertiary (~65 Ma) when the modern Southern Rocky Mountains were born. While these weaknesses likely existed prior to the Pennsylvanian Period, the scenario proposed by these geologists is problematic when confronted with more recent evidence.

An Alternative Hypothesis: Compression from the Southwest

Baars and Stevenson attribute the rise of the Uncompahgre highlands, in part, to strike-slip faulting along the southwest side of the ancient range. In this scenario, where bends in the fault occur, localized compression produced vertical uplift (Fig. 2.4). They suggest, however, that most of the uplift in the Uncompahgre, and dramatic subsidence in the adjacent Paradox basin, was on normal faults and therefore caused by extension. However, recent drilling and geophysical imaging suggest that both the ancestral Uncompahgre and Front Range highlands were heaved upwards along large thrust faults. So the Ancestral Rockies must have been raised by compression. Furthermore, the geometry of these faults strongly suggests compression from the southwest. The problem with a southwest compression source has been the lack of evidence for other features that might indicate such a stress direction. This stemmed from a dearth of information on the geology of the interior of Mexico. Research in the last two decades has considerably improved our understanding of this region, providing new data relevant to the creation of the Ancestral Rockies.

The identification of Pennsylvanian-Permian volcanic rocks in northeast Mexico raises the possibility of compression from the southwest as a mechanism for Ancestral Rockies uplift. Although no actual volcanoes have been identified, blocks of volcanic andesite and rhyolite up to 100 m in length(!) have been recognized in the Chihuahua basin. These landslide blocks, apparently tumbled from some long-disappeared volcanic highland, are mixed with limestones that contain Pennsylvanian fossils. These unusual deposits point to the collapse of a high-relief volcanic arc that must have stretched along what was the southwest margin of Mexico at that time. The presence of a Pennsylvanian volcanic arc of andesite/rhyolite composition leads to the inevitable conclusion that a subduction zone existed offshore of Mexico (see Fig. 2.3). Subduction zones typically are a source of intense compression.

The association of undated volcanic rocks and fossiliferous Pennsylvanian limestone emphasizes relationships that sometimes must be used to infer the age of rocks and, thus, the events that formed them. The recognition of similar volcanic deposits in the area, but of Permian age, lends credence to these inferences.

These relationships were first reported in 1988 in a paper spearheaded by Edwin H. McKee, a geologist with the U.S. Geological Survey (USGS). Although the implications of these isolated volcanic rocks may have a great bearing on the origin of the Ancestral Rockies far to the northeast, many geologists are reserving judgment until more information from Mexico is available. As we will see time and again, old hypotheses die hard in the geologic community.

More recently another group of geologists from the Massachusetts Institute of Technology (MIT) have tackled this problem. In a report by Hongzhuan Ye and

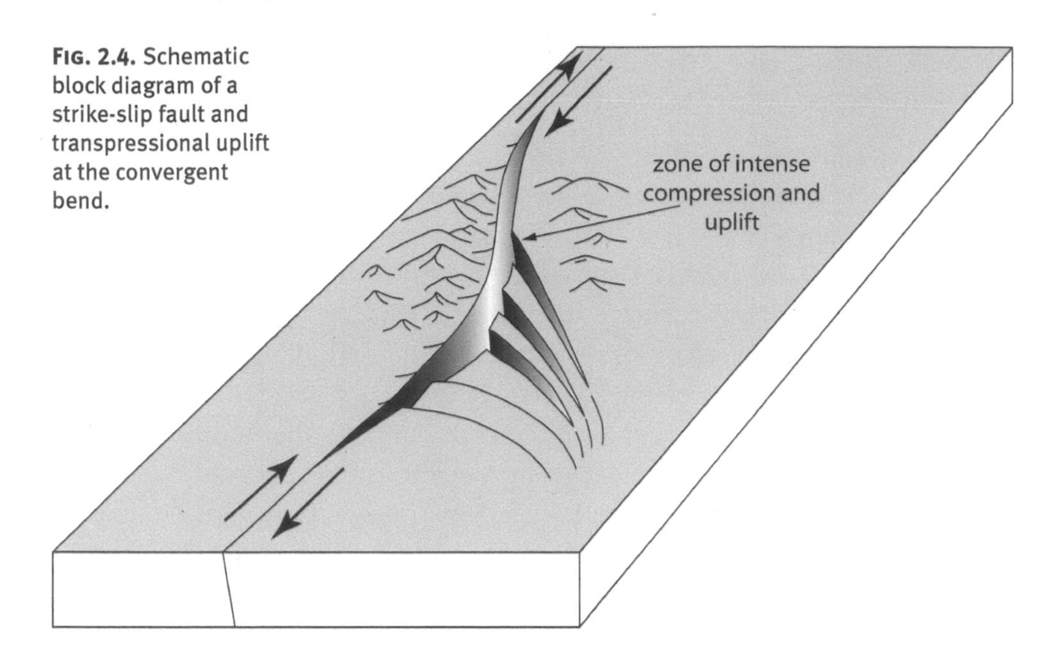

Fig. 2.4. Schematic block diagram of a strike-slip fault and transpressional uplift at the convergent bend.

several others (1996), the timing and intensity of uplift in the Ouachita-Marathon belt were carefully compared to those in the Ancestral Rockies. They concluded that the collision to the southeast was an unlikely source and also called on a compressional source to the southwest, probably the volcanic arc setting documented by McKee and his coworkers. Careful reexamination of earlier relationships and new studies of poorly known regions and rocks will be required before any consensus is reached on the origin of the Ancestral Rockies.

Regardless of their origin, the obvious result of the Ancestral Rockies was the related birth of the adjacent sedimentary basins: great troughs that received and preserved the detritus washed from the highlands. Although those mountains have long since been worn flat, it is the sediment--the remnants of the rock that made up these mountains—that contributes so much to the modern landscapes of the Colorado Plateau and Southern Rocky Mountains.

Paradox Basin

The ancestral Uncompahgre highlands contributed sediment to the Paradox basin to the west and the Central Colorado trough to the east (see Fig. 2.2). These two basins share a similar history, although the larger Paradox basin opened to the sea to the west. The narrower CCT was bounded on both sides by mountains; the ancestral Front Range lay to the east and also supplied large amounts of sediment to the linear basin. The following in-depth discussion concentrates on the Paradox basin.

The Paradox basin is broadly defined by geologists as the lateral limit of evaporite rocks of the Pennsylvanian Paradox Formation (Fig. 2.5). Most of the basin

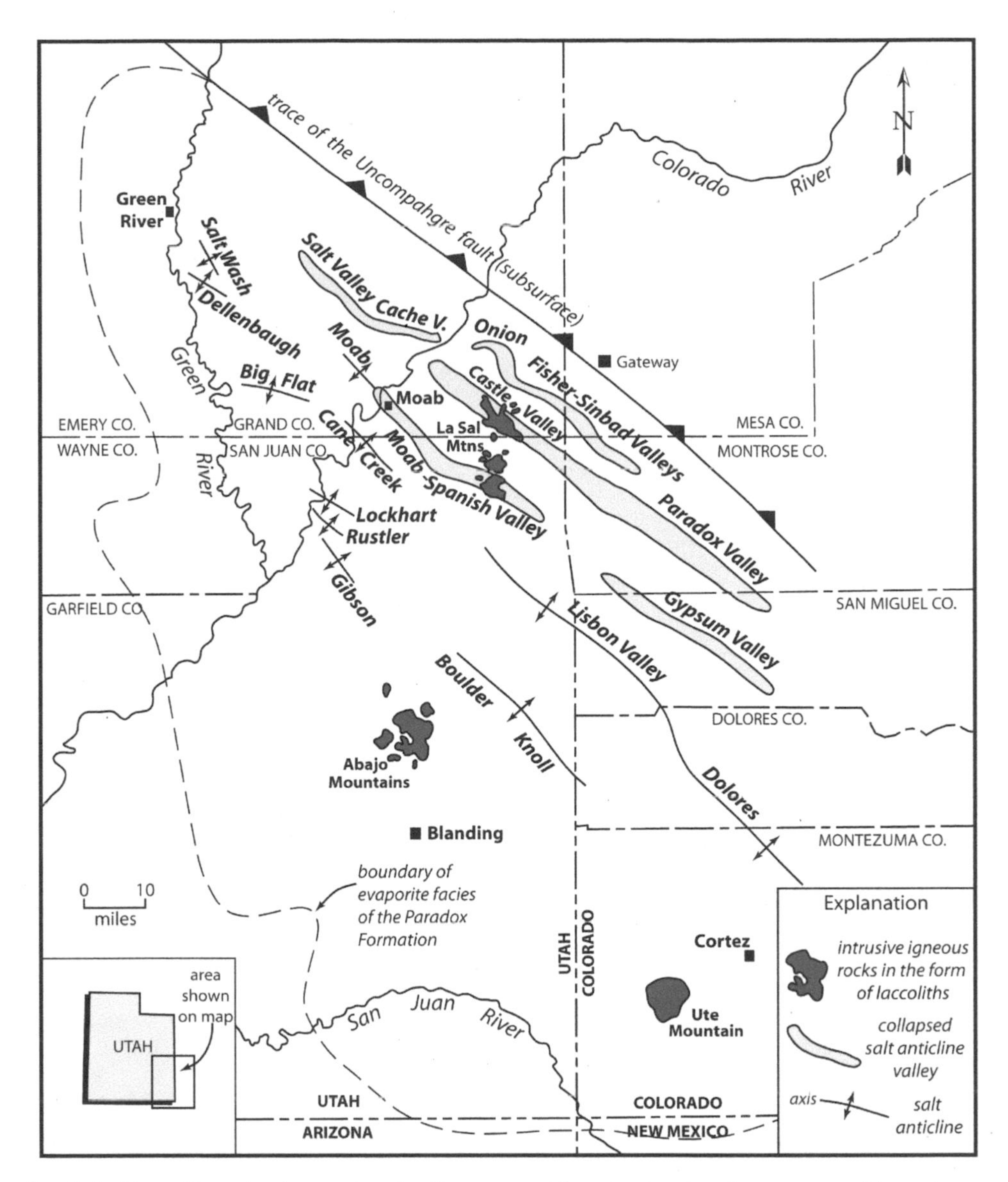

Fig. 2.5. Map of Paradox basin showing its geographic extent and various geographic and geologic features, including the salt deformation features of salt anticlines and collapsed salt anticline valleys, laccolithic intrusions, and various towns and rivers. Modified from Doelling 1985.

lies in southeast Utah, but it extends into a large part of southwest Colorado, and into the northwest corner of New Mexico and northeast Arizona. The basin gets its name from the formation which, in turn, was named for exposures in Paradox Valley in western Colorado. The Paradox Valley was named for the strange course of the Dolores River through the northwest-trending valley. The Dolores flows northeast across the valley, slicing through the steep walls of Mesozoic sandstone on either side. Thus the course of the river and the origin of the valley were a paradox. As we will see later in this chapter, that paradox is a direct consequence of the evaporite deposits in the Paradox Formation.

A

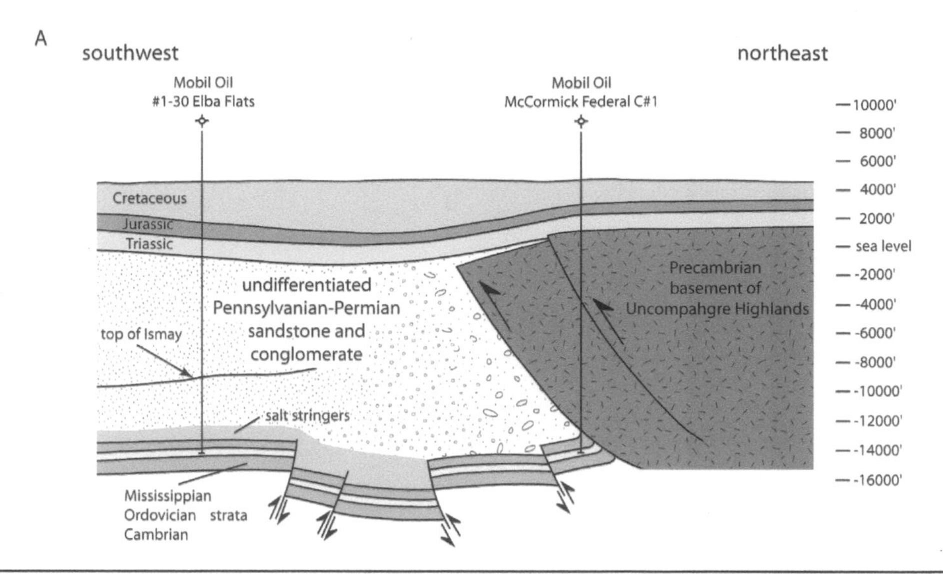

B

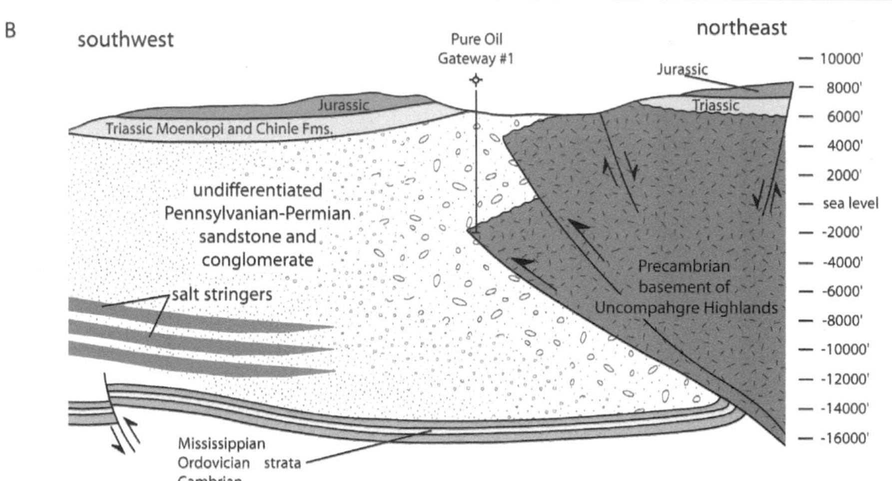

Fig. 2.6. Southwest to northeast cross sections showing the modern subsurface geology across the eastern Paradox basin and the front of the Uncompahgre highlands. The geology is known from a combination of seismic reflection data and drill hole data (locations and depths shown on cross sections). Cross section A parallels Interstate 70. The northeast part is just west of the Utah-Colorado border, and the southwest termination is near Salt Wash, immediately north of Arches National Park. Cross section B is located completely in the state of Colorado, just south of the town of Gateway. Both cross sections are modified from White and Jacobson 1983.

The elongate, northwest-trending Paradox basin began to subside in the Early Pennsylvanian, propelled by the rise of the Uncompahgre highlands. The boundary between the basin and the highlands is marked by large-scale reverse faults. Drill hole data and seismic imaging of the boundary near Cisco, Utah, and to the south near Gateway, Colorado, show that movement on the main fault, or collectively along several smaller faults, shoved Precambrian basement rocks southwestward, up and over younger Paleozoic sedimentary rocks (Fig. 2.6). Mobil Oil geologists Claudia Frahme and E. B. Vaughn, in a 1983 paper, showed that

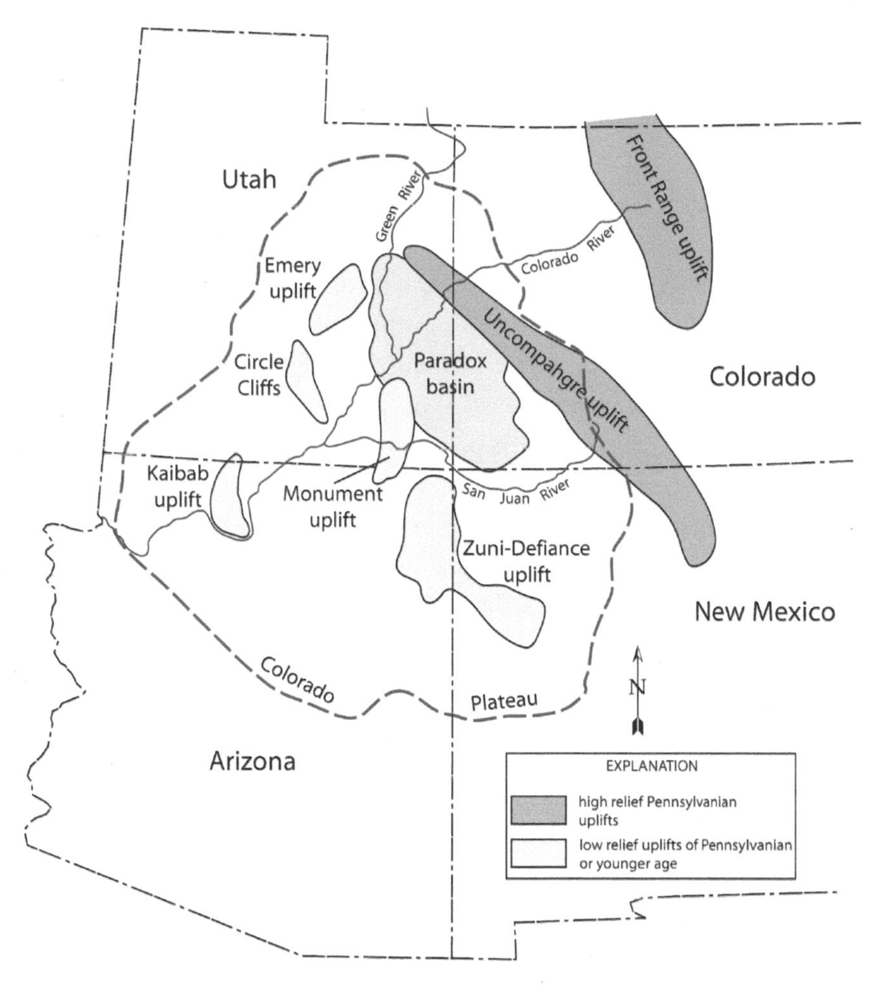

Fig. 2.7. Map of the Four Corners area showing several types of Pennsylvanian tectonic elements. The Ancestral Rockies (Uncompahgre and Front Range uplifts) were high-relief features that acted as sources for abundant sediment. The lighter-shaded features (e.g., the Emery and Defiance-Zuni upwarps) were lower-relief features that may have acted as barriers to the Paradox basin but did not necessarily contribute any sediment during this time. Note that not all of these features were active during the Pennsylvanian Period. For instance, the Circle Cliffs upwarp and Monument upwarp may not have been elevated until later in the Permian or Triassic-Jurassic periods.

the fault dips 30° to the northeast, and that horizontal displacement of at least 6 miles is coupled with ~20,000 feet of vertical uplift. The amount of movement along this fault zone is huge by any standard.

It was the lure of petroleum-bearing rocks that spurred Mobil and Chevron, among several others, to explore this part of the Paradox basin. Although exploratory wells were dry, much new information on the geology was obtained. The lack of petroleum in this area probably allowed the important proprietary data eventually to be released by these companies, adding significantly to the scant knowledge of this subsurface feature. Similar situations exist around the world where geologic data are shrouded in secrecy. Competition between oil companies is fierce in some prospective regions. If the possibility of petroleum reserves exists, costly data may remain secret for many years. Because academic researchers and even government research groups cannot afford the price tag of large-scale subsurface investigations, oil companies are a vital resource for subsurface data and the advancement of the science of geology.

Although the Uncompahgre is the major structure associated with the Paradox basin, there are several mild upwarps that form the other basin margins and periodically affected sedimentation. The Zuni-Defiance uplift bounded the basin to the south, the Monument uplift formed part of the southwest margin, and the Emery uplift lay to the northwest (Fig. 2.7). None of these were as lofty as the Uncompahgre, and during much of the late Paleozoic they were covered by a shallow sea. They did, however, act as restrictions to an influx of sea water during times of low sea level, controlling the precipitation of evaporites in the deeper central and eastern parts of the Paradox basin. Like the faults that the Uncompahgre rose along, these structures have been intermittently active throughout geologic history.

To the southwest of the deep basin center the basin gradually shallowed, and the distal basin margin was encountered. During the Pennsylvanian this margin was defined by a shallow marine shelf and carbonate deposition. Unlike the evaporite rocks in the basin center, which today are deeply buried or poorly exposed, these shelf carbonates are spectacularly displayed in the deeply incised canyons of the San Juan River and along the Colorado River in Cataract Canyon. Because of these continuous exposures, geologists have not had to rely so much on subsurface data. In fact, abundant detailed information is available in scores of reports and articles. This does not mean that petroleum geologists are uninterested in these rocks. Quite the opposite! Where these shallow-water carbonates are present in the subsurface, they make up the numerous small oil fields that dot the Four Corners region, including the small field at Mexican Hat. In fact the limestone outcrops in the shadowed depths of the lower San Juan canyon make this the number one destination for petroleum geologists interested in exploring the Paradox basin. It is likely that an additional attraction to examining these rocks is getting out of the office and hiking through spectacular scenery in a remote setting. Many students have become geologists for this reason.

Thickness patterns for Pennsylvanian sediments in the northwest-trending Paradox basin are strongly asymmetric when viewed along a northeast to southwest transect and provide some insight into subsidence patterns during this period (Fig. 2.8). Locally, along the northeast basin margin that abuts the Uncompahgre highlands, an astounding 18,000 feet of coarse sandstone and conglomerate was deposited. A short distance southwest, in the basin center, cyclic successions of deep-water evaporites, black shale, and carbonate accumulated to a thickness of ~7,000 feet. Farther southwest, in the vicinity of the modern San

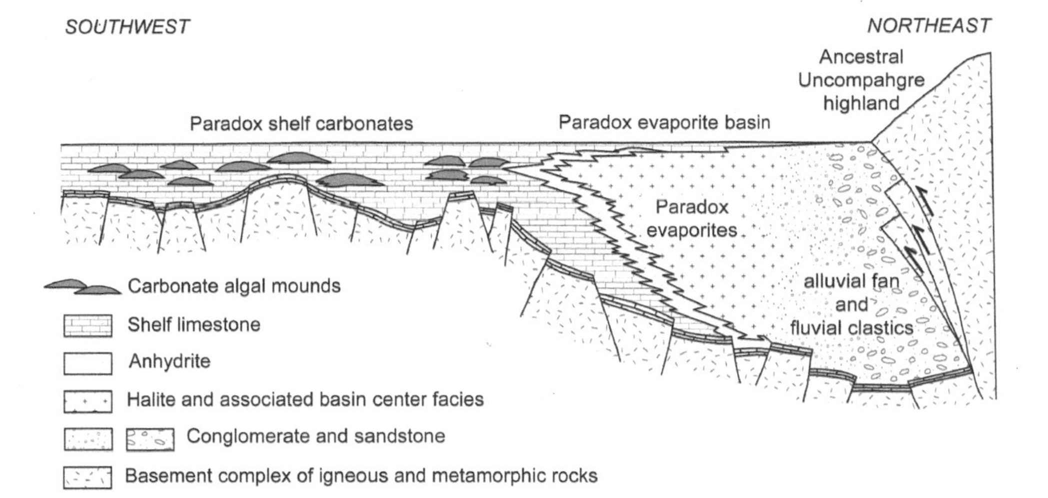

Fig. 2.8. Schematic cross section across the Paradox basin during Late Pennsylvanian time showing the relations between the carbonate shelf facies to the southwest, the evaporite facies in the basin center, and the eastern clastic facies against the Uncompahgre highlands source terrane that bounded the basin along its northeast margin. Modified after Stevenson and Baars 1986.

Juan River canyon and the settlement of Mexican Hat, Utah, ~1,000 feet of cyclic, shallow-water carbonates were deposited.

The asymmetry of the basin, with the pronounced deepening toward the northeast margin, suggests a compressional tectonic setting. Foreland basins, as these types are known, form when the load of a mountain belt causes the underlying crust to flex in the form of downwarping. Maximum subsidence occurs directly beneath and adjacent to the mountain belt, in this case the Uncompahgre highlands and the northeast margin of the Paradox basin, respectively. Subsidence decreases with distance from the mountain belt. In the Paradox basin this culminates with the distal southwest margin, which was a shallow marine shelf.

The modern interpretation of the Paradox basin as a foreland basin contrasts with earlier interpretations of it as an extensional basin (Baars 1976), or as a pull-apart basin in a dominantly strike-slip fault setting (Stevenson and Baars 1986). But these early interpretations cannot be faulted; much of the detailed subsurface data now available had not yet been collected or were unavailable when these interpretations were made. Thus the science of geology progresses not only because of better technology, but also because of larger numbers of scientists working in an area and studying the rocks in greater detail.

Pennsylvanian Stratigraphic Evolution of Paradox Basin

The initiation and evolution of the Paradox basin can be traced by examining regional Pennsylvanian depositional environments and thickness patterns. In the preceding Mississippian Period (~360–320 Ma) western North America was blan-

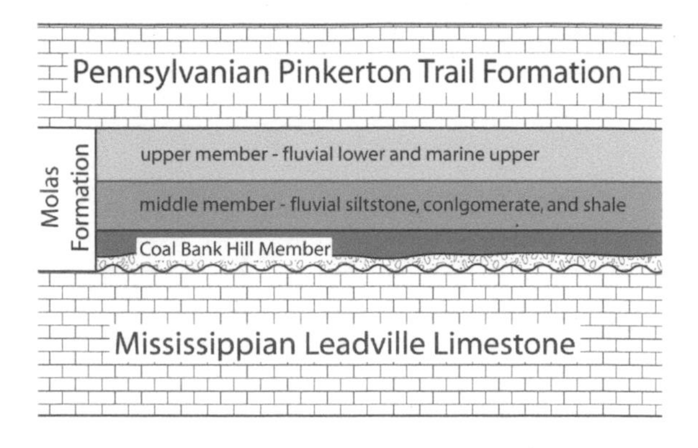

FIG. 2.9. Stratigraphy of the Mississippian-Pennsylvanian Molas Formation as exposed in the San Juan Mountains of southwest Colorado. Here the Molas consists of the basal Coal Bank Member and the two informal overlying members defined by Merrill and Winar (1958). The Molas rests unconformably on the Leadville Limestone and is conformably overlain by the Pinkerton Trail Formation, the lowest part of the Hermosa Group.

keted by a warm, shallow sea in which up to 1000 feet of limestone and dolomite were deposited. The period closed with a westward retreat of this sea, leaving the Mississippian Leadville Limestone high and dry. Carbonate rock, which makes up the bulk of the Leadville, slowly dissolves when exposed to even slightly acidic water. All meteoric water (rain, snow, etc.) and most surface water is slightly acidic. The exposure of thick Mississippian limestone for several million years created a deeply weathered surface pocked with sinkholes as dissolution formed shallow caves and their rocky roofs collapsed. Mississippian limestones throughout western North America are characterized by this rugged karst topography, named for the Karst region in Yugoslavia. It was on this deeply eroded surface of Leadville Limestone that the basal Pennsylvanian deposits of the Paradox basin were laid down.

Molas Formation

The Molas Formation, which overlies the Mississippian Leadville Limestone, spans the boundary between the Mississippian and Pennsylvanian periods. The Molas is named for exposures at its type locality near Molas Lake between Durango and Silverton, Colorado, in the San Juan Mountains. The unit is not exposed in southeast Utah, but its occurrence there is documented from drill hole data. Since these data document little more than unit thickness, the following description comes from the type locality. Merrill and Winar (1958), in one of the few studies of the Molas Formation, recognized three members near Molas Lake (Fig. 2.9).

The basal Coalbank Hill Member is composed of red-brown mudstone with angular boulders and cobbles of limestone and chert. This unit, which cloaks the erosional surface, represents residual material left behind as the Mississippian limestone was being weathered. Limestone and chert fragments are remnants of the once underlying Leadville Limestone. The silt and clay have recently been interpreted as loess, fine sediment blown into the region by wind to fill in the cracks and crevices of the irregular karst surface (Evans and Reed 1999).

The middle member, which is not formally named, consists of ~40 feet of stratified shale, siltstone, and conglomerate. Limestone and chert pebbles are better rounded than fragments in the underlying member. Stratification and rounded pebbles point to a **fluvial** (deposited by rivers) origin, indicating that streams cut across the underlying Coal Bank Hill Member and reworked its upper sediments.

The upper member also is dominated by stratified shale, siltstone, and conglomerate but contains in its upper part interbedded sandstone and limestone. Fossils such as brachiopods, bryozoans, and echinoderms are abundant in the limestone beds and indicate an Early Pennsylvanian age. These uppermost beds herald the return of the sea to the region.

The middle and upper members are informally designated and thus have not been given proper names. The boundary between the two is within a fluvial succession and is not recognizable in the field. Merrill and Winar (1958) defined this boundary based on a change in clay minerals within this otherwise continuous sequence. Different clay minerals can only be identified by laboratory analysis, so the criteria that formal members be recognizable in the field cannot be met. This explains the informal designation.

The three members of the Molas Formation mark the evolution of the region from a low-relief, weathered surface on which residual material slowly accumulated to the establishment of a regional drainage network and fluvial system across the vast landscape. Finally, the sea returned, drowning the rivers and dominating the Paradox basin for tens of millions of years.

The consistent regional thickness of the Molas Formation suggests that the Paradox basin had not yet developed into the rapidly subsiding trough that it would soon become. As sea level began to rise toward the end of Molas deposition, the region was gradually inundated, and the earlier fluvial deposits were reworked by the shifting shoreline. The continued climb in sea level allowed a variety of marine organisms to move in, and shallow-water carbonate deposition marked the top of the Molas Formation.

Pinkerton Trail Formation

The Pinkerton Trail Formation is a succession dominated by gray fossiliferous marine limestone with gray-black shale and ranges up to 200 feet thick. It is named for exposures just north of Durango, Colorado. The formation conformably overlies the Molas Formation, and fossils indicate a Middle Pennsylvanian age. The formation is not exposed in southeast Utah but is known to extend throughout the Paradox basin based on abundant drill hole data. Mountain-building activity has provided excellent outcrops along the west side of the San Juan Mountains in western Colorado, which is where much of the following description is taken from.

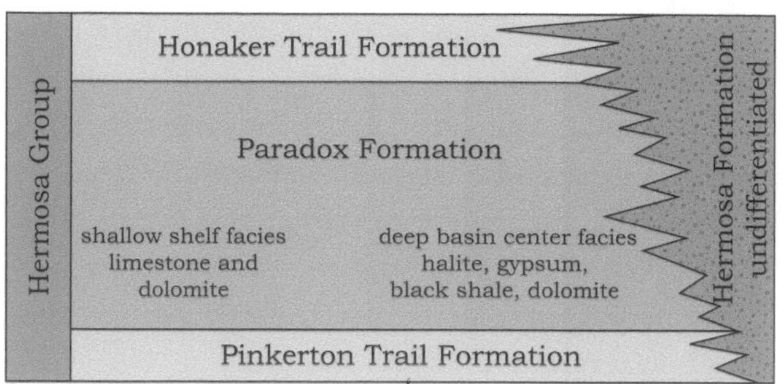

FIG. 2.10. Stratigraphy of the Pennsylvanian Hermosa Group in the Paradox basin showing the change from its eastern extent, adjacent to the Uncompahgre highlands, west to the shallow shelf environment on the basin's west margin.

The formation is the lowest unit in the Hermosa Group, which includes, in ascending order, the Pinkerton Trail, Paradox, and Honaker Trail formations (Fig. 2.10). Collectively these rocks provide a record of the Pennsylvanian Period in the Paradox basin.

Although the Pinkerton Trail consists of several rock types, it is the fossiliferous limestone that provides details on the shifting conditions in the shallow sea. According to a detailed study of the formation near Hermosa, Colorado, by Karen Franczyk and her colleagues at the USGS (1995), vertical changes in limestone composition reflect changes in sea level. Basal limestones in the formation consist of broken crinoid stems mixed with quartz sand. These deposits represent a high-energy shoreline which crept into the area from the west as sea level began to rise. Overlying limestone is different, containing a diverse fossil fauna of brachiopods, bryozoans, sponges, and crinoid stem fragments. The abundant organisms suggest deepening water out of the high-energy shoreline area, but still mildly affected by wave activity. In contrast, succeeding limestone beds contain more carbonate mud, with fossils limited to sponges and thin-shelled brachiopod fragments. Thin shells indicate an adaptation to low-energy conditions, places where the organism would not be tumbled about on the sea floor by strong currents. Likewise, the presence of lime mud suggests a low-energy setting, allowing the fine mud particles to settle out of the undisturbed water column without being wafted away by currents. These muddy limestones point to deposition in deeper water, probably below wave base, a term that refers to water deep enough that waves do not commonly scour the sea floor, except possibly during storm events. Finally, uppermost carbonate beds exposed in Colorado contain a diverse fossil assemblage, similar to the middle part of the formation. This suggests that at the close of Pinkerton Trail deposition, sea level was dropping.

Detailed analyses of cores taken from a Department of Energy drill hole in San Juan County, Utah, a few miles east of the Canyonlands Needles District park boundary, provides insight into changes in the Pinkerton Trail Formation toward the center of the Paradox basin. These cores indicate that the uppermost part of the formation contains several thick beds of anhydrite. Evaporite deposits indicate an increasingly arid climate and isolation of the basin from the open sea. As sea level dropped, surrounding upwarps restricted the circulation and influx of seawater. High evaporation rates in the restricted basin eventually triggered the precipitation of evaporite minerals from the chemical-rich brines that filled the basin. These were the controlling factors in sedimentation in the overlying Paradox Formation, which is dominated by cyclic evaporite deposits.

Paradox Formation

The Middle Pennsylvanian Paradox Formation and equivalent rocks vary markedly with location in the basin. It was during Paradox deposition that fault-induced subsidence along the northeast side of the basin accelerated, producing the deep northwest-trending trough. Strata along the northeast margin consist of coarse sandstone and conglomerate derived from the rising Uncompahgre highlands. This succession reaches up to 18,000 feet thick--more than three vertical miles of sandstone and conglomerate! This fringe of clastic sediments is not considered a formal part of the Paradox Formation. The problem in this area lies in defining the lower and upper boundaries, which are fuzzy to nonexistent due to a seemingly endless flood of coarse sediment off the adjacent highlands. A short distance to the southwest, away from the influence of this uplift, vertical changes in strata conveniently allow for the division of the Pinkerton Trail, Paradox, and Honaker Trail formations, which combined form the Hermosa Group. The clastic sediments that line the northeast basin border prohibit such a division and are simply referred to as "undifferentiated" Hermosa Group (Fig. 2.10).

In the central part of the basin the Paradox Formation consists of cyclic, evaporite-dominated sequences. The formation reaches a maximum depositional thickness of 6000 feet with 33 separate cycles recognized from drill hole data. Individual cycles are dominated by halite but also contain interbeds of anhydrite, dolomite, and organic-rich calcareous shale. The study of these rocks over the last fifty years has been stimulated by petroleum exploration in the Paradox basin. The cyclic evaporites were deposited while subsidence was rapid and pronounced. Simultaneously, worldwide sea level was rising and falling over regular time intervals. When sea level dropped, the shallow shelf areas that defined the basin margins to the south and southwest inhibited circulation and partially isolated the basin from the open sea to the west. This created a giant evaporative pan from which evaporite minerals precipitated. When sea level rose, fresh seawater again flowed into the basin to dilute the briny waters. Through many such events the multiple cycles of the Paradox Formation were deposited.

Evaporite rocks of the Paradox Formation are exposed in their namesake Paradox Valley and other salt anticline valleys in the Utah-Colorado borderlands. These large structures were formed by the mobilization of salt in the Paradox Formation by viscous flow as overlying sediments were deposited. Such movement of salt also is responsible for the evolution of the scenic Needles area in Canyonlands and Arches National Park. The development of salt anticline valleys are discussed at the end of this chapter.

Paradox evaporites are also exposed along the Colorado River in Cataract Canyon, where the river has cut deep enough to exhume these older rocks. Another glimpse of these rocks is provided at the Raplee anticline, where the San Juan River has cut into this uplift just east of Mexican Hat. Here it is the spectacular uparching of the anticline that brings these rocks to the surface.

Moving southwest, out of the deeper central part of the basin, carbonate-dominated shallow shelf deposits are encountered. Like the evaporites, these rocks were deposited in cyclic successions but are dominated by limestone with black shale, dolomite, and siltstone. These rocks interfinger with evaporites in the transition zone between the shelf and the deep basin center. Collectively the carbonate cycles reach a maximum thickness of ~1000 feet, reflecting a decrease in subsidence with distance from the abutment of the Uncompahgre highlands.

Depositional setting of the basin center evaporite facies of the Paradox Formation

The cyclic deposits of the Paradox Formation have long been attributed to regular fluctuations of sea level during the Pennsylvanian; however, the magnitude of change and the degree of isolation of the basin during sea level lowstands have been a recurrent source of controversy. These problems are best understood by looking in detail at a characteristic evaporite cycle from the basin center and a carbonate interval from the shelf area. In doing so, the depositional processes throughout the basin become more apparent.

The most detailed study of Paradox evaporite cycles comes from USGS geologists Omer Raup and Robert Hite (1992, 1996). Hite has studied the Paradox Formation for the last forty-five years. Their most recent work focuses on cores taken from the upper Paradox Formation by companies interested in associated potash salts. One drill hole was located in the Shafer basin, near the mining complex of Potash, 8 miles southwest of Moab along the Colorado River. Another core was obtained from a drill hole ~7 miles to the south, on the opposite (southeast) side of the Colorado River. Both are close to the east boundary of Canyonlands National Park.

A typical, complete evaporite cycle consists, in ascending order, of basal anhydrite, dolomite, carbon-rich black mudstone, dolomite, anhydrite, and finally, halite (Fig. 2.11). When potash occurs in a cycle, it may be interbedded with the upper part of the halite or overlie it to form the top of the cycle. Individual cycles are marked, top and bottom, by sharp contacts interpreted as disconformities. A disconformity is a line of contact between two rock units, in this case halite and overlying anhydrite, that represents an unspecified interval of missing time. Thus,

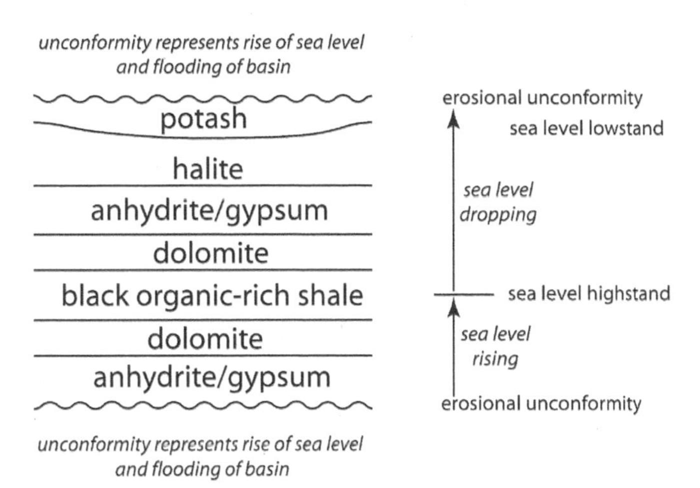

FIG. 2.11. Idealized evaporite cycle from the center of Paradox basin showing vertical sequence of lithologies and the unconformities that bound the cycle at top and bottom. Note that in any given cycle the potash may be absent, in which case the unconformity would be at the top of the halite facies. The thickness of a lithology in any given cycle will vary, and some cycles may not contain all lithologies.

while rock units represent deposition during some period of time, a disconformity marks an interval that is not represented by rock, due either to removal by erosion or a period of nondeposition. The cycle-bounding disconformities in the Paradox evaporites are thought to represent a short time period relative to the cycles of deposition. Contacts between units within the cycles are gradational, suggesting continuing deposition as gradual changes in water depth and chemistry took place.

Not all of the thirty-three recognized evaporite cycles contain the complete sequence outlined above. Additionally, the thickness of individual cycles, as well as units within the cycles, varies considerably. Cycles and the units that comprise them tend to thicken toward the basin center, and thin or pinch out (thinning to zero thickness) toward the margins. For example, the thickness of various halite units ranges from 20 to 800 feet.

Any discussion of the Paradox Formation must address the causes of cyclicity. Recent studies by Raup and Hite (1992, 1996) and Bruce Rueger (1996) have documented a regular rise and fall in sea level, coupled with related climate changes from warm and humid to cool and dry. Because the Paradox basin lay in the equatorial region, it was affected differently than eastern North America, which was located in the more northerly paleolatitudes (see Fig. 2.1). While both regions record cyclic sedimentation associated with pulse-like sea level fluctuations, the Paradox was relatively arid; however, coal deposits in eastern North America point to a humid climate. As we shall see, the trough-like configuration of the Paradox basin was also an important contributing factor.

To get a snapshot of what was happening during sedimentation, a description of a single, idealized evaporite cycle as proposed by Hite and Buckner (1981) follows. This succession was subsequently documented in previously mentioned cores. The description will begin at the base of a cycle, at the disconformity with underlying halite, and continue upward through the succession to where it is bounded by another disconformity at the top.

Erosion by dissolution marks the beginning of a new cycle. As sea level again began to rise, the water in the shallow shelf area that separated the isolated Paradox basin from the open sea to the west began to deepen. This allowed fresh seawater to pour into the basin and dilute the brines that had concentrated from earlier evaporation. These diluted brines quickly became undersaturated in sodium and chlorine, allowing the new mixture to dissolve rather than precipitate halite. This influx of water also replenished the brines with calcium and sulfate ions. Anhydrite ($CaSO_4$), a chemical sediment precipitated from ion-rich water, forms the base of each cycle and rests on an erosion surface that defines the top of the underlying cycle (Fig. 2.11). The brines had been depleted of these ions during anhydrite deposition of the previous cycle. This renewed stock of Ca^{2+} and SO_4^{2-} ions caused anhydrite to precipitate out of the water and be deposited on the basin floor on top of the erosion surface.

Overlying the anhydrite is silty dolomite [$MgCa(CO_3)_2$], also precipitated from the water column (Fig. 2.11). As the basin brine was further diluted by the open sea, HCO_3 ions were reintroduced to the mixture, and calcium and magnesium ions bonded instead with CO_3 to form dolomite.

Black mudstone overlies dolomite and forms the midway point of the cycle (Fig. 2.11). Units may reach 160 feet thick, but most are much thinner. Shale is composed of roughly equal amounts of carbonate minerals (calcite and dolomite), silt-sized quartz, and clay minerals. Organic carbon content may be as high as 13 percent and gives the shale its dark gray-black color. This relatively large organic component is, to petroleum geologists, the most intriguing aspect of the Paradox Formation and is one reason that so much exploration effort has been put into the basin. Sparse fossil fragments within shale units indicate minor biological activity on the basin floor during deposition, something that was not likely in the brine that earlier evaporite rocks were deposited from.

When considering the origin of black shale, several questions must be addressed. Most important, where did the organic material come from, and why so much in that part of the sequence? Deposition of black shale ultimately was controlled by sea level rise, but this rise also brought about other changes, for instance in climate and biological production. Black shale was deposited during the culmination of the rising sea, or highstand, during peak circulation and exchange of basinal water with the open sea. As the inflow of fresh seawater increased, evaporite precipitation halted, and bacteria and algae growth may have been stimulated, contributing to the organic sediment. Furthermore, a growing body of evidence suggests that as sea level rose and expanded over low-lying continental areas, the climate became wetter. This would have produced more vegetation, thus organic detritus, on the adjacent Uncompahgre highlands to be washed into the basin by increased runoff. The analysis of kerogen (insoluble organic material) from black shale beds points to a terrestrial plant source for most of the organic sediment.

Sugar-textured dolomite overlies black shale and heralds the return of evaporative conditions as sea level turned around and began to drop (Fig. 2.11). This dolomite contains small amounts of organic material and appears to preserve burrows of organisms that lived on the sea floor.

The contact between dolomite and overlying anhydrite in the upper part of the cycle is gradational. Anhydrite was deposited as salinity further increased in the basin, and calcium and sulfate ions became further concentrated. Careful inspection of anhydrite layers reveals some interesting textures. Crystal forms are of the mineral gypsum rather than anhydrite, suggesting that it was originally gypsum that precipitated from the Pennsylvanian brines, and it was subsequently altered to anhydrite. Evaporite minerals are very reactive with fluids, and this is one example of a common chemical reaction. The chemical formula of gypsum is $CaSO_4 \cdot H_2O$, whereas anhydrite is simply $CaSO_4$. When precipitating from brines, gypsum incorporates two water molecules into its crystal structure. This explains why it is such a soft mineral that it can be easily scratched with your fingernail. Anhydrite, as its name implies, consists of the same calcium and sulfate ions, but without the water. It is generally thought that anhydrite originates as gypsum. When gypsum is buried by subsequent sediments and put under pressure, water is slowly squeezed from the crystal structure to form the much harder, water-free anhydrite. On the other hand, anhydrite is rarely exposed at the surface of the Earth. If uplifted and exposed to moisture, it becomes rehydrated and reverts back to gypsum. Where Paradox evaporites are exposed, for instance, in the Paradox Valley, Colorado, and Onion Creek, eastern Utah, it is gypsum that we see.

Halite (NaCl) marks the culmination of a single cycle over most of the basin and grades upward from underlying anhydrite deposits. Halite—or, where present, potash—is sharply overlain by deposits of the succeeding cycle. Halite marks the maximum drop in sea level within a single cycle, and therefore the greatest isolation of the Paradox basin from the open sea. How much sea level was lowered during these lowstands remains debatable, but evidence in correlative shelf carbonate rocks to the southwest suggests local emergence and the beginning of soil-forming processes.

Potash occurs at or near the top of eighteen of the thirty-three halite beds in the Paradox Formation. Potash is a general term for potassium-bearing compounds such as potassium carbonate (K_2CO_3), potassium oxide (K_2O), and potassium hydroxide. Most of the potash in the Paradox basin is in the form of the mineral sylvite (KCl). Potash in the Paradox is confined to halite, at the top of the cycle, where it is either dispersed throughout the upper part of the halite or concentrated into discrete beds above it. Potash beds range more than 100 feet (30 m) thick.

Potash represents precipitation only after the most extreme salinity conditions had been reached in the final stage of the cycle as sea level dropped to an all-time low and the basin became a giant evaporative pan. The degree of isolation of the Paradox basin at this point in the cycle remains controversial. Recent studies by S. C. Williams-Stroud (1994) suggest that the basin was completely cut off from the open sea at this stage. The dogmatic view maintains that such large-scale isolation is unlikely. As a geologist, I find such unknowns and recurring controversies refreshing. These arguments spur researchers to continue to experiment, and to analyze and interpret the rock record with increasing detail in an attempt to find answers to such questions.

Basin isolation and the concentration of brines was cut short by a rise in sea level. As fresh seawater again flooded into the basin, evaporite precipitation was interrupted; in fact, the uppermost part of the halite/potash deposits was dissolved. Thus, one cycle concluded and another began, their boundary marked by a dissolution surface—an abrupt reminder of the dramatic change.

Next we move to the shallower shelf of the Paradox basin, located south and southwest of the evaporite-dominated basin center. Here we will see how cyclic sea level changes influenced the carbonate factory that was in full production. However, before we can effectively understand the limestone successions in the Paradox Formation, some insight into the components of limestone and what they tell us is essential. The following is a general introduction to the products of the carbonate factory in a shallow marine environment.

The making of limestone

Limestone, though always composed chemically of $CaCO_3$, is built of sediments of varied origins. In the simplest view, limestone ingredients are grouped into two categories: the particles or grains that form the framework of the rock, and the material that fills in the spaces between the grains, holding them together. All the components that ultimately make limestone provide information that can be used for interpreting conditions of deposition such as water depth and chemistry, climate, and current energy. Taken together, these can yield a detailed reconstruction of any ancient marine environment. Such interpretations of ancient settings are well founded on detailed studies of modern carbonate environments, particularly in the Bahamas and neighboring areas, and the Persian Gulf. While the specific organisms that play a large role in carbonate production may have changed through time, the processes are the same.

Framework grains. Four basic types of carbonate particles are recognized by geologists: fossils, ooids, peloids, and intraclasts. Each provides vital information in the interpretation of limestone-forming environments.

Fossils in limestone usually consist of the calcified hard parts of marine invertebrate organisms, either protective shells such as of brachiopods and molluscs, or skeletal parts that support an organism (e.g., corals and bryozoans). Limestone components in this category include complete fossils or fragments that have been broken up by waves and tidal currents. Many fossil organisms can be traced back to specific environmental conditions, making them valuable for such interpretations, as we will see in the subsequent section.

Ooids are the dominant component in oolitic limestone and are small, sand-sized spherical or ovoid concretions of $CaCO_3$. They form in shallow, agitated water with a high calcium carbonate content as grains are rolled in the surf. The inorganic calcium carbonate precipitates from the water onto some nucleus, typically a shell fragment or sand grain. The vigorous current activity rolls these growing grains along the sea floor, ensuring ooids of uniform size and shape. Oolitic limestone in the rock record is a sure indicator of a shallow, high-energy marine environment.

Peloids are sand-sized, ovoid particles composed of carbonate mud with no internal structure. Modern particles of this type are fecal pellets of worms, molluscs, crustaceans, and other organisms that feed on organic-rich mud. After they absorb nutrients from the mud, pellets of residual carbonate mud are excreted. Pellets harden rapidly due to the high calcium carbonate content so are likely to be preserved. They are common in modern quiet-water lagoons, so even though they occur in shallow water, they are not likely to be destroyed by waves or strong tidal currents. Modern particles of definite fecal origin are called pellets, whereas similar grains in ancient limestone are termed peloids because of the less certain nature of their origin.

Intraclasts are mostly larger particles derived from the local erosion of partially consolidated sediments. They are especially common in carbonate environments because carbonate sediments are so readily cemented. Calcium carbonate is the cement, so even a small amount of $CaCO_3$ precipitated from the sea water hardens the sediments soon after deposition. If exposed and dried out, carbonate mud will curl up into hardened mud chips that may become intraclasts. Storm waves and other energetic currents also rip partially consolidated carbonate sediments from the shallow sea floor and redeposit them nearby. The internal composition of intraclasts varies greatly and depends on the makeup of the original sediments.

Interstitial material. There are only two possible compositions when considering the matrix or interstitial material in a limestone. Each reflects very different conditions. The first is microcrystalline calcite or carbonate mud, shortened

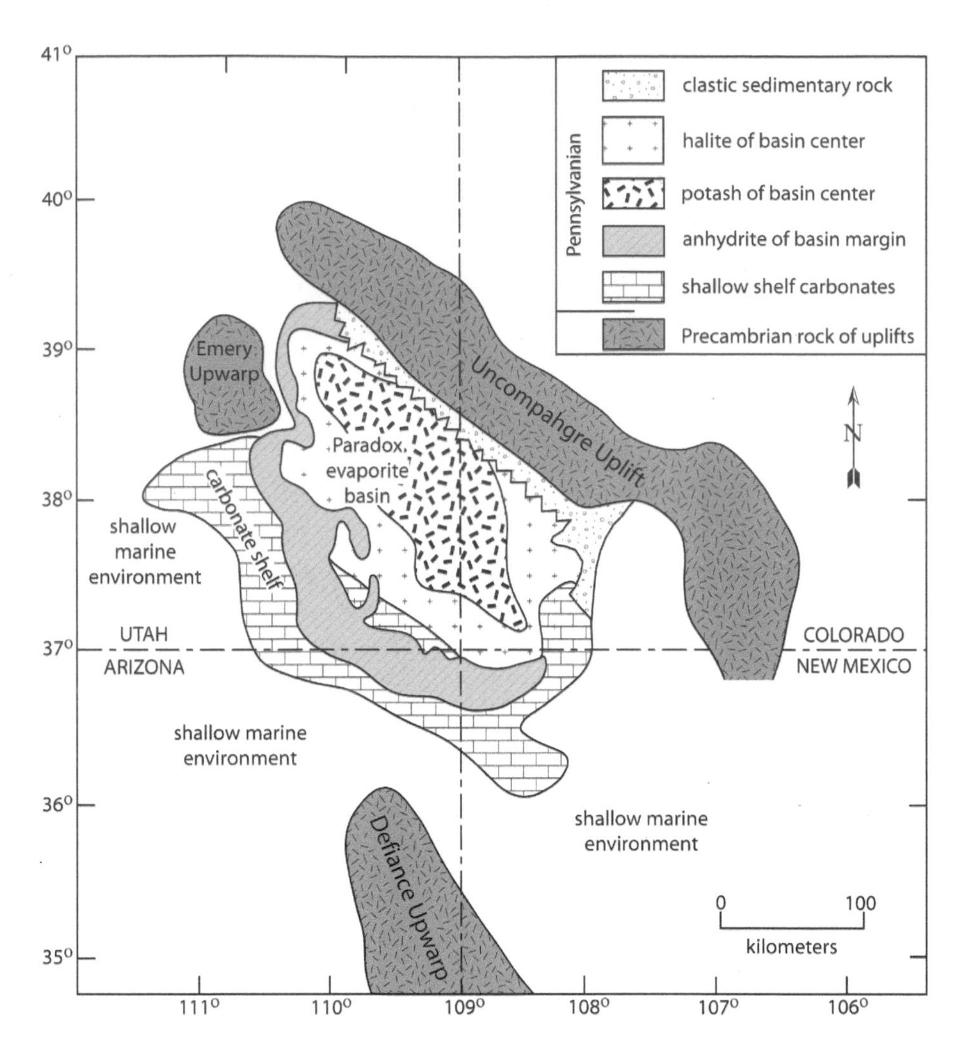

FIG. 2.12. Map of Paradox basin showing approximate limits of facies in the Pennsylvanian Paradox Formation and various surrounding features that influenced Pennsylvanian deposition. Map is modified after Williams-Stroud 1994.

by geologists to micrite. The second is sparry calcite, or spar, which consists of coarse, clear or white calcite crystals that fill spaces between the framework grains.

Micrite accumulates simultaneously with the grains that it encompasses. It is derived primarily from the breakdown of calcareous algae and marine plants, although some is generated by microscopic organisms that bore into shells and other carbonate grains. Whatever its origin, micritic limestone implies a low-energy depositional environment. Although micrite may be produced in a high-energy, shallow marine setting, it is easily winnowed seaward into deeper water, where it is more likely to come to an undisturbed rest. Micritic sediment is most common on deeper parts of the shelf, well out of the range of wave activity, and in lagoons or any other protected, shallow water setting.

Interstitial sparry calcite signifies very different conditions than those of micrite. Clear calcite crystals are precipitated slowly in the small voids between the framework grains as $CaCO_3$-rich fluids filter through the porous sediments. Precipitation may occur soon after deposition or later, after burial by subsequent sediments. The extreme mobility of $CaCO_3$ is one aspect of the compound that we will see again and again. It is easily dissolved by natural water, but also precipitates readily from water that contains dissolved $CaCO_3$. Pore-filling sparry calcite is characteristic of limestone deposited in high-energy settings where currents are too swift to allow carbonate mud to settle onto the sea floor. The voids between the irregularly shaped carbonate grains accommodate the subsequent crystal growth.

Associations between framework grains and interstitial material reinforce our interpretations of current energy and water depth. For instance, ooids and fragmental fossil grains are interpreted to form in high-energy currents. Limestone dominated by these particle types is usually cemented by sparry calcite. On the other hand, peloids and whole fossil shells commonly occur in a micritic matrix, all components attesting to a low-energy environment, one in which peloids and shells were not reduced to unrecognizable rubble, and mud was not winnowed away. While these interpretations generally hold true, it would be dangerous to interpret depositional environments strictly on the components of a rock. Instead geologists rely on a combination of field relations such as sedimentary structures, relations with laterally equivalent and over- and underlying rocks, as well as the composition of the unit in question. These methods will become evident in the following discussion of the shelf carbonate successions in the Paradox Formation.

Depositional setting of the shallow shelf carbonate facies of the Paradox Formation

While evaporite deposition was taking place in the basin depths, carbonate was accumulating on the broad, shallow shelf that defined the south and southwest basin margins (Fig. 2.12). At times, shallow water on the shelf was teeming with carbonate-producing organisms, notably the phylloid algae Ivanovia that grew into large mounds, or bioherms, but also brachiopods, crinoids, sponges, and bryozoans. The Paradox carbonate succession has been informally divided by petroleum geologists into four intervals that include, from bottom to top, the Barker Creek, Akah, Desert Creek, and Ismay intervals (Fig. 2.13). These subdivisions are defined by marker beds at their base and/or top, unique units that are easily recognized and traced in the subsurface through drill hole data. Each of these intervals records multiple cyclic sea level changes, and each correlates to several evaporite cycles (Fig. 2.13). Like its evaporite counterpart, the carbonate sequence was profoundly affected by the ups and downs of the sea.

Water depth on the shelf regulated both sediment type and the thickness of its accumulation. The broad nature of the shelf resulted in a large variation in car-

FIG. 2.13. Pennsylvanian stratigraphy of the Paradox basin, southeast Utah, showing the zones of the Paradox Formation established by workers in the petroleum industry. These zones are traceable throughout the subsurface through drill hole data and recognition of various marker beds, notably thick, black shale beds that were deposited during sea level highstands, but also laterally continuous dolomite and anhydrite beds. Note that the Molas Formation actually straddles the Mississippian-Pennsylvanian boundary.

Pennsylvanian Period	Honaker Trail Formation	
	Paradox Formation	Ismay zone
		Desert Creek zone
		Akah zone
		Barker Creek zone
		Alkali Gulch zone
	Pinkerton Trail Formation	
	Molas Formation	

bonate composition over short distances at any given time. An important factor in this diversity was preexisting topography on the shelf floor, a condition that was sometimes generated by the rapid buildup of Ivanovia and sponges. The main control, however, was true sea level change (eustatic sea level changes) that periodically resulted in subaerial exposure and erosion on the shelf. The restriction of overall thickness to 1000 feet in the shelf strata contrasts noticeably with the ~7000 feet of time-equivalent basin center deposits. The much reduced subsidence rate on the shelf severely limited the space available for deposition of sediment (known as accommodation space). Essentially, rapid lateral changes in sediment during any given time interval, coupled with a constantly fluctuating sea, produced a patchwork of rock types that defies simple description. Still, there are some dominant facies that can be related to eustatic changes in the same way as the previously discussed evaporite cycles. Thus, the representative carbonate shelf sequence in the following discussion correlates with the basin center evaporite cycle outlined earlier.

The following discussion traces the evolution of a single carbonate-dominated cycle as the sea rose from its lowest level to a highstand. This was followed closely by a slow drop, until the sea again reached a lowstand. The sea level lows are marked by disconformities so that each shelf cycle is sandwiched between erosion surfaces.

The cycle begins with the lowstand erosion surface that formed when sea level dropped below the shelf, exposing it to the forces of weathering. This was followed by a sea level rise as the open sea that lay to the south and west overtopped the shelf and flooded the basin. This event is marked by the deposition of black, organic-rich mudstone—the only unit in the carbonate cycles that can consistently be traced into the basin center. Once the restricted Paradox basin received this influx of fresh seawater, the carbonate machine turned on. Deep-water carbonate mud (micrite) rich in sponge fragments was deposited on the black shale. Succeeding units record increasingly shallow water as a stable carbonate platform became established. Sponge-bearing limestone grades up to limestone with a diverse shallow marine fossil fauna and a decrease in mud. This is overlain by

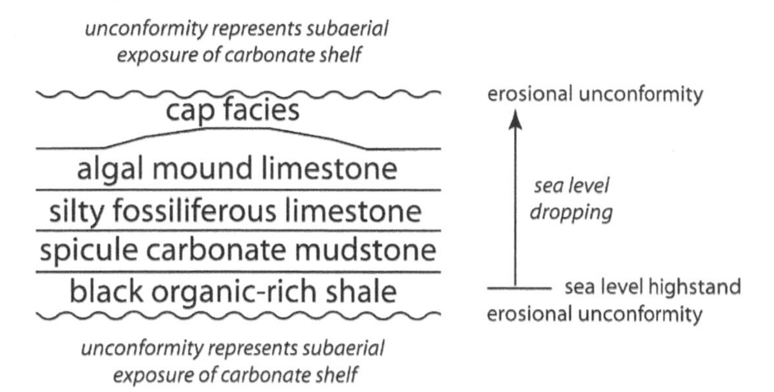

FIG. 2.14. Idealized carbonate cycle from the shallow shelf setting Paradox Formation on the southwest margin of the Paradox basin showing the carbonate facies present and the cycle-bounding unconformities.

bioherms composed of phylloid algae. These are capped by limestone composed of broken fossil fragments (grainstones) and ooids, both indicative of a very shallow, high-energy shoal setting. It is within these shallow water grainstones and algal mounds that the cycle-terminating disconformity developed as the sea again receded to expose the shelf.

Like the evaporite succession, shelf carbonate cycles record a regular rise and fall of sea level (Fig. 2.14). It is there, however, that the similarities stop. Besides the obvious differences in rock types between the two areas, there is no symmetry in the shelf cycles. Additionally, the erosion surface on the shelf has a different origin and developed over a greater period of time.

Lowstand erosion surface. Erosion surfaces that bound each cycle above and below formed when sea level dropped below the shelf floor, exposing the carbonate sediments to the elements. Calcium carbonate is extremely susceptible to alteration by meteoric water such as rainfall. All natural waters in the form of rain and snow are slightly acidic due to mixing with CO_2 in the atmosphere (to produce carbonic acid—H_2CO_3). Calcium carbonate dissolves when exposed to such waters. Thus, when sea level dropped, one of the dominant weathering processes was dissolution. An important result was the development of a porous zone along this surface as meteoric water trickled down through the unprotected sediments, enlarging preexisting voids. This type of alteration is vitally important in the search for petroleum since it is in these pore spaces that the petroleum occurs. Larger pore spaces may mean more petroleum.

A product of this dissolution was caliche, a residual type of $CaCO_3$ that forms when $CaCO_3$-rich water evaporates from the soil. As meteoric waters percolated through the carbonate sediments, the waters quickly became enriched in $CaCO_3$. In the arid climate that is believed to have dominated the Paradox basin, moisture easily evaporated, abandoning the $CaCO_3$ to the sediments to form a crust just below the surface.

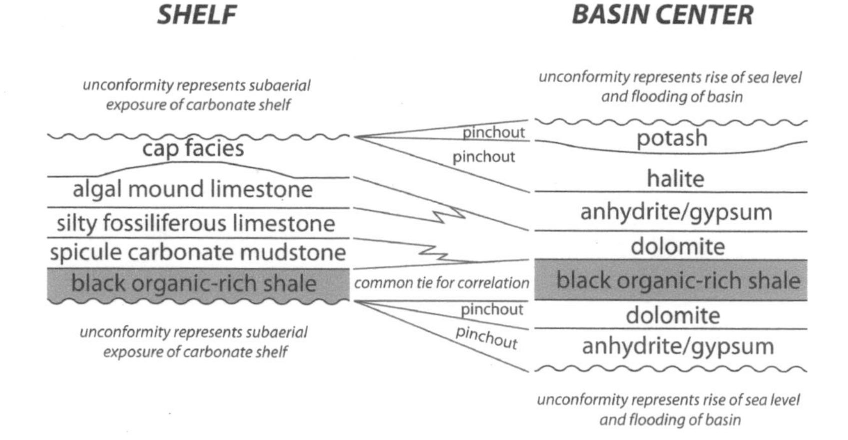

Fig. 2.15. Correlation of cycles in the Paradox basin showing a single cycle from the evaporites in the basin center with a cycle from the shallow, carbonate-dominated shallow shelf. Note that while both cycles are bounded by unconformities, they form through very different processes and at different times.

Other significant, but less common features that aid in recognizing erosion surfaces are thin localized soil zones (paleosols) marked by calcified traces of plant roots (rhizoliths) that extend into underlying units. Rarely these are associated with thin lignite horizons, brown-black carbonaceous layers formed by the preservation and compaction of vegetation that eked out a tenuous existence in this short-lived and hostile environment.

In a study of the Barker Creek and Akah intervals exposed in the San Juan River canyon, geologists Gary Gianniny and Toni Simo (1996) of Fort Lewis College in Durango, Colorado, have recognized at least one instance of fluvial deposits nested within an erosion surface. In this case, large, localized, lens-shaped bodies of coarse, pebbly sandstone are incised into rocks of the underlying cycle. These are interpreted as large river channels. The development of a drainage system on the emergent shelf suggests exposure over a broad area for a lengthy period of time.

As the sea dropped below the level of the shelf, eventually to reach a lowstand, the emergent area became a barrier that restricted the exchange of water between the basin and the open sea. While the shelf was exposed, successive beds of anhydrite, halite, and potash were deposited in the basin center. Thus, the evaporites that make up the upper part of each basin center cycle are the time equivalents of an erosion surface and a period of nondeposition on the shelf (Fig. 2.15).

During some intervals of shelf exposure, eolian sand was blown onto the shelf from the southwest. The sand was trapped in depressions on the dessicated shelf and is mostly preserved in cracks that formed in the carbonate sediments as they dried out. In some cases, sand was blown basinward to interfinger with anhydrite that was deposited from the basin water. During the subsequent sea level rise, eolian sands were reworked into marine shoreline deposits.

Organic-rich mudstone. Black organic-rich shale was the first widespread unit to be deposited on the lowstand disconformity. Organic-rich clay and carbonate mud were deposited as sea level rose and the basin received a fresh influx of water. These shales correlate with the basin center shales that mark the "turnaround" point of the symmetric evaporite cycles. These beds are easily traced across the basin, even in the subsurface, and provide a much needed common tie between the very different lithologies of the basin center and shelf cycles. Shelf shales range 3 to 10 feet thick (1–3 m) and expand considerably basinward to 30 to 50 feet (10–13 m). Black shale on the shelf, as elsewhere, is laminated and has a high organic content. In the subsurface it is believed to be the source rock for petroleum that occurs in overlying porous limestone. The sparse fossil content and scarcity of burrows, features that would indicate the presence of organisms, coupled with the preservation of organic matter, suggest oxygen-deficient, possibly toxic, hypersaline conditions. Exxon geologist Robert Goldhammer and colleagues (1994) estimate that black shale on the shelf was deposited at depths of more that 100 feet (30 m). Thus, organic-rich shale in the Paradox represents deposition during the later stages of sea level rise as it reached its maximum depth.

Spiculitic carbonate mudstone. Black shale typically is overlain by gray, chert-rich carbonate mudstone. Wavy laminations and abundant siliceous sponge spicules are important elements of this facies, while delicate bryozoan and thin-shelled brachiopods are rare but also significant. Needle-like sponge spicules form the framework structure of sponges and are the part that is likely to be preserved after death. While the hard parts of most marine organisms are composed of $CaCO_3$, these spicules are siliceous, meaning they are composed of microcrystalline quartz (SiO_2). The presence of chert, also microcrystalline quartz, likely is related to the abundant sponge spicules.

Deposition of carbonate mudstone is interpreted by Goldhammer and his colleagues (1991) to mark the gradual transition from sea level highstand to slightly shallower depths, around 50 to 100 feet (15 to 30 m). These depths are based on the preservation of thin laminations and the thin-shelled fossils. Fine laminations suggest a low-energy setting in water that was deep enough to be unaffected by most shelf currents. Thin-shelled fossils also indicate low-energy conditions because the fragile shells could not have withstood high- or moderate-energy currents without being broken. Sponges can withstand almost any conditions. The low-diversity fauna and scant burrows in the sediment suggest that environmental conditions remained far from ideal. This is probably due to continued oxygen deficiency and/or hypersaline conditions.

Carbonate mudstone is transitional between underlying nonfossiliferous black mudstone and overlying fossiliferous limestone. Conditions had improved enough that sponges were able to tolerate these waters and take advantage of this vacant niche, but not so much that anything else was able to thrive.

Fossiliferous silty limestone. Silty limestone grades up from spiculitic mudstone and tracks the slowly shallowing shelf. These units are laterally continu-

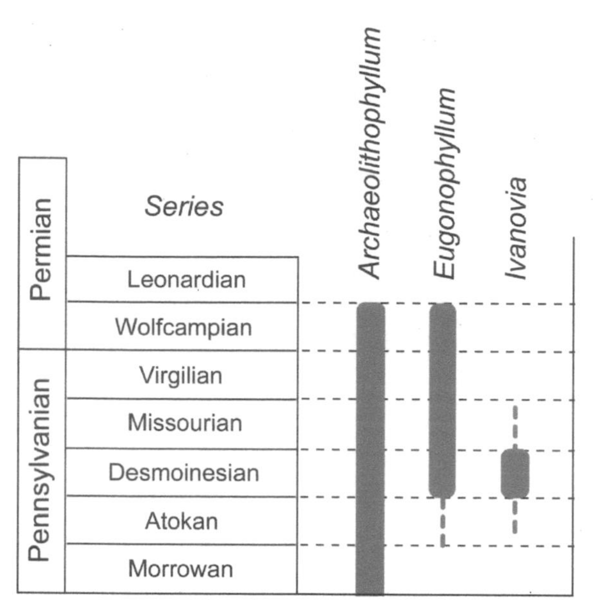

Fig. 2.16. Time range of Ivanovia compared to other common Pennsylvanian-Permian phylloid algae. Modified after Ginsburg and others 1971.

ous and range > 100 feet thick (> 30 m). Limestone beds are constructed of coarse skeletal fragments and whole fossils in a silt and carbonate mud matrix. Fossils are abundant and comprise a diverse assemblage of normal marine organisms, including an assortment of brachiopods, bryozoans, robust corals, fusulinids, large crinoids, and sparse phylloid algae (Fig. 2.16). Burrowed sediment is pervasive, also indicating a healthy and robust fauna that lived and fed in the muddy sediments.

Fossiliferous limestone heralds a change to normal marine salinity and well-oxygenated shallow marine water. As these conditions evolved, abundant and varied organisms moved onto the shelf, wasting no time in colonizing this previously barren niche. The fossil fauna and carbonate sediments suggest deposition in water depths of ~15 to 30 feet (5 to 10 m). Shallow, normal marine water on the shelf paved the way for the revival of the carbonate platform on which the succeeding algal mound facies was deposited.

Algal limestone. The algal mound facies consists of light gray limestone composed mostly of platy, leaflike fragments of the phylloid algae Ivanovia. Individual algal plates are irregularly shaped and measure up to several centimeters across. Stacked plates are cemented by micrite or sparry calcite. Common additions to the mud include peloids and skeletal fragments of normal marine organisms such as bryozoans, crinoids, fusulinids.

Mounds formed by the growth and accumulation of algae are well documented from excellent exposures of the Desert Creek and Ismay intervals along the spectacular canyon walls of the lower San Juan River (Pray and Wray 1963, Goldhammer and others 1991, Grammer and others 1996, Gianniny and Simo 1996). On this part of the shelf, mounds mostly are discrete, isolated features with flat bottoms and convex-upward tops. The flanks of these mounds reached angles of 25° to 30°. This is believed to represent the geometry during deposition. Individual mounds ranged 20 to 40 feet thick and up to 90 feet long. The lower parts are dominated by micrite, with scattered algal and skeletal fragments. Micrite decreases upwards as the algal chips become the primary component.

Algal facies in the subsurface ~70 miles (110 km) east of the San Juan River exposures are also well studied due to their tendencies to contain petroleum. In fact, Pennsylvanian algal limestones are the dominant petroleum producers in the

Four Corners region. The largest of these are the Aneth field in the southeast corner of Utah and the Ismay field, which straddles the Utah-Colorado border. While algal limestone in this area is very similar to that exposed along the walls above the San Juan River, these accumulations are much more extensive. Algal limestone units still range up to 40 feet thick, but they may extend laterally for several thousand feet, blanketing the shelf on which they were deposited. This large area of algal limestone is believed to be the result of the lateral overlap of individual mounds; they amalgamated into what appears to be, from the viewpoint of drill hole data, a single, widespread unit. These overlapping mounds tend also to be stacked vertically, forming multiple extensive units in the shelf deposits in this area. These relations suggest that conditions in this particular part of the shelf were consistently ideal for the growth of Ivanovia.

Ivanovia was a dominant Mid-Pennsylvanian mound-building algae but disappeared from the seas by the end of this period (Fig. 2.16). These algae apparently thrived in depths as shallow as 10 to 15 feet (3 to 4 m). Water was deepest in the early stage of mound construction, possibly up to 50 feet (15 m). This is told by the muddy lower mound deposits in which algae were relatively sparse. Micritic algal limestone represents the onset of shelf depths and water conditions that were suitable for colonization by Ivanovia. Initially the shelf was deep enough that mud remained unaffected by wave currents, but shallow enough to be within the photic zone required for algal growth. As sea level continued its slow drop and the mounds began to build vertically, wave energy increased and mud was more easily winnowed into sheltered areas between mounds or onto deeper parts of the shelf. Upper mound deposits consist dominantly of stacked algal plates cemented by later-stage sparry calcite cement. During deposition, large void spaces between the irregular "potato chip"–shaped plates were porous, but have since been partially filled with calcite. These rocks still contain good porosity, which is what makes them such a promising target for petroleum exploration.

Researchers have suggested that Ivanovia plates originated as leaves on a stalked, upright-growing algae that rose several centimeters off the sea floor, similar to the modern calcareous algae Halimeda. During the best of times they formed lush meadows on the shelf, their broad leaves swaying in the currents. Upon death (the worst of times), the leaves detached from the stalks and settled to the sea floor, carpeting it with a layer similar to that found on the floor of a deciduous forest in the fall. Mud and skeletal fragments that washed in on wave currents accumulated in the hollows of the leafy plates. The roots, or holdfasts, that connected the stalks to the substrate probably helped to stabilize the mounds and even contributed to their construction by inhibiting erosion during high-energy events such as storms. Additionally, the upright stalks and leaves likely acted as baffles and sediment traps by dampening the increasingly energetic currents as the mounds grew up into shallower water.

The algal mound facies is the most intriguing and important because, of all the facies in the Paradox Formation, it is the most likely to hold petroleum. It is for this reason that the shelf carbonates in the Paradox basin have been exhaustively studied and are so well known from both outcrop and subsurface data. This relationship with petroleum will be revisited in a later section of this chapter.

Cap facies. The importance of the algal mound facies is emphasized by the term "cap facies" that is used to describe the numerous and diverse rock types that directly overlie the mounds. Collectively, the rocks of the cap facies represent the shallowest water and highest-energy conditions recorded in the shelf limestones, although some lower-energy deposits accumulated in more protected parts of the broad shoal that formed on the shelf.

The high-energy cap facies rocks are mud-poor and dominated by sand-sized carbonate fragments of various origins including ooids and fossil fragments of crinoid stems, fusulinids, brachiopods, bryozoans, molluscs, and rare phylloid algae. All skeletal fragments show signs of abrasion, and the limestone units they comprise typically are cross-bedded, indicating that vigorous currents swept across the shelf. Some skeletal sands occupy the broad troughs and hollows that separate the mounds. These deposits are interpreted by Michael Grammer and colleagues (1996) to outline channels through which sand was threaded as forceful currents washed across the extensive shelf. These deposits slowly filled the hollows, lapping onto the steep mound flanks and eventually blanketing them. According to analogues with modern skeletal sands, these high-energy limestones were deposited in water less than 15 feet deep (< 5 m) and record the lowest sea level on the shelf except for the subsequent lowstand exposure.

Moderate- to low-energy cap facies are represented by peloidal limestone and laminated anhydrite. These contrasting lithologies shared a common element during deposition—namely, a position on the shallow shelf that was sheltered from the high-energy currents that washed across it.

Besides containing peloids and micrite, peloidal limestone has several features that provide clues to its depositional setting. Dessication cracks indicate periodic exposure as the muddy carbonate sediments shriveled in the dry heat of the bare shelf. Intraclasts occur in this facies and likewise point to exposure. Intraclasts typically are composed of fragmented mud chips that curled up during exposure and dessication. These are subsequently shuffled around and redeposited as the sea sweeps back across the mudflat. Dessication cracks and intraclasts are good indicators of a tidal flat setting. The conditions required to form these features are provided by the alternating flood and withdrawal of the tidal water across the shelf. The final piece of evidence for a tidal mudflat origin is the presence of wavy laminations in the micritic limestone. These are interpreted as mats of blue-green algae, based on similar wavy laminations observed in modern algal mats on modern tidal flats in the Persian Gulf and the Bahamas. Blue-green algae are one of the few organisms, past and present, that can tolerate the harsh and varying conditions of an arid climate tidal flat.

Laminated anhydrite overlying algal mound carbonate is recognized in the subsurface, along the southern shelf, in the Four Corners area. Laminations indicate quiet water deposition, but its association with algal limestone indicates accumulation in shallow water, a stark contrast with anhydrite deposited in the deep basin center. These relations have led Grammer and coworkers (1996) to suggest that anhydrite was precipitated in restricted lagoons that evolved within and adjacent to the mound buildups. As sea level dropped, intermound depressions and nearby low-lying areas became quiet evaporative pans, shielded from the high-energy tidal currents and waves. Additionally, the mounds inhibited the influx of fresh seawater into the lagoonal areas, promoting the concentration of brines. Interbedded with anhydrite are wavy laminated carbonate interpreted as mats of blue-green algae, attesting not only to a shallow water origin, but also to the adaptability of blue-green algae to conditions that would be lethal to most organisms.

Petroleum in the Paradox basin

More than 100 small oil fields have been discovered in the Four Corners area along the southern shelf of the Paradox basin (Fig. 2.17). Most of these fields produce from the algal mounds and related facies of the Ismay and Desert Creek zones in the Paradox Formation; however, minor production is from the Mississippian Leadville Limestone or other Pennsylvanian and Permian rocks. The largest field, by far, is the Greater Aneth field, which covers an area of 75 square miles in southeast Utah. The Aneth is surrounded by smaller fields that extend into Colorado, New Mexico, and Arizona. Collectively, these fields have produced more than 400 million barrels of oil and one trillion cubic feet of natural gas.

Early exploration of Paradox shelf carbonates revealed all the essential ingredients for the possible occurrence of oil and gas, including organic-rich source rocks, porous reservoir rocks, and impermeable caprocks. Source rocks for the hydrocarbons consist of the black, organic-rich shale beds that punctuate every carbonate and evaporite cycle. When the shale reached a temperature of ~435°C, organic matter began to transform to petroleum. If the temperature rose to 460°C or higher, the hydrocarbons were converted to natural gas. Thus, the window of temperature for petroleum generation is small. This depends partly on depth of burial because the temperature in the Earth's crust increases at a regular rate with depth. As part of a large government initiative to study the Paradox basin in detail, USGS geologists Vito Nuccio and Steve Condon (1996) reconstructed the burial history of the Ismay and Desert Creek zones throughout the Paradox basin. According to their study, these rocks in southeast Utah were buried at a maximum depth of 13,400 feet about 25 million years ago—soon after the large-scale

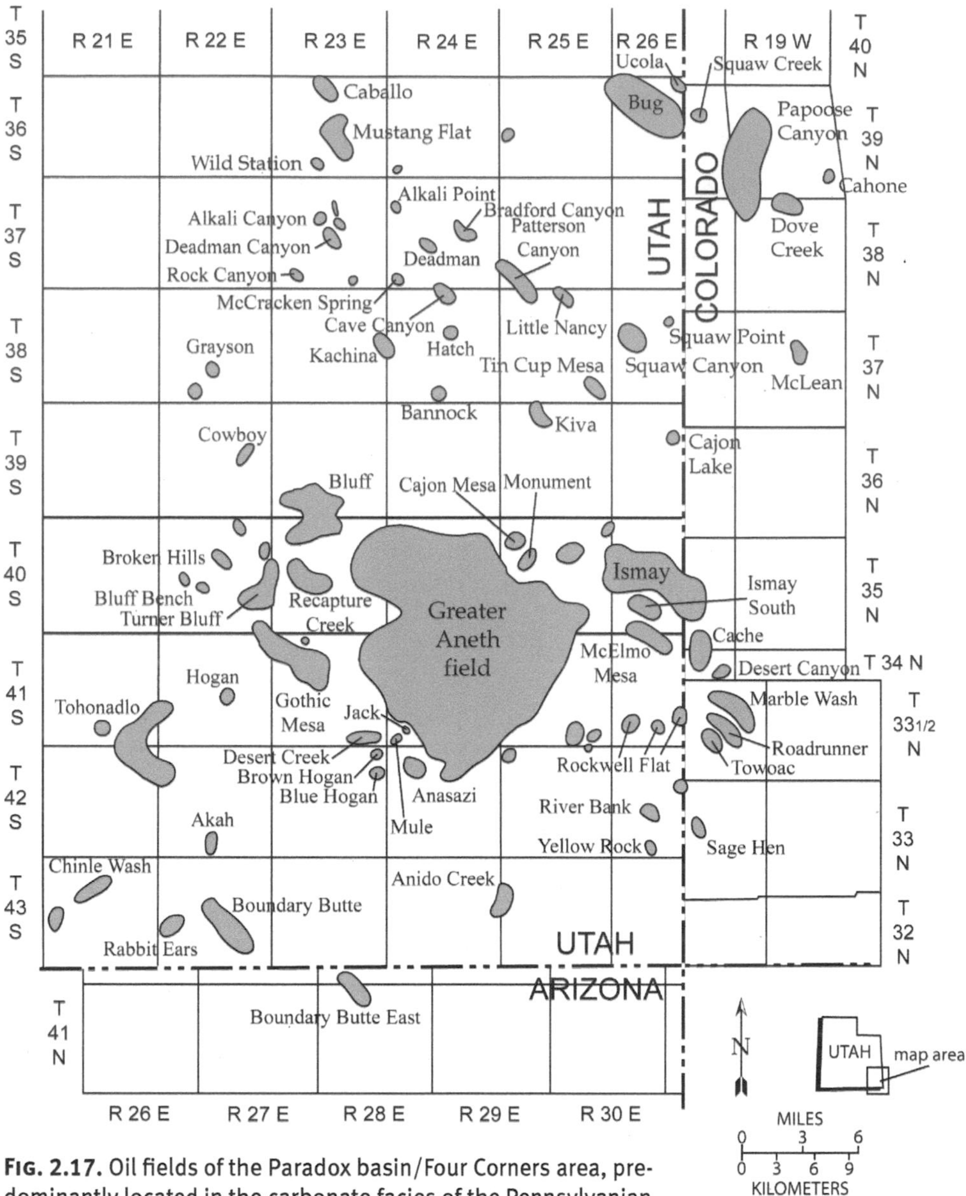

FIG. 2.17. Oil fields of the Paradox basin/Four Corners area, predominantly located in the carbonate facies of the Pennsylvanian Paradox Formation.

eruptions of the San Juan volcanic field in southwest Colorado. At least 1000 feet of the sediments in the Paradox basin that laid on top of the Ismay and Desert Creek zones is believed to have been volcanic detritus shed from the east. Much of this material has since been stripped away by erosion. In fact, in the Aneth field the Desert Creek zone presently lies ~5500 feet below the surface, indicating that 7900 feet (a mile and a half!) of sediments have been scoured away in the last 25 million years.

As liquid petroleum was released from the source rock, it slowly migrated upwards toward porous rock and an escape from the pressure of thousands of feet of overlying material. This condition was met by porous algal limestone and the carbonate sand of the cap facies. Dissolution of calcium carbonate during lowstands undoubtedly improved the porosity and permeability. Pore spaces must be interconnected to create a permeable passageway through which petroleum can flow. This is vital for the entrance and exit of oil from the rock. Upon drilling into a porous reservoir, the petroleum must be pumped to the surface, sucked out through the drill pipe much like lemonade in an ice-filled glass is sucked through a straw.

The last requirement for the generation and preservation of petroleum is an impermeable caprock above the reservoir rock. In the Paradox Formation carbonate mud or anhydrite seals the top of the porous reservoir, halting the upward migration of petroleum. This allows the petroleum to accumulate in one place in economic amounts and keeps it from rising to the surface, where the volatile material would escape to the atmosphere, rendering it to a useless residue.

A common problem upon discovering oil in rock is its actual removal. In the Paradox oil fields the recovery rate typically is less than 20 percent of the oil that is present in the reservoir. While all the oil in a reservoir can never be recovered, rates can be improved. Since each Paradox field was discovered and developed, production has decreased. This is a common pattern with oil fields; however, scientists with the Utah Geological Survey, through funding by the U.S. Department of Energy, are intensely studying several Paradox fields to determine the character of the reservoirs and to design appropriate methods for enhanced recovery. Possibilities include flooding the reservoir with water or injecting it with carbon dioxide. Both approaches would wrest some of the unrecovered petroleum from the tortuous rock passageways. It is estimated that another 200 million barrels of oil could be extracted through these methods (Chidsey and others 1996).

Sea level changes and cyclic sedimentation

Cyclic sedimentation in Pennsylvanian rocks has been recognized around the world, allowing us to confidently relate it to periodic eustatic sea level changes. This also rules out local effects such as tectonic activity and changes in sedimentation rates. The growing consensus is that sea level changes were caused by regular climate fluctuations which, in turn, triggered alternating accumulation and melting of glacial ice in southern Pangea. During the Pennsylvanian the southernmost part of this supercontinent sprawled across the South Pole, providing an anchor for a great thickness of ice. In this scenario, cooling generated growth of the ice cap, locking up a substantial part of the planet's water and dropping sea level. Subsequent warming provoked melting, generating a sea level rise, up to 115 feet by some estimates.

While the waxing and waning of glaciers at the Pennsylvanian South Pole has long been suspected as the source of cyclic sea level changes, the cause of the cli-

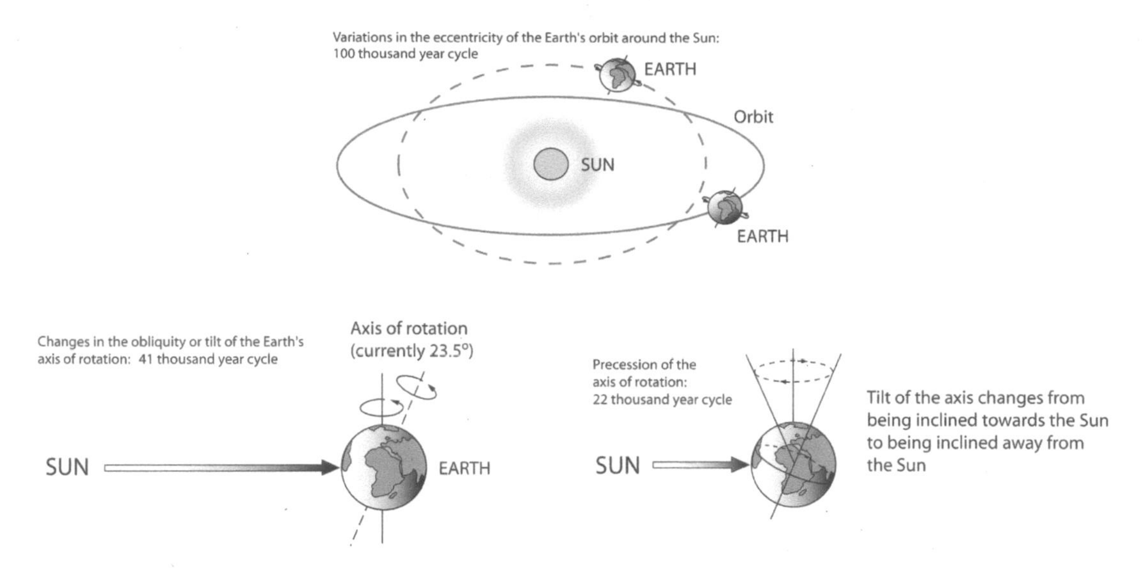

FIG. 2.18. Factors in the rotation of Earth around the Sun that are part of the Milankovitch cycles, which were responsible for relatively short term glaciation and deglaciation cycles. Modified after Leeder 1999.

mate fluctuations that sparked it remained elusive. New discoveries and ongoing research into climate change of the last few million years have been applied with some success to this ancient problem.

In the 1920s and 30s, Serbian geophysicist Milutin Milankovitch worked out three variations in Earth's orbital geometry and hypothesized the occurrence of regular global climate change based on the compounded effect of these variations. These changes have since become known as Milankovitch cycles (Fig. 2.18). The first variation is in the eccentricity of the Earth's orbit around the Sun, which changes from more circular to elliptical over a period of 96,000 years. The second factor, which varies over a 41,000-year period, is the inclination or tilt of the Earth's axis, on which it spins. This spin axis is not vertical, but varies between 21.4° and 23.4°. Like the eccentricity, this changes the amount of heat received from the Sun. Finally, the Earth wobbles on its axis like a giant spinning top. This regular precession, as it is called, has a periodicity of 23,000 years. In concert, these variations may be enough to nudge the climate across the threshold of glaciation in the polar regions. It might not take a dramatic temperature change—perhaps just a small variation in the length of winter or summer, or a change in seasonal precipitation patterns—to cross this line. For instance, a shortened summer could result in the incomplete melting of the previous winter's snow and, over a long period, lead to the evolution of glaciers. Similarly, a change from dominantly summer precipitation to a winter-dominated precipitation pattern might lead to the development of glaciers.

Milankovitch's hypothesis couldn't be tested at that time because there were no detailed climate data spanning the amount of time he was proposing. It was not until the 1970s that quality climatic data for the past few million years began to be obtained from cores of deep sea sediments. These sediments, which have accumulated undisturbed for hundreds of thousand years, hold the most detailed paleoclimate record available. In comparing this record with Milankovitch's hypothesis, an agreement with recent geologic history has become apparent. Many geologists who study Pennsylvanian rocks now suspect that the same processes were influencing sedimentation in the Paradox basin and elsewhere at this time. While Milankovitch cycles likely have been occurring over a vast period of geologic time, conditions during the Pennsylvanian were optimal for recording the changes. Conditions that collaborated to leave this record include the location of southern Pangea on the South Pole, a climate that straddled the boundary between glaciation and melting, and shallow marine conditions over large parts of the continent that were dramatically altered by small, regular changes in sea level.

A recent study by Bruce Rueger (1996) has answered some of the larger questions concerning the Pennsylvanian climate in the Paradox basin. Rueger analyzed palynomorphs, microscopic plant pollen, from the evaporite-dominated basin center rocks. Palynomorphs were blown into the basin from adjacent landmasses and provide a record of the flora that occupied the terrestrial realm during deposition of a given rock type. Essentially, Rueger found recurring variations of species within the strata that can be related to cyclic changes in climatic conditions.

In an analysis of several cycles from a drill core of the Paradox Formation taken immediately east of Canyonlands National Park, Rueger found that dolomite and shale beds contain the same palynomorphs as those found in Pennsylvanian coal beds in the midcontinent of North America. These suggest warm, moist conditions during deposition of the middle part of the cycles, when sea level was high. In contrast, these species are absent from halite beds. Palynomorphs from halite instead suggest a cool, arid climate during times of low sea level. Anhydrite units contain only fragmented bits of organic material, believed to be unidentifiable remnants of palynomorphs that were mashed during the transition from gypsum to anhydrite.

To summarize the relationships between sea level, climate, and deposition, the cyclic growth of glaciers in the Southern Hemisphere removed water from the planet's hydrologic budget, triggering a drop in sea level and creating an arid climate in the equatorial regions. Reconstructions of Pangea for the Pennsylvanian place the Paradox basin well within this equatorial desert. As the sea approached its lowest level, the shelf emerged and the basin became a restricted inland sea or lake in which thick successions of halite accumulated. On the adjacent Uncompahgre highlands, the cool, dry climate limited vegetation to plants that were tolerant of a low-moisture setting. At the other end of the spectrum, interglacial melting released huge volumes of water to the sea, enough to raise sea level worldwide and increase humidity globally. As marine water flooded over the shelf and into the basin, the brines became diluted, and dolomite and shale deposition replaced the earlier evaporite setting. Palynomorphs from these deep-water deposits reveal a warm, wet climate. Although the plants that populated the Uncompahgre landmass at this time were the same as those that flourished in coal swamps of the midcontinent and elsewhere, they never grew in sufficient amounts to form coal. However, the additional runoff from the Uncompahgre highlands during this time, coupled with the abundant vegetation, did contribute a substantial volume of organic detritus to the black shale that was blanketing the basin. As we saw earlier, this became the source for the hydrocarbons that are sought after today. In the next section we revisit the strata of the Paradox Formation for a final look at the most recent interpretations of their origin.

Sea level changes and deposition of the Paradox Formation

In the Paradox basin, the simultaneous deposition of different lithologies on the shallow shelf and in the deep basin center are deciphered by tracking sea level changes. Except for the black mudstone that blanketed the entire basin, the two settings share no rock types. Even the unconformities that so clearly delineate the cycle boundaries in both parts of the basin formed at different times and by different processes. For example, while the shelf was exposed during the lowstand, evaporite deposits of halite and potash were accumulating in the basin center. It is here, at the end of a cycle, that we will begin our inquiry into the evolution of the next one.

As sea level approached its minimum depth, the shelf area emerged from the sea, and the top of the preceding carbonate succession was set upon by dessication and erosion. This marks the initiation of the disconformity on the shelf. Meanwhile, in the basin center, evaporite deposition continued and likely was intensified. The exposure of the shelf further isolated the already restricted basin, possibly cutting it off completely to form an enormous hypersaline lake. It was probably at this point that the brine became concentrated enough so that halite and, in extreme events, potash dropped out of the toxic solution. In correlating individual units within the cycles from basin to shelf, halite and potash are the time equivalents of the cycle-bounding unconformity on the shelf (see Fig. 2.15).

The next stage was a gradual sea level rise. Following an interval of exposure the shelf was slowly inundated. Judging from the rock record, deposition on the shelf during this time was minimal. An exception was the redistribution by water of eolian sand that had blown onto the emergent shelf during the prior lowstand. The basin center facies, however, were extensively affected by the inpouring of fresh seawater. As the incoming water mixed with the brine, it became diluted, and halite precipitation ceased. In fact, the diluted mixture easily dissolved the uppermost part of the thick halite interval, generating the disconformity that marks the cycle boundary here.

As fresh seawater continued to enter the basin and dilute the brine, a predictable succession of evaporite deposits was laid down. Above the irregular halite-dissolution surface, first gypsum and then dolomite were precipitated from the increasingly dilute solution. This predictable sequence is the reverse order of mineral precipitation that has long been recognized from normal seawater as it becomes increasingly concentrated by evaporation. Since experiments reported in 1849 by Usiglio, it has been known that through concentration by evaporation, a well-ordered succession of evaporite minerals precipitate from seawater. The first to precipitate are calcite and dolomite, followed by gypsum. Gypsum is later altered to anhydrite during burial with a loss of water from its crystal structure. It is only after the depletion of magnesium, calcium, and sulfate ions from the brine that halite begins to precipitate. Finally, in cases of extreme brine concentration, potassium-bearing salts may crystallize.

We will see soon enough how this process relates to the upper part of the basin center cycle, but how does it apply to the lower half? Essentially, the lower part of the cycle is a reversal of the sequence outlined above. It begins with the hypersaline brines that were precipitating halite of the underlying cycle. As the incoming seawater mixed with these brines and halite sedimentation was cut short, the modified fluid instead dissolved the upper halite layers. As further mixing and dilution progressed, gypsum precipitated, followed by a gradual change to dolomite formation. As the extremely concentrated brines slowly became diluted, a reversed succession of evaporite minerals accumulated in the basin center. While gypsum and dolomite were forming in the basin depths, the shelf area apparently was still adjusting to being submerged again, and no deposition was taking place.

As sea level neared a maximum, deposition of black, organic-rich mudstone began throughout the basin, providing a valuable basin-wide reference point. In fact, it is these mudstones that this entire discussion hinges on. Mudstone deposition continued past the sea level maximum and into the following regressive phase. Organic material came from both marine and continental sources. Lowered salinity in the basin likely promoted algal blooms and high rates of bacterial reproduction. Additionally, a switch to a humid climate spurred vegetation growth on adjacent landmasses with a concurrent increase in runoff. This washed abundant organic material and the very fine detritus of weathered rock into the basin. The floor of the basin must have been oxygen deficient at this time given the amount of organic material that was preserved. (Organic carbon decomposes rapidly in oxygenated water.)

It was only after the sea level began its steady descent that normal marine water began to blend with the basin water on a large scale. As this transition progressed, sponges populated the muddy shelf. Eventually a more diverse fauna of brachiopods, bryozoans, crinoids, and corals colonized the broad shelf, marking a return to optimal conditions for healthy marine organisms. Although it was slowly shallowing, the shelf was deep enough that carbonate mud remained the dominant sediment type.

While carbonate-producing organisms gained a toehold on the shelf, basinal mudstone gave way to dolomite. Some researchers have suggested that the dolomite originated in shallower water and was flushed basinward by currents to settle to the basin floor. Others, however, maintain that it precipitated directly from the water as the influx of fresh marine water again was reduced. The gradational contact with underlying mudstone, and with overlying anhydrite, suggests that dolomite deposition was forced by sea level changes and a related increase in salinity. On the other hand, the high silt content of the dolomite suggests a detrital origin, as both dolomite and silt-sized quartz were possibly flushed into the basin from the northeast by runoff that emanated from the highlands. It is likely that a combination of dolomite precipitation from the water column and detrital material washed from the basin margin contributed to these deposits.

As sea level continued to drop, the shelf shallowed. When the sea became shallow enough that sunlight could penetrate to the shelf floor, the shrub-like, phylloid algae Ivanovia began to grow on the muddy substrate. With time and the accumulation of sediments, mounds of algae and mud built up from the sea floor. This shoaling eventually put the mound tops into the zone of strong wave and tidal currents, which swept the mud away to leave stacked plates of Ivanovia.

Just before the sea dropped below the level of the shelf, water was at its shallowest and, because of waves and tidal currents, its most energetic. This resulted in deposition of the cap facies over much of the shelf: clean carbonate sand composed of ooids and skeletal fragments that rolled back and forth in the currents that plowed across the shallow rim of the Paradox basin. Towards the basin center the cap facies is onlapped by a thin wedge of anhydrite that thickens northeastward into the upper anhydrite of the evaporite cycle (Fig. 2.19). This relation provides a much-needed tie between the shelf and basinal cycles.

While the algal mound and cap facies were accumulating on the shallow shelf, dolomite in the basin center was probably giving way to gypsum (eventually to become anhydrite). As sea level diminished, the restricted basin grew increasingly saline--to the point that gypsum precipitation became favored over dolomite. The dolomite-gypsum transition marks the beginning of the classic evaporite sequence that we saw in reverse for the lower part of the cycle.

For many years geologists believed that the growth of algal mounds on the shelf was accompanied by halite deposition in the basin center. This was based on the interpretation that the dissolution surface at the top of the halite formed at the same time as the exposure surface that marked the top of each carbonate cycle. As is common in geology (and all sciences), new discoveries and more detailed work by subsequent researchers may cast a new eye on an old problem. New interpretations of the Paradox Formation, built on a strong foundation of

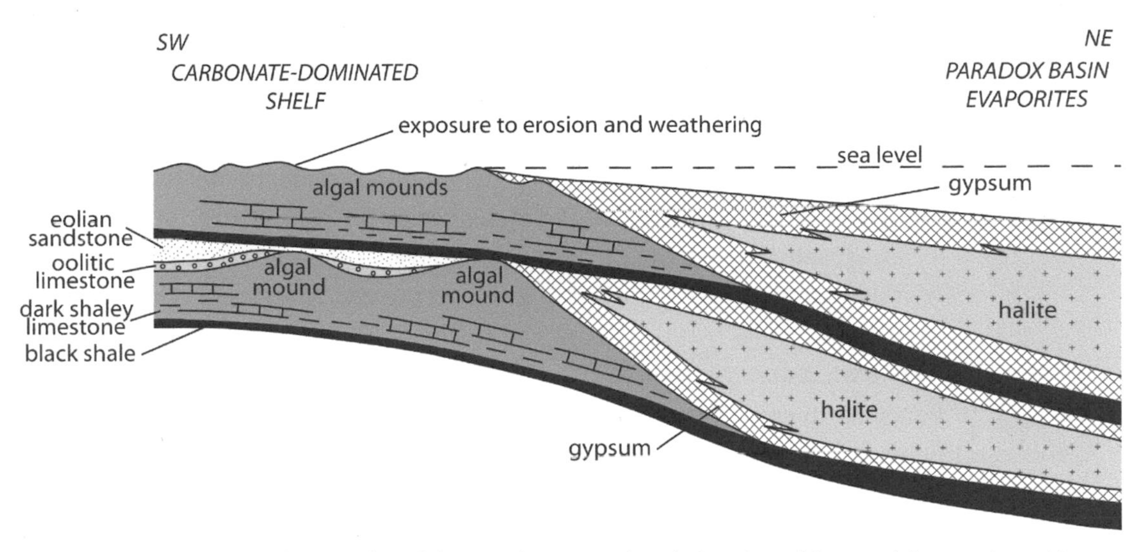

FIG. 2.19. Cross section of two cycles of the Paradox Formation during deposition, and the stratigraphic relations between the basinal evaporites (northeast) and the shelf carbonates (southwest). Note that the anhydrite onlaps the algal mounds on their basinward side. The only common tie between the two different facies is the black shale bed deposited during the sea level highstand. Also note that the basinal evaporites are somewhat simplified, with only the halite and gypsum/anhydrite facies shown.

earlier studies, suggest that it was actually exposure of the shelf that prompted halite precipitation in the basin center.

When sea level dropped well below the shelf, the Paradox basin became completely isolated from the open ocean that bordered it to the south and southwest. This transformed the restricted inland sea, at least temporarily, into a giant saline lake (Williams-Stroud 1994). As the inflow of seawater became obstructed by the emergent shelf, evaporation far exceeded the influx of water from precipitation and runoff. This situation was exacerbated by the accompanying shift to aridity, further increasing salinity and creating an ideal setting for halite precipitation. The relatively great thickness of individual halite beds (up to 800 feet) has led many researchers to conclude that precipitation at this stage was rapid.

Part of the argument for the complete separation of the Paradox basin from the open sea is the concentric map pattern of the successive evaporite deposits that took shape as each subsequent rock type occupied a smaller area of the basin. This systematic decrease in map area suggests that as evaporation concentrated the brine, the volume of liquid decreased. This is most obvious along the southwest basin margin, where the sea floor rose gradually toward the shelf. For instance, anhydrite—which precipitated as gypsum early in the succession, when the water volume was high—laps onto the edge of the shelf. Overlying halite contracted in area to the middle part of the basin, where the trough-like geometry was more pronounced. This has been interpreted to reflect the area that was actually submerged during halite deposition. Potash, which in some cycles overlies halite, is even more limited, typically to a narrow, northwest-trending zone.

Potash, mostly in the form of sylvite, has been recognized at the top of seventeen of the thirty-three evaporite cycles. Where present, it grades up from the underlying halite and reflects deposition during the highest salinity reached for any given cycle. It has been found experimentally that potassium salts begin to precipitate from brines of original seawater composition only after 98 percent of the original water has evaporated. Potash represents the final depositional event before the flooding and dissolution that signals the end of a cycle.

The economic aspects of potash

Potash is an economic mineral that today is mainly used in fertilizer. Although it is widespread in the subsurface of the Paradox basin, it is mined on a large scale only in the Cane Creek anticline area, along the Colorado River about seven miles southwest of Moab. The mine, which was opened in 1964, is owned and operated by Texasgulf Inc. Initially potash was extracted by normal underground mining methods, but the work was difficult due to the high underground temperatures (100°F) and the folded evaporite beds that made following the deposits a daunting task. In 1970 Texasgulf decided to try solution mining, at first pumping water into the mine to circulate through the salts, and then pumping the saturated brines out. The plan proceeded more rapidly than anticipated when an exploratory drill hole encountered groundwater, and the mine was flooded. Eventually the leaks were sealed, and today the solution operation is well controlled. After the saturated brine is recovered from the ground, it is placed into giant solar evaporation ponds to remove the liquid. This stage of the operation is essentially a smaller-scale replay of the original depositional processes of the evaporite deposits ~295 million years ago! These ponds cover ~400 acres and are easily seen from the overlook at Dead Horse Point State Park, where they look like giant mirrors tucked into the red, crenulated landscape.

After the water is evaporated, the salts are taken to the nearby processing plant (shown as "Potash" on maps) along the Colorado River. Here the powder is refined to 60 to 62 percent K_2O. A railroad spur from the plant transports the material north to Thompson, where it continues by railroad to various plants.

Honaker Trail Formation

The Pennsylvanian Honaker Trail Formation conformably overlies the Paradox Formation and is the uppermost unit of the Hermosa Group. Like the Paradox, the Honaker Trail records cyclic sea level fluctuations. This formation, however, contains no evaporites and instead consists of alternations of marine carbonate and shale interbedded with fluvial and eolian sandstone.

The Honaker Trail Formation, originally defined by Wengerd and Matheny (1958), is marked by a color change from the gray and black of the Paradox to the red-, brown-, and buff-colored strata of the Honaker Trail. This change proved difficult to recognize from drill hole data, but after several rounds of revision, petroleum geologists now place the contact at the top of the uppermost halite bed in the Paradox Formation. The contact is defined as the top of the Ismay zone near the basin center, but it is the top of the Desert Creek or even the Akah zone to the southwest, where first the Ismay and then the Desert Creek halite beds pinch out (Fig. 2.20).

The formation is named for exposures at the top of the Honaker Trail on the east side of the San Juan River canyon, a few miles northwest of Mexican Hat. Here, in 1904, numerous switchbacks were blasted into the steep limestone cliffs by a crew of men being supervised by a local Mormon cowboy named Henry Honaker. The motivation for this undertaking was the discovery of gold in a nearby sandbar along the San Juan River. After completing the trail, the prospectors found that the gold was too sparse to be worthwhile. Today, however, the trail provides one of the few routes in this stretch of deeply incised canyon to reach the banks of the San Juan. It also offers the most convenient way to see the shelf carbonates of the upper Paradox Formation and the Honaker Trail Formation up close.

The limestone, sandstone, and shale of the Honaker Trail Formation are exposed sporadically throughout the Paradox basin. Along the northeast basin margin, strata equivalent to the formation are exposed at Hermosa Mountain, a few miles north of Durango, Colorado. Halite is absent here. Because it is the halite that allows for the subdivision of the Upper Pennsylvanian strata, the Paradox and Honaker Trail formations are lumped into undifferentiated Hermosa Group. The formation is also exposed in Fisher Valley, Utah, near the Colorado-Utah border. The most complete and laterally extensive exposures are to the west, along the Colorado River and its tributaries in Canyonlands National Park. The Honaker Trail lines the precipitous canyon walls from just north of the confluence of the Green and Colorado rivers, southward through Cataract Canyon. Here, in the lower tributaries, notably Gypsum and Dark canyons, are some of the most accessible (relatively speaking!) exposures in the region. In fact, the inviting waterfalls in lower Dark Canyon spill over the resistant limestone beds in the formation. Finally, the Honaker Trail is well exposed in the San Juan River canyon where uplift of the Monument upwarp and incision by the San Juan have conspired to create a spectacular gallery of Upper Pennsylvanian strata.

The most detailed study of the Honaker Trail Formation on the northeast margin of the basin comes from the work of Karen Franczyk and others (1995) at Hermosa Mountain, Colorado. Although the absence of a clearly recognizable lower boundary prohibits the definition of the formation here, they informally recognize the upper 830 feet of the succession as the equivalent of the Honaker Trail Formation. This is based on lithologic correlation with drill hole data a short distance to the west. The Hermosa Mountain strata consist of cyclic alternations of

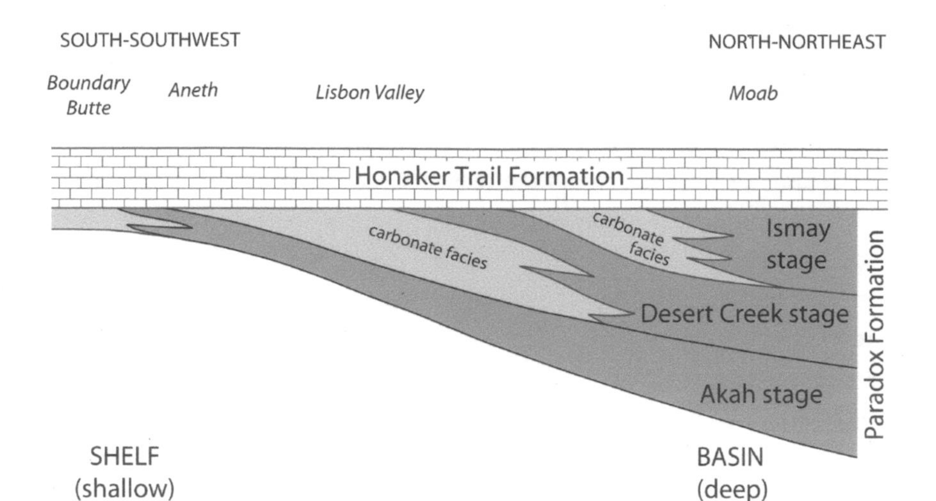

FIG. 2.20. Schematic cross section of Pennsylvanian strata in the Paradox basin showing the relationship between the various stages of the upper Paradox Formation and the overlying Honaker Trail Formation. The Paradox cycles are defined by shale beds that separate evaporite-dominated successions. The dark shading of the labeled stages denotes halite-dominated strata. These grade southwestward into shallow-water shelf carbonate facies. Note that the Honaker Trail overlies increasingly older Paradox cycles as they pinch out to the southwest onto the shelf.

carbonate and clastic rock that record a continuation of glacially driven sea level fluctuations. Because the area lay at the foot of the Uncompahgre highlands, the succession is dominated by clastic rock composed of detritus washed from this ancient range. Cycles at Hermosa Mountain begin with marine carbonate and grade upward into clastic rocks of various origins that become progressively coarser. These cycles reflect an initial sea level highstand followed by a slow drop.

Carbonate beds that signal the beginning of each cycle host an assortment of fossils, including brachiopods, bryozoans, fusulinids, and sponge spicules, encased in a lime-mud matrix. Taken together, these components point to deposition on a muddy carbonate shelf under normal marine conditions. Higher in the succession, as carbonate beds thin, fossil diversity diminishes and sand content increases. Franczyk and colleagues suggest deposition in an intertidal setting into which sand was washed from adjacent highlands. The overall vertical trend in these limestone units suggests initially moderate depths--deep enough that mud was not winnowed away. Subsequent limestone beds indicate a gradual drop in sea level, and an incremental drop with each fluctuation. The succession probably records both a filling of accommodation space along this margin as sand influx began to exceed basin subsidence, and a large-scale westward retreat of the sea, even as its shoreline oscillated on a finer scale.

Clastic sequences that succeed each carbonate bed become coarser-grained upwards, typically grading from siltstone into sand and, in some instances, conglomerate. As we have seen in the Paradox Formation, cyclic sequences track the west-retreating sea through each oscillation. As the sea receded following limestone deposition, a delta trailed it, first dropping clay and silt at the delta front, followed by sand—all of it derived from rivers that tapped the tireless sediment generator, the Uncompahgre highlands. As deltas extended vast tracts of land westward, over the limey sea deposits, rivers stretched their channels across the delta plain, etching it with ribbons of coarse sand and gravel. Still, after all that erosion, deposition, and infilling, the sea ultimately rose to regain its former position, clearly marking it with another limestone. Thus the Pennsylvanian cycles continued.

The best and most spectacular place to examine the Honaker Trail Formation is along the walls of the Colorado and San Juan rivers. Here, geologists Mark Williams (1996) and S. C. Atchley and David Loope (1993) have studied in detail the cyclic alternations of eolian and fluvial sandstone, and marine shale and limestone. Large-scale depositional patterns in these continuous exposures, augmented with drill hole data from the subsurface immediately to the east, recount a marked change in basin geometry over this period. In the early stages of Honaker Trail deposition the basin had shallowed, but had maintained its earlier trough-like geometry with a marine-dominated setting in the deeper central part. The shelf that lay to the southwest was periodically blanketed by eolian sand when emergent. During the later stages of deposition the trough became filled with sediment and its surface expression was lost to a gentle slope that extended unbroken from the Uncompahgre highlands farther westward to a fluctuating shoreline.

Honaker Trail cycles consist of the now-familiar pattern of an erosion surface overlain by sediments that first record a sea level rise, followed closely by a drop. This commonly incomplete record culminates in another erosional exposure surface, punctuating each cycle like bookends. Rather than exhaust the possibilities of rock types that could be sealed between these erosion surfaces, we turn to a case study of a single cycle in the Honaker Trail Formation, documented by Mark Williams (1996) from the lower part of Dark Canyon. Unlike the underlying Paradox Formation, there is no typical or "standard" cycle for the Honaker Trail Formation. The following example, however, serves to illustrate some of the differences between the two formations, as well as some of the processes involved in Honaker Trail deposition.

In this cycle the basal erosion surface developed on marine limestone as it was exposed during the lowstand. It is generally agreed that erosion was carried out dominantly by eolian processes as coastal winds scoured the former sea floor. At this point erosion presided over deposition, and the winnowed detritus was swept downwind. It was not until the sea began to rise that the eolian sediments became damp and stationary. The disconformity is covered by up to 30 feet of large-scale, cross-stratified sandstone deposited as the windblown sand was molded into large dunes. Eolian sandstone grades up into bioturbated sandstone that tracks the advancing shoreline into the coastal dune field. Abundant burrows indicate that marine organisms found this new environment to their liking. As the sea deepened, thick deposits of fossiliferous limestone were laid over the sandstone. At this point, water was shallow enough to provide ideal conditions for a diverse fauna, but deep enough that carbonate mud remained in place, eventually to entomb the abundant and varied organisms. Fossils include bryozoans, brachiopods, horn corals, fusulinids, and crinoid stems, all signifying normal marine conditions. Thin carbonate shale beds overlie the limestone and mark the deepest water deposit within the cycle. This "turnaround" point was followed closely by the regressive phase of the cycle as the sea began to recede from the area. This relatively deep water shale is succeeded by silty carbonate mudstone with a limited fauna of ostracodes, clamlike crustaceans that can tolerate most conditions. This restricted fauna, coupled with mudcracks and salt crystal impressions, suggest a shallow intertidal setting in which high evaporation rates created isolated saline pools. The abrupt jump from highstand shale to mudcracked tidal flat sediments almost certainly involved an interval of erosion, possibly during the passage of a high-energy shoreline as sea level dropped. An interesting feature of this facies is sand-filled fissures that are interpreted by David Loope and Zsolt Haverland (1988) as large mudcracks that opened up as the tidal sediments dried. Subsequently, some of the windblown sand that bounced across this barren surface during the ensuing exposure fell into these fissures, eventually filling them. This wind-scoured surface marks the top of the cycle. Overlying strata of the next cycle consists of shallow marine sandstone. Thus, continental deposits are missing from the base of the next cycle, even though evidence for the existence of such a setting is present. Such are the whims of erosion and sedimentation.

The mélange of nonmarine facies in the Honaker Trail facies ranges from eolian sandstone to fluvial conglomerate, to lacustrine (lake) limestone. The variations in these deposits are understandable, as they are largely controlled by local conditions such as sediment supply and local topography. Variation in marine deposits is attributed to the shallow nature of the Late Pennsylvanian sea, and the ease with which erosion erased parts of the record. A shallow sea means a greater variety of carbonate facies since even minor changes in depth can dramatically change the amount of sunlight that penetrates to the sea floor, and therefore the type and rate of biological productivity, the current energy, and salinity of the water. Smaller-scale oscillations of sea level are always superimposed on the larger, glacially driven cycles. In a shallow sea these minor variations produce more obvious changes in lithology. Additionally, Atchley and Loope (1993) found that most of the Honaker Trail cycles had been "beheaded"—that is, the upper part of the cycle had been removed by erosion. This was attributed to a combination of exposure and erosion during lowstand periods, and to a lack of early cementation in the limestones. This was convincingly demonstrated in their microscopic analysis of the eolian sandstone. Although most of the grains were composed of quartz, some units contained up to 30 percent fossiliferous limestone grains that had obviously been stripped from nearby shallow marine deposits during lowstand exposure and piled into the dunes downwind.

The upper part of the Honaker Trail Formation marks the filling of the trough that defined the Paradox basin over most of the Pennsylvanian Period. As Honaker Trail deposition came to a close, the rate of sediment coming into the basin exceeded the subsidence rate, and as the trough filled, the fingerlike seaway that had once occupied this depression was pushed out of the region. By the close of the Pennsylvanian a gradual, low-relief slope extended uninterrupted from the apron of coarse clastic rock that skirted the Uncompahgre highlands westward to the shoreline in central Utah. The trough that had earlier defined the basin center was filled, its topographic expression gone.

This is not to imply that tectonic activity in the form of uplift in the highlands or subsidence in the basin had ceased. In fact, subsequent Permian rocks provide a clear record of continued activity in this mountain belt/basin pair. This record, however, is dominated by eolian and fluvial deposition as the shoreline was nudged westward. Additionally, ongoing, cyclic sea level fluctuations continued to affect the nature of the sediments.

Salt Anticlines and Related Salt Deformation Features

Soon after the thick evaporite succession of the Paradox Formation was deposited and began to be buried, halite was mobilized. The lateral and vertical migration of salt was a response to the increasing load of overlying sediments. Halite has unique properties compared to other geologic materials. It is a very low density mineral and under pressure has the ability to easily flow in a plastic manner, with a consistency reminiscent of cold molasses. Movement typically is upward due to its considerable density difference with surrounding rock. In the Paradox basin, most of the vertical salt movement occurred along elongate, northwest-trending zones where the salt pushed up to form regional-scale salt anticlines: large, uparched folds of sedimentary rocks cored by rising salt. Such activity is common wherever thick salt deposits are covered by later sediments. It is currently taking place along the Gulf Coast of Texas and Louisiana, where Jurassic-age salt is rising up through more recent Gulf Coast sediments.

The scenic wonders of Arches National Park and most of Canyonlands National Park owe their existence to the mobility of salt. Regionally, salt anticlines affect a large portion of southeast Utah and southwest Colorado: virtually the entire region is underlain by the thick evaporite facies of the Paradox Formation. In Utah, important salt anticline valleys include the Fisher-Professor Valley, where the mazelike Fisher Towers are found; Castle Valley, at the north end of the La

Sal Mountains; Salt Valley, which forms the landscape of Arches; and Spanish Valley, occupied by the sprawl of Moab. Along the western border area of Colorado lie Paradox Valley, Gypsum Valley, and the Lisbon-Dolores Valley structure (see Fig. 2.5). All these salt anticlines have collapsed along their crests. This occurred as salt rose toward the surface, only to be dissolved by percolating meteoric water. Thus what was once the crest or axis of these linear folds is now a valley. The evolution of these unique features is discussed in detail below.

To the southwest, in Canyonlands National Park, the movement of salt takes on some different forms. The Needles fault zone, which is responsible for the labyrinthine topography of the Needles District, is attributed to salt beneath the area. In this region the underlying salt acts as a slide surface for overlying rocks as they slowly collapse toward the Colorado River canyon (Cataract Canyon). This colossal shift was initiated by the methodical excavation by the river of great volumes of rock and was coupled with a gentle regional tilt of the strata to the west. The removal of support by erosion triggered the collapse of overlying strata into the resulting void by sliding on a viscous surface of salt. Faulting within the overlying brittle plate of Pennsylvanian-Permian strata produced the jumble of downdropped and uplifted blocks that define the Needles District. These blocks have been split into a colorful array of fins and pinnacles that give the area its name.

The area north of the Needles is marked by several small salt anticlines that have not been breached by erosion or salt dissolution. These include Gibson dome, Rustler and Lockhart anticlines, and the Cane Creek anticline, which is easily viewed from the overlook at Dead Horse Point State Park. The potash plant and its large evaporative pools are situated along the axis of the Cane Creek anticline (Fig. 2.21).

One of the more intriguing salt-related structures in Canyonlands is the Meander anticline. The axis of this salt-induced fold follows the trace of the Colorado River through Canyonlands, even curving around bends in the canyon, from the south part of the Needles, northward almost to Cane Creek anticline, a distance of more than 25 miles (Fig. 2.21). There are several hypotheses for the origin of this unique, snakelike fold. These features will be addressed in detail in a following section.

By far the most controversial, potentially salt-related feature in the region is Upheaval dome, located in the Island in the Sky part of Canyonlands. This circular, dome-shaped structure is ~4 miles (6 km) across. The central portion, however, is in reality a crater. The origin of this unique structure has been variously interpreted as a volcanic-related feature, a salt flow feature, and a meteorite impact crater. Geologists have debated the origin of Upheaval dome since the 1930s, and today we are no closer to a consensus than we were seventy years ago. In fact, recent research has brought the debate back to the forefront, focusing

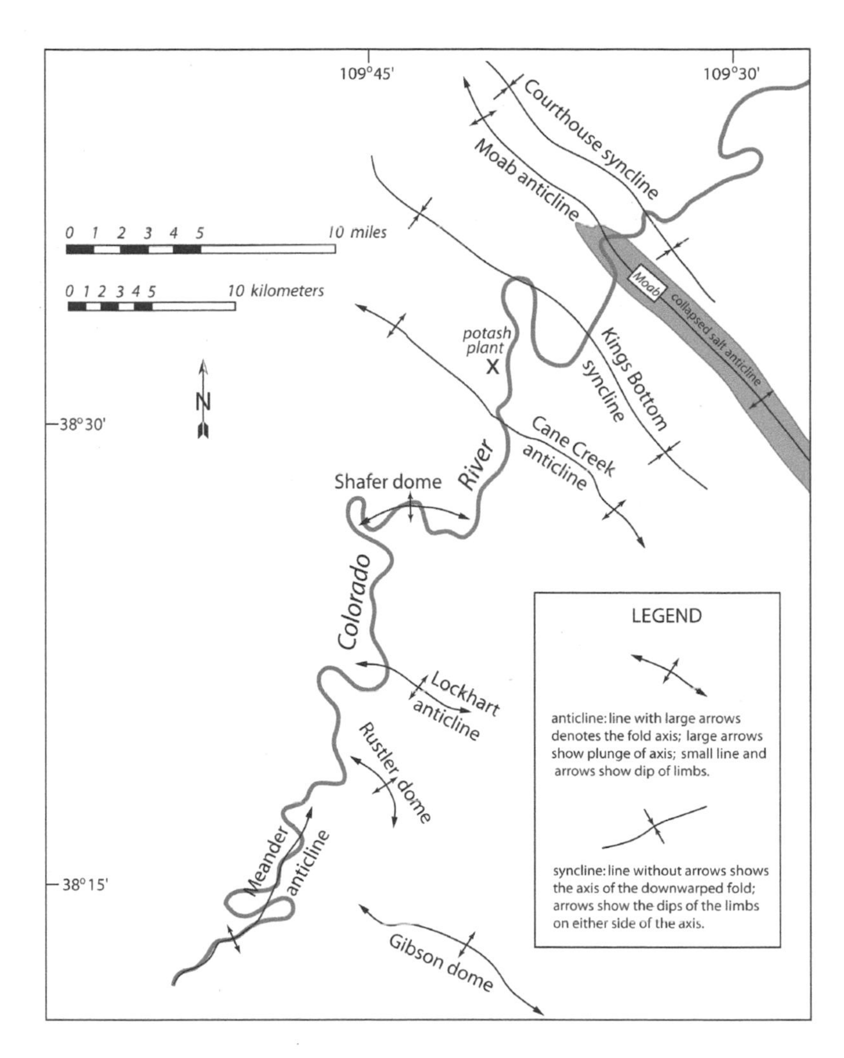

FIG. 2.21. Map of the unbreached salt anticlines in northern Canyonlands and adjacent areas. Note the relationship between the potash plant and the Cane Creek anticline. The collapsed Moab–Spanish Valley anticline, the town of Moab, and the Colorado River are shown for reference. Data are from Williams 1964.

dominantly on the salt and meteorite hypotheses. The following section will look at arguments for both sides.

Salt Anticline Valleys

The large salt anticline valleys along the Colorado-Utah border consistently trend northwest-southeast, parallel to the main bounding fault that separates the Uncompahgre highlands from the Paradox basin, and parallel to the elongate geometry of the basin. In fact, most of the structural features in the region follow this orientation (see Fig. 2.5). This has long been attributed to the rejuvenation of a system of regional northwest-trending fractures in the Precambrian basement rocks that underlie the southwest part of North America. The genesis of these fractures has most recently been related to extension ~1.0 billion years ago as the Precambrian supercontinent of Rodinia broke up. During this breakup, the block that today makes up Antarctica pulled away from southwest North America, generating the system of large-scale, northwest-trending fractures (Marshak and others 2000, Timmons and others 2001). When later tectonic stress rippled through the region, these fractures acted as faults, accommodating vertical motion. Initial reactivation during the Pennsylvanian was in response to the rapid subsidence in the Paradox basin, and likely prompted the first phase of salt movement.

Hellmut Doelling of the Utah Geological and Mineral Survey, who has studied in detail the Salt Valley anticline that makes up most of Arches National Park, has recognized four periods of salt-related activity in its evolution. Although his work is centered on Arches and the adjacent Moab area, a similar origin can be inferred for the salt valley anticlines throughout the region.

According to Doelling (1988), the period of most active salt movement began during the Pennsylvanian Period, soon after the first halite bed was deposited. This was a time of tectonic convulsion, as the Ancestral Rockies pushed skyward and the fractured blocks that floored the Paradox basin began jostling for position during subsidence. Rapid but sporadic regional salt movement continued to the end of the Triassic Period, a time span of about 75 million years.

Evidence for the initiation and continued rise of salt anticlines is contained in the sediments laid down during this period. Pennsylvanian, Permian, and Triassic deposits thin dramatically over the crests of these great folds, and locally they disappear completely from the axial region (Fig. 2.22). These wedge-shaped sediment packages indicate that the rise of salt was already creating topographic relief along the axes of the anticlines, inhibiting sedimentation in these areas. These relations are much less pronounced in units deposited after the Late Triassic Chinle Formation.

The result of this salt deformation is clearly displayed along the Colorado River, on Highway 128 between Castle Valley and Moab. Here, lower Chinle strata have been tilted and are overlain by flat-lying upper Chinle deposits, producing an obvious angular unconformity. Soon after deposition of the lower Chinle sediments, the rise of underlying salt locally domed these strata upwards. This domed surface was quickly beveled flat and covered with upper Chinle sediments, creating the angular unconformity within the Chinle Formation. Angular unconformities typically are assumed to represent long periods of unrecorded time—that is, a prolonged period of erosion rather than sedimentation. This is based on the idea that millions of years are required to fold the rocks, then erode them, and finally bury the erosion surface with sediment. They are relatively common between formations, especially in tectonically active regions. In this case, however, the unconformity occurs within the formation, suggesting the processes responsible for this surface occurred over thousands, rather than millions, of years. These localized, angular unconformities occur throughout the region and underscore the rapid rate of salt movement during this period.

Adjacent to the salt anticlines are complementary synclines, less obvious downwarped folds that initiated during the lateral subsurface flow of salt away from the areas between the anticlinal rises. The thickening and upward move-

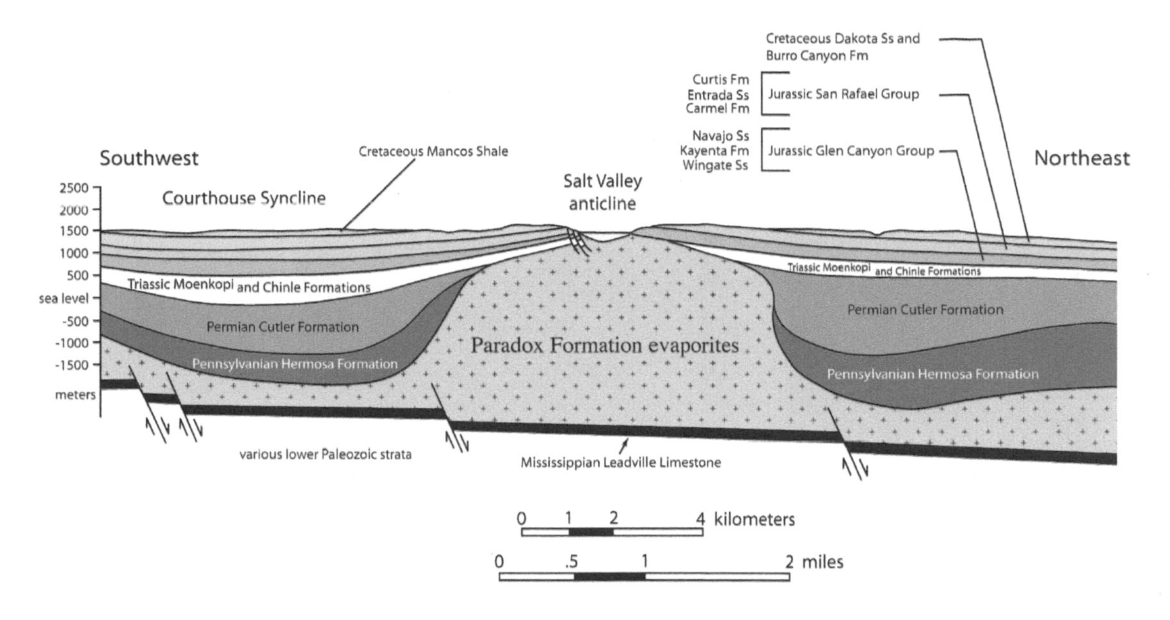

FIG. 2.22. Cross section through the Salt Valley anticline immediately north of Arches National Park showing the lateral flow of the Paradox Formation evaporites and their rise into the Salt Valley anticline. Note the thickening of overlying Pennsylvanian, Permian, and Triassic strata southwest of the anticline and the downwarping in this area due to salt withdrawal from beneath. This forms the Courthouse syncline. Note also the thinning and, in some cases, pinchout of these same strata on flanks of the Salt Valley anticline. This indicates salt movement during deposition of those units. Units that show no thickness changes across these structures were likely deposited during periods of no salt movement. Fm = Formation; Ss = Sandstone. Modified after Doelling 2001.

ment of salt along the anticlines required salt withdrawal from adjacent areas. The result was synclinal troughs between the anticlinal highs. Just as sedimentary units thin across the anticlines, the same units thicken in the intervening synclines (Fig. 2.22). For example, the Chinle Formation thins to 334 feet on the southwest flank of Salt Valley anticline, but thickens to more than 800 feet in the axis of Courthouse syncline, less than five miles away (Doelling 1988).

Courthouse syncline, located in Arches National Park, is bounded on the east by the Salt Valley anticline and to the west by the Moab anticline (Fig. 2.23). The syncline axis hosts Courthouse Wash, probably the most accessible and scenic canyon in Arches. Drainages coincident with syncline axes are common in folded regions because synclines form natural valleys. Runoff tends to drain off the limbs of the fold, where it is then concentrated in the axis, where the main drainage becomes rooted. Another nearby example of this type of relationship is the Salt Wash syncline, named for Salt Wash, which faithfully follows the syncline axis until draining southward into the Colorado River. Salt Wash syncline lies immediately east of Salt Valley anticline (Fig. 2.23).

The first period of salt movement was only the formative stage of what is seen today, and while these structures initiated during this time, they continued to evolve. The second period of salt activity was marked by a slowdown in deforma-

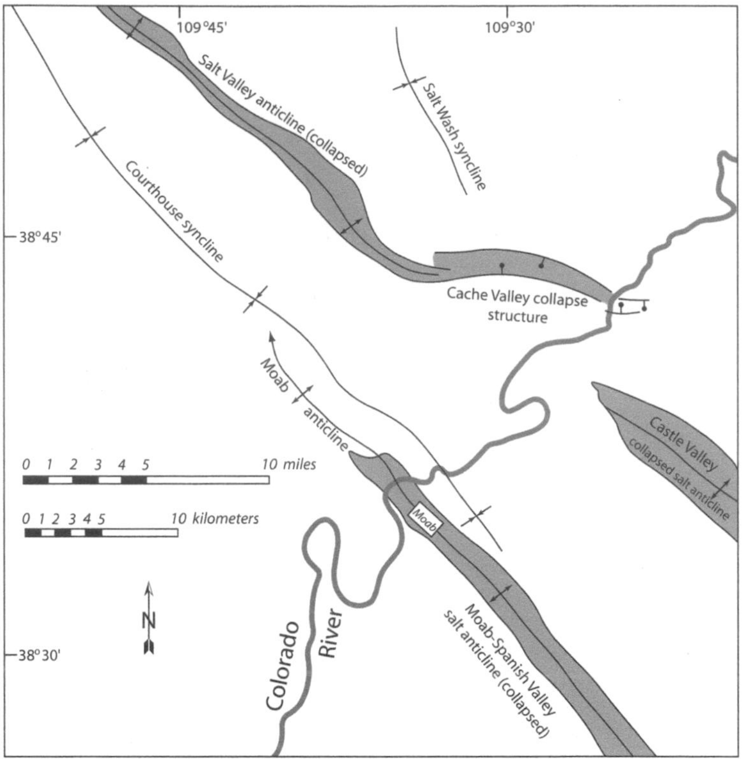

Fig. 2.23. Map of the major salt structures around the town of Moab and Arches National Park (the Salt Valley anticline). Note particularly the synclines between the major anticlines. As underlying Paradox salt flows laterally and rises to form anticlines, the withdrawal of salt from adjacent areas results in downwarping and the development of synclines.

tion, with only localized activity. This period extends through the entire Jurassic Period and into the Early Cretaceous, a span of ~125 million years (Doelling 1988). Movement was concentrated in the anticline crest areas where salt was thickest, and mostly was in the form of its continued rise. Sediments deposited during this time thin over the fold crests but do not thicken appreciably in the syncline areas, suggesting the lateral flow of salt had ceased. According to Doelling, even the rise of salt had halted by the end of this period.

During these first two periods of salt activity, deformation was not limited to simple folding. As the salt rose, punching up through the overlying units, faults and fractures formed in the brittle caprock. These weaknesses acted as conduits for meteoric waters, and as salt reached closer to the surface, it was easily dissolved.

The next period was dominated by regional subsidence and the deep burial of salt structures beneath thousands of feet of sediment. During this period, which lasted from Late Cretaceous to Late Tertiary time (~90 million years), salt mostly was dormant. The Early Tertiary in western North America was a time of renewed mountain building as compression again surged through the region. This orogeny forced up the monoclines that today characterize the Colorado Plateau and formed the modern Rocky Mountains of adjacent Colorado. It is likely that the salt anticline region was at least mildly deformed by these viselike stresses, probably through the rejuvenation of the salt-induced folds and faults. However, compelling evidence for this is lacking, and probably has been masked by subsequent salt dissolution, which triggered the large-scale collapse of most of the anticlines.

The final stage of salt anticline activity began about 10 million years ago and continues today. Rather than movement, this period is dominated by dissolution of salt that earlier had risen near the surface. As the salt crept upward during this time, mostly through the erosion of the thick succession of overlying sediments, any water that seeped down through the fractured caprock quickly dissolved whatever salt it came into contact with. This efficiently excavated the salt from beneath the overlying, more resistant, but fractured rock, causing blocks to collapse into the voids. What began as minor dissolution and localized collapse evolved into extensive salt anticline valleys tens of miles in length. It is these valleys that characterize the region today.

The salt anticline valleys today are wide, flat-bottomed features rimmed on both sides by linear escarpments of resistant sandstone. The strata that bound either side of these valleys tilt noticeably away from the valley center, reflecting the precollapse anticlinal structure. It is easy to imagine these colorful strata, at some previous time, extending upward above the present valley to form the apex of a great upwarp.

Although it is responsible for most of the modern landscape, salt is never seen at the surface. Even in today's arid climate it is readily dissolved. What is exposed, however, are light gray mounds of more resistant gypsum dotting the valley floors. Gypsum is the alteration product of the anhydrite that was interbedded with the salt. As it rose closer to the surface with the salt, it became rehydrated in the presence of shallow groundwater. Upon close inspection, gypsum is interbedded with thin, black shale layers, another less soluble remnant of the basinal Paradox Formation.

The collapse of the salt anticline crests and conversion into a valley probably occurred through multiple processes, though all were related to the dissolution of underlying salt. One mechanism was the downdropping of large portions of the crest along extensive normal faults that marked the valley margin. Movement on these faults, while sporadic, would have affected a large area. The alternative method was similar, but instead collapse occurred on smaller blocks riddled with closely spaced fractures that were generated earlier. In this case, movement at any time was localized and distributed across smaller, fracture-bounded blocks. The blocks eventually foundered into the salt and were buried by the thick blanket of detritus that covers the valley floors today. It is likely that in the evolution of any salt anticline valley in the region, both processes were important.

Unbreached Anticlines

While the salt anticline valleys to the east are lined by abrupt cliffs and floored by wide, flat-bottomed valleys, uncollapsed anticlines reside in the complicated jumble of rocks that make up the northern part of Canyonlands. Here the gentle dips of the strata may not even be detected by the casual observer. However, this subtle procession of anticlines and synclines from above resembles an undulating carpet of red sandstone, but one sliced by the ribbon of the Colorado River.

Unbreached salt anticlines that are concentrated west of the Moab–Spanish Valley structure provide an example of these features before collapse, although they are on a smaller scale. From northeast to southwest, the more pronounced of these are Cane Creek anticline, Shafer dome, Lockhart anticline, Rustler dome, and Gibson dome (see Fig. 2.21). Like their larger, more dramatic relatives to the east, these also are bounded by intervening synclines.

Although the geometry, history, and origin of these smaller western anticlines are similar to the larger salt valleys, size differences and the conspicuous lack of collapse structures must be explained. One important difference is that the salt in these anticlines has not ascended to the shallow levels found to the east. For instance, the Paradox Formation beneath the Cane Creek anticline lies 1500 feet below the surface, covered by the overlying Honaker Trail Formation and the lower Cutler Formation. This thick section of caprock obstructs the downward trickle of groundwater, prohibiting the salt dissolution that promotes collapse.

An influential factor in this is the original thickness of the underlying salt. In this western part of the basin, thickness decreases to ~1000 feet, compared with 4000 feet in the salt valley anticline region. There simply is not enough salt to generate the force needed to push up near the surface, nor to form anticlines as massive as those to the east. A thinner Paradox Formation also means less pressure in the early history of the salt, when rapid sedimentation to the east was already mobilizing the lower salt beds. It is likely that pronounced salt movement did not begin until much later, after enough overlying sediment had accumulated to generate the pressure needed to set it in motion.

An exception to the lack of salt-related collapse features west of the anticline valley region is Lockhart basin, situated 18 miles southwest of Moab. According to a hypothesis advanced by Peter Huntoon and Henry Richter (1979), this small (~5 mile diameter) circular basin developed when strata overlying the Paradox Formation settled into a cavernous void that formed when salt, far below the surface, was dissolved. Dissolution was driven by groundwater that gained access to the top of the salt by deep fractures that formed in overlying rocks as they were arched upwards. As groundwater slowly ate into the salt, the void grew. Initially the collapse of overlying rocks was localized, with rocks immediately above the void dropping first. As progressively higher rocks were left without support, the collapse worked its way upwards. Eventually, the circular zone that defines Lockhart basin experienced wholesale collapse, and overlying strata dropped downward along flexures that today bound the basin. Although Lockhart basin is undoubtedly a salt collapse feature, salt was never near the surface. Unlike the

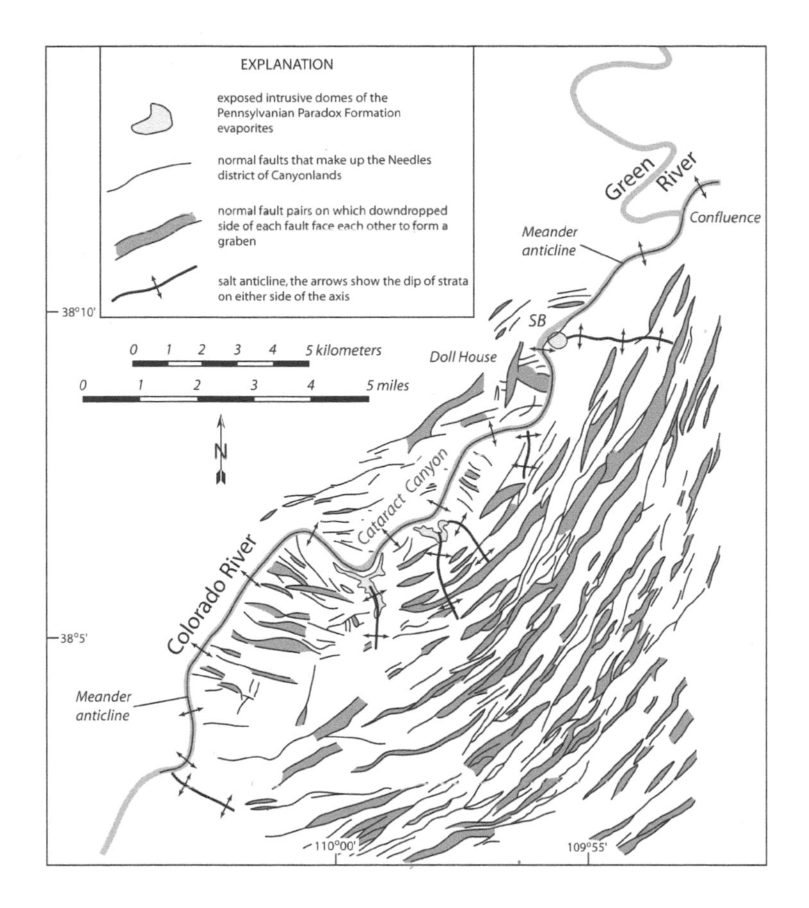

Fig. 2.24. Tectonic map of the Needles District of Canyonlands National Park showing the fault traces in the horst and graben area on the east side of the Colorado River. Also shown are related salt-induced anticlines, including the Meander anticline, which follows exactly the trace of the Colorado River canyon through the region, and the smaller side canyon anticlines with axes oriented perpendicular to the Meander anticline. SB denotes the location of Spanish Bottom for geographic reference. Modified from Huntoon and others 1982.

large anticline valleys to the east in which collapse occurred along faults and fractures, Lockhart basin formed by the downward flexure of strata that overlie the Paradox salt.

Needles Fault Zone

The Needles fault zone is a salt-related feature that encompasses the Needles District of Canyonlands National Park. This faulted and jumbled region covers a 90 square mile area along the east side of Cataract Canyon (Fig. 2.24). The tectonic evolution of this spectacularly convoluted landscape is closely related to incision by the Colorado River in the last 5 million years.

The Needles fault zone is defined by a system of north-trending arcuate faults that parallel the trace of the river (Fig. 2.24). These normal faults slice the Pennsylvanian Honaker Trail Formation, lower Cutler beds, and the Permian Cedar Mesa Sandstone, which caps the district. Numerous geologists have interpreted the faults to terminate in the subsurface in the Paradox salt, which underlies the area. The mobility of the salt, which is being squeezed toward Cataract Canyon, is also driving the overlying Pennsylvanian and Permian rocks westward into this void. Movement takes place on a low-angle glide surface of salt. The wholesale west-directed movement of the Needles is accommodated incrementally on the

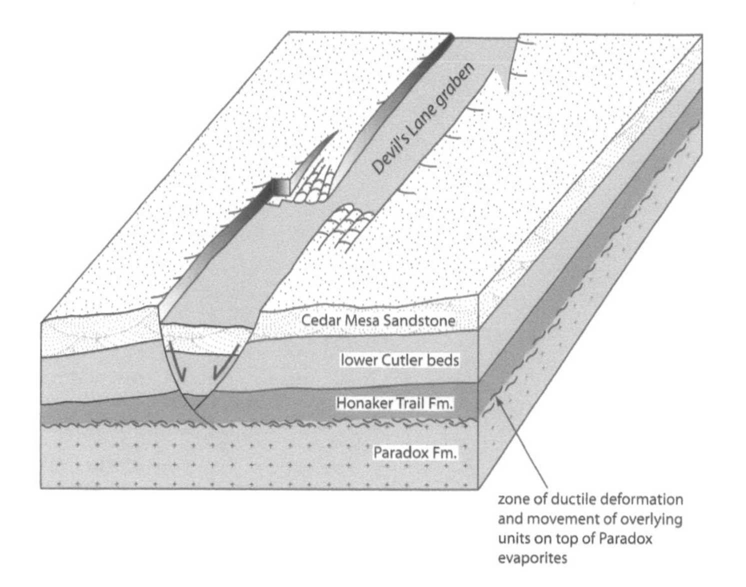

‹ **Fig.** 2.25. Block diagram of Devil's Lane graben in the northwest part of the Needles District, Canyonlands National Park. The cross-sectional view shows the strata involved and their relationship to the underlying evaporites of the Pennsylvanian Paradox Formation. The view is approximately to the north. Modified after Trudgill and Cartwright 1994.

⌄ **Fig.** 2.26. Oblique view aerial photo of the Needles District of Canyonlands looking north toward the Colorado River canyon. The sandy wash on the right is Lake Canyon graben, which turns west at a right angle to drain into the Colorado via Lower Red Lake Canyon.

numerous faults that cut the zone into a series of blocks. As these blocks jostle for position, some are dropped down while others rise, creating a classic horst and graben setting (Figs. 2.25, 2.26, and Plate 1).

The slow westward slide of Pennsylvanian-Permian strata toward the abyss of Cataract Canyon is also thought to have spawned the Meander anticline, which is discussed in the next section. Because development of the fault zone and anticline were triggered by this colossal excavation, they are probably the youngest salt-related features in the region. Geologically speaking, the Colorado River is young, its drainage network established only within the last 5 million years. Since much downcutting must have preceded the initiation of these structures, their maximum age is limited to 2–3 million years. Movement continues today. In fact, Peter Huntoon (1979, 1988) proposes that the rate of deformation is actually accelerating. He suggests that the numerous faults that permeate the Needles area act as conduits, focusing the sparse surface water downward to mix with the impermeable, but easily dissolved salt. At the interface between water and salt, at the top of the Paradox, a less viscous mixture is concocted. This creates a lubricated surface along which the overlying strata, under the gentle nudge of gravity, easily move.

Meander Anticline

The Meander anticline follows the trace of the Colorado River through most of Canyonlands over a distance of 26 miles. Regionally, the fold trends northeast. It extends northward from Gypsum Canyon at the terminus of Cataract Canyon, traversing the Needles fault zone and through the Confluence, and dies out south of the Cane Creek anticline (Fig. 2.27). Like the closely related Needles fault zone, the fold involves the Pennsylvanian-Permian strata that overlie the Paradox salt.

Although the regional fold trend is northeast, the fold axis in detail is very sinuous, winding faithfully along the circuitous route of the river (Fig. 2.27). It is from this sinuosity that the anticline gets its name. The curious course of the axis affirms a close relationship between canyon cutting and the genesis of this unique fold. The tilt of the strata on the flanks of the fold locally reaches 50°, but this decreases dramatically a short distance from the river. Combined, these features require that any explanation for the origin of the Meander anticline consider the role of the Colorado River.

Most geologists agree that uplift of the Meander anticline is linked to downcutting by the Colorado River; the exact match between the course of the river and the trace of the anticline provides an unmistakable connection. However, a dispute arises when attempting to decipher the exact origins of this peculiar uplift. Early interpretations called on excavation by the Colorado River to remove enough material to release vertical pressure on the underlying strata, triggering the lateral flow of Paradox salt into this zone. At this stage the processes resemble those responsible for the salt anticline valleys to the east. The salt ascended into the zone of reduced pressure, pushing the overlying Pennsylvanian-Permian strata upward into a narrow anticline, but one that happened to be guided by the path of the river.

More recent investigation of the Meander anticline led Peter Huntoon (1982, 1988) to propose an origin for the fold that conveniently explains the horsts and grabens of the Needles as well. The fieldwork of Huntoon and his associates (1982) when mapping the geology of the entire Canyonlands region undoubtedly took them to areas that had not previously been studied in such detail, and thus

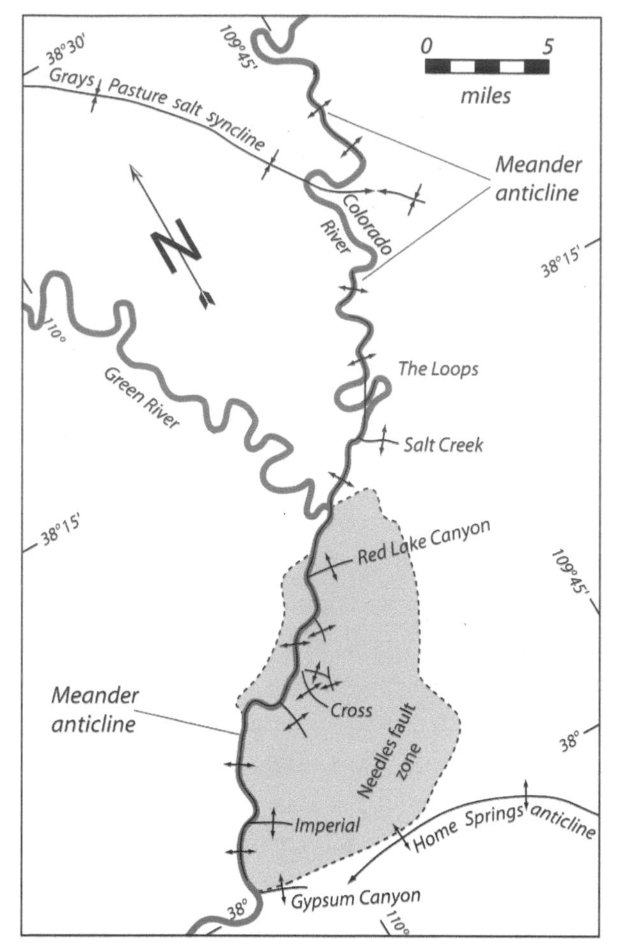

FIG. 2.27. Map of the Colorado River through Canyonlands National Park area showing the Meander anticline and associated small tributary canyon anticlines in the Cataract Canyon area south of the confluence of the Green and Colorado rivers. After Huntoon 1982 and Huntoon and others 1982.

were not considered in previous interpretations. Although Huntoon agrees that the Meander anticline is controlled by downcutting of the Colorado River, he maintains the actual flow of salt into the core of the anticline is an incidental occurrence; that is, salt flows because of the anticlinal uplift, rather than driving it. Instead, the anticline and the horst and graben system of the Needles are the direct result of a brittle sheet of overlying Pennsylvanian-Permian strata gliding toward the canyon on a surface of Paradox salt. The anticline is pushed up by the horizontal compression generated by the gliding mass, much like a wrinkle in a rug. As rocks on either side of the canyon crowd toward each other at the river, they bulge upwards. It is only this buckling that allows the Paradox salt to rise into the fold.

Part of Huntoon's argument for horizontal compression stems from his discovery of thrust faults along the axis of the anticline. The orientation of these wedge-like faults provides clear evidence for horizontal compression perpendicular to the anticline axis. The crucial exposures are in the Pennsylvanian-Permian strata that connect the Loops, a trace of the river marked by two large meanders that almost form a figure eight (Fig. 2.27). The fin of rock that connects these meander bends is bisected by the fold axis. This is the only place in the entire length of the fold that rocks in its core are preserved; elsewhere the axis coincides with the gap of the canyon. The fold axis in this fin is cut by numerous small thrust faults with spacing between faults ranging from one foot to several hundred feet. Maximum offset on any single fault is 50 feet, but most show much less. Collectively the numerous faults that slice these rocks produce a significant amount of offset. These faults provide the most reliable stress indicators and document horizontal compression, which, according to Huntoon (1982), is generated by the deliberate glide of a 2000 foot thick sheet of rock toward the chasm of the Colorado on a surface of viscous salt.

A closer look at the Needles District along Cataract Canyon reveals some interesting smaller anticlines with axes oriented at high angles to the Meander anticline and the trace of the river (Fig. 2.27). These folds are confined to the mouths of deeply incised side canyons that empty into the Colorado, and over their relatively short extent, their axes follow these drainages. Strata on the limbs of these folds locally dip up to 50°, but this tilt diminishes rapidly toward the interior of the Needles, where the folds die out. Like the larger Meander anticline, numerous small thrust faults riddle the core of these folds, suggesting a common origin. According to Huntoon's model, these folds also are generated by the methodical movement of rock atop the salt toward the deeply cut cavities. So, while movement mostly is focused toward the spectacular gorge of the Colorado, side canyons that have been breached as deeply also trigger the lateral movement into their vacated space. The disappearance of these folds up the side canyons makes a strong case for downcutting and subsequent horizontal compression as a major control on these features.

Distributed along the margin of Cataract Canyon, where the deepest side canyons enter from the east, are four small evaporite plugs (Fig. 2.24). These plugs are less than a kilometer in diameter and rise up to 100 m off the canyon floor. Their rise was driven by the mobility of the Paradox salt, which underlies the area at a shallow level. Although such features are common in the larger salt anticline valleys to the east, their occurrence in the vast Canyonlands region is limited to this deeply incised segment along the river. These plugs occur at river level at the mouth of Cross Canyon, in a small, unnamed tributary immediately to the south, and in Lower Red Lake Canyon to the north, near Spanish Bottom (Fig. 2.24). All of these diapirs, as they are known, occur at the intersection between the Meander anticline and the smaller tributary anticlines, within the Needles area.

Although the formation of these dome-like features is driven by the low density and mobility of salt, no salt is exposed in the plugs. Instead they are composed of gypsum, with lesser limestone and shale. The salt that was most certainly in these plugs was easily dissolved as the rising diapirs encountered shallow groundwater near the surface. According to Huntoon (1982), the Paradox in this area is ~85 percent halite. This suggests that the plugs originally contained a much greater volume than their present mass indicates, but through dissolution have been reduced to less soluble residual material.

The ascent of these evaporite plugs was influenced by several cooperating factors. The diapirs are limited to the area that has been most deeply excavated, placing the Paradox closer to the surface here than anywhere else in the region. Additionally, the zones where these plugs breach the surface contain the weakest rocks in the region. The crests of smaller tributary anticlines and the Meander anticline are riddled with fractures, so they easily succumbed to the push of the evaporites that so forcefully sought release from their restraints. The position of plugs at the intersection of these fold axes provides a strong case for this as a major control. Finally, all occur at the toe of the gliding mass of rock that comprises the Needles, probably placing greater than normal pressure on this salt and squeezing it upward like toothpaste from a tube.

Huntoon's proposed origin for the Meander anticline and smaller tributary folds differs in several ways from those of earlier workers (e.g., Harrison 1927, Prommel and Crum 1927, Stokes 1948, Baars and Molenaar 1971). Early explanations called on the vertical rise of salt into the zone that today is the Colorado River canyon, with deformation driven by vertical compression--forces generated by the upward push of Paradox salt as it ascended through overlying strata. The interpreted timing of this movement ranges from Pennsylvanian-Permian to recent. These particular interpretations derive from detailed study of the large salt valley anticlines to the east, where such an origin has been accepted and where there is abundant evidence for Pennsylvanian to recent deformation. Huntoon, in contrast, calls on horizontal compression, generated in the last 2 to 3 million years by the slow glissade of overlying strata along the viscous upper surface of Paradox salt. This movement toward the recently excavated terrain of the Colorado River confines the resulting folds and thrust faults to the narrow zone of the canyon (Fig. 2.28). Huntoon's interpretation has the advantage of explaining in simple, easy to visualize terms the extensional horst and graben system of the Needles and the compression required to produce the Meander anticline and associated folds. However, no consensus has yet been established, and it is possible that both processes have had a hand in the evolution of this fascinating part of Canyonlands.

The age of deformation along Cataract Canyon is necessarily linked to the age of the river that cut it. Most evidence suggests that the present Colorado River drainage has only existed for the past 5 million years. This implies that incision through the Permian rocks that line the upper walls of Cataract Canyon probably occurred 3–4 million years ago. Huntoon (1988) estimates that extensional faults in the Needles initiated less than 2 million years ago. Longtime researchers in the Needles, George McGill and Albert Stromquist (1975), used the average downcutting rate of the river to extrapolate the time that the canyon had reached a sufficient depth to spur development of the Needles fault zone. Although they admitted it was a crude estimate, they concluded that faulting began ~500,000 years ago. While the exact time of initiation will likely remain fuzzy, most studies agree that deformation is a continuing process, and the Needles District is a work in progress.

Researchers in the Needles have long reported signs of ongoing deformation. Geologist A. A. Baker (1933) documented several features indicating recent fault movement. These include earlier stream courses that had been abruptly truncated by a graben, causing the present drainage to be diverted at a right angle into the

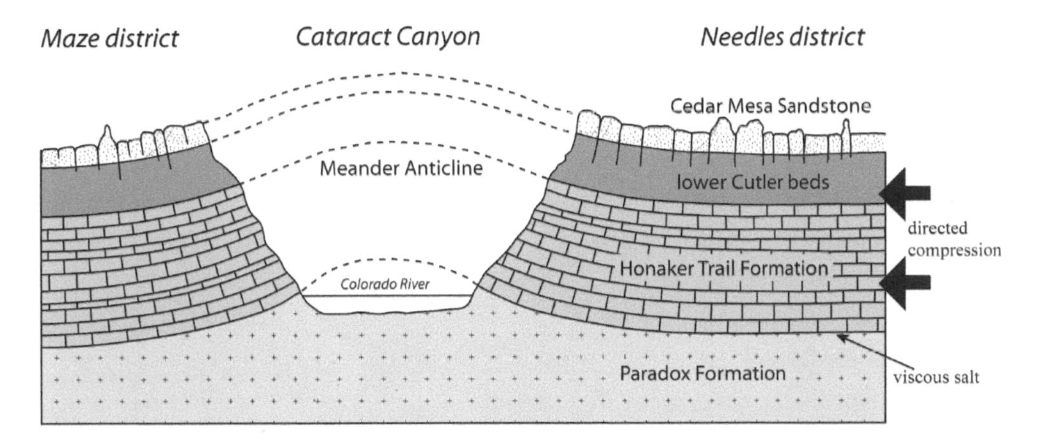

FIG. 2.28. Schematic cross section of the Meander anticline from the Needles District (southeast) across the Colorado River and Cataract Canyon, into the Maze District (northwest). Huntoon's (1982) hypothesis proposes that the excavation of Cataract Canyon by the Colorado River has allowed the northwest-dipping Needles District to glide northwestward on the underlying horizon of viscous Paradox salt into the void of Cataract Canyon. This slow glide has produced compression in the canyon that has driven anticlinal uplift. In this way the axis of the Meander anticline faithfully follows the bends and curves of the Colorado River (see previous map). Modified from Stromquist 1976.

more recently downdropped graben. In other parts of the Needles, grabens such as Cyclone Canyon are so young that no drainage network has yet been established.

Modern extensional deformation is also indicated by large open holes on the canyon floors at the edge of the grabens (Biggar and others 1981). These "swallow holes" mark the graben-bounding fault and are opened during movement on these faults. Fissures in the form of large, gaping fractures on the floors of graben and on top of horst blocks have been noted to open up in recent times as well. Both features reflect the regional stretching that is taking place as the Needles block continues its deliberate slide into the depths of Cataract.

Finally, we turn to the river itself. Upstream and downstream from the churning water of Cataract Canyon, the Colorado River is tranquil, with a gradient of less than 1 foot/mile (pre–Glen Canyon dam). This translates to an average of a 1 foot drop in the elevation of the river for every mile of horizontal distance. Through the chaos of Cataract, however, the gradient increases to 8 feet/mile. Normally, the increased energy of a river driven by such a steep rise in gradient would, over time, prod the river to cut downwards in its steeper course and establish a gradient closer to that of the up- and downstream stretches, eventually reaching a state of equilibrium. This would be in the form of a continuously low gradient through this region. The intense turbulence through Cataract Canyon indicates that the river is straining to correct its current state of disequilibrium. However energetic, the river is unable to keep pace with the continued rise of the anticlines and the encroachment of the toe of the mass that makes up the Needles as it slides into the canyon. The river also steepens in the canyon in an effort to flush out the debris that collapses from the canyon walls into the river. Thus, the river is being assaulted on all fronts: from below by uplift of the Meander anticline, from the side by the westward translation of the Needles block, and from above by the collapse of unstable canyon walls. So, while the spectacular turrets and spires of the Needles are ultimately the product of salt deformation, so too are the violent rapids that define Cataract Canyon.

Upheaval Dome

There is probably no geologic feature on the Colorado Plateau that has generated more controversy than Upheaval dome. Structurally it is a dome, with surrounding strata dipping radially away from a projected, but now absent central high point. Topographically, however, Upheaval dome is a crater, with a structure that is ~3.5 miles in diameter. This peculiar geologic feature is embedded into Mesozoic rocks on the high plateau of the Island in the Sky District in northern Canyonlands National Park. This plateau is confined by the Green River canyon to the west and the Colorado River on the east.

Upheaval dome is an amazingly circular feature in map view and is defined by a central crater about 1 mile in diameter and a surrounding sandstone rim that extends from the crater for another mile (Fig. 2.29 and Plate 2). The center of the crater is marked by a conspicuous high point composed of uplifted, intensely deformed slivers of Permian White Rim Sandstone and Triassic Moenkopi Formation, the oldest rocks exposed in the complex. This high point is encircled by a strip of colorful shale and sandstone of the Triassic Chinle Formation, which forms the rugged crater floor. Prominent orange cliffs of Jurassic Wingate Sandstone rim the crater and tilt away from the center at angles up to 80°. Outward, away from this rim, are cliffs and domes of the overlying Kayenta and Navajo formations. Within these units the tilt direction of the strata changes, and the dip (up to 20°) is toward the crater center, forming a mild but pronounced syncline that rings the dome (Fig. 2.30). Traversing still farther from the crater, strata begin to flatten and within a 2 mile radius of the center revert to undeformed, flat-lying beds.

Upheaval Canyon cuts through the northwest margin of the dome, opening the central crater to the nearby Green River. This canyon offers a spectacular cross-sectional view of the dome that is vital to unraveling the mystery of its origin.

Various hypotheses for the origin of Upheaval dome have been proposed over the past century, but none has been unanimously embraced by the geologic community. One early explanation proposed a volcanic origin that called on some type of eruption, but a complete absence of volcanic rocks discounted this pos-

FIG. 2.29. Aerial view of Upheaval dome in the Island in the Sky District of Canyonlands. The orange, cliff-forming sandstone that bounds the central crater is the Jurassic Wingate Sandstone. The outer ring of resistant sandstone is the Jurassic Navajo Sandstone. The light-colored and red strata in the central crater are the Triassic Moenkopi (red) and Chinle (light-colored) formations. Upheaval Canyon drains the central crater on the upper left side of the photo. Note the road to the trailhead in the lower left corner.

sibility (McKnight 1940). The presence of a thick succession of Paradox evaporites beneath this region, coupled with the abundance of surrounding, salt-related features, has long made a salt dome origin a favorable explanation (e.g., Jackson and others 1998). Although this hypothesis has its detractors, it has until recently endured attempts to rule it out.

Another recurring explanation is a meteor-impact origin. Because geologists are loath to call on extraterrestrial origins for geologic features, this proposal has gone in and out of favor through the years. Recent work, however, by the late Eugene Shoemaker and his coworkers has brought this idea to the forefront (Huntoon and Shoemaker 1995, Kriens and others 1999). Most recently, an international team of scientists has recovered unequivocal evidence for a meteor-impact origin for Upheaval dome, laying the controversy to rest (Buchner and Kenkmann 2008). This indisputable evidence is discussed below, following a brief overview of the recently discounted salt dome hypothesis.

Salt dome hypothesis

Early explanations of Upheaval dome as a salt-related structure called on the simple ascent of Paradox evaporites through the overlying Permian and Mesozoic strata (McKnight 1940, Mattox 1968). Instead of rising along a linear zone, as seen in the salt anticlines to the east, it was proposed that Upheaval dome formed from the escape of salt via a single point source, generating the dome-shaped fold in which the axial region, essentially a point at the culmination, eventually col-

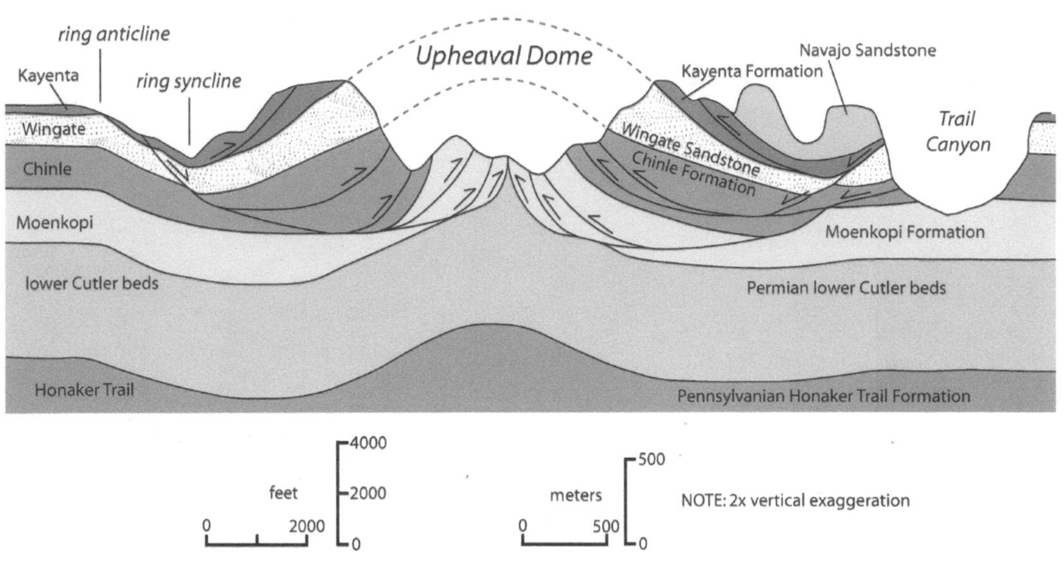

Fig. 2.30. Simplified cross section of Upheaval dome through its center. View of cross section is to the northwest, with an orientation of approximately northeast-southwest. Note the transition of normal faults at the outer part of the dome to converging thrust faults toward the center. The vertical scale has been exaggerated 2X to show clearly the stratigraphy and faults. Modified after Kriens and others 1997.

lapsed. For many years this scenario largely was accepted by the geology community. Concurrence was based on the thick succession (~2300 feet) of Paradox evaporites beneath the area and the variety and abundance of other salt deformation features that dominated the regional landscape. Thus, it was a case of guilt by association.

Doubt was cast on the salt dome hypothesis when Louie and others (1995) reported the results of a seismic survey across Upheaval dome. Seismic surveys are complex but effective methods for obtaining information on the geometry of subsurface features such as rock units, faults, and folds. These surveys are conducted by sending seismic waves, or vibrations, into the Earth from some source, typically a large "thumper truck" designed for this purpose. As the waves pass into the subsurface, they encounter rock units of different densities. At each change in density (and rock type) some waves are reflected back toward the surface. Upon reaching the surface, the waves are recorded by a line of sensors, or geophones, planted in the ground. The recorded signals are then processed by computer, and the end result is a cross-section of the subsurface along that line. A good seismic line is invaluable for interpreting subsurface geology. The findings of the experiment at Upheaval dome shows that the Paradox salt remains 1500 feet below the surface, not immediately beneath the surface, as early proponents of the salt dome mechanism suggested. This is corroborated by a drill hole through the syncline that rings the crater that hit the Paradox at a depth of ~1500 feet.

In 1984 Gene Shoemaker and K. E. Herkenhoff presented convincing evidence for Upheaval dome as a meteor-impact crater, and the salt dome hypothesis seemed headed for the geological dustbin. However, a later, detailed analysis of the structure by a team of scientists from the Texas Bureau of Economic Geology and Exxon Production Research resuscitated the idea. In their report (Jackson and others 1998), they propose a series of detailed hypothetical events induced by the rise of salt for the evolution of Upheaval dome. According to this chain of events, deformation began as early as the Pennsylvanian and continued into the Jurassic.

The proposed salt-induced movement began in the Pennsylvanian Period as the Paradox was buried by later sediments (Fig. 2.31). During the early phase of deposition of the Late Triassic Chinle Formation, continued extrusion kept the salt at or near the surface. Around this time the regional blanket of sediment thickened enough to push from the dome more rapidly and spread glacier-like over the immediate Chinle landscape (Fig. 2.31). The sediments of the upper Chinle Formation would have abutted this pancake-shaped salt mass as it swelled up and outwards. It should be emphasized that there is nowhere today that salt extrudes onto the surface of the Earth: it is simply too soluble to survive at the surface, even in arid conditions.

The Jurassic Wingate Sandstone that overlies the Chinle also contains features attributed by Jackson and others (1998) to salt activity. They suggest that wind-blown sand of the Wingate never made it onto the crest of the salt dome, so the Wingate never existed on the central part of the structure, which is now pocked by the crater.

During deposition of the overlying Jurassic Navajo Sandstone, salt extrusion again accelerated, and the arid climate is interpreted to have allowed the emerging mass to swell and spread laterally (Fig. 2.31). The soft sand was tilted as the amoebic salt blob bulldozed into the surrounding eolian sand pile that today is Navajo Sandstone.

The final phase of this hypothesis is the constriction and eventual closure of the conduit that fed salt to the surface. Jackson and his colleagues suggest that closure occurred during, or soon after, Navajo deposition (Fig. 2.31). A similar pinch-off geometry, coupled with overlying salt masses, has been imaged in subsurface salt provinces around the world, including the Gulf of Mexico, and the North Sea and Red Sea regions. Because all of these observations were made through seismic imaging, geologists are unable to get a detailed look at the rocks and structures. Thus, the interpretation of Upheaval dome as a pinched-off salt structure was significant as an exposed example of an otherwise inaccessible feature.

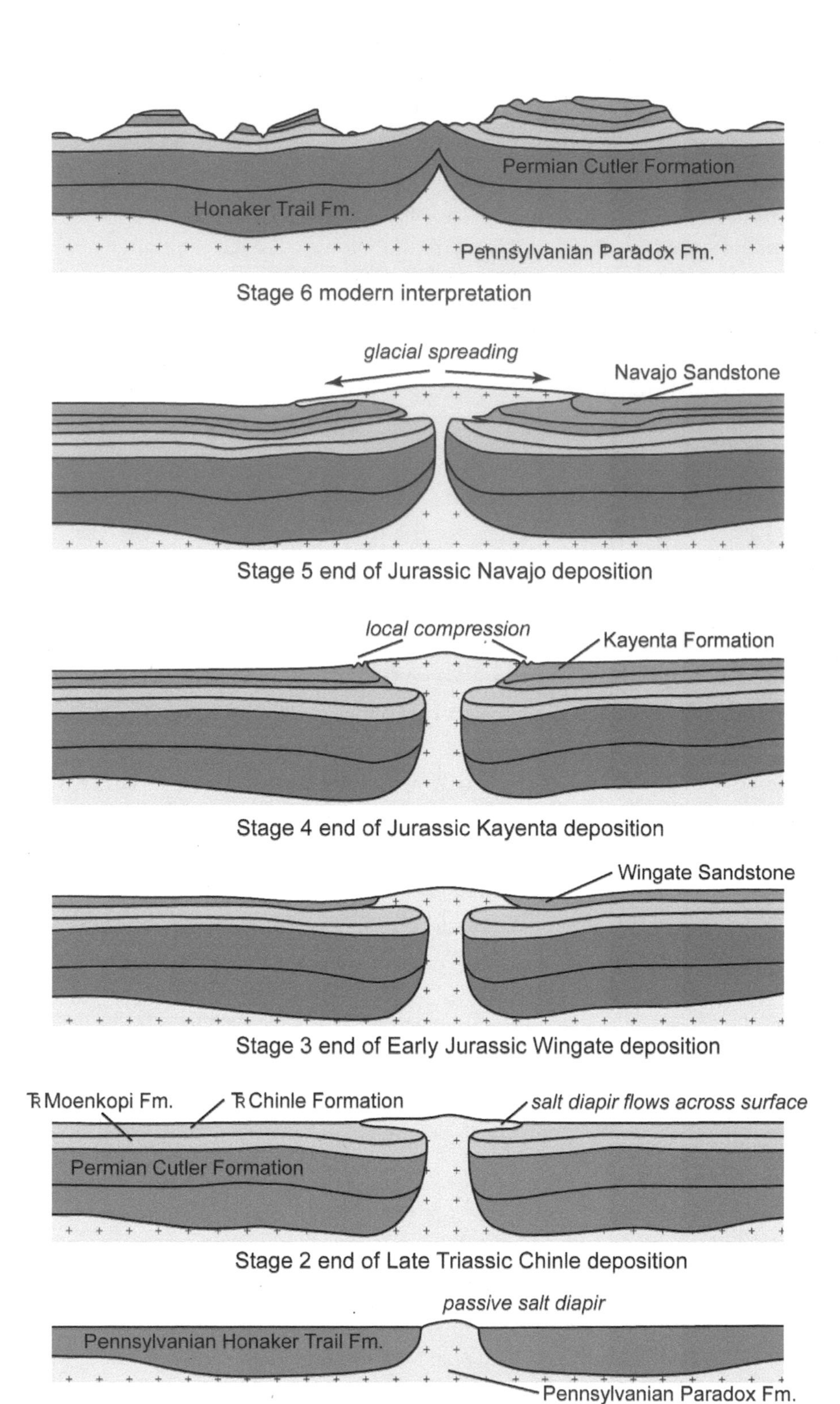

Fig. 2.31. Model of Jackson and others (1998) for a salt dome origin for Upheaval dome. The model shows the stepwise evolution of Upheaval dome from the salt's initial movement in the Pennsylvanian Period to the modern setting.

The remaining obstacle in this model is explaining the step-by-step processes involved in the evolution of these structures. Even Jackson and others concede that the exact mechanisms of pinch-off are unknown. Most interpretations of ancient geologic processes are firmly grounded in the detailed observations of modern processes. The disconnection of rising salt diapirs from their source by pinch-off simply has not been observed in a modern setting. As will be seen again and again, the answer to every question (or simply the attempt to answer any question) brings up a multitude of new ones. Thus, 4.6 billion years of Earth history has provided geologists with a mind-numbing job security.

Although the salt dome hypothesis for Upheaval dome has recently been disproven, this model outlines the most detailed scenario to date for such an origin. At this time, however, it enters the category of a rejected hypothesis, one that will no longer be considered. The following section reviews the history of the meteor-impact origin and the abundant evidence for it.

Meteor-impact hypothesis

Probably the greatest obstacle to acceptance of the meteor-impact origin for Upheaval dome is geologists' aversion to explaining any natural phenomenon by catastrophic processes or extraterrestrial sources. This stems from the formative years of geology in Europe in the 1600s, when the Church had a strong and oppressive influence on the sciences. During this time, any interpretation of the Earth was required to adhere strictly to the Bible, regardless of what the evidence might suggest. For instance, based on the genealogy outlined in biblical stories, the Earth could be no greater than 6000 years old. In one of the more outrageous pronouncements of the time, Irish archbishop James Ussher, after "careful" calculations, declared that the world had been created on October 23, 4004 BC! This view dominated for 200 years. Today, thanks to modern analytical techniques, radiometric dating, and a multitude of geologists looking at rocks around the globe, the age of the Earth is estimated to be 4.6 billion years.

Interpretations of the rock record during the early years of science were also required to follow the Bible, and the single event of "the great flood" became the unfortunate explanation for all the rocks on Earth. From this repressive view of nature emerged the notion of "catastrophism," the belief that the Earth has not evolved over time, but came to its present state—mountains, canyons, and all—through a single catastrophic event. Needless to say, this belief stifled the advance of geology dramatically. Finally, after reversing this repressive, religion-based pseudoscience through careful investigation and conservative interpretations, the science of geology stepped out of the shadow of the Church. Yet the scars remain. The resistance of modern geologists to accept any type of catastrophic process to explain geologic features is a by-product of these past events. In fact, in cases where such processes are called on to explain geologic features and events, the burden of evidence is much greater than would be expected for a noncatastrophic interpretation. With this history in mind, it is time to return to the case of Upheaval dome.

The interpretation of Upheaval dome as an impact structure hinges on the combination of observed faults, folds, and other features similar to those found in well-documented impact craters. Equally important is the large-scale geometry of Upheaval dome and its similarities with both known impact craters and experimental models. According to these comparisons Upheaval dome is a small, complex impact crater, meaning it has a characteristic geometry and features that formed through a specific sequence of events. The sequence initiates with the impact of the meteor and intense compression, immediately followed by excavation and crater formation. Finally, the modification stage is defined by rim collapse into the crater center, while the center is forced upward in an elastic response to the preceding compression. Geologically speaking, the entire process is instantaneous, occurring within seconds or minutes of the initial impact. The structures discussed in this section are clearly and easily attributed to these specific stages.

Faulting is the dominant mode of deformation in Upheaval dome and consists of the unusual combination of normal faults and thrust faults. This association is at odds because normal faults indicate extensional stress, whereas thrust faults form under compression. These two stress types are in opposition to each other, yet here they are found acting simultaneously and in the same direction. This occurs only in the rarest instances; one such instance is the formation of complex meteor-impact craters.

The different fault types are confined to specific zones of Upheaval dome and are easily explained by the proposed evolution of the crater. Normal faults are concentrated in the outer rim of the crater, chiefly in exposures of the Wingate, Kayenta, and Navajo sandstone. Movement on these low-angle faults was down toward the crater center; the faults had the effect of thinning the units. In contrast, thrust faults are clustered in the center of the crater and mostly are seen in the White Rim Sandstone and Moenkopi Formation, which make up the central uplift. These faults have the opposite effect of structurally thickening these units (see Fig. 2.30). The central high part of the crater was formed by the convergence of these strata toward the crater center, forcing the numerous fault slivers upward along the closely spaced thrust faults.

The relationship between the outer normal faults and inner thrust faults is that they very likely are physically connected (see Fig. 2.30). Most of these faults formed during the final modification stage, as the outer rim of the crater collapsed inward. At the same time, the crater center was rebounding upward in an elastic response to the initial impact and compression stage. Earth behaves like any other material when violently and intensely compressed: it rebounds to a limited extent upon removal of that stress. As the crater center rebounded, the normal fault-bounded sheets that were collapsing into this maelstrom were dragged upward to become thrust faults. Thus, if erosion had not removed the Wingate and Kayenta sandstone from the central high, the normal faults that cut them around the rim of the crater would likely continue toward the center, where, due to a change in fault orientation, they would become thrust faults (see Fig. 2.30).

Folds in Upheaval dome are best defined in the outer part of the crater, in the Wingate-Kayenta-Navajo succession. They vary in size from small, centimeter-scale folds to those with wavelengths measured in kilometers. The orientation of fold axes also varies, but dominantly is one of two types: radial or circumferential to the crater center. Radial fold axes extend outward from the crater center like the spokes of a wagon wheel. Circumferential folds are circular in map view, with axes that trace the outline of the crater and, in this case, define its outer perimeter. The folds detailed below are large, kilometer-scale folds, but it is important to understand that small-scale, subsidiary folds with the same orientation have a similar origin. The small-scale mimics the large-scale; this rule can be applied to most folded terrains.

Circumferential or ring folds mark the outer limits of Upheaval dome. A peripheral monocline defines the outline of the structure by clearly separating unaffected strata on the fringe of the crater complex from folded and faulted rocks of the interior. Outside the monocline strata are flat-lying, with no discernible tilt. The circular monocline tilts outermost deformed strata toward the crater center at angles up to 20°, much like the upturned rim of a dinner plate, but one with a diameter of ~4 miles! In some places the fold is cut by normal faults, but in other places the faults are folded. This suggests that the folding and faulting were formed at the same time and thus are related. As the inward-dipping strata of the monocline dive down toward the crater center, they meet the outward tilting rocks of the main dome structure. Where these oppositely tilted strata meet, they form the axis of the ring syncline (see Fig. 2.30).

The ringlike structural depression that lies inside the outer monocline and encircles the central uplift forms a circular synclinal trough. The inner limb of this fold consists of uplifted strata flanking the central peak, while the outermost monocline of inward-dipping strata forms the outer limb (see Fig. 2.30). The massive Navajo Sandstone occupies much of the fold axis. The syncline also acts as a boundary between the outer zone of normal faulting and the inner thrust-faulted chaos of the central uplift. According to Bryan Kriens and others (1997, 1999), the ring syncline formed while the central peak was uplifted along the converging thrust faults. As the central part of the crater vaulted upwards by the violent recoil of elastic rebound, adjacent material was sucked into the vacuum, creating

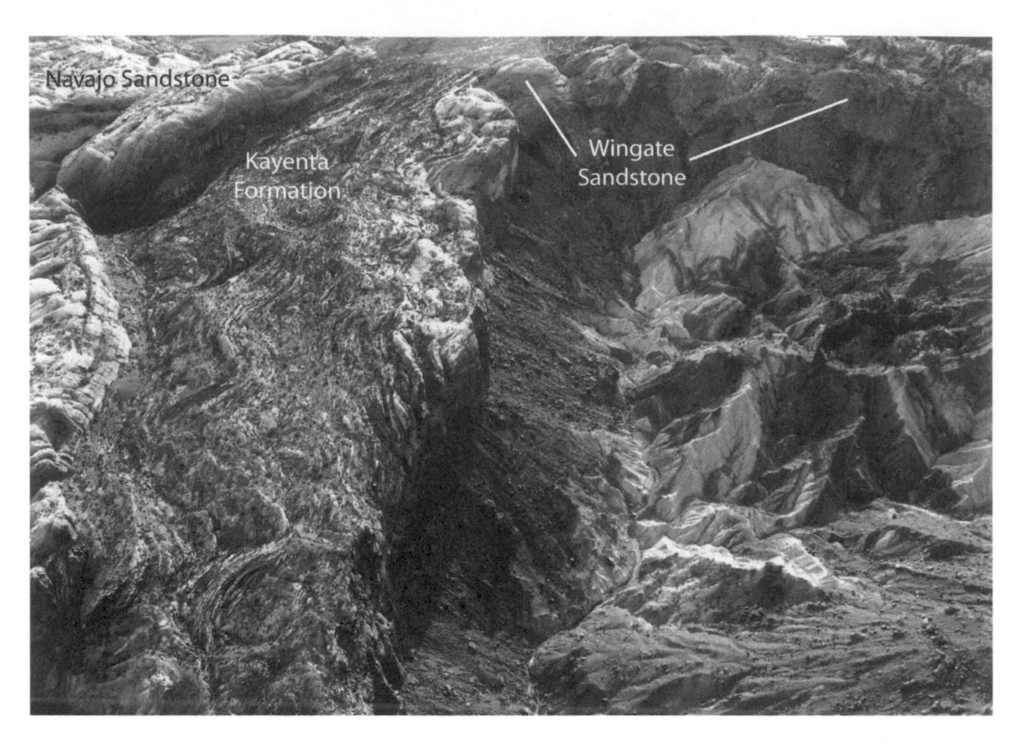

Fig. 2.32. Aerial view of Upheaval dome showing the radial folds in the Kayenta Formation and Wingate Sandstone.

an outer concentric depression. Thus the central uplift generated the ring syncline, and the two structures are synchronous. This structural pairing is a well-documented characteristic of complex impact craters.

Smaller folds with axes that radiate from the center of the structure have wavelengths measured in tens of meters rather than kilometers. The folds are clearly displayed in the panoramic Wingate cliffs that encircle the crater (Fig. 2.32 and Plate 3). Besides diverging from the center, the folds die out with distance from the center (Fig. 2.33). Like the other fold types, some are cut by faults, while some faults are affected by folding. This suggests that faulting and folding were simultaneous and occurred over a very short time period. Smaller, meter-scale radial folds appear to be the result of soft sediment deformation—possibly suggesting that folding took place before the sediments had been cemented into rock.

Radial folds are likely related to the ring syncline and formed as the sedimentary layers, recoiling from the initial compressional shock, were pulled into the crater center. As material converged in this central zone, space became limited, and the layers were constricted into radial undulations. Convergent flow, and the possibly unconsolidated nature of Wingate sand, is supported by its pronounced thickening toward the central zone.

Clastic dikes are a common feature throughout Upheaval dome, particularly in the center. Dikes consist of narrow sheets of clastic sediment, mostly quartz sand, that form fingerlike intrusions into surrounding rock units. Width ranges from a centimeter to several meters, and they typically extend a short distance. Dikes in the crater center are white and were fed by the underlying White Rim Sandstone. Some follow the planes of thrust faults, suggesting a close association with them. In the outer rim areas of the crater, dikes are composed of orange sand and are easily traced into their Wingate, Kayenta, or Navajo source rock. Along the sharp contact between the Chinle Formation and overlying Wingate Sandstone, the Chinle locally pushes up into the Wingate; in other places veins of Wingate sand snake downward, invading the Chinle. Clastic dikes are common features in settings where unconsolidated or fluid-saturated sediments suddenly are pushed over some threshold level of compression. The abrupt loading of sediment results in the injection of sediment-fluid mixtures into surrounding material, commonly through fractures generated by that same stress. Crosscutting relations between clastic dikes and previously discussed folds and faults follow a familiar pattern. Some dikes are cut by faults, while others, especially in the center, follow fault planes. Such relations suggest the features formed at the same time—and very rapidly.

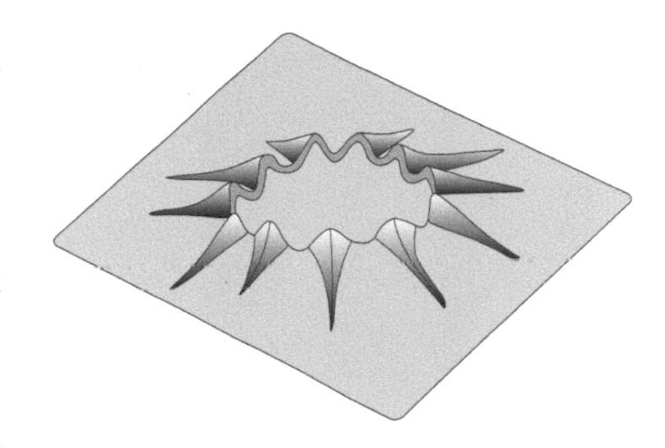

Fig. 2.33. Diagram illustrating the inward flow of the Wingate Sandstone during impact of the meteor. As the material in the crater rebounded, the Wingate was pulled inward, creating a space problem that was accommodated by the development of folds with axes radiating from the central crater. Kriens and others (1999) have suggested that prior to erosion the overlying Navajo Sandstone likely had the same geometry. Other exposed units in Upheaval dome reacted to impact by faulting with minor folding. After Kriens and others 1999.

A significant aspect of the dikes in the central peak area, which are easily traced into underlying White Rim Sandstone, is the nature of the sand grains. When viewed through a microscope, grains within the dikes are very angular and fragmented, suggesting they were shattered during emplacement. This is a stark contrast to sand grains elsewhere in the White Rim, which are well rounded. This rounding reflects their earlier history of ricocheting about in the gusty Permian wind and tumbling in the nearby surf. Dikes likely formed during the violent expulsion of water from pore spaces in sandstone during the impact. This abrupt event would have entrained sediments from the saturated rock, crushing the fragments as they smashed violently into each other during their upward propulsion.

A similar but geographically isolated feature known as Roberts rift has been interpreted as a large clastic dike that may be related to the Upheaval dome impact (Huntoon and Shoemaker 1995). This linear, sediment-filled fissure is located 22 km northeast of Upheaval dome and can be traced for another 10 km (Fig. 2.34). The narrow fissure is on a northeast trend that radiates from the dome. Although the fissure is less than a meter wide, the white, bleached zone that sur-

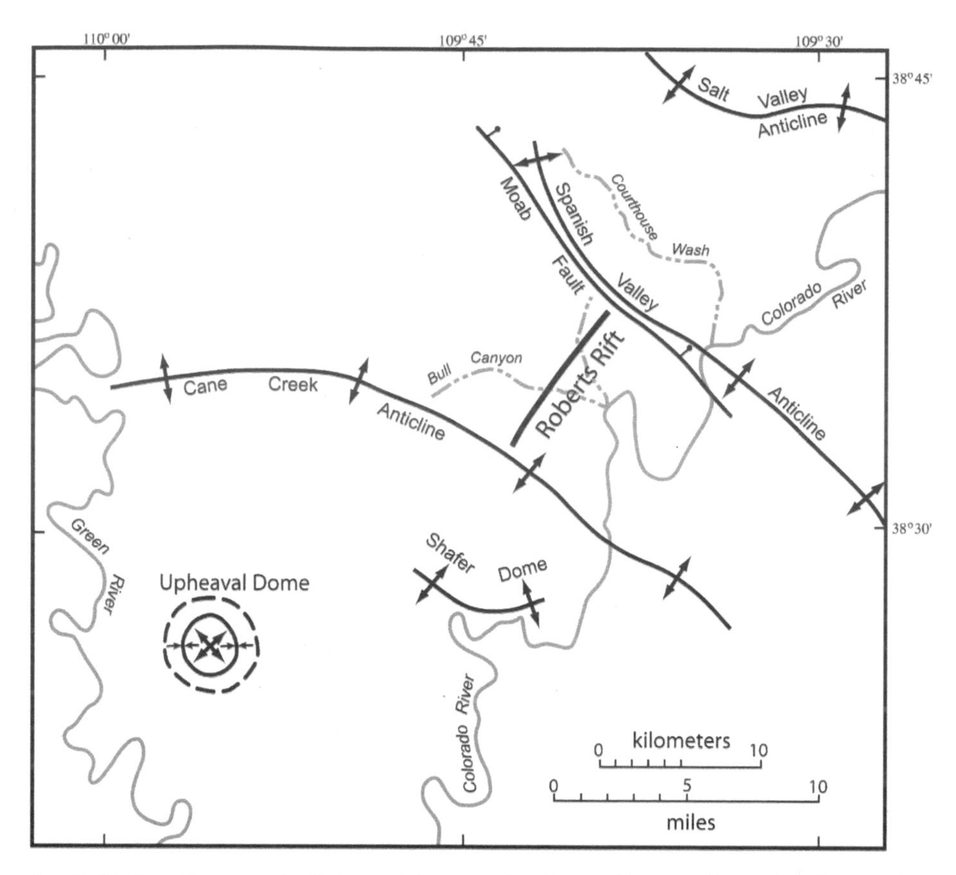

Fig. 2.34. Location map for Upheaval dome and Roberts Rift, as well as salt-related folds and geographic information. After Huntoon 2000.

rounds it is ~20 m wide. Host rocks include the varicolored Chinle Formation and the red Wingate and Kayenta sandstones, so the bleached zone stands out. Roberts rift is well exposed along the walls of the upper reaches of Little and Bull canyons, which are easily accessed via the road to Potash that follows the Colorado River (Utah Highway 279). A well-cemented mixture of large rock fragments and sand fills the fissure. Fragments include black shale and blocks of fossiliferous limestone up to 30 cm. Limestone can be confidently identified as pieces of the Honaker Trail Formation, which lies buried 3000 feet below the present surface. Black shale is believed to be from the Paradox Formation, which underlies the Honaker Trail Formation. Before the interpretation of Upheaval dome as an impact-related structure, the origin of these blocks was difficult to imagine. According to Huntoon and Shoemaker, these blocks were propelled upward by the force of fluids that were rapidly expelled from the subsurface, probably in response to the shock waves that were transmitted through the regional subsurface as a small asteroid or comet collided with Earth.

In the hypothesis proposed by Huntoon and Shoemaker (1995), the shock wave generated by the impact event at Upheaval dome had far-reaching consequences. As the initial wave of compression swept into surrounding rock, it affected units that were saturated with fluid. Many of the fluids contained in deeper strata are under intense pressure. As the shock wave rippled through these overpressured reservoirs, they may have literally exploded due to the rapid crush of the pore spaces. Such a "popping" of an overpressured reservoir would have resulted in the rapid and violent eruption of fluid, reservoir rock, and anything else caught in the path as the mixture blasted toward the surface.

The source for fluid, based on fragments identified in the Roberts rift, is probably the Paradox or Honaker Trail formations. Oil wells in the vicinity that have drilled into these units have a history of pressure-induced "blowouts." These occur when the drill bit penetrates an overpressured reservoir packed with fluid or gas, and material blows out of the drill hole until the pressure is relieved. This may take a few hours or several months. In 1925, a nearby well drilled into the Cane Creek anticline along the Colorado River blew out, spraying burning oil 300 feet into the air. Although the fire was extinguished, for six months several thousand barrels of oil per day gushed from the well into the river.

The pronounced bleaching of rock adjacent to the fissure provides some insight into fluid composition. Fluids in the Paradox and Honaker Trail reservoirs typically consist of salt water with various amounts of natural gas, hydrogen sulfide gas, and petroleum. The passage of these fluids through iron-rich red rock would easily bleach the rock, removing the red color that characterizes the host rocks at the surface.

An impact origin for Upheaval dome certainly provides a convenient and plausible mechanism for ejecting rock fragments more than 3000 feet upward through solid rock. Any other mechanism is difficult to imagine, although impact by an extraterrestrial body admittedly was a bizarre and unacceptable explanation a few tens of years ago.

Problem solved: An impact origin is confirmed

Indisputable evidence of a meteor-impact origin in the form of shocked quartz from the crater has long eluded geologists. This changed in 2008 with a report by two German researchers, Elmar Bucher and Thomas Kenkmann, who had been conducting detailed studies on Upheaval dome. Their discovery of shocked quartz sand grains in the Kayenta Formation within the crater conclusively confirms a meteor-impact origin for Upheaval dome, laying to rest one of the major long-running controversies in Colorado Plateau geology.

Quartz, in general, is much more resistant to deformation than most minerals. It is harder and has a strong crystal structure with no inherent weaknesses. In contrast, shocked quartz is riddled through with closely spaced, regular sets of planar fractures that are formed only at sudden, high pressures at relatively low temperatures. Its occurrence is known from only two distinct settings—known meteor-impact sites and beneath nuclear bomb detonations—both places with overwhelmingly large, instantaneous pressures. It is found in no other settings.

So finally the "smoking gun" has been discovered. Upheaval dome was formed by a meteor impact, validating the numerous structural and geophysical studies that suggested, but could not prove, such an origin. The remaining question is concerned with when the impact occurred.

Prior to the discovery of shocked quartz, Kenkmann and his coworkers conducted a variety of detailed studies of Upheaval dome (Kenkmann 2003, Kenkmann and others 2005, Scherler and others 2006). These investigations had already convinced them of an impact origin, causing them to think beyond its origin and consider instead the type of impact and the timing of the event. Their studies include detailed mapping, structural analysis, and computer modeling using all these data. Based on the asymmetric distribution of fault densities and the orientations of deformed strata, among other things, they concluded that the meteor struck at an oblique angle, coming from the northwest, dragging and shearing material toward the southeast.

Kenkmann and others (2005) also determined, based on the degree of deformation of the Permian White Rim Sandstone, that the impact likely occurred when it was beneath 6600 feet (2000 m) of other sediments. This suggests a Late Cretaceous impact, probably during deposition of the deep marine Mancos Shale. This provides a starting point in the search for evidence in surrounding areas.

A final, intriguing line of reasoning for the age of the impact at Upheaval dome also comes from outlying areas. According to Walter Alvarez and coworkers (1998), widespread and pronounced deformation in Middle Jurassic strata in Arches National Park and nearby areas may have been induced by regional shaking during the impact. Deformation, mostly in the form of folds, is confined to the Dewey Bridge Member of the Carmel Formation and, locally, the lower part of the overlying Slick Rock Member of the Entrada Sandstone (Fig. 2.35 and Plate 4). The Navajo Sandstone that underlies these units is undeformed, as is the overlying Upper Jurassic Morrison Formation. Folds vary greatly in style and orientation over a short distance, lending a chaotic appearance to the affected units. Other features include plugs and pipes of intrusive sandstone, contributing to the discordant nature of the deformation.

Folding and sand intrusion occurred while sediments were unlithified, probably during deposition of the Slick Rock Member. At this time these sediments would have held shallow groundwater. As the shock wave of the impact pulsed outward from the crater, the saturated sediments became liquified. This rapid conversion to quicksand took place as the water sought release from the confines of the pore spaces. In such events the mobilization and upward escape of water convulses the soft sediments, throwing the strata into chaotic undulations. This is a common process during large-magnitude earthquakes, which the impact-

FIG. 2.35. Convolutions in the Dewey Bridge Member of the Carmel Formation in Arches National Park and surrounding areas may be related to shock waves that rippled through the region during the meteor impact that formed Upheaval dome in the Island in the Sky District of Canyonlands National Park.

generated shock wave likely resembled. Alvarez and others estimate that the energy required to deform these sediments was equivalent to an earthquake magnitude of more than 8.0.

Alvarez and others concede that the association between deformation and impact is tentative, but propose it as food for thought. If it turns out to be related to the impact, it could provide much needed constraints on the timing of the event, which is currently limited to "younger than the Navajo Sandstone." Moreover, somewhere near the top of the deformed strata should lie the surface of the region at the time of impact. This would be an obvious place to begin the search for material ejected from the crater during impact.

Perhaps the greatest obstacle to uncovering such evidence is erosion; however, due to uncertainties about its age, the amount since the birth of Upheaval dome remains obscure. Whatever its origin, age has a great bearing on the thickness of sediment that has been removed. The present age constraint of younger than Early Jurassic is not very enlightening. If the structure is Mid-Jurassic in age, most of the affected strata are preserved, and clues to its origin should be present. But if it formed later—for instance, during the Cretaceous, as some have suggested—then thousands of feet of sediment have been erased, along with the most compelling evidence for its origin. Thus, determining the age of Upheaval dome would be an important step in ultimately deciphering its origin.

The Permian Period

Rivers, Ergs, and Shorelines

The Permian Period, which spans 300 to 248 Ma, was marked by global climate change. The dominant environment of the massive Pangean landmass shifted from waterlogged Pennsylvanian coal swamps to a Permian world of extreme aridity marked by evaporite deposition and blankets of **eolian** (wind blown deposits such as sand dunes) sand. In fact, the name *Permian* stems from the salt deposits of the Perm province in Russia, where rocks of this age were first recognized.

Because the Paradox basin already was arid during the Pennsylvanian, the Permian climate change had little effect on the region. The exact boundary between the two periods over much of the region is difficult to pinpoint because there is no distinct break, and much of the strata are fluvial or eolian deposits, so they are barren of useful, age-diagnostic fossils. The boundary is discernible in the marine deposits just to the west, where it is marked by a change in species of fusulinids; however, it takes a specialist to recognize these different species. To most people fusulinids look like grains of wheat or some form of petrified breakfast cereal.

In the Permian Paradox basin, it was business as usual. The lofty Uncompahgre highlands continued to rise and were simultaneously broken down into fragments that were then funneled westward into the Paradox basin. Major adjustments included a filling of the deep trough that had characterized the basin during the Pennsylvanian, and a westward migration of the shoreline into central Utah. Southeast Utah became a constantly changing patchwork of terrestrial environments that included alluvial fans, high- and low-energy rivers, and eolian dune fields of variable extent. Continued fluctuations in sea level periodically inserted a shallow marine shoreline into the mix and, as we shall see, affected sedimentation in the coastal dune fields by raising and lowering the water table.

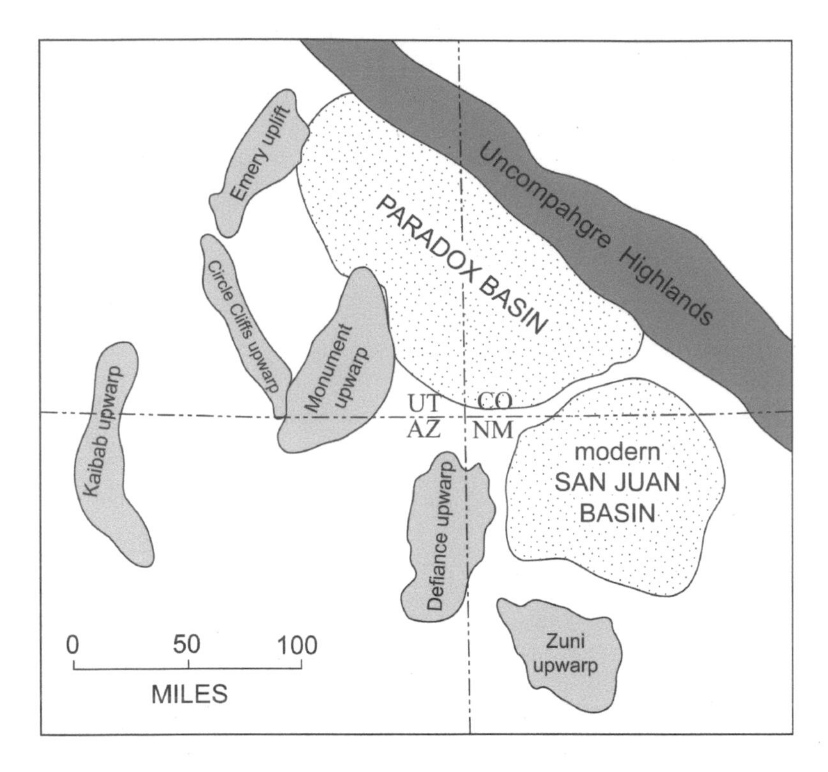

Fig. 3.1. Map showing Paradox basin, the Uncompahgre highlands source area, and various upwarps of moderate relief. Although the upwarps were not significant sediment sources, they affected sedimentation patterns in the Permian and later times as they were reactivated. The typical evidence for their activity during any given time is the thinning of sedimentary units across the tops of the upwarps. All these upwarps are expressed in modern geography as they were uplifted during the Tertiary Laramide orogeny.

Tectonic Setting

The Permian tectonics of the Paradox basin and the Four Corners region was a continuation of events that began in the Pennsylvanian. The Uncompahgre highlands continued to rise on the east-dipping reverse faults that bounded its steep southwest flanks. The already ancient Precambrian igneous and metamorphic rocks that composed the core of the range were pushed an estimated 6000 to 7000 feet above the broad plain of the Paradox that lay immediately to the southwest. Sitting on the summit of the highlands looking west, one would have seen numerous ribbons of water cutting westward across a plain broken only by sporadic dune fields. On a clear day, as most of them probably were, the sea possibly could have been seen on the far horizon, just over the intervening coastal dune complex.

The Paradox basin was bordered to the south-southwest by several mild uplifts (Fig. 3.1). The largest and most pronounced was the Monument upwarp—a feature that was recurrently active from the Pennsylvanian into more recent times. The most recent incarnation, uplifted ~60 million years ago, is defined by the area west of the Comb Ridge monocline in southern Utah and northern Arizona. The modern Monument upwarp includes the present-day areas of Monument Valley, Cedar Mesa, and the deeply incised canyons of the San Juan River.

Other Late Paleozoic positive elements in the region include the Kaibab arch, which is the site of the modern Kaibab Plateau on the north rim of the Grand Canyon, northern Arizona. Just east, in northern Arizona and New Mexico, lay the Defiance uplift. All these features were reactivated during the most recent episode of tectonism, the Laramide orogeny (~65–50 Ma), and thus represent recurrently active, large-scale weaknesses in the crust.

It should be emphasized that although these were positive elements, the topographic relief was subtle, and these features did not serve as large sources for sediment like the Uncompahgre highlands did. Instead, their activity at any given time is told by a thinning of sedimentary units across these features (Fig. 3.2).

The Permian saw a substantial decrease in subsidence in the Paradox basin. This can be demonstrated by comparing thickness of strata and basin geometry for the Pennsylvanian and Permian periods. Although the thickness of both successions increases dramatically toward the fault-bounded Uncompahgre mountain front, Pennsylvanian strata reach up to ~18,000 feet thick, whereas Permian strata reach a maximum of ~8000 feet. While both are exceptionally thick due to tectonically driven subsidence, the thinner Permian strata suggest waning tectonic activity in the Ancestral Rockies and associated basins.

By the end of the Permian the Uncompahgre highlands and the rest of the Ancestral Rockies were inactive; movement on their bounding faults had concluded for the time being. However, it was another 80 million years before erosion was able to wear the extremely resistant Precambrian rocks down to a level that they could be buried by sediment.

Permian Stratigraphic Evolution of the Paradox Basin and Surrounding Areas

The Permian rocks of the Paradox basin and adjacent areas have a long and controversial history among geologists who have worked on them. Much of this disagreement has to do with the nature of the rocks. Rapid lateral and vertical changes in rock type and the lack of fossils have made correlating these rocks across the region a somewhat hazardous occupation. Part of the dispute can be attributed to the obstinate nature of geologists and their resistance to change, even when more detailed observations show earlier work to be in error. Several of these controversies will be recounted briefly as modern interpretations are reviewed.

Like the underlying Pennsylvanian-age Hermosa Group, the Permian rocks of southeast Utah make up the "undifferentiated" Cutler Formation to the east; to the west, equivalent strata are elevated to the Cutler Group and are subdivided into several different formations based, as always, on obvious differences in rock type. Along the northeast basin margin, where it was bordered by the tectonically active Uncompahgre highlands, the Cutler Formation consists of a homogeneous succession, up to 8000 feet thick, of coarse conglomerate and sandstone. Farther west, away from the overwhelming influence of this high-relief source area, Perm-

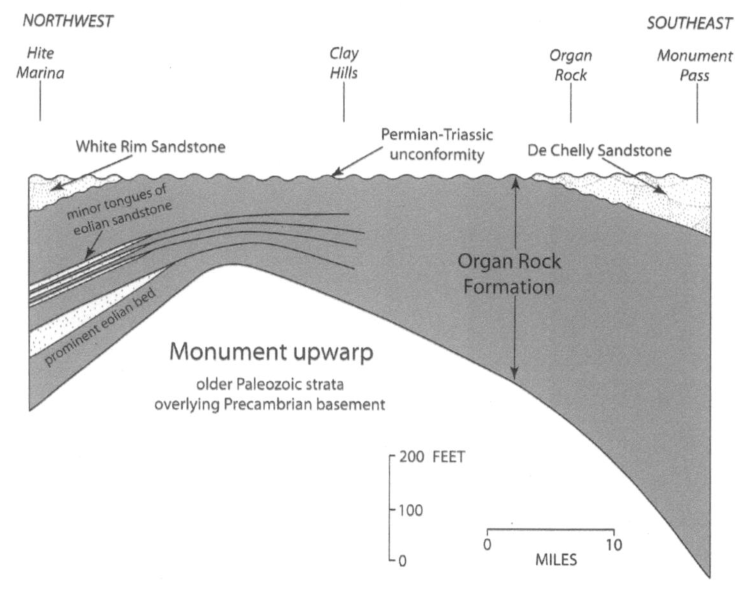

FIG. 3.2. Cross section of the Monument upwarp showing stratigraphic relations in the Permian Organ Rock Formation as it was deposited across the rising upwarp. Note the thinning and eventual pinchout of the eolian deposits across the crest of the upwarp. The diagram illustrates a well-documented example but also can be generally applied to the thinning of strata across any low-relief upwarp. After Stanesco and others 2000.

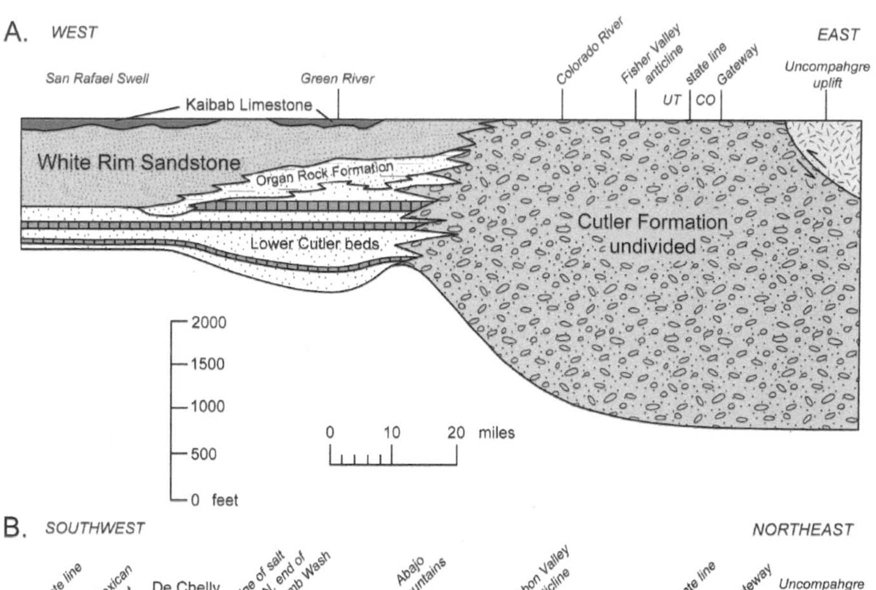

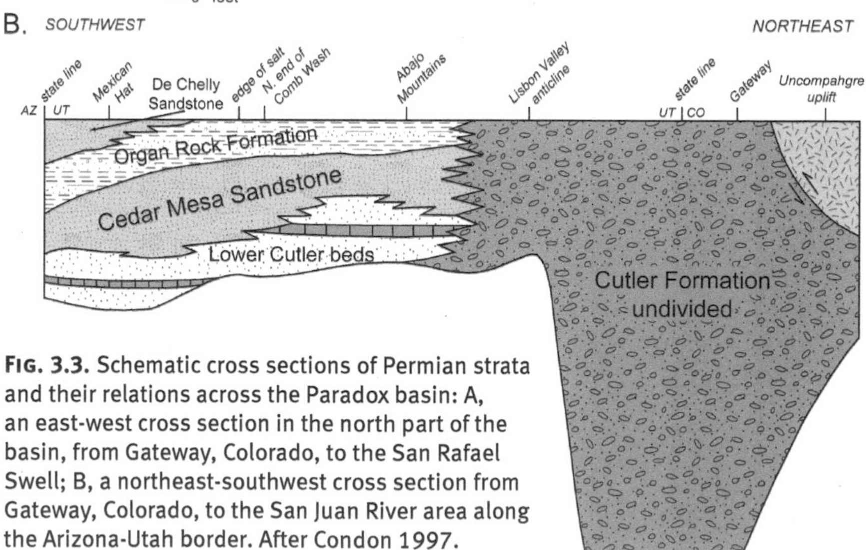

FIG. 3.3. Schematic cross sections of Permian strata and their relations across the Paradox basin: A, an east-west cross section in the north part of the basin, from Gateway, Colorado, to the San Rafael Swell; B, a northeast-southwest cross section from Gateway, Colorado, to the San Juan River area along the Arizona-Utah border. After Condon 1997.

ian strata are easily divided into formations. Throughout most of the basin these include, from bottom to top, the lower Cutler beds, Cedar Mesa Sandstone, Organ Rock Formation, White Rim Sandstone, and the Kaibab Formation (Fig. 3.3a). All but the Kaibab are part of the Cutler Group. To the south, along the Utah-Arizona border, the Organ Rock is overlain by the De Chelly Sandstone (Fig. 3.3b); here the White Rim and Kaibab are not present. Collectively these formations represent eolian, fluvial, alluvial fan, and shallow marine environments. While sediment influx from the Uncompahgre highlands reigned as the dominant influence on the various systems at any given time, sea level fluctuations and climate grew increasingly important to the west.

The Permian sedimentary rocks that fill the Paradox basin and make up such a vital part of the scenery are discussed in detail in this section, beginning with the undifferentiated Cutler Formation that occupies the northeast basin margin along the Colorado-Utah border. Traversing west into the time-equivalent Cutler Group, discussion begins at the base of the succession, followed by a unit-by-unit explanation of each formation. This will provide, in chronological order, an understanding of the events that led to the evolution of these varied and colorful rocks.

Undifferentiated Cutler Formation

The Cutler Formation exposed immediately southwest of the Uncompahgre highlands consists of red to purple sandstone, conglomerate, and siltstone deposited dominantly by west-flowing rivers that gushed energetically from their mountainous source. Three different environments are recognized in these rocks based on grain size, sedimentary structures, and to a limited extent, location relative to the Uncompahgre mountain front. Represented in these rocks are alluvial fan, fluvial, and eolian environments. Alluvial fan deposits are the coarsest and are confined to a narrow belt adjacent to where the steep mountain front was located. Fan deposits change over a short distance westward into fluvial deposits. On the wide arid plain across which these rivers coursed, fine, windblown sand accumulated, sometimes piling into small dune fields.

The only exposures of the very coarse, most proximal (meaning closest to the source) Cutler Formation occur around the settlement of Gateway, Colorado, just east of the Utah-Colorado border (Fig. 3.4a). Here the formation is dominated by two facies; both are characteristic of deposition on modern, arid-climate alluvial fans. The first and most impressive facies consists of thick beds of unsorted chaotic boulder conglomerate. Boulders are composed of granite and range *up to 25 feet in diameter* (Fig. 3.4b). These boulders are moderately rounded and water polished, indicating that runoff of colossal proportions emanated, at least sporadically, from the Permian Uncompahgre highlands. The precipitous mountain front probably lay less than a mile to the east.

FIG. 3.4A. Outcrop of the spectacular alluvial fan facies of the Cutler Formation in Casto Draw near Gateway, Colorado, and the Utah/Colorado border.

FIG. 3.4B. Large granite boulder in debris flow deposits at the top of the Cutler Formation, Gateway, Colorado. Six-foot-tall student for scale.

Deposition of boulder conglomerate probably was triggered by exceptional rainfall events. On modern, arid-climate alluvial fans, such as in Death Valley, sediment ultimately comes from weathered rock that accumulates on the steep slopes and canyon floors within the mountains. During infrequent cloudbursts large volumes of debris and water wash down the slopes to be funneled into the steep-walled canyons. Eventually these viscous sediment-water mixtures burst from their mountainous confines onto the relatively flat valley floor. Here at the foot of the mountains the gradient decreases abruptly, and the gravity-driven mass spreads into a lobe-shaped deposit of mud and boulders. Over time, numerous debris flow lobes coalesce to build a radial, fanlike feature whose apex is the canyon that feeds it debris, hence the name **alluvial fan**.

The second facies consists of thinner beds (less than 1 foot) of sandstone and pebble conglomerate. Grain size within a single bed decreases gradually, but noticeably, from base to top, typically from pebble conglomerate to sandstone or siltstone. Each "fining-upward" sequence, as this pattern is known, is deposited from a single sheetflood event.

Sheetfloods are water-rich flows that occur on alluvial fans when water spills out of the channel at the head of the fan to spread over its surface. The unconfined flow loses its ability to carry sediment as the water disperses into a thin sheet. The largest particles drop out first, followed in rapid succession by increasingly finer sediment. This produces a single, fining upward bed that is a record of a single sheetflood event.

Like debris flows, sheetfloods are discrete, sporadic events that occur during exceptional runoff. The main difference between the two is that sheetfloods contain more water and less sediment, so they are fluid, turbulent flows from which deposition takes place particle by particle as flow velocity decreases. In contrast, debris flows contain large volumes of mud, which makes them viscous and gives them strength. Deposition is by a "freezing" of the viscous mass as the slope down which it flows drops below some critical angle required for movement. The high strength of the mud-rich matrix actually allows large boulders to be supported and carried *within* the flow rather than sinking to the bottom, as in a more fluid mixture.

The question that comes to mind when considering "sporadic" or "episodic" debris flow and sheetflood events is, how often did they occur? In a study of Cutler sediments near Gateway, geologists Greg Mack and Keith Rasmussen (1984) from New Mexico State University documented evidence for the development of soil horizons at the top of some sheetflood and stream deposits. Soil formation implies a hiatus in deposition since it takes a considerable length of time to break the sediments down into soil and for plants to become established. Evidence includes horizons of calcified traces of plant roots and small nodules of calcium carbonate known as **caliche** that typify arid climate soils. Based on rates of modern soil formation in similar settings, these features take at least 100 years to develop. Thus, some of these **paleosols** (ancient soils) represent at least a century between depositional episodes.

Increasing distance from the ancient Uncompahgre source area is revealed in numerous ways. The Cutler Formation becomes progressively finer-grained in exposures a short distance to the southwest, specifically in Fisher Valley, Utah, and the San Miguel River canyon and along the Dolores River in western Colorado. Bedding in the sediments becomes more obvious and better defined. Besides red sandstone and conglomerate, thin beds of purple shale become more common and create obvious breaks in the thick sandstone/conglomerate successions that dominate the Cutler.

Clasts that compose the conglomerate also become smaller to the southwest. Pebbles and cobbles become increasingly abundant at the expense of boulders, which are rare to the southwest. Geologists who specialize in sedimentary rocks precisely define conglomerate clasts based on their size. **Pebbles** range from 2 to 64 mm, whereas **cobbles** are from 64 to 256 mm in diameter. Anything more than 264 mm (~10 inches) is a boulder. Do geologists actually walk around and measure the size of clasts? When size is important to what they are investigating, the answer is yes. Distinction between clast sizes is particularly relevant when studying ancient river deposits because it reflects the energy of that long-vanished river. Clast size decreases with distance from a river's source for several reasons. First, the ability of a river to move larger clasts lessens with distance

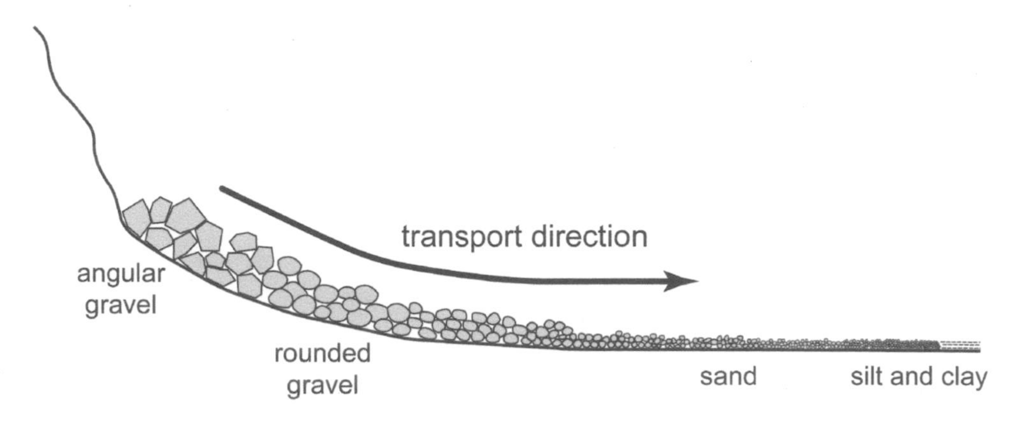

Fig. 3.5. Diagram showing the progressive downslope rounding and fining of clastic sedimentary particles with transport distance.

Fig. 3.6. Photo of the Cutler Formation at Fisher Valley, Utah, showing the interbedded nature of the eolian and fluvial strata.

from its source because the gradient, or slope of the channel, diminishes and the water does not have as much energy. Rivers are steepest in and near their mountainous sources and are reduced to their lowest gradient as they approach the sea. Another factor is the abrasion and physical breakdown of rock fragments, which increases with transport distance. As these particles bang against each other when rolling and bouncing along the river bed, they are broken and smoothed into increasingly smaller fragments (Fig. 3.5).

The regional distribution of grain size in a rock unit is important to reconstructing ancient geographies because sediments coarsen toward the mountains and become finer toward the sea. The Pennsylvanian-Permian Uncompahgre highlands are long gone, but early geologists in the Paradox basin knew of their existence from observations of regional grain size distribution in the Cutler Formation and related strata. Since then, numerous other lines of evidence have confirmed this interpretation, but it was changes in grain size that originally brought the Ancestral Rockies to the attention of geologists. So while mountain ranges inevitably are reduced to rubble and disappear through time, their ghostlike presence can be demonstrated through a careful look at the sedimentary record.

Excellent and widespread exposures of the Cutler Formation occur in the canyons of the Fisher Tower/Onion Creek area west of Gateway and east of Moab. In this area the formation consists of red, coarse conglomeratic sandstone interbedded with fine-grained sandstone (Fig. 3.6). These have been interpreted by various geologists as the deposits of braided rivers and eolian dunes, respectively (e.g., Campbell 1980).

Fluvial sandstone is distinguished by coarse angular fragments of granite and conglomeratic beds. Clasts of granite and metamorphic rocks range in size from common pebbles and cobbles to sparse boulders.

A variety of fluvial cross-stratification types ornament the red canyon walls. The abundant and diverse designs chronicle alternating periods of erosion and deposition that characterize braided rivers. Trough cross-stratification dominates and consists of concave-upward basal surfaces that formed as high-energy currents scoured into loose sediment on the channel floor (Fig. 3.7). These erosional hollows were filled with cross-stratified pebbly sandstone during subsequent lower-energy flows. Horizontal stratification is also a common component of these braided river deposits (Fig. 3.7b).

The term "braided" is applied to the rivers responsible for this part of the Cutler and describes the multichannel nature of the rivers in plan view. Braided rivers form today where stream gradient drops and the flow loses energy, causing it to deposit its sediment load in the channel. This occurs relatively close to the mountainous source, but beyond the alluvial fans. The abundant sand and gravel from the upper reaches is deposited, yet the river must cut through it to continue seaward. This is accomplished by molding the sediment into sand and gravel bars through which the shallow watercourses must thread their way. During high flow the wide channel is flooded bank to bank. As flow wanes and water level drops, bars are sculpted from the sediment and emerge, forcing the flow into multiple courses as it is diverted around these self-made obstructions.

At Fisher Towers fine-grained eolian sandstone forms interbeds tens of feet thick within coarse-grained fluvial deposits. Eolian deposits are recognized by uniform grain size, large-scale cross-stratification, and by an orange hue that stands

Fig. 3.7. Examples of trough cross-stratification in braided fluvial deposits from the Cutler Formation in the area of Fisher Valley, Utah: A, coarse-grained conglomeratic strata with multiple troughs scouring into underlying troughs; B, a single trough of sandstone filling a scour at the base of a channel and grading up into horizontal stratification. Note that this trough also erodes into horizontal stratification.

out from the red and purple fluvial deposits. These beds represent small, isolated dune fields that developed as blowing sand piled up on the arid floodplain. In some cases, especially farther west, they may represent fingers of Cedar Mesa Sandstone, an eolian unit that is part of the time-equivalent Cutler Group (see Fig. 3.3a). According to geologist Jack Campbell (1981), who conducted a regional study of Cutler Formation exposures, eolian beds decrease in abundance a short distance south, near Lisbon Valley, and are absent in exposures still farther south.

While fluvial sediments were carried from the Uncompahgre highlands westward to the sea, eolian sand has a different transport history. Based on dominant southeast dip directions in eolian cross-strata, it appears that fine sand was recycled from shoreline deposits to the northwest and blown to the southeast onto the vast plain that separated the mountains from the sea. In this process sand originally was transported west by rivers, then back east by wind, possibly several times before submitting to a final resting place. Such an erratic path undoubtedly contributed to a diminished size and more rounded grains. In fact, eolian sand grains are more spherical than fluvial grains, attesting to recycling and a greater travel distance.

Braided stream deposits shift into low-energy, meandering stream deposits farther to the southwest, notably in exposures in Lisbon Valley and near La Sal, Utah, and around Durango, Colorado. Meandering stream deposits differ in that they are dominated by thick sequences of mudstone that may make up to 85 percent of the formation (Campbell 1979, 1981). This type of stream deposit is characterized by isolated lenses several tens of feet wide that delineate the actual stream channels. Sandstone bodies are completely encased in red mudstone, which represents floodplain deposition.

Meandering rivers are low-gradient, low-energy rivers with a noticeably sinuous single channel that loops lazily across a broad, mud-dominated floodplain. They invariably occur downstream from braided parts of the system where the gradient continues to decrease with distance from the source. Meandering river channel deposits may contain minor amounts of conglomerate, typically at the base of the channel, but chiefly are sandstone. Because most coarse sediment drops out in the braided reaches, it is mostly sand and mud that move in the channel, but mud is deposited primarily on the floodplain.

Floodplain deposition occurs, appropriately, during floods. When a meandering river rises to flood stage, the high velocity of the water increases the turbulence. This stirs up the fine sediment and muddies the water. While clay and silt are suspended in the water column, the water overtops the banks and spills across the flat floodplain. As the once tumultuous flow spreads across the flat expanse, it loses all energy, and the mud is deposited. In this way a thin veneer of clay and silt builds up the floodplain incrementally. Over thousands or millions of years, accumulations easily add up to hundreds of feet of mudstone.

The Cutler Group

Shifting to the western part of the Paradox basin, there is enough variation in the rock type or lithology that the Cutler can be subdivided into several easily recognized units. In this area the Cutler is elevated to a higher status: it becomes the Cutler *Group* and is composed of several formations. Formations are the fundamental rock units in the science of geology. The designation of formations, which is governed by a rigorous set of guidelines, enables geologists and anyone else to discuss specific rock units with the confidence that everybody involved knows exactly which rocks are being discussed. In short, a formation is recognized by its specific rock type or types, has a definable lateral extent, and well-defined lower and upper boundaries. It is worthwhile at this point to briefly discuss some aspects of formations.

A formation is a body of rock defined by lithology and its position within a vertical sequence. It must be recognizable in outcrops and in the subsurface through drill hole data. While other features in the rock may be important, the most basic aspect of the unit, the type of rock that makes it up, is of primary importance. A formation may be defined as a single lithology, such as sandstone or shale, or it may be a heterogeneous unit with several rock types. Those made of a single rock type may have that lithology as part of its formal name—for instance, the White Rim Sandstone that is part of the Cutler Group. If the unit is composed of several rock types, the term "formation" becomes a formal part of the name. An example of this is the Paradox Formation, which consists of limestone, shale, and evaporites.

Formations are named for places where they are first recognized and/or most completely exposed. Such places are known as the **type locality** for that formation. Many formations were identified by early geologists in the region. The Cutler Formation was named by Whitman Cross in 1899 for exposures along Cutler Creek a few miles north of Ouray, Colorado. The Cedar Mesa Sandstone was named by A. A. Baker and J. B. Reeside Jr. in 1929 for spectacular exposures along the rim of Cedar Mesa and in the canyons that dissect it.

A problem with the designation of new formations that occasionally arises is their creation for erroneous or invalid reasons. This may be caused by a geologist that does not have a complete regional knowledge of the unit, or possibly the unit had already been named, but in an adjacent area, so that the unit ends up with two names. This was more common in the late 1800s and early twentieth century, when geologists were more isolated while in the field. In other cases later, more detailed work may show that relations between units are not as originally thought, and adjustments or redefinition must be made. Such revisions are common; it is the rare formation that remains exactly as it was originally defined, especially if it was first identified more than twenty years ago. Do not get the impression that geologists race to establish new formations; in fact, it is the opposite. New formations are now established only when absolutely necessary, and there are several stratigraphic problems in southeast Utah that were recognized years ago that remain uncorrected. Geologists are hesitant to formally solve some of these problems, especially if the original designations have become entrenched in the geologic literature. In most cases, when changes are backed up by careful fieldwork and good evidence, geologists accept and even welcome these changes because they clarify stratigraphic relations and remedy long-standing problems. In rare instances, however, obstinate geologists have refused to accept changes to their designations, leaving it up to the geology community to decide what will and will not be used. The basal strata of the Cutler Group are one such problem, and it is discussed in the following section.

Lower Cutler Beds

Strata at the base of the Cutler Group have been informally designated the "lower Cutler beds." These rocks are well exposed along the Colorado River and its tributaries in southeast Utah and are known from drill hole data. The lower Cutler beds overlie the Pennsylvanian Honaker Trail Formation of the Hermosa Group and are overlain by the Cedar Mesa Sandstone or the Organ Rock Formation (see Fig. 3.3). As defined by Steve Condon of the USGS (1997), the lower part of this informal unit is of Pennsylvanian age, whereas the upper part is Permian. The current informal standing of the lower Cutler beds emphasizes the controversy that has plagued these rocks for the last century. An account of this unit is outlined below.

The varied history of these rocks begins with Whitman Cross, the pioneering geologist who, with several colleagues, first mapped the geology of southwest Colorado in the late 1800s. Cross and Spencer (1900) named the Rico Formation for the Rico Mountains of southwest Colorado and defined it as a sequence of limestone, sandstone, and shale that was transitional between underlying marine strata and overlying continental deposits of the Cutler Formation. Marine fossils indicated a Pennsylvanian and Permian age. The name "Rico Formation" was later applied to similar rocks in the San Juan River canyon to the west, where it cuts through the Monument upwarp. The name became confusing when various geologists working during the Four Corners oil boom of the late 1950s and 60s began to use the name for any transitional sequence associated with Pennsylvanian-Permian strata. If correlation with the original Rico Formation in Colorado was uncertain, the informal term "Rico facies" was used, adding to the confusion. Eventually the term "Rico" became a confusing obstruction to understanding these strata.

In 1962 geologist Don Baars, in an attempt to clarify the Permian stratigraphy of the Colorado Plateau, called for the abandonment of the name "Rico." Baars defined a new Permian unit for the upper part, calling it the Elephant Canyon Formation for its type locality in a Colorado River tributary 3 miles north of the Confluence, where the Green and Colorado rivers come together. The lower part of the Rico was assigned to the underlying Honaker Trail Formation. This redefinition was based on Baars's interpretation of an unconformity separating the Pennsylvanian and Permian successions in this area. According to Baars, this gap in time took the form of an angular unconformity in the Cataract Canyon and Confluence areas. In Baars's view the Pennsylvanian Honaker Trail, as he redefined it, was folded into an anticline in latest Pennsylvanian time, and the crest of the fold had been beveled flat. According to Baars, this was followed closely by deposition of Permian sediments on the erosion surface (Fig. 3.8). Folding was triggered by the local upward flow of the underlying Paradox salt in response to the pressure

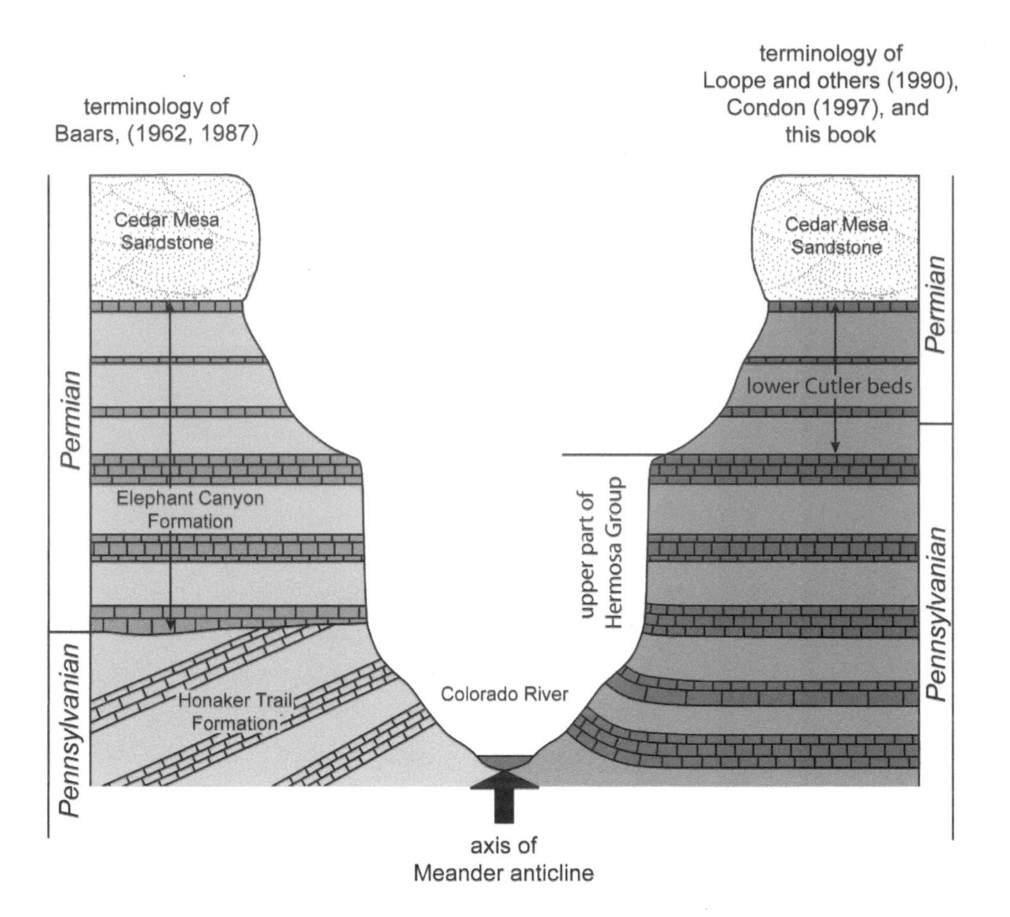

FIG. 3.8. Stratigraphic column showing the two conflicting interpretations of Upper Pennsylvanian and Permian strata along the Colorado River through Canyonlands National Park. The column on the left is the interpretation of Baars (1962, 1987), which was based on biostratigraphic evidence and an interpreted angular unconformity. The column on the right is the more recent interpretation based on lithology and geologists' ability to recognize the units in the field, which are the proper criteria for establishing such units. This is the nomenclature currently used by most geologists. Note that the "lower Cutler beds" comprise an informal unit that straddles the Pennsylvanian-Permian period boundary. Modified after Loope and others 1990.

of overlying sediments. Moving away from the Meander anticline, as this fold is called, the angular relationship disappears, and the unconformity changes into a type known as a disconformity, in which beds above and below are flat-lying. In many cases such cryptic unconformities can only be recognized by using fossils to determine the age of the strata that bound the surface. In this case, Baars called on paleontologists from Shell Oil Company to examine tiny fusulinid fossils in limestone beds to identify the disconformity. In the field, however, there is no other way to tell the presence of the unconformity because rocks of the upper Honaker Trail and the overlying Elephant Canyon Formation are identical.

There are several problems with Baars's definition of the Elephant Canyon Formation. First and foremost, formations are defined by rock type, and their boundaries should mark changes in lithology. The basal contact with the Honaker Trail

Formation is practically impossible for other geologists to identify because of the similarities between the rocks above and below the boundary. Even by Baars's own admission, the contact, where it is not an angular unconformity, must be identified by fossils. This breaks one of the rules for defining a formation. Additionally, subsequent geologists who chose to accept the Elephant Canyon as a valid formation placed the lower contact at a different level than Baars's original definition, illustrating the difficulties in recognizing it; this includes the geologic map of Canyonlands National Park created by Peter Huntoon and others (1982). USGS geologists who worked in the area never accepted the name because of an inability to recognize the formation boundaries.

Another problem lies in defining a formation based solely on the presence of an unconformity. This is not a valid criterion for a formation boundary, particularly where rocks are identical across that boundary. While many well-accepted formations are indeed separated by unconformities, the rock type changes recognizably across these boundaries. After studying these strata extensively in Canyonlands, geologist David Loope strenuously objected to the Elephant Canyon Formation (Loope 1984, Loope and others 1990). Loope argues that there is no angular unconformity or identifiable disconformity within this succession. Along with paleontologists George Sanderson and George Verville, Loope reexamined Baars's type section in Elephant Canyon. A detailed look at fusulinids throughout the succession suggests that the lower 450 feet, as defined by Baars, are actually of Pennsylvanian age. Their data produced no evidence for a disconformity at the type locality as well. Similar results were obtained from a study of the alternate type section that Baars designated. An alternate type section was proposed because of the remote nature and difficulty of access to the original type locality in Elephant Canyon. These rocks came from a subsurface section obtained from a drill hole core along the Green River, north of the Confluence (Sanderson and Verville 1990).

Recent publications have avoided Baars's terminology, instead resorting, at least for now, to "lower Cutler beds" to refer to these strata (e.g., Dubiel and others 1996, Condon 1997). It is likely that some time in the near future, once the Pennsylvanian-Permian has again been deciphered, another formal designation will be proposed. At that point the saga will be over, or it will plunge into yet another chapter.

The "lower Cutler beds" were originally defined by Loope and others (1990) in their call for the abandonment of the Elephant Canyon Formation. The unit overlies the Honaker Trail Formation of the Hermosa Group and is overlain by the Cedar Mesa Sandstone, although the uppermost part also grades laterally into the Cedar Mesa. The uppermost part of the lower Cutler beds also includes the former Halgaito Formation, which has been abandoned as a formal unit. As currently used, the base of the lower Cutler beds is the same as the base of the Elephant

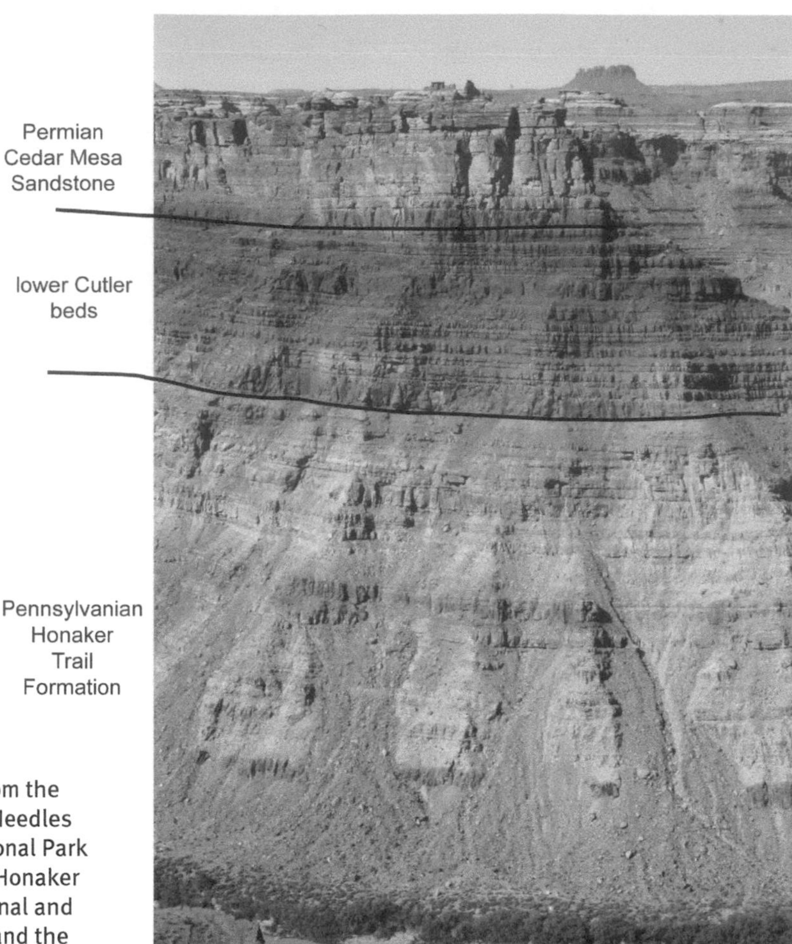

FIG. 3.9. View to the west from the Confluence Overlook in the Needles District of Canyonlands National Park showing the Pennsylvanian Honaker Trail Formation, the transitional and informal lower Cutler beds, and the Cedar Mesa Sandstone.

Canyon Formation as mapped by Huntoon and others (1982) on the Canyonlands National Park geologic map. This suggests that the base of the lower Cutler beds is a more logical and recognizable boundary.

In its northwestern reaches in the San Rafael Swell area, the lower Cutler beds lie unconformably on Mississippian rocks. Strata of the Pennsylvanian Hermosa Group are completely absent, due either to erosion or nondeposition. Regardless of why they are missing, this relationship indicates Pennsylvanian tectonic activity in the form of mild uplift in the area, probably on an early incarnation of the San Rafael Swell. This is not an isolated event; there is abundant evidence for similar activity on this structure during the subsequent Mesozoic Era as well.

The lower Cutler beds are exposed in a belt that extends along the Colorado River from the Cane Creek anticline in the north to the San Juan River canyon in the south. Throughout this region the lower contact is at the top of a conspicuous thick cliff band of gray limestone and white sandstone of the Honaker Trail Formation. The lower Cutler beds consist of thinner beds of fossiliferous limestone, sandstone, and red mudstone that weather into a steep red slope broken by thin ledges. The unit is well displayed along the precipitous walls of the Colorado River canyon where it is bound below by the Honaker Trail cliffs and above by blocky cliffs of the Cedar Mesa Sandstone, rendering it largely inaccessible except in side canyons (Fig. 3.9).

Lower Cutler beds reach a maximum thickness of ~1000 feet in the Green River canyon just north of the Confluence, although this has been determined from a combination of outcrop and drill hole data. Outcrops along the Colorado River expose ~400 to 600 feet of the lower Cutler. The unit thins to the northeast as it grades into undifferentiated Cutler Formation. To the southwest it grades into shallow and deep marine deposits that appear to have been outside the influence of sediment influx from the Uncompahgre highlands.

Although the great variation in rock type is typical of any given succession of the lower Cutler beds, the relative percentages of conglomerate, sandstone, mudstone, and limestone also vary greatly, reflecting different depositional environments. The environments represented in these successions include fluvial, deltaic, eolian, and shallow marine settings. Rapid lateral changes in thickness and rock types are attributed to the interplay of such factors as sea level fluctuation, changes in sediment input from the Uncompahgre highlands, and variable subsidence rates within the basin.

Lower Cutler beds exposed in the north part of the outcrop belt, notably the Cane Creek anticline and Shafer dome areas, are dominated by two distinct types of sandstone: coarse-grained arkose and fine-grained, quartz-dominated sandstone. Each reflects a different mode of transport and source area. Arkosic sandstone, or **arkose**, is rich in feldspar fragments and is derived from high-relief granitic mountains, in this case the ancient core of the Uncompahgre highlands. Such mountains gradually decompose to fragments, producing abundant, feldspar-rich sand which is then whisked away by high-energy rivers, deposited, and quickly buried by more sediment. Rapid breakdown, transport, and burial shields the feldspar from a prolonged attack by water, which normally would break it down to clay. Conglomerate within the arkose—in conjunction with the lens-shaped geometry of these units, which mimics the outline of relict river channels—attests to deposition in the west-flowing rivers that originated in the adjacent Uncompahgre highlands. Many generations of these rivers met the sea in this area to form deltas.

Fine- to medium-grained quartz sandstone ranges from maroon to orange and forms small cliffs. Beds reach up to 20 feet thick, with thicker beds clearly showing large-scale cross-bedding. These deposits originated as eolian dunes and as eolian sand sheets that skirted the small dune fields. The sand accumulated on the fluvial floodplain, locally blanketing the area enclosed by active river channels, and also along the coastal plain that lined the shallow seaway that lay immediately west. Cross-bedding dominantly dips to the southeast, indicating a regional wind blowing in that direction. Sand came from the northwest and west. Most geologists who have examined these strata agree that quartz sand was blown inland by winds coming off the sea. Sand was derived from the reworking of shallow marine and shoreline deposits as they were intermittently exposed during constant sea level fluctuations.

Fossiliferous, shallow marine siltstone and sandstone are interbedded with fluvial and eolian deposits. Fossils are varied but consist mostly of gastropods and bivalves that could tolerate fluctuations in salinity as seawater mixed with rivers. These organisms mostly lived and fed by burrowing through the sediments. The clamlike bivalve *Wilkingia*, which grew to more than 10 cm long, is commonly found preserved in its burrow. Most of the sediments that host these fossils have been churned up by the excavation activity of the numerous organisms that lived in this environment. Geologist Forrest Terrell (1972), who studied these rocks and fossils in detail, interpreted the deposits to be the product of small deltas that stretched out into the shallow marine environment wherever sediment-laden rivers met the sea. Mud and fine sand accumulated sporadically in low-lying areas between the numerous delta channels. As sediment dropped out of the river at its interface with the sea, the sluggish rivers had to force their way through their self-imposed barrier. The branching of a single river channel into multiple, deltaic distributary channels is the natural response of a low-energy watercourse spreading out to find the easiest outlet.

Terrell (1972) and botanist William Tidwell (1988) have reported the presence of a thin coal bed in the lower Cutler beds near Potash that is encased in green-gray siltstone. Coal provides compelling evidence for a swamp environment and supports Terrell's interpretation of a deltaic setting. Most coal swamps are closely related to delta environments, and delta/coal swamp deposits are especially significant in the later geologic history of the region, during the Cretaceous Period. It appears, however, that the Pennsylvanian-Permian climate in the Paradox basin was too arid for large-scale coal swamp development. Coal deposits require hundreds of thousands of years to accumulate to an economically beneficial thickness. It takes ~100 feet of vegetation, stacked and compressed, to produce just one foot of coal. At this locality, the swampy environment appears to have endured long enough to develop a noticeable thickness of carbonaceous sediment, preserving some material to the extent that many plant species can be identified. In addition to this thin coal seam, Terrell discovered a 60-foot-long petrified conifer log at the base of an arkose-filled river channel deposit. So while the climate was arid and likely hindered the large-scale growth of plants, limited areas such as swamps and riparian zones that slashed across the sand-cloaked plain supported a healthy flora.

Limestone beds are thin, gray, and fossiliferous and commonly contain chert. Although limestone beds are interspersed throughout the lower Cutler, they are most abundant near the base and top. Collectively, limestone beds have a shallow marine origin, but individually they represent an astoundingly diverse range of environments, from high-energy offshore shoals that were pounded by waves to quiet, isolated estuaries. Some beds also contain coarse, angular granitic fragments and arkosic sand, probably propelled seaward by the jetlike currents of rivers in flood. Other units have a component of red silt, likely mixed into the carbonate sediments by longshore drift—powerful marine currents that moved parallel to the convoluted shoreline, winnowing the fine sediment from the delta front and redistributing it over a wider area. Fossils are commonly encountered in these beds, and they often litter the slopes below limestone beds as they weather from the long-held confines of their matrix. Fossils include a variety of brachiopods, gastropods, bryozoans, sponges, bivalves, crinoids, fusulinids, and rarely, trilobites. The specific associations of these various fossils, in conjunction with their encasing sediment, are what allowed Terrell (1972) to recognize the different subenvironments of the lower Cutler beds.

The rapid changes and variety of rock types at the Cane Creek anticline near Moab was attributed by Terrell to the constant shift of delta lobes as they built out into the shallow marine environment. This was undoubtedly coupled with continued, regular sea level changes as well. To the south, in the Cataract Canyon, Dark Canyon, and San Juan River areas, lithologic changes in the lower Cutler are similar, but the influx of sediment from the east is considerably reduced. Here, various workers have suggested similar environments, but they were dominantly affected by sea level changes and eolian processes.

Cedar Mesa Sandstone

The Cedar Mesa Sandstone, which overlies the lower Cutler beds, consists mostly of red and white cross-bedded eolian sandstone (see Fig. 3.3). The formation is especially well known for its spectacular exposures in the Needles District of Canyonlands, where the colorful red- and white-striped sandstone has been sculpted into a maze of pinnacles with mushroom-shaped summits. To the south, where the sandstone caps its namesake Cedar Mesa, the deeply cut meanders of White Canyon host Natural Bridges National Monument. Cedar Mesa Sandstone also makes up the steep walls of the remote Grand Gulch, as well as Lime Canyon, Road Canyon, and the connected system of Owl and Fish canyons. These deep incisions into the Cedar Mesa Sandstone characterize the rugged beauty of southeast Utah canyon country.

The Cedar Mesa Sandstone is widely distributed through southeast Utah. It extends from the Confluence southward through the Needles and Cedar Mesa area, and southwest into the Four Corners area, where it is recognized in the subsurface from drill hole data. North and northeast of the Confluence, the Cedar Mesa intertongues with age-equivalent strata of the Cutler Formation undiffer-

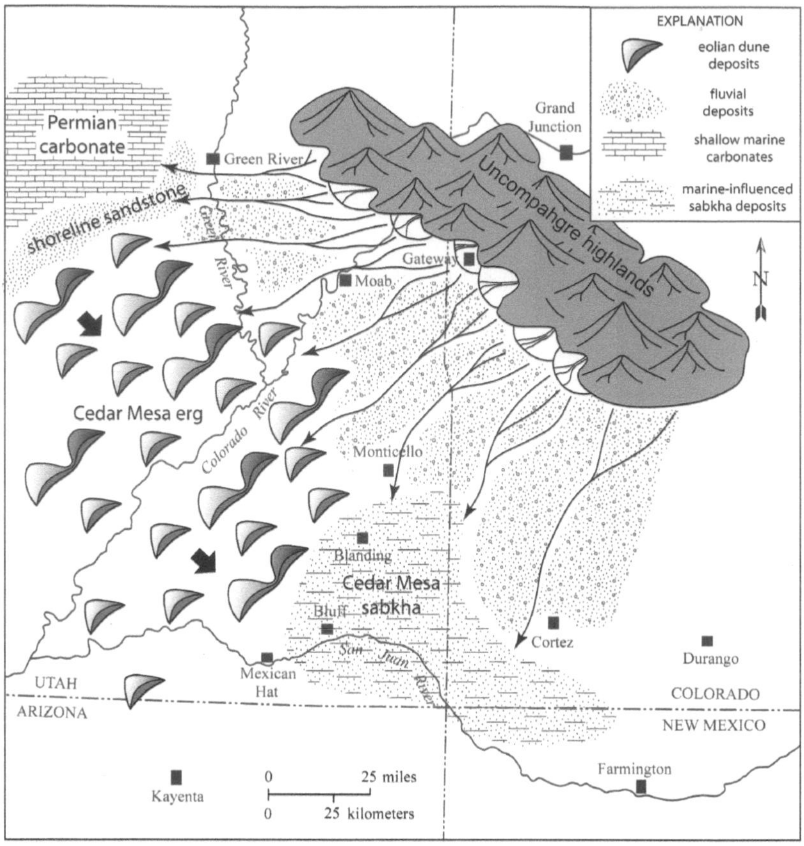

Fig. 3.10. Paleogeographic reconstruction of southeast Utah and surrounding regions showing depositional environments of the Cutler Formation adjacent to the Uncompahgre highlands and the eolian and sabkha facies of the Cedar Mesa Sandstone to the west. Immediately west of the area shown was a shoreline and shallow marine environment. Modified after Condon 1997.

entiated. The sandstone reaches an exposed thickness of ~1000 feet in the Dark Canyon area, and drill holes in the Glen Canyon/Lake Powell area have penetrated more than 1200 feet of Cedar Mesa Sandstone.

The depositional setting of the Cedar Mesa Sandstone was dominated by eolian processes that piled sand into a vast dune field. The fringe areas, and even the interior of this dune field, were regularly invaded by other, water-driven systems. A shallow fluctuating sea bounded the dunes to the west and south, while rivers that continued to emanate from the Uncompahgre highlands modified the northeast margin (Fig. 3.10).

Eolian sandstone of the Cedar Mesa consists dominantly of quartz sand with large-scale crossbedding that typifies wind-blown dune deposits. Crossbedding is a pattern common to sedimentary layering when viewed in cross-section, such as along a canyon wall or cliff face. It consists of multiple sloping layers of sand: each layer slopes steeply at the top and sweeps gracefully downward in an arc to the base where the layers flatten tangentially to a common, flat surface (Fig. 3.11a). While crossbedding in the Cedar Mesa is eolian in origin, it can form under any conditions in which currents—whether wind or water--are moving sediments. The most useful aspect of crossbeds is that the direction that these layers slope—or **dip,** as geologists say—indicates the direction that the current was moving

Fig. 3.11a. Cross-stratification sets in the eolian Cedar Mesa Sandstone in White Canyon. Note the consistent dip direction to the right, representing the wind direction. Note also the tangential relations at the base of each set of cross strata.

Fig. 3.11b. Oblique view of eolian dune deposit in the Cedar Mesa Sandstone in the Needles District of Canyonlands showing cross-stratification and the actual dune front. Note that the wind was blowing, and the dune was migrating in the dip direction shown in the photo.

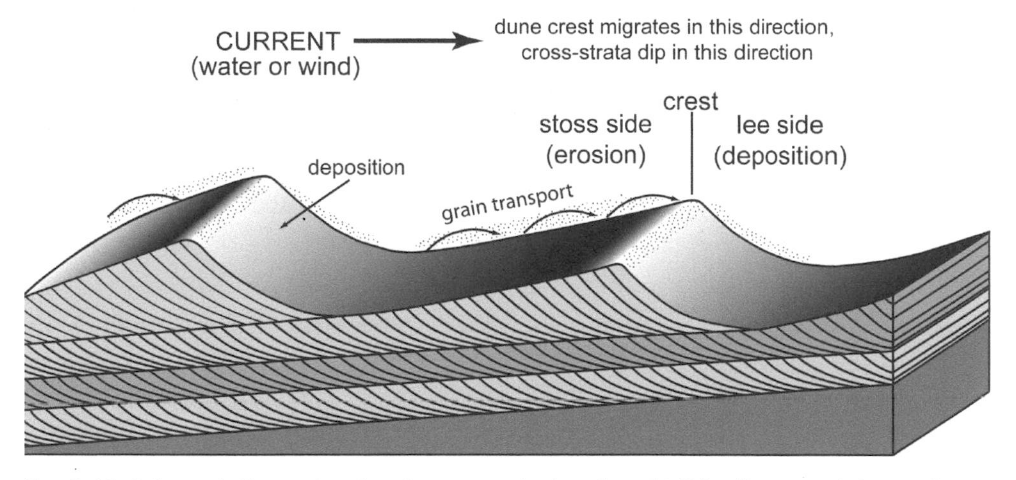

Fig. 3.12. Schematic figure showing downcurrent migration of 3-D bedforms and the production of cross-stratification. As the current erodes the stoss (upcurrent) side of the bed form, the moving sand migrates over the crest and out of the flow of the current to be deposited on the lee side. Each lee side layer that builds downcurrent produces an increment of downcurrent bed form migration. Cross strata dip in the direction of the current. Note that this activity can occur in eolian sand dunes on a large scale (tens of meters), as medium-scale bed forms on the bottom of a river (tens of centimeters), or as small-scale ripples (millimeter scale) in water or wind. Regardless of the scale or origin of the current (wind or water), the pattern of migration and cross-stratification development is the same.

when the sediments were deposited (Figs. 3.11b and 3.12). For this reason, crossbedding is an important element in the reconstruction of ancient landscapes. It informs us of the direction that long-extinct rivers flowed—something that can be used to infer where the mountainous source for the river was located, as well as the direction of the sea.

Crossbedding in the Cedar Mesa Sandstone consistently dips toward the southeast, indicating that the wind, by modern coordinates, was blowing from northwest to southeast. Thus, the main source area for the large volumes of sand in this ancient dune field was the fluctuating shoreline that bordered it to the west.

Besides crossbedding, among the more obvious features in the Cedar Mesa Sandstone are the thin, laterally extensive layers that separate the crossbedded strata. These typically differ in color from the sandstone and form horizontal recesses that cut across the buttes, pinnacles, and cliff bands of the formation. The shadowed recesses that mark these layers also host many remnants of the Anasazi civilization. According to detailed field investigations by John Stanesco and John Campbell (1989), many of these layers can be traced for more than 200 km.

Rocks that comprise these thin layers range from limestone to siltstone, shale, and fine-grained sandstone. Varied rock types reflect the numerous possibilities for the origin of these layers. While the exact genesis of these deposits remains controversial, geologists agree they are related to an invasion of the arid, sand-covered region by water. The question is, what was the source of the water? Was it surface water that originated in the sea to the west or rivers to the east, or was it groundwater that rose from below during a shift to a wetter climate? The answers may lie in relations between the various lithologies and the many features found within them.

Within these layers localized lenses of micritic limestone grade laterally into siltstone and structureless sandstone. These rocks contain abundant burrows and root casts, and localized mudcracks. Taken together, these rocks and their features represent deposition in interdune areas, which are troughs and low-lying areas between the high-relief dunes. Limestone formed in ephemeral ponds in the deeper hollows during wet intervals. Burrows and rhizoliths indicate that moist conditions lasted long enough to be colonized by invertebrate organisms and allow plants to become established. In the Maze part of Canyonlands, root casts range up to 20 cm in diameter and have laterally radiating root branches that can be traced for several meters. The size of these rhizoliths indicates the growth of tree-sized plants—a feature that suggests relatively long lasting wet conditions and the development of a soil zone.

The lateral continuity of these layers indicates that deposition was not confined to interdune areas. In fact, the geometry of these beds requires sedimentation on a broad, flat plain. Thus, the dunes must have been absent. The entire process—the erosion of sand dunes and colonization of the resulting flat, sandy surface—is simply explained by a rise in the groundwater table. As the water table rose, low-lying interdune areas were the first to be flooded and transformed into ponds in which limestone was deposited. While the northwesterly winds drove the dunes to the southeast, the lack of new, incoming sand caused erosion to cut down to the damp, sandy surface of the water table. The erosive wind, however, was unable to winnow the water-secured sand. This wet and stable surface became an ideal setting for invertebrate organisms and plants, generating the laterally extensive layers seen today in the rock record.

The main unknown in this interpretation is the cause of the water table rise. There are two possible scenarios. The first concerns the feeding of a shallow aquifer by precipitation in the Uncompahgre highlands to the east. As rivers from these mountains headed seaward across the arid fluvial plain, much of the runoff likely seeped into the sediments. During wet periods, increased runoff may have raised the water table. As we have seen during Paradox deposition, sea level rose and fell with a stunning regularity, resulting in wetter and drier periods, respectively, in adjacent highlands. There is abundant evidence from around the world that such sea level oscillations continued into the Permian. In this scenario, dune deposition occurred during dry periods, while wet periods were marked by the complete erosion of dry sand down to the wet surface of the shallow water table. It was this surface that controlled the depositional process until the water table again dropped and the dunes became reestablished. According to geologist David Loope (1985), who has worked extensively on the Cedar Mesa Sandstone, the crossbedded sands that make up the bulk of the formation represent a much shorter time span than the thinner and less voluminous deposits of the wet interval.

The second scenario involves an elevation of the water table triggered by sea level rise. The shoreline of the sea, which bordered the dune field along its west and northwest margin, was the source for the voluminous quartz sand in the Cedar Mesa erg. As the wind swept onshore from the northwest, it plucked the fine sand from its shoreline moorings, blowing it back inland to feed the dunes. Once sea level rose, the sandy shoreline was inundated, cutting off sediment to the erg. The rising sea would also have seeped laterally into the porous sand of the dune field, raising the water table. Sea level rise would have had the dual role of cutting off the sediment supply and elevating the water table, both promoting erosion down to a flat, unbroken plain.

It is possible, even likely, that both scenarios had an influence on non-eolian sedimentation in the Cedar Mesa—with identical results. The strata in the east part of the erg would have been more easily influenced by increased precipitation and runoff from the Uncompahgre highlands. The western part, however, would more likely have been affected by sea level fluctuations. As we have seen from evidence in the Paradox Formation, a regional sea level rise promotes increased precipitation in nearby highlands.

Geologists Richard Langford and Marjorie Chan (1988) of the University of Utah documented a localized process for producing non-eolian strata in the Cedar Mesa Sandstone. Their study area was centered on the Colorado River area between Moab and the Confluence. Here, shale and fluvial sandstone of the undifferentiated Cutler are interbedded with eolian sandstone of the Cedar Mesa. These intertonguing relations are along the northeast erg margin (see Fig. 3.3b) and suggest that the rivers that emitted from the highlands to the east repeatedly flooded the erg fringe, depositing shale in the interdune lows and reworking the eolian sands. Langford and Chan also found that the thickness and number of these flood surfaces decreased toward the erg interior, suggesting that the water did not invade far into the erg.

In these same fluvial deposits in the lower Indian Creek area of the Needles District, John Stanesco and John Campbell (1989) noted the occurrence of a variety of plant fossils. Here, interdune pond limestone hosts several large petrified conifer logs, the largest one being a meter in diameter, with an exposed length of 8 m. It is likely that these logs were washed into the erg during floods. Limestone interfingers with coarse arkosic fluvial sandstone, suggesting that ponds were fed by streams. Nearby, shale associated with fluvial sandstone contains impressions of fossil horsetail plants and ferns. The fragmented nature of the fossils led Stanesco and Campbell to suggest that these also were washed in from outside the erg on a rising flood.

South of its namesake Cedar Mesa, in the Comb Ridge area, the formation transforms rapidly into shallow marine gypsum, limestone, shale, and sandstone. These rocks were deposited in a sabkha setting—an arid-climate tidal flat that was alternately covered by a shallow, fluctuating sea and exposed, stagnant evaporative pools and dessicated sediments (see Fig. 3.10). According to Ron Blakey (1980) of Northern Arizona University, these sabkha deposits extend southward for at least 125 km, although they are nowhere exposed at the surface.

Originally there was some question regarding the origin of these beds; that is, were they deposited in an extensive but isolated interdune wetland fed by a fluctuating water table, or was it a shallow marine embayment controlled by sea level fluctuations? This was settled by Stanesco and Campbell (1989), who conducted chemical analyses on sulfur in the thick gypsum beds. In these evaporite deposits they discovered an isotopic sulfur composition similar to that found in known Permian marine deposits elsewhere. These sulfur isotopic signatures reflect the distribution of various isotopes of sulfur present in the Permian seawater, providing a fingerprint for its source.

FIG. 3.13. The Chocolate Drops, part of the Land of the Standing Rocks in the Maze District, Canyonlands. The dark, stratified part of these pinnacles and their base is in the Organ Rock Formation. The pinnacles are capped by the white eolian White Rim Sandstone, and the white rock in the canyons below is Cedar Mesa Sandstone.

Organ Rock Formation

As we step upward through the varied Permian strata of southeast Utah, the Cedar Mesa Sandstone abruptly gives way to the Organ Rock Formation. The dark, brown-red slopes of the Organ Rock are a stark contrast to the light-colored cliffs and overhangs of the Cedar Mesa Sandstone. The Organ Rock is composed mostly of sandstone and siltstone, with lesser amounts of conglomerate and mudstone. The thin beds of these various lithologies give the formation its distinctive horizontal outcrop pattern. The formation is an important component of the classic southwestern landscape throughout its exposure.

The extent of the Organ Rock roughly parallels that of the underlying Cedar Mesa Sandstone. Exposures of the Organ Rock are numerous but confined to the deeply incised canyons of the Colorado/Green river system and its tributaries, and the Monument upwarp to the south. The formation rims Monument basin, in the northern part of Canyonlands, where it is overlain by the White Rim Sandstone. Monument basin lies just south of the White Rim Trail, a popular bike and four-wheel-drive route, and can be clearly seen from the overlook at Grandview Point in the Island in the Sky part of Canyonlands. To the southwest, across the Green and Colorado rivers, isolated remnants of the Organ Rock form the slender pinnacles for which the Land of the Standing Rocks was named by John Wesley Powell (Fig. 3.13). These towers sit at the head of the intricate canyon system of

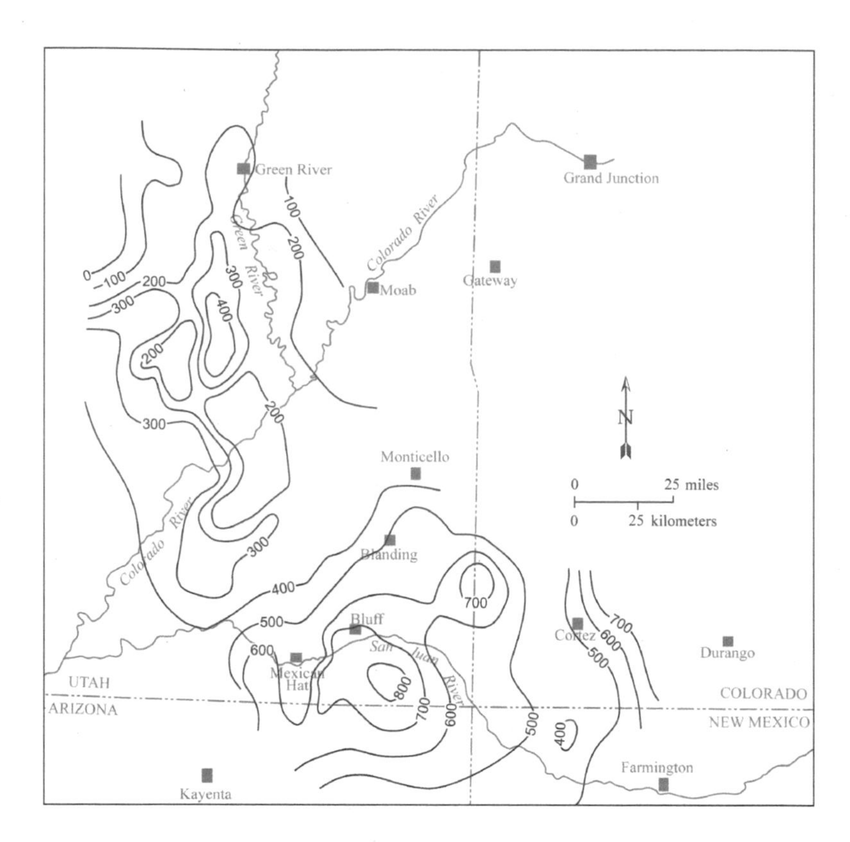

Fig. 3.14. Isopach map of the Permian Organ Rock Formation. Lines and numbers show the total thickness of the formation throughout its extent. The numbers refer to thickness in feet. Thickness data come from a combination of well data and outcrop measurements. Note the abrupt thinning trend northwest of the towns of Blanding and Monticello. The northeast trend of this thinned zone reflects the location of the Monument upwarp at this time. After Condon 1997.

the Maze, the remote western part of Canyonlands National Park. To the southeast, on Cedar Mesa, the Organ Rock makes up the apron of deep red slopes at the base of Jacob's Chair and the Cheesebox, two conspicuous buttes that look down on White Canyon and its fingerlike tributaries. Finally, along the Utah-Arizona border, one of the icons of the southwest, Monument Valley, is floored by the Organ Rock, upon which sit the buttes and pinnacles composed of the Permian De Chelly Sandstone.

The Organ Rock attains a maximum thickness of ~800 feet in southeast Utah, although it varies dramatically over its extent. Using a combination of outcrop and drill hole data, USGS geologist Steve Condon (1997) produced an **isopach**, or thickness, map of the Organ Rock Formation. Isopach maps look like ordinary topographic maps at first glance, but the curving contour lines represent the thickness of a unit rather than elevation. The most notable trend on this map is a pronounced thinning of the formation, locally down to 100 feet, in a triangular zone bounded to the south by the San Juan River and to the northwest by the Colorado River (Fig. 3.14). This zone of thinning coincides with the modern Monument upwarp, a mildly uplifted area that has been intermittently active through much of the region's geologic history. According to a detailed study by John Stanesco, Russell Dubiel, and Jacqueline Huntoon (2000), the Monument upwarp was rising during Organ Rock deposition, contributing to variations in thickness and, as we shall see, depositional setting.

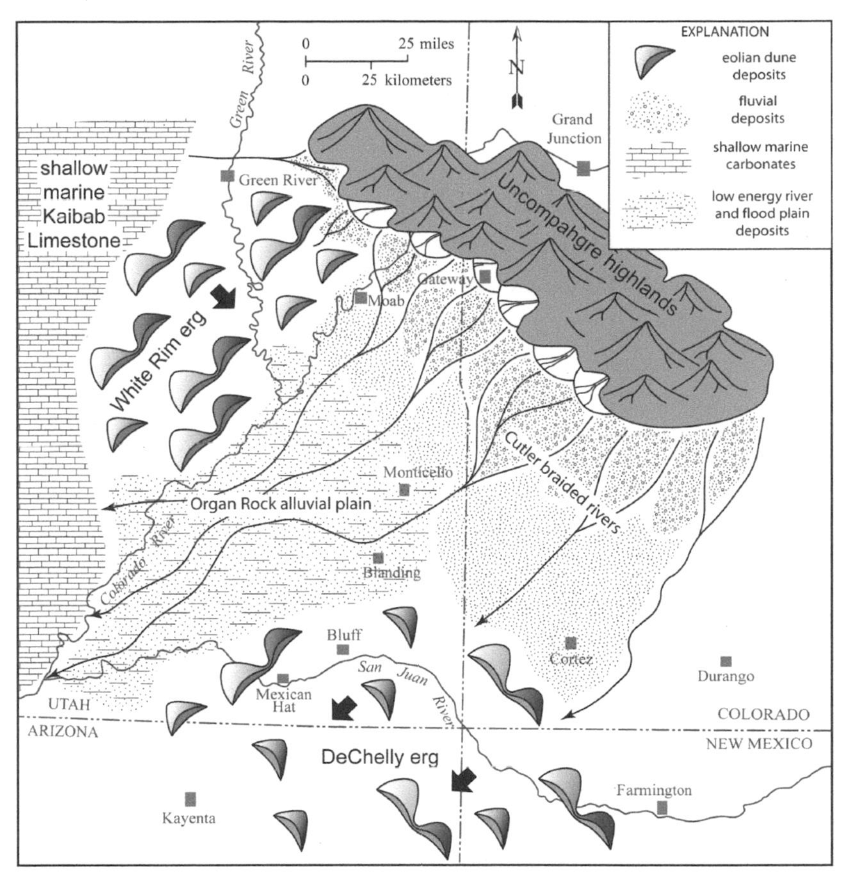

FIG. 3.15. Permian paleogeography of southeast Utah and surrounding areas during deposition of the Organ Rock Formation, White Rim Sandstone, and De Chelly Sandstone, all parts of the Cutler Group, as well as the Cutler Formation to the northeast. Figure is after Condon 1997 and Stanesco and others 2000.

Although the stratigraphy of the Organ Rock Formation and associated Permian strata in southeast Utah have been exhaustively treated over the last seventy years, the depositional setting of the formation, until recently, remained unstudied and speculative. Because of this, the following discussion relies heavily on the recent, detailed work of Stanesco, Dubiel, and Huntoon (2000).

The Organ Rock Formation consists of a variety of fluvial and eolian deposits that differ in style and relative abundance with location. Exposures in Monument basin to the northeast, for instance, are dominated by conglomeratic sandstone representing west-flowing, high-energy braided rivers. Conglomerate clasts consist of granite and metamorphic rocks that were part of the continuing slurry of sediment shed from the deeply incised core of the Uncompahgre highlands. Farther west, in the Maze area, fluvial deposits give way to a fine-grained eolian sandstone. Here the westward path of the rivers was hampered by the sporadic sand piles. Fluvial sandstone interbedded with eolian deposits indicate that rivers eventually found their way through, eventually emptying into the shallow sea on the west side of the sandy barrier (Fig. 3.15).

To the south, in Monument Valley and the White Canyon area, fluvial deposits dominate Organ Rock strata, but they differ from those to the northeast in several ways. First, these deposits are finer grained. Lens-shaped river channel deposits consist mostly of sandstone with sparse limestone pebbles confined to the basal part of the units. Second, these discrete river channel deposits are isolated from each other by laterally extensive siltstone and mudstone, interpreted as floodplain sediments. Both channel and floodplain deposits represent sedimentation in a low-gradient, meandering stream system. Isolated, channel form sandstone encased in floodplain siltstone and mudstone is a pattern that typifies these highly sinuous, low-energy rivers.

Organ Rock floodplain deposits preserve an important record of river activity and evidence for climatic conditions. Siltstone commonly contains vertical tubes that branch downward. These are fringed by pink-purple zones, where the typical dark red color has been slightly bleached. These cylinders are interpreted as rhizoliths, remnants of plant roots that extended down into the Permian floodplain soils. Abundant rhizoliths indicate a stable floodplain that locally was protected from erosion by clots of plants, probably living along the well-watered river bank areas. Another significant feature in these sediments is zones of limestone nodules that formed within the soil zones. These caliche nodules, as they are known, range up to several centimeters in diameter and are an indicator of stable, arid-climate soils. Additionally, it is these caliche nodules that make up the limestone pebbles at the base of the channel deposits. The nodules were dropped into the river as it cut into the floodplain during normal lateral migration of the winding channel or during floods. As the river cut into the floodplain, cohesive chunks of soil collapsed into the river. Most of the soil mass broke up into silt and clay particles; however, the more-resistant limestone nodules became coarse, pebble-sized sediment in an otherwise fine-grained river system.

Paleogeography during Organ Rock deposition was strongly influenced by regional-scale features. From east to west these are the enduring Uncompahgre highlands that bounded the basin its northeast margin; the Monument upwarp, which occupied the center of the basin; and the sea, lying to the west, in central Utah (Fig. 3.15).

Organ Rock deposition was largely controlled by the Uncompahgre highlands, which supplied sediment and attracted precipitation that fed the westward drainage system that traversed the arid plain.

The Permian version of the Monument upwarp was not so dramatic as today, but still had a substantial effect on Organ Rock sedimentation (Figs. 3.2 and 3.14). This northeast-trending platform sat in the middle of the Organ Rock basin but was not so high as to act as a sediment source. Instead, its presence is told by thinning and facies changes in Organ Rock strata across its crest and around its margins. For instance, fluvial systems in the northeast part of the basin that came from the Uncompahgre highlands should have flowed southwest or west. At the northeast end of the upwarp, however, rivers were deflected northwestward, around the nose of the upwarp (Fig. 3.15). Stanesco and others (2000) have also found that eolian deposits, which make up a large volume of the formation to the northwest, thin and eventually pinch out southeastward across this subtle high point (see Fig. 3.2).

The sea to the west acted as a receptacle for fluvial sediment that made the westward journey from the highlands, but, more important to the Organ Rock Formation, it also acted as the main source for eolian sand. Windblown sediments dominate westernmost Organ Rock deposits and represent a coastal dune field. Crossbedding in these strata consistently dip to the southeast, indicating a dominant wind direction from northwest to southeast. This points to a sediment source in the sandy shoreline, with a surface area that contracted and expanded with the ups and downs of sea level.

Finally, low-energy stream deposits in the Organ Rock to the south, in Monument Valley, came from a different sediment source than strata to the north. These streams meandered lazily toward the north-northwest out of modern-day northern Arizona. The highlands that drove these rivers were subtle, probably a precursor to the modern Kaibab arch or Zuni-Defiance uplift to the southeast. Both of these are modern topographic expressions that were re-uplifted ~60 million years ago as monoclinal folds. Like the Monument upwarp, many of these modern structures have a long and sporadic history of uplift. As we will see, most were also active through much of the ensuing Mesozoic Era.

White Rim Sandstone

Not surprisingly, the White Rim Sandstone, the uppermost unit in the Cutler Group, consists of fine-grained white sandstone. For the most part it is confined to a region west of the Colorado River, and where exposed, it forms a prominent bench known as the White Rim, for which it was named. It reaches a maximum thickness of 800 feet in its western extent near Hanksville, although this is determined from drill hole data since it is not exposed in the area. Where it is continuously well exposed—along the White Rim north of the Green/Colorado Confluence and in Elaterite basin in the Maze—it ranges 200 to 400 feet thick. Eastward, along the Colorado River, the formation thins and pinches out (Fig. 3.16). However, along the west wall of Castle Valley, well east of this pinchout, a band of white sandstone separates the underlying, undifferentiated Cutler Group from the overlying Triassic Moenkopi Formation. This sandstone likely is an isolated outlier of eolian sand that accumulated farther inland than the main body of the White Rim Sandstone.

Over most of its extent the White Rim overlies the Organ Rock Formation, although locally it rests on lower Cutler beds or Cedar Mesa Sandstone. To the

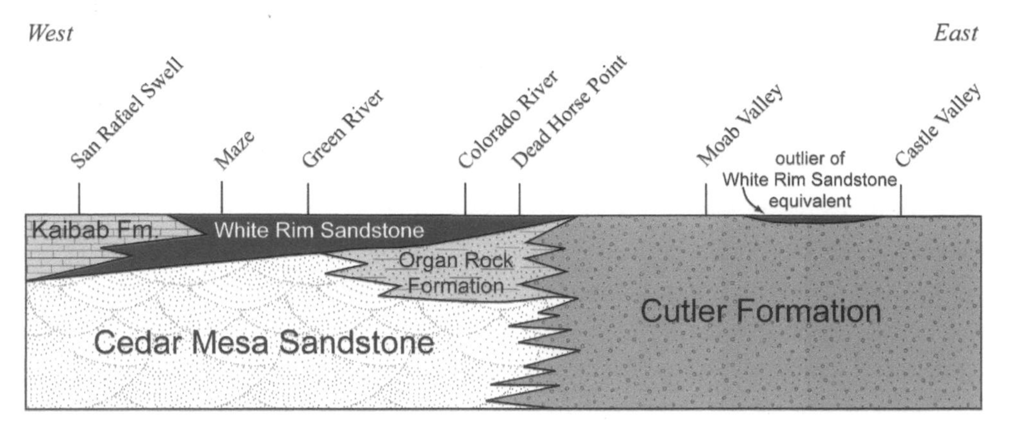

FIG. 3.16. Schematic stratigraphic cross section of Permian strata associated with White Rim Sandstone showing the pinchout to the southwest and northwest of the main erg, and the isolated outlier of eolian sandstone exposed in the west wall of the Castle Valley salt anticline, which is likely the equivalent of White Rim Sandstone.

west, in the Capitol Reef and San Rafael Swell area, White Rim Sandstone grades into shallow marine deposits of the Toroweap and Kaibab formations. To the east the White Rim pinches out and is replaced by fluvial strata of the Cutler Group.

Although the White Rim is dominated by eolian strata, other types of deposits have been recognized. The formation consists of dune, interdune, sabkha, and even marine deposits (Steele 1987, Huntoon and Chan 1987). Dune deposits contain features identical to those in the underlying Cedar Mesa Sandstone, notably large-scale crossbedding, well-sorted quartz sand, and wind-ripple laminations that are unique to wind-formed ripples. Wind ripples are very low relief compared to waterlain types and, in cross section, form laminations that become coarser grained upward over a thickness of a few millimeters. Even in the absence of more obvious features, such as large-scale crossbedding, these small-scale laminations can be used to confidently identify sediment deposited by eolian processes.

Interdune deposits in the White Rim were laid down in isolated pockets between the sinuous, wind-blown ridges of sand. In contrast to the thick, sheet-like dune sands, interdune sediments form thin, lens-shaped deposits. Features in these isolated bodies include wavy laminations, and polygonal bedding plane structures that are interpreted as dessication cracks that formed as wet sediment dried. Taken collectively, interdune features point to a variety of wet and dry conditions that shifted with the vagaries of sea level. When sea level was high, the groundwater table also was high as seawater filtered laterally through the porous sand to saturate the hollows between the dunes. When sea level dropped, the once wet interdune areas dried out, and the wind again took control.

The easternmost belt of White Rim Sandstone consists of sabkha deposits: essentially the product of an extensive, arid-climate wetland. The sabkha was bordered to the east by rivers that continued to roll off the Uncompahgre highlands. West of this sabkha lay the White Rim dune field. These sabkha deposits are similar in lithology to the interdune deposits, but instead of occurring in isolated lenses, they form a laterally extensive, north-south-trending belt. This suggests similar processes, but over a far greater area. This sabkha differs from most in that it apparently had no connection to the sea. Instead, it appears to have been sustained by groundwater, which probably rose steadily as sediment accumulated, maintaining a wet surface (Steele 1987). The lack of a fluctuating water table suggests that it was fed by a more reliable, consistent source, probably the adjacent river system. This is also suggested by the greater distance from the coast, making the sea an unlikely source. It is possible, however, that the sea was a minor contributor to the groundwater supply.

To the west, in Elaterite basin of Canyonlands and in Capitol Reef National Park, the uppermost White Rim has an obvious marine influence. As White Rim sedimentation progressed, sea level rose and advanced over the western edge of the dune field, modifying the upper few feet of sand to leave a clear record of the incursion (Kamola and Chan 1988).

A vital part of this marine record is the feeding and dwelling traces of invertebrate organisms preserved in the sandstone. These **trace fossils**, as they are called, are classified into different species, just as fossil organisms are. Trace fossil species are defined by the geometry and orientation of the tracks and burrows. Although we may not know what organism actually left a trace, hundreds of different species have been recognized, and they are a significant part of the rock record. Particular species have been reliably linked to specific environments and, in the case of some marine trace fossils, to specific depths beneath the sea. The trace fossils *Chondrites* and *Thalassinoides* that are reported from the White Rim Sandstone indicate a shallow marine environment (Kamola and Chan 1988). *Thalassinoides* has been interpreted as a horizontal dwelling burrow, while *Chondrites* is a feeding trace. Both are confined to an upper veneer of sandstone in the White Rim.

Geologists Diane Kamola and Marjorie Chan (1988), who studied the White Rim Sandstone in detail at Capitol Reef, documented ripple-laminated sandstone interbedded with thin limestone beds in the uppermost part of the formation. These oscillation ripples are symmetrical in cross section, an indicator of back and forth currents in shallow water (Fig. 3.17). In contrast to the very low relief eolian ripples, these waterlain features are characterized by high relief and sharp crests. Taken in concert, ripples, limestone, and trace fossils provide convincing evidence for a shallow marine genesis for these strata. As sea level rose to inundate the margin of the coastal dune field, waves and currents remolded the eolian sand, stamping their own imprint into the top of the formation.

The White Rim Sandstone has not been without controversy. Early geologists considered the formation to be of eolian origin, based mostly on the large-scale cross-strata (e.g., Baker and Reeside 1929, McKnight 1940, Kunkel 1958). However,

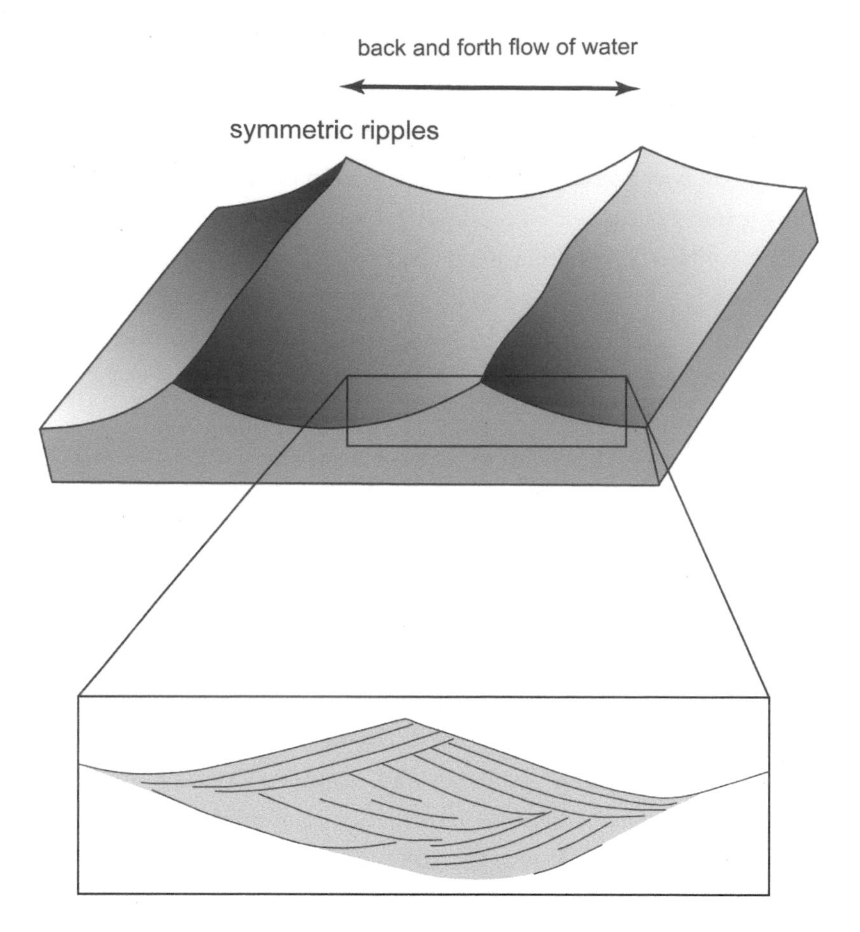

Fig. 3.17. Origin of symmetric ripples in bidirectional subaqueous flow, such as occurs on lake margins or marine shorelines. Note that the angle of dip from the ripple crest down both sides is approximately the same. Inset shows the complex dip directions that represent flow direction at various times, but ultimately back and forth.

a later, detailed study of the formation by geologists Don Baars and W. R. Seager (1970) interpreted the formation to be wholly marine in origin. Based on meager evidence, Baars and Seager pronounced that the White Rim was the product of shallow marine, offshore sandbars. Subsequent study has since unequivocally shown the formation to be an eolian deposit with only minor marine strata to the west (Steele 1987, Huntoon and Chan 1987, Kamola and Chan 1988, Chan 1989). Unfortunately, many popular publications on the regional geology of the Colorado Plateau have continued to champion a marine origin, continuing a misperception among the general public. Today the White Rim in the Canyonlands area is considered by most geologists to be a classic eolian dune deposit.

A unique and fascinating aspect of the White Rim Sandstone in Elaterite basin is the numerous tacky, black tar seeps that ooze from the canyon walls, especially during the hot summer months. Locally, in Elaterite basin and in the subsurface farther west, this sandstone is saturated with heavy hydrocarbons. Here freshly broken surfaces of the normally buff-colored sandstone are dark gray to black, resembling some type of granular charcoal. An initial encounter with these seeps in this remote area is disconcerting, as the long drips of sticky tar are strangely reminiscent of filled cracks in streets. They simply don't fit the setting.

The term **elaterite,** for which the area is named, is defined as "a brown asphaltic pyrobitumen, soft and elastic when fresh but hard and brittle on exposure to the air" (Jackson 1997)—in a word, tar. The White Rim Sandstone and, to a limited extent, the Cedar Mesa Sandstone hold the largest tar sand deposit in the United States (Campbell and Ritzma 1979). It has been estimated that between 12.5 to 16 billion barrels of oil reside in these rocks—if the oil could be put in barrels; however, these hydrocarbons are unrecoverable from both a technological and economic standpoint.

Several conditions in the White Rim Sandstone combined to make it, at one time, a receptacle for petroleum. First, eolian processes responsible for deposition of the sandstone generated a sheet of sand with grains that were well-rounded and uniform in size. This resulted in a sandstone body with high porosity and permeability—an ideal rock for the entry of petroleum migrating from some organic-rich source rock. Originally the White Rim contained little to no organic material. So what was the source for petroleum in the White Rim? Baars and Seager (1970) pointed to the adjacent marine Kaibab Limestone. It contains a moderate organic content and was deposited next to the White Rim dune field, so oil migration would have followed a very short and simple path. Although he considered a Kaibab source, Richard Sanford (1995) found that the organic content of the Kaibab was insufficient to generate the volume of hydrocarbons found in the White Rim. Sanford favors a source in Precambrian Chuar Group rocks, a deeply buried, organic-rich succession that underlies much of the region. Studies on black shales in the Chuar Group in the only place where it is exposed, in the bottom of the Grand Canyon, and drill hole data to the north indicate that it is a suitable source (Cook 1991, Uphoff 1997), but it has yet to be determined with certainty as the source. This would require geochemical fingerprinting of the organic material from both the Chuar rocks and the White Rim for comparison.

The final requirement for a petroleum reservoir was an impermeable caprock over the White Rim to block vertical migration. This trap was provided by the overlying Triassic Moenkopi Formation, which is dominated by very fine grained, impermeable shale and siltstone. For a time the Moenkopi retarded the flow of petroleum, and it accumulated in the White Rim reservoir. This confinement, while probably long-lived, came to an end sometime in the last 5 million years as the modern Colorado River drainage cut downward, eventually breaching the White Rim reservoir. At this time the volatile gases and liquid in the petroleum were released, and the hydrocarbons were altered to a solid, tarry substance, rendering it unusable.

The top of the White Rim Sandstone heralds the end of Permian deposition in the Paradox basin. On a larger scale, it also marks the end of the Paleozoic Era. Because erosion rather than deposition dominated the Colorado Plateau 245 million years ago, our view of the Permian-Triassic boundary is unclear. Erosion is attributed to a westward recession of the sea at this time.

The Permian-Triassic Extinction Event: A Segue into the Mesozoic

Although it is not obvious in the rocks of southeast Utah, the Permian-Triassic boundary represents a profound change. It marks the most severe extinction event in Earth history, one in which 90 to 95 percent of the marine species died out. It is the destruction of this marine fauna on which the Paleozoic-Mesozoic boundary is defined. These devastated fauna had ruled the marine realm since the dawn of the Cambrian, when shelled vertebrates first appeared. Almost all the fossil organisms seen in abundance in the previously discussed Pennsylvanian and Permian strata were wiped out. Gone were the fusulinids, and most species of bryozoans, brachiopods, crinoids, and corals. These organisms, dominant for 300 million years, left a void that would not be filled until well into the Triassic. The marine environment would recover, but it was to be a vastly different population. In fact, the Triassic recovery gave rise to what is known today as the "modern fauna"; that is, the marine organisms that evolved from this Permian-Triassic devastation gave rise to the organisms that populate the seas today.

The exact cause of the terminal Permian extinction remains debatable, but several lines of evidence point to multiple rapid climate changes. Extinction does not appear to have been triggered by any single catastrophic change. Rather, the fossil record indicates it was spread over the last 7 million years of the Permian Period. Recent work suggests that the Earth cooled dramatically during the Middle Permian. Remember that the Pangean supercontinent stretched from the north to south polar regions, a configuration that stalled ocean circulation. Immense ice sheets blanketed both polar landmasses. This was the first time that ice was widespread enough to accumulate at the North Pole. Extinction patterns suggest that as the Earth cooled, the tropical climate belt that girdled the equatorial area contracted and eventually disappeared, to be replaced by a temperate climate belt. Hardest hit were the tropical animals. Temperate-climate organisms that survived were forced to adapt to cooler water or migrate toward the equator, where they could find a suitable climate. The organisms that endured this rapid cooling were weakened by the stress of the change.

The latest Permian Earth, in contrast, was besieged by rapid global warming. The striking nature of this change is underscored by the presence of Late Permian petrified trees in Antarctica, where only a few million years earlier thick blankets of glacial ice had scoured the landscape. It is speculated that this extreme shift further stressed surviving populations, particularly those that earlier had adapted to a cooler climate, pushing large numbers over the brink of extinction.

Other potential triggers for the Permian extinction that are perpetually under investigation include a major regression and the possibility of an extraterrestrial impact on Earth. A large-scale regression would have substantially reduced the area of shallow marine habitat around the world. A problem with the regression hypothesis is that an even more pronounced sea level drop took place earlier in the Permian, and it had only a minor effect. The specter of a terminal comet or meteor-impact event recurrently rises as a cause of this extinction. This is one of the most intensely investigated possibilities in recent times, especially since the terminal Cretaceous extinction, which defines the Cretaceous-Tertiary boundary, has been shown to be due to such an impact. So far, diagnostic evidence for a Permian impact has been elusive, but investigations continue wherever a complete record for the Permian-Triassic transition is preserved.

Chapter 4

The Mesozoic Era

Rivers, Dunes, Red Rocks, and Dinosaurs

The Mesozoic Era (245 to 66 Ma) includes the Triassic, Jurassic, and Cretaceous periods, respectively. From a global perspective, it was a time of great change. Terrestrial plants and animals, including the dinosaurs, gained a firm foothold on the continents, a relatively new environmental niche. The Pangean supercontinent, which dominated the planet through the Late Paleozoic, began to break up, and by the end of the Mesozoic, Earth began to assume a configuration that resembles that seen today. As the Pangean jigsaw puzzle split apart, seaways opened between the pieces, narrow and constricted at first, but gradually widening to large ocean basins.

This activity affected western North America only indirectly because the region was far removed from the rifting margins of the breakup. Still, Mesozoic western North America saw many changes in the landscape. The Ancestral Rocky Mountains that dominated the scene through the Late Paleozoic slowly disappeared under the assault of erosion and regional subsidence. They endured as low hills through the Triassic, sporadically submitting small amounts of fine-grained sediment to the west. By Mid-Jurassic time they were covered by sediment and were no longer an influence. Instead, chains of mountains rose far to the west, first in central Nevada; by the Late Cretaceous, the locus of uplift had migrated eastward into what is today western Utah. These mountainous highlands loomed as an important ingredient in the evolution of the Colorado Plateau, providing abundant sediment, but also driving subsidence so that the sediments were preserved as the colorful strata that create the splendid landscape today. These new mountains also reversed the Late Paleozoic east-to-west drainage patterns. By the Mid-Jurassic these highlands drove rivers northeastward.

The sea came and went on the Mesozoic Colorado Plateau, but it was a minor component of the harsh desert landscape that dominated the Triassic and Jurassic periods. The seas that were present were mostly shallow fingers of water that extended into adjacent dune fields. In the Late Cretaceous, however, a worldwide sea level rise, coupled with a locally high subsidence rate, brought a deep seaway onto the Plateau and much of western North America, blanketing the region with a thick layer of deep marine mud. By the end of the Cretaceous the sea had receded from the area, leaving the mud exposed to the elements.

Geologic eras and periods are defined by the appearance and extinction of certain groups of organisms. Evolution and extinction are an integral part of geologic history, and the basis of its division into increments of time. As we have seen, the Paleozoic-Mesozoic boundary is marked by the extinction of much of the marine fauna at the end of the Permian. The ensuing Triassic Period dawned with a nearly deserted marine environment. The devastation of reef-building corals in the Permian left a void that remained unfilled until the Middle Triassic. The scleractinian, or "modern," corals appeared in the shallow seas at this time, apparently evolved from some so far unrecognized ancestor. The numerous species of brachiopods that dominated the shallow sea floor throughout the Paleozoic were replaced by the bivalves, which include modern clams. Only a single genus of ammonites survived the Permian extinction. From this surviving stock, however, numerous species evolved throughout the Mesozoic. In fact, ammonites changed so rapidly and obviously over this time span that they are the dominant index fossil for Mesozoic marine strata. An **index fossil** is a fossil organism used as a time marker. It represents, or "indexes," a specific time interval; thus, when found, it establishes the age of the rock that contains it. Index fossils are also vital to establishing time equivalence between two geographically separated rock units. For instance, if a Middle Triassic ammonite species recognized in Nevada is later discovered in western China, it can be stated with a fair amount of certainty that the strata in China are also of Middle Triassic age.

Large Permian reptiles that survived into the Mesozoic gave rise to the first dinosaurs by the Late Triassic. Dinosaurs and their close reptile relatives quickly diversified into a great variety of large animals that ruled the land, air, and sea. Mammal-like Permian reptiles also evolved into the first mammals in the Mesozoic. The mammals, however, were limited in diversity and consisted only of small, rodent-like animals throughout the era. The giant reptiles dominated all environmental niches, and it would take their destruction for mammals to really take hold. The dramatic extinction event at the end of the Cretaceous in which all the dinosaurs were wiped out provided the mammals their opportunity. The following Cenozoic Era is characterized by a rapid diversification of mammals as they took over the niches previously held by the giant reptiles.

The large, herbivorous dinosaurs experienced a great diversification in the Cretaceous Period. This was a direct result of the arrival of flowering plants, or **angiosperms.** Once they appeared, these seed plants expanded and evolved at a furious pace. Their expansion was boosted by the numerous methods of seed dispersal, but they also grew and reproduced more rapidly than the earlier gymnosperm plants that had grown on land since the Mid-Paleozoic. Insects were an important ally in the spread of angiosperms and enjoyed an evolutionary leap at this time as well. Insects pollinated the flowers, speeding seed production, but also cross-pollinating plants, quickly and efficiently producing numerous new species. The importance of angiosperms to other organisms—from the great herbivorous dinosaurs, the largest animals ever to exist on Earth, to tiny insects—illustrates the weblike complexity of ecosystems and the reliance of one form of life on others as evolution progressed and life became increasingly diverse.

The Triassic Period

Tidal Flats and Forests

The record of the Triassic Colorado Plateau opens with shallow marine deposits of the Moenkopi Formation and closes with the sluggish river and swampy lake deposits of the Chinle Formation. A period of erosion separated these vastly different environments. While the Colorado Plateau at this time was mostly a low-relief, featureless region, localized, small-scale tectonic convulsions wracked the area throughout the period, locally affecting sedimentation patterns. Both units were supplied with sediment from the east, in part from the slowly disintegrating remnants of the Ancestral Rockies, but dominantly from some source farther to the southeast, probably highlands in the northern Mexico borderlands and the Oklahoma/Texas region.

The Triassic tectonic setting for the eastern part of the North American continent was one of rifting and extension. As the Pangean supercontinent split up, the landmasses of Europe, Africa, and South America pulled away from North America. Where mountains stood earlier, narrow seas were born. As the Mesozoic progressed, the narrow seas widened, eventually forming the modern Atlantic Ocean and Gulf of Mexico.

In western North America the only activity in the Ancestral Rockies was the rasplike action of erosion that had already reduced these once lofty peaks to low hills. At the beginning of the Triassic, however, a new mountain-building event, the Sonoma orogeny, was taking place in central Nevada. The driving force behind these mountains was the collision of a moderate-sized volcanic island with the west coast of the continent, which prior to that time *was* central Nevada (Fig. 5.1). As much of western Nevada and northeastern California were added to the continent, these new mountains were squeezed up in the intervening area. Such additions continued throughout the Mesozoic, continually renewing the mountains

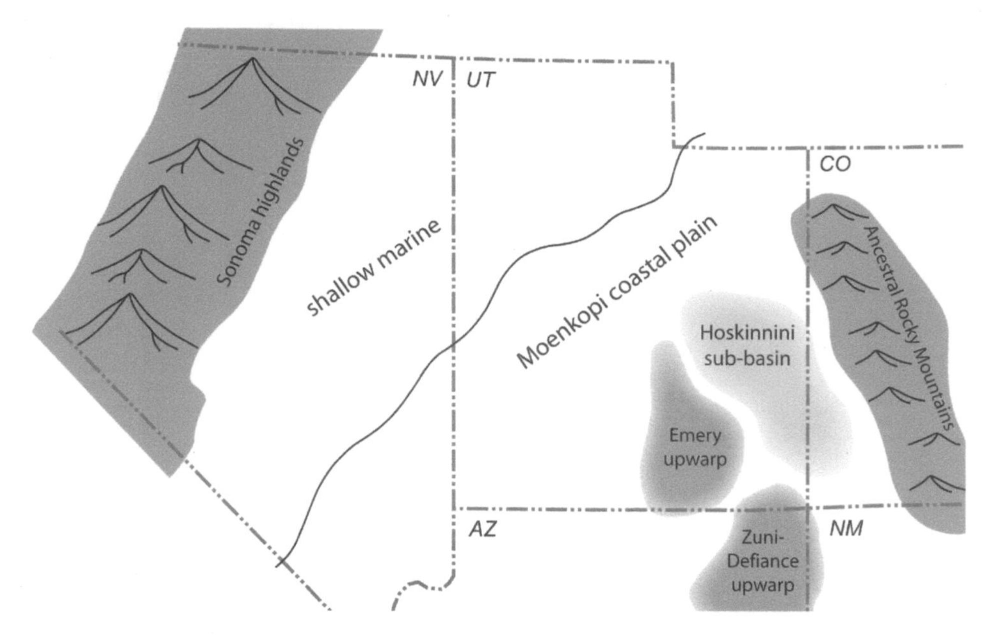

FIG. 5.1. Paleogeography of highlands and the Hoskinnini sub-basin during the Early Triassic at the beginning of Moenkopi Formation deposition. The Monument (not shown here) and Zuni-Defiance upwarps were low-relief features, but high enough that they acted as sediment sources during initial Moenkopi deposition. The Ancestral Rocky Mountains were much reduced in relief by this time and were no longer tectonically active. They did, however, continue to shed sediment westward. The Sonoma highlands had little effect on deposition but likely acted as a barrier to moisture, creating an arid climate across the Four Corners region during this time.

and driving them farther eastward. Initially these mountains had only the indirect effect of influencing the climate of the Colorado Plateau region. It was not until later in the Jurassic when drainage directions and other geographic features of the plateau were affected.

The arid climate that endured through the Late Paleozoic persisted into the Early Triassic. This extreme aridity was due to the low latitudes of the region during this time. An additional influence was the new Sonoman highlands that had risen to the west, effectively blocking any moisture that may have come off the vast sea farther west. The Middle Triassic is represented by an unconformity between the earlier Moenkopi Formation and the Late Triassic Chinle Formation, so there is no record for this time; however, by Chinle time the region had become much wetter, with a monsoon-type climate. By the end of the Triassic the climate again was trending toward extreme aridity, which would continue through most of the following Jurassic Period.

Moenkopi Formations

The Early Triassic Moenkopi Formation throughout the Colorado Plateau is easily identified by its deep red color, fine, horizontal lines, and its tendency to weather into steep slopes and benches. The formation is dominated by thin alternations of sandstone, mudstone, and shale, which contribute to its horizontally striped appearance. Limestone is present to the west, in the Capitol Reef area, but is very thin to absent in southeast Utah. The Moenkopi was deposited in a variety of environments depending on location, yet its appearance in outcrop is unchanging. Easternmost exposures in the salt anticline region along the Colorado-Utah border represent fluvial channel and floodplain deposits. Westward, in Canyonlands and the Cedar Mesa/White Canyon area, the Moenkopi gains a marine tidal influence. Traversing farther west to Capitol Reef, deltaic sandstone and marine limestone enter the mix. Finally, in Nevada and northern Utah, where the strata thicken dramatically, the Early Triassic deposits split into three distinct marine formations. From bottom to top these include the Woodside, Thaynes, and Mahogony formations.

In a general sense, the Moenkopi to the east records the passing of low- to moderate-energy rivers that traversed westward across a featureless, west-sloping coastal plain. The shoreline that met these rivers varied in location, shifting great distances in an east-west direction with only minor variations in sea level. The very low relief of the coastal plain allowed the smallest sea level changes, or even tidal fluctuations, to shift the shoreline tens of kilometers, alternately exposing and inundating vast tracts of muddy sediments.

Thickness of the Moenkopi ranges from a feather edge in the east, where it pinches out against the Uncompahgre highlands, to more than 2000 feet in the southwest corner of Utah. The westward thickening mostly is gradual, except for the salt anticline region, where locally exaggerated thickening and thinning chronicles Early Triassic salt movement. In this district the Moenkopi is conglomeratic, indicating that the Uncompahgre highlands remained a positive element that was still capable of delivering coarse sediment. However, the formation rapidly becomes finer-grained to the west, suggesting the rivers that emitted from these diminished highlands had a relatively low gradient and were thus unable to carry a coarse load very far.

The Moenkopi is subdivided into numerous members that change names completely in an east-west traverse. The four members recognized in the salt anticline region share no names with the six members a short distance west, in the Canyonlands and Cedar Mesa/Lake Powell areas. This is due largely to the effect of four transgressive/regressive events to the west that deposited several tongues of limestone as convenient marker beds. The record of these events is not easily recognized in the dominantly continental deposits to the east.

The following discussion of the Moenkopi opens with the origin of the extensive Permian-Triassic unconformity that underlies the formation and is an integral part of its history. Basal Moenkopi strata throughout southeast Utah will be addressed regionally, as they have a great bearing on the genesis of the unconformity. Once this foundation is established, strata of the eastern Moenkopi will be discussed, followed by a treatment of the western members.

Permian-Triassic Unconformity (Tr-1) and Basal Moenkopi Strata

The regional extent of the Permian-Triassic unconformity (Tr-1 unconformity) indicates far-reaching erosion beginning in the final phase of the Permian Period and continuing well into the Early Triassic, a span of ~22 million years. This boundary is distinguished by downcutting into the Permian surface and subsequent filling by Triassic conglomerate throughout Utah and adjacent parts of Nevada, Arizona, and New Mexico. Most geologists attribute this erosion to a significant drop in sea level, stimulating the evolution of a regional drainage network. The subsequent infilling phase of Moenkopi deposition likely was induced by a sea level rise that reduced the gradient of the streams, lowering the energy level.

While sea level changes certainly played a role in the development of the unconformity, tectonic activity locally had a hand in its creation. In eastern Nevada, conglomerate-filled erosional channels in Moenkopi-equivalent strata are attributed to the rise of the Sonoman highlands in central Nevada. In southeast Utah the Tr-1 erosion surface cuts more deeply than anywhere else over its extent. Channels carved into Permian strata range up to 60 feet deep and are filled with locally derived, Triassic chert pebble conglomerate. This basal conglomerate is limited in extent and outlines a broad area of mild uplift in southeast Utah.

Geologists Jacqueline Huntoon, Russell Dubiel, and John Stanesco (1994) traced the Tr-1 unconformity through southeast Utah in an effort to delineate the early Triassic upwarp and document the basal Moenkopi sediments that came off it. According to their work, the Emery high, as it is called, was a mild upwarp centered on the modern Circle Cliffs region (Fig. 5.1). This may have been an earlier incarnation of the monoclinal Waterpocket Fold that defines Capitol Reef in south-central Utah today. The exact origin of the Emery high remains enigmatic, but it initially divided the broad Moenkopi coastal plain into two sub-basins. The upwarp shed coarse sediment to the north into the larger, main Moenkopi basin and eastward into the Hoskinnini sub-basin, which was also bounded on its east margin by the degraded Uncompahgre highlands.

Basal Moenkopi conglomerate on the east side of the Emery high, in the Lake Powell and White Canyon region, is an unnamed unit that rests on various Permian units, depending on location. The chert pebble conglomerate, which ranges up to 40 feet thick, fills channels carved into Kaibab Limestone along the edge of the Emery high in the modern Circle Cliffs. Traversing east, the conglomerate scours into White Rim Sandstone in the Lake Powell area and is well exposed on the cliffs above the confluence of the Colorado and Dirty Devil rivers. It cuts into the Organ Rock Formation in its easternmost extent, along White Canyon on Cedar Mesa

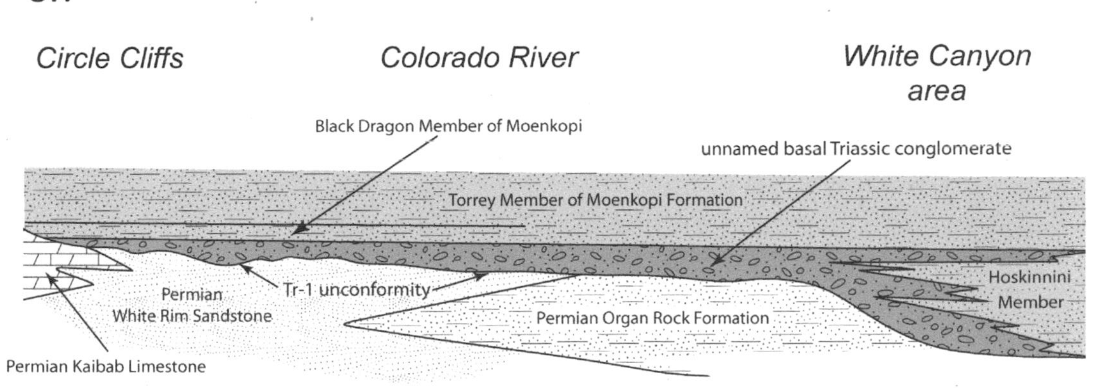

FIG. 5.2. Schematic stratigraphic cross section showing basal Triassic conglomerate derived from the Emery upwarp to the southwest and shed eastward into the Hoskinnini sub-basin. Note that the Figure 5.1 unconformity cuts into varying Permian strata. After Huntoon and others 1994.

(Fig. 5.2). At this eastern limit the conglomerate grades into the basal Hoskinnini Member of the Moenkopi, which is dominated by sandstone and siltstone. Within this transition zone these basal strata rapidly thicken to more than 80 feet. For a short distance the fine-grained Hoskinnini is sandwiched between thin fingers of chert pebble conglomerate at its base and top. Farther east the conglomerate disappears completely.

This chert pebble conglomerate was deposited in energetic, braided rivers that descended eastward off the Emery uplift. Chert was derived from erosion of the chert-rich upper Kaibab Limestone that capped the Emery high. The White Rim Sandstone contributed the sandy matrix that encases the pebbles. In the White Canyon area the vigorous Triassic rivers were immediately subdued as they met the low-energy environment of the restricted Hoskinnini basin. Sandstone and siltstone of the Hoskinnini Member have been interpreted by many workers to represent a sabkha setting (e.g., Blakey and others 1993, Huntoon and others 1994). As sabkha deposits filled the Hoskinnini basin, the sea was pushed out, and rivers from the west spilled across the surface, blanketing it with a veneer of conglomerate and coarse sandstone.

An identical chert pebble conglomerate occupies the base of the Moenkopi in the San Rafael Swell, some distance northwest of the White Canyon locality. Similarly, it fills channels scoured into Kaibab Limestone and White Rim Sandstone. Clearly it is the equivalent of the White Canyon conglomerate; however, in the San Rafael Swell it makes up the base of the Black Dragon Member of the Moenkopi, named for Black Dragon Wash.

In northwest Canyonlands, between the San Rafael and White Canyon localities, conglomerate is absent. Instead, interbedded sandstone and siltstone of the Black Dragon overlie a thin wedge of Hoskinnini sandstone or Permian strata. The

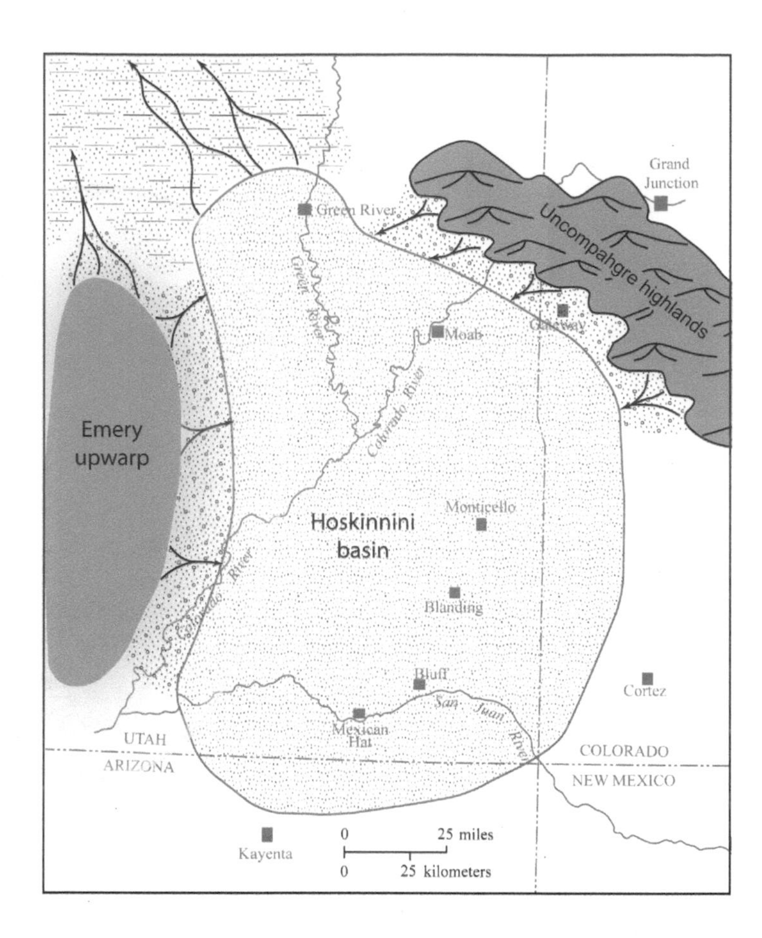

FIG. 5.3. Paleogeography of southeast Utah and surrounding areas during deposition of the Hoskinnini and Black Dragon members of the Lower Triassic Moenkopi Formation.

Black Dragon and Hoskinnini members both overlie the Tr-1 unconformity and share some distinctive rock types, including the basal chert pebble conglomerate and shallow marine sandstone and siltstone. This suggests they were deposited during the same interval, but in geographically separated basins and under different conditions.

University of Utah geologists Steffen Ochs and Marjorie Chan (1990) studied the Black Dragon in the San Rafael area in detail. Here they discovered a greater variation in rock types compared to the basal strata around Lake Powell and White Canyon. This difference stems from a closer proximity to the sea in the San Rafael region. Because of the very low relief of the Triassic coastal plain, even minute sea level changes altered the landscape dramatically.

Basal conglomerate in the Black Dragon was laid down by north-northwest-flowing rivers that initiated on the Emery high, which rose slightly to the south (Fig. 5.3). Rivers rolled across the broad coastal plain, arcing gently northwest to rendezvous with the shoreline in northern Utah. In their study, Ochs and Chan (1990) identified with certainty a Kaibab source for the chert clasts; this is based on fossils contained within chert and limestone cobbles. Like the basal strata at White Canyon, the abundant sand is attributed to erosion of the White Rim Sandstone.

Conglomerate in the San Rafael is overlain by thin but laterally continuous interbeds of red sandstone, siltstone, and shale. Ripple marks are especially prominent features of bedding surfaces and mudcracks are evident locally. Although fossils are absent, some units contain traces of the passage of invertebrate organisms. Collectively, these features led Ochs and Chan to interpret these rocks as the product of an intertidal marine environment. This indicates a postconglomerate rise in sea level, intermittently inundating the San Rafael area. In summary, fluvial conglomerate deposition was closely followed by a sea level rise, transforming the vast coastal plain into a broad tidal flat.

Tidal flat deposits are blanketed by a sheet of sandstone ranging up to 25 feet thick. The sandstone contains a variety of cross-stratification types, all of which dip northward. Mud pebbles line the base of some crossbed sets. No body or trace fossils that might disclose a marine setting have been recovered from this unit. Instead, these strata are believed to mark a regeneration of the earlier north-flowing river system, probably provoked by a retreat of the sea. The northern flow direction indicates the continued presence of the Emery high, but it must have been a lower-relief feature than before, as it was unable to generate the coarse pebbles, cobbles, and boulders that characterize the basal conglomerate. This drainage system consisted of sand-dominated, braided streams that pushed their way across the restored coastal plain.

This second phase of fluvial activity in the Black Dragon was interrupted by another incursion of the sea, drowning the coastal plain as it reverted back to an immense intertidal zone. While alternations of sandstone, siltstone, and shale again were deposited in thin layers, the water continued to deepen. This ongoing sea level rise drove the shoreline farther eastward, and the San Rafael area became a subtidal environment, so low tide no longer exposed the sediments. As the sea deepened, the zone of clastic deposition migrated eastward with the shoreline, allowing carbonate sediments to dominate in the San Rafael area. Algal and oolitic limestone characterize these uppermost Black Dragon strata, although limey sandstone, siltstone, and shale make sporadic appearances. These upper Black Dragon limestone beds grade upward into fossiliferous limestone of the widespread Sinbad Limestone Member of the Moenkopi.

Basal Moenkopi strata in the Elaterite basin and Island in the Sky areas of Canyonlands, situated between the previously discussed White Canyon and San Rafael areas, differ from both these localities. In this area of northwest Canyonlands National Park, basal conglomerate is absent, and locally less than 10 feet of Hoskinnini Member siltstone lies on the White Rim Sandstone. The Hoskinnini, of sabkha origin, is overlain by a thick succession of red sandstone, siltstone, and shale assigned to the Black Dragon Member. Where the Hoskinnini is absent, the Black Dragon rests on the White Rim. Various workers interpret this succession to represent a prograding delta (e.g., McKnight 1940; Stewart and others 1972a, 1972b;

Blakey 1974). According to recent work by Shaun Baker and Jacqueline Huntoon (1996), the delta records a gradual sea level rise in the Hoskinnini basin, and the filling and leveling of the erosional Tr-1 topography with sediment. The Black Dragon thins over residual highs and thickens over the deeper erosional hollows.

The deltaic succession documented in the Black Dragon built northeastward into the Hoskinnini basin, creating land where previously a shallow inland sea existed. Because the main shoreline, the one that marked the open sea, was located to the northwest, the northeast-directed growth of the delta goes against what would be expected. The Hoskinnini basin, however, must have been deep enough that rivers emitted from the Emery high in this area were pulled northeast to drain into the sub-basin.

The distinctive red delta deposits are overlain by a thin yellow bed (less than 5 feet thick) composed of ripple-laminated micritic limestone and sandstone. According to Baker and Huntoon (1996), this is the eastern tongue of the Sinbad Limestone, which represents the maximum sea level for this transgressive phase of the Moenkopi. At this point all earlier topography that had affected sedimentation, including the Emery high, were blanketed with sediments.

The areal extent of the Hoskinnini Member outlines the Hoskinnini sub-basin, which covers a north-south-trending oval area covering much of southeast Utah and part of northern Arizona (Fig. 5.3). The basin stretches south to Monument Valley and north to Moab, where it grades into the laterally equivalent Tenderfoot Member of the salt anticline region. In the Maze area, along the northwest margin of the basin, the Hoskinnini quickly thins and pinches out. To the southwest, along White Canyon, it also thins suddenly and grades into the much thinner conglomerate discussed previously. In fact, USGS geologist John Stewart (1959) found that the Hoskinnini reaches a maximum thickness of 126 feet at Clay Hills, along the lower reaches of the San Juan River, but thins to a feather edge just a short distance to the west. Apparently the west margin of the Hoskinnini basin had an abrupt, bowl-like geometry. On the other side of the basin, along the east margin, the Hoskinnini disappears below younger strata, where it gradually thins and pinches out against the slopes of the Uncompahgre highlands.

The Hoskinnini Member consists of red-brown siltstone and fine-grained silty sandstone. Unlike the rest of the Moenkopi, the Hoskinnini is well cemented and resistant to erosion, so it tends to weather into steep, smooth cliffs. A defining aspect of the member is the presence of sparse pockets and lenses of well-rounded fine to coarse sand grains, as well as scattered discrete sand grains, all encased in siltstone. Internally the member is characterized by indistinct, wavy laminations. This part of the Moenkopi contains no ripple marks, a feature that is diagnostic of the rest of the formation.

Geologists who have encountered the Hoskinnini agree that it was deposited in a sabkha setting, in this case a low-lying area that was rarely inundated by water, yet was often wet. The lack of standing water is told by the absence of ripples or other structures that might indicate subaqueous deposition or currents. Instead, water probably entered the basin from below via groundwater, from the sea to the northwest, or as runoff from the adjacent Uncompahgre highlands and the Emery high. In such a setting alternating episodes of precipitation and dissolution of gypsum within the near-surface sediments may have generated the wavy laminations that characterize the Hoskinnini. When gypsum crystallizes near the surface, the growth of the crystals pushes the surrounding sediments apart. If new water passes through the sediments, the gypsum is dissolved and the sediments are compacted. The end product of this seesaw evaporite activity is a disruption of the original stratification, transforming it into irregular, wavy laminations.

The Hoskinnini sub-basin was an inland basin, unique in its genesis because it formed by erosion rather than subsidence. Evidence for this can be seen in thickness relations between the Hoskinnini Member and underlying Permian strata, across the Tr-1 unconformity. Where the Hoskinnini thickens, it does so at the expense of the more deeply incised Permian strata beneath it. Further evidence is provided by the steep western basin margin, which resembles the edge of an erosional valley. Moreover, subsiding basins rarely fill completely because as sediment is deposited, the basin continues to slowly sink. This provides continual accommodation space for the vertical accumulation of sediment. In contrast, the Hoskinnini sub-basin filled with sediment until the erosional topography was buried, concealing it as a separate entity. Apparently the 22 million year period of erosion across the Permian-Triassic boundary was a time of pronounced downcutting in southeast Utah, forming the oval-shaped depression of the Hoskinnini sub-basin.

Moenkopi Formation of the Salt Anticline Region, Utah and Colorado

Basal Moenkopi strata in the salt anticline region along the Utah-Colorado border differ enough from the Hoskinnini Member to warrant a different name. The basal **Tenderfoot Member** is named for exposures on Tenderfoot Mesa, which overlooks the Dolores River southeast of Gateway, Colorado. The contact between the Tenderfoot and the underlying Cutler Formation everywhere is an angular unconformity. In most exposures the Cutler was tilted only a few degrees prior to Tenderfoot deposition; however, along the flanks of the salt anticlines the Cutler locally has been tilted up to 90° relative to the Moenkopi. These regional angular relations highlight the profound influence of salt movement during Moenkopi sedimentation. Extreme thickness and lithologic changes in the Moenkopi of the region likewise record the effect of salt flow.

The Tenderfoot Member ranges up to 290 feet thick, but is missing in some places. Four distinct units have been recognized in the member, but rarely are all four well developed at a single locality, and in many places one or several units are absent. The base of the Tenderfoot consists of sandstone and conglomerate that was obviously reworked from the underlying Cutler Formation. This is overlain by up to 7 feet of gypsum. According to USGS geologists Eugene Shoemaker and William Newman (1959), who worked intensively in the region during the uranium frenzy of the 1950s, gypsum is found in approximately half of the Tenderfoot exposures.

The overlying third unit consists of massive red to brown, cliff-forming mudstone and silty sandstone. Upon close examination this unit reveals wavy laminations and contains scattered coarse, rounded quartz grains. All these features are properties of the Hoskinnini Member, which led Shoemaker and Newman, and most researchers since, to correlate the two. This Hoskinnini-equivalent is topped by a unit of similar lithology, but with well-defined bedding due to thin intervening shale layers. This capping unit also contains ripple marks and minor gypsum. Along the flanks of several salt anticlines, notably Sinbad and Paradox, the Tenderfoot contains sporadic conglomerate beds with coarse, angular limestone fragments. This sediment is believed to have been shed from the core of the large folds as carbonate interbeds within the Paradox salt were dragged up with the ascending salt mass.

Conglomeratic sandstone of the basal Tenderfoot consists of weathered sediment from the uppermost Cutler Formation that was reworked by Triassic rivers. The streams may have originated in the residual Uncompahgre highlands or on the rising salt anticlines, or both. The regional drainage pattern must have been complicated by the salt anticlines and the discrete basins that formed between them. Present data make it difficult to evaluate the various possibilities, but overlying gypsum suggests ponded water and at least partially closed basins.

Hoskinnini-equivalent mudstone and sandstone, like the Hoskinnini Member, were deposited in a groundwater-fed sabkha environment that likely evolved as the underlying units filled the deeper parts of the basins. Ripple marks and gypsum in the uppermost fourth unit of the Tenderfoot signal a recurrence of ponding, caused either by increased runoff into the basins or a regional rise in the water table above the sediment surface.

The **Ali Baba Member** of the Moenkopi lies on the Tenderfoot or, where the Tenderfoot is missing, the Cutler Formation. This contact is an angular unconformity that is particularly distinct in the Sinbad, Paradox, and Fisher valleys, again demonstrating the power of moving salt. The member is named for exposures on Ali Baba Ridge, one of the more colorfully named features in Sinbad Valley, Colorado. Although the member reaches up to 290 feet thick, it is absent in the Gateway area, along the margin of the Uncompahgre highlands. Like the rest of the Moenkopi in the salt anticline region, its thickness is irregular, varying greatly over short distances.

Fig. 5.4. Abundant cross-stratification in the fluvial Ali Baba Member of the Moenkopi Formation in the upper Fisher Valley area. This type of cross-strata reveals the direction of flow for the river system that deposited this member. Note the consistent dip direction to the left.

The Ali Baba Member is dominated by red conglomeratic sandstone with red siltstone beds sandwiched between. Conglomerate clasts range up to cobble size (up to 10 inches or 25.6 cm) and consist of granite and metamorphic rock ferried westward from the adjacent hills of the Uncompahgre. Sedimentary structures within coarse clastic units such as crossbedding and scour surfaces indicate powerful rivers as a transporting mechanism (Fig. 5.4). A unique feature in these sandstone beds are strange-looking knobs and ridges that protrude downward from their undersides (Fig. 5.5). These **load structures** formed as coarse sediment was rapidly deposited on unconsolidated, wet, clay-rich silt. The weight of the coarse sediment locally forced bulbs and ridges of sand down into the easily displaced silt. These strange features and the coarse-grained nature of the sandstone are defining characteristics of the member, allowing for its easy recognition in the field.

Along a northeast to southwest transect across the salt anticline region, the Ali Baba Member becomes appreciably finer grained. In a comprehensive study of the Moenkopi Formation, a USGS team of geologists led by John Stewart (Stewart and others 1972a, 1972b) documented a southwestward decrease in conglomerate that was augmented by a concurrent increase in siltstone. For instance, in its easternmost extent along the Dolores River, where the basin abutted the ancient Uncompahgre highlands, the Ali Baba is 80 percent conglomeratic sandstone and contains abundant cobbles. Approximately 20 miles east, in the expansive, escarpment-rimmed Richardson Amphitheater along the Colorado River, conglomeratic strata decline to ~55 percent, and clasts are only a few inches in diameter. Where it rims the floor of Castle Valley only 6 miles farther west, con-

FIG. 5.5. Blocks from the base of a sandstone bed in the Ali Baba Member of the Moenkopi Formation. The characteristic irregular knobs and ridges form on the underside of a sandstone bed that overlies silty claystone. Upon deposition, the load of the overlying sand exerts pressure on the underlying water-saturated clay. As the clay deforms, the sand sinks into it, producing these irregular features. This type of soft sediment deformation typically occurs at the contact between some upper sediment type and underlying clay. For scale, the lens cap is 67 mm in diameter.

glomeratic sandstone dwindles to less than 20 percent. Finally, in the Moab area the member is reduced to a single thin sliver of ledge-forming sandstone enveloped in shale and siltstone. A short distance west of Moab the Ali Baba Member disappears completely, overwhelmed in a mass of siltstone and shale.

More recent work on the Ali Baba Member has shown that much of the fluvial sediment came from the erosion of strata exposed in the surrounding salt anticlines, which were actively rising during its deposition (Fillmore 2006). A detailed analysis of the grains that make up the Ali Baba show a dominance of arkosic sand, but with a component of pebbles and sand composed of limestone. Some of these clasts contain telltale Pennsylvanian and Permian marine fossils. This indicates that much of the Ali Baba sediment derives from the erosion of the Permian Cutler Formation and the Pennsylvanian Honaker Trail and Paradox formations as they were pushed upward and eroded in the rising salt anticlines.

In addition, paleocurrent data obtained from abundant cross-strata in the Ali Baba Member (see Fig. 5.4) indicate a very different flow direction for these rivers than predicted by a simple paleogeography. These data show a north- to northwest-directed flow, at a right angle to the west-southwest-direction expected if the rivers had simply drained from the Uncompahgre highlands to flow unobstructed to the sea that lay to the west. The northwest paleoflow directions for the

Ali Baba throughout the salt anticline region indicate that rivers were confined to the synclinal downwarps between the anticlines and were guided northwestward by these parallel, northwest-trending uplands. The few rivers that did come off the remnant Uncompahgre highlands were quickly deflected into northwest-running rivers, caught between the Uncompahgre highlands to the east and the Fisher Valley–Sinbad anticlines to the west. Eventually these rivers, after completing an end run around the north tip of the salt anticlines, drained westward to the shallow sea. My ongoing work will further refine the fluvial setting of the Ali Baba Member.

The **Sewemup Member** of the Moenkopi succeeds the Ali Baba throughout the salt anticline region. The member is named for its type locality at Sewemup Mesa, which forms the eastern rampart of Sinbad salt valley in Colorado. In contrast to the contacts between the Moenkopi members so far discussed, this basal contact is conformable and gradational over most of its exposure. The gradational contact is distinguished by a color change: from a distance the underlying Ali Baba is dark brown, whereas the Sewemup is light brown. The change is caused by a subtle increase of gypsum in the Sewemup, slightly lightening its hue.

The Sewemup Member reaches up to 500 feet thick and consists of thinly bedded red brown to gray red siltstone and shale with interspersed sandstone beds. The siltstone contains horizontal and ripple laminations. Gypsum occurs in the lower part as cement, holding silt and clay particles together, and as thin white veinlets that cut through the red rock at various angles, locally creating an irregular weblike appearance. As gypsum content increases upward, it forms thin beds of nodules near the top. The Sewemup is the only member in the salt anticline area in which fossils have been found. Gastropods and *Meekoceras* ammonites recovered from the Sewemup in the Salt Valley anticline suggest, at least in this westernmost exposure, a marine influence.

Numerous features in the Sewemup Member point to a tidal influence on sedimentation. Thin interbeds of siltstone and shale form the "classic" tidal flat couplet. Siltstone probably was deposited by relatively high energy tidal currents as the tide was rushing in or out. Shale was deposited at the zenith of the tide as water deepened and current energy waned, allowing clay to settle out of the water column. Ripples and horizontal laminations also record fluctuating energy levels. Gypsum likely was laid down after the tide had receded, leaving pools of seawater stranded in low-lying areas to evaporate under the intense Triassic sun.

The uppermost member of the Moenkopi in the salt anticline area is the highly localized **Parriott Member**, named for Parriott Mesa, which looms above the La Sal Mountain Loop road at the northern entrance to Castle Valley. The basal contact appears to be conformable and is defined by a change from the light brown of the underlying Sewemup to the multihued brown, red, purple, and orange strata of the Parriott. Exposures are confined to the Richardson Amphitheater, Castle

Valley, and an isolated outcrop 5 miles to the southwest, along a large meander of the Colorado River called the Big Bend. The only other documented exposure in the region is an outlier in Sinbad Valley that originally was considered part of the overlying Chinle Formation.

The Parriott Member is characterized by numerous lithologies and a variety of colors. Sandstone beds may be conglomeratic and form red brown to purple brown ledges. Interbedded mudstone, siltstone, and shale range in color from chocolate brown to orange and red, and tend to form steep slopes.

Variations in thickness of the Parriott are particularly important in deciphering the conditions under which it was deposited. According to work by Stewart and others (1972a, 1972b), the member at its type locality, Parriott Mesa, is 135 feet thick. Across Castle Valley, only a short distance west, it thickens to several hundred feet, and at Sinbad Valley it reaches 252 feet. This rapid thickening was caused by an increased subsidence rate in areas peripheral to the salt anticlines. This was driven by the gradual withdrawal of several thousand feet of salt from beneath these areas as it flowed into the rising anticlines. Thus, the thinning, or even pinchout, of Moenkopi strata across the anticlines is amplified by a noticeable thickening between them.

The multiple rock types in the Parriott Member contain no evidence of a shift in depositional environment from underlying members. The abundance of sandstone indicates a recurrent influx from the Uncompahgre highlands and possibly the salt anticlines. Siltstone, mudstone, and shale contain ripples and horizontal laminations that suggest ponding between positive areas or a possible tidal influence. Rather than any change in depositional environment, the Parriott appears to combine all the previous settings, producing the most heterogeneous mixture of rock types seen in the Moenkopi of the salt anticline region.

The limited occurrence of the Parriott, coupled with its deposition in low-lying synclinal basins, led Stewart and others (1972a, 1972b) to suggest that it is much younger than the top of the Moenkopi elsewhere. If true, the Parriott was laid down during a hiatus in deposition that affected the entire Colorado Plateau. While this regional unconformity is also present throughout the salt anticline district, where the Parriott Member is preserved, it may represent a shorter span of time. The impetus for deposition during an otherwise quiet period was active subsidence due to salt withdrawal in the synclines, concurrent with uplift on adjacent anticlines. All the necessary ingredients were present in these areas: a ready sediment source in the form of the anticlines, and subsidence that created space for sediment to accumulate and eventually be preserved. In the salt anticline area the Moenkopi/Chinle unconformity is mostly angular due to the salt tectonics that repeatedly convulsed the region through the Triassic.

The Moenkopi Formation as a whole ranges 350 to 450 feet thick in the Utah-Colorado borderland (Stewart and others 1972a, 1972b). The salt anticline region,

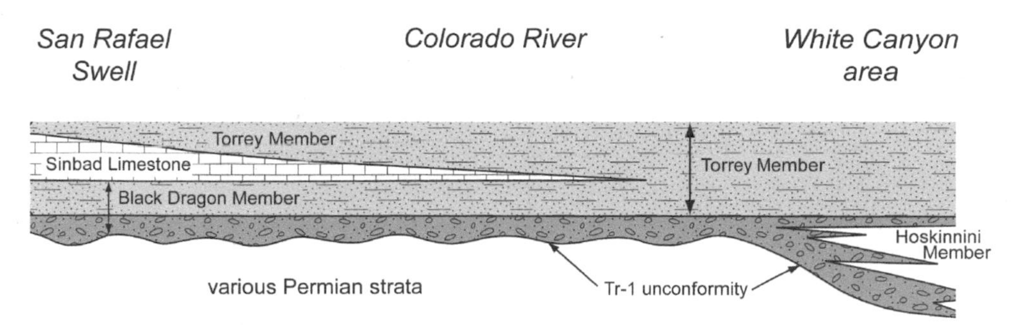

Fig. 5.6. Stratigraphic cross section of the lower part of the Moenkopi Formation emphasizing the relations between the Torrey Member to the east and the Black Dragon and Sinbad Limestone members to the west. After Huntoon and others 1994.

however, is a different story. On the flanks of Sinbad Valley it reaches a maximum measured thickness of 1264 feet (Shoemaker and Newman 1959). This immense sequence is attributed to subsidence driven by salt withdrawal. Conversely, immediately south of Sinbad, along what was the crest of the Paradox salt anticline, the Moenkopi is missing, and the Chinle rests on the Paradox salt. Similar relations are present in other salt anticlines, notably in Gypsum and Lisbon valleys, Colorado. This absence may be explained by nondeposition in these crestal areas and/or erosion of the Moenkopi prior to Chinle deposition. In either case, these relations indicate active uplift of the salt anticlines during the Triassic Period. As we will see, this activity continued well into Chinle time.

For the following discussion of the Moenkopi we traverse westward, out of the realm of salt tectonics and into a vast, low-relief region influenced more by the coming and going of the sea. Here also the members of the formation change completely, and in some cases their relationships with members in the salt anticline region remain fuzzy.

The Moenkopi Formation in the Canyonlands and Glen Canyon Regions

The **Sinbad Limestone Member** of the Moenkopi succeeds the basal Black Dragon Member in Capitol Reef, San Rafael Swell, and the Orange Cliffs, which rim the Maze in westernmost Canyonlands National Park. The long-recognized member is named for exposures in the Sinbad region of the San Rafael Swell (an area with no relation to Sinbad Valley in Colorado) (Gilluly and Reeside 1928). Where it is present, the yellowish Sinbad serves as a valuable marker unit that separates the red delta deposits of the underlying Black Dragon from the similar red beds of the overlying Torrey Member. Sinbad Limestone forms a yellowish-tan cliff in what is otherwise a red landscape etched with fine creases of bedding. This limestone thins to the southeast and pinches out along the Colorado River in Canyonlands and along the Circle Cliffs uplift farther to the southwest (Fig. 5.6).

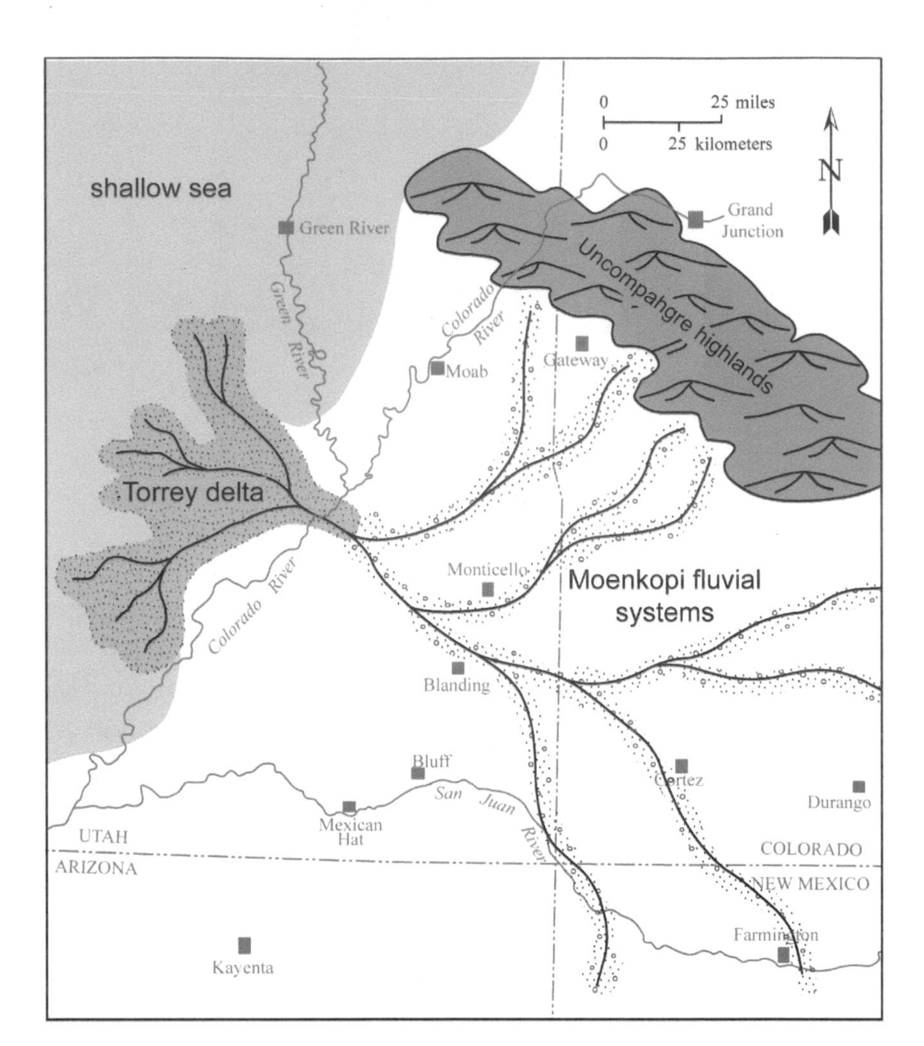

FIG. 5.7. Paleogeographic reconstruction of the Colorado Plateau and surrounding areas during deposition of the Torrey Member of the Moenkopi Formation. Note the disappearance of the earlier Emery upwarp and Hoskinnini subbasin due to infilling and burial by deposition.

Southeast of this pinchout the Moenkopi becomes a thick, indivisible succession of red slope- and ledge-forming siltstone, shale, and sandstone.

The contact between the Black Dragon and the Sinbad is a gradual one, from red deltaic deposits upward into fossiliferous limestone. This points to a progressive rise in sea level. Sinbad limestone was deposited in a shallow marine environment amid high- to moderate-energy conditions. Limestone components varied with energy, from crossbedded ooids and abraded shell fragments to well-preserved fossils entombed in lime mud. Fossils include gastropods, pelecypods, and *Meekoceras* ammonites, which document an Early Triassic age and suggest normal marine conditions.

The top of the Sinbad is a mirror image of the base, as it grades up into the red sandstone and siltstone of the Torrey Member. This change heralds the retreat of the shallow sea and the resurrection of the delta.

As the sea receded to the northwest, it was closely tracked by a flush of fine sediment from the east that was forged into large deltas wherever the shoreline lingered. These clastic sediments constitute the **Torrey Member**, named for the small town of Torrey that lies at the western border of Capitol Reef. At the onset of Torrey deposition, carbonate sedimentation continued in the western third of Utah (Blakey and others 1993). As the sea continued to recede, however, Torrey red beds blanketed all but a small corner in the northwest part of the state. At this point, most of Utah was a vast plain that sloped imperceptibly to the northwest. Only narrow ribbons of water interrupted the plain—rivers that labored to ferry their sediment load westward to the muddy deltas (Fig. 5.7).

Sandstone beds in the Torrey are common but irregularly distributed throughout the member in the form of thick lenses and discontinuous beds. The resistant sandstone gives the Torrey a distinctive ledgy appearance, and prior to being formally designated as the "Torrey Member," the unit was informally recognized as the "ledge forming member" (e.g., Stewart and others 1972a, 1972b; Blakey 1973). Interbedded siltstone is abundantly rippled and weathers into slopes and recesses beneath the sandstone ledges. The Torrey ranges up to 100 feet thick in its eastern extent and thickens to ~300 feet westward.

Like the rest of the Moenkopi, the Torrey Member is generally red. In parts of the San Rafael Swell, however, many Torrey sandstone bodies are olive gray due to residual petroleum held in the pore spaces. These tar sands, as they are known, are mostly confined to the sand-rich delta-front deposits (Blakey 1977). In some places the entire lower Torrey has been bleached white, probably by the migration of petroleum through the rock. Clearly, the Torrey Member in the San Rafael Swell, before being breached by erosion, was at one time a substantial petroleum reservoir.

Overlying the Torrey Member is the **Moody Canyon Member**, which comprises the top of the Moenkopi over most of southeast Utah. Its type locality is Moody Canyon, a tributary to the lower Escalante River that originates in the Circle Cliffs uplift. Both the Torrey and Moody Canyon members were defined by geologist Ron Blakey (1974) of Northern Arizona University, whose work on the Moenkopi has spanned 30 years. Prior to Blakey's designation, the Moody Canyon Member was known informally as the "upper slope forming member" due to its nonresistant nature.

The Moody Canyon Member is dominated by slope-forming siltstone and mudstone with minor amounts of gypsum, dolomite, and sandstone, all of which are thinly bedded. Siltstone units occasionally display ripples, and mudstone is mostly laminated. Gypsum and dolomite occur together in very thin beds. A particularly interesting aspect of the Moody Canyon is its incredible lateral continuity and the regular thickness of its units (Blakey 1974).

The basal contact of the Moody Canyon Member has long been defined as the top of the highest ledge-forming sandstone in the underlying Torrey Member. This works well in the San Rafael Swell, where the slope-forming Moody Canyon contrasts with the underlying ledge-forming strata. To the east, however, in the Cedar Mesa and Stillwater Canyon (Green River) areas, the Moody Canyon is sandier, and the two members interfinger, obscuring the contact (Blakey 1974).

Thickness of the Moody Canyon ranges to less than 100 feet at its eastern limit, near Moab. Westward the member gradually thickens to ~400 feet in the San Rafael Swell and the Teasdale uplift, just west of Capitol Reef. This thickening trend is disrupted locally in parts of the San Rafael Swell, Circle Cliffs, and Monument upwarp, where a large part of the upper Moody Canyon was eroded before the Chinle Formation was laid on top. The significance of this erosion is discussed in the following section on the unconformity between the Moenkopi and the Chinle.

The Moody Canyon is interpreted as the product of a vast, low-relief tidal flat (Blakey 1974). This is based on the numerous thin siltstone-mudstone alternations, abundant ripples, and the incredible persistence of individual beds. Gypsum and dolomite couplets suggest local ponding in low-lying areas followed by rapid evaporation. According to Blakey's work, the Moody Canyon tidal plain sloped westward, with a gradient of less than 6 inches per mile; thus a rise or drop in sea level of only 5 feet would have shifted the shoreline up to 10 miles. The rushing ebb and flow of tides must have been spectacular! These regular, wide-ranging oscillations account for the exceptional lateral continuity of beds in the member.

Correlations between Moenkopi strata in the salt anticline region and the various members to the west have been difficult to establish with confidence due to the paucity of fossils and the very different factors controlling deposition in the adjoining regions. Nevertheless, good correlations between the disparate sections have been made based on position within the sequences, lithologic similarities, and sparse but useful fossil evidence. As mentioned earlier, the Hoskinnini and Black Dragon members are correlated with each other and with the basal Tenderfoot Member in the salt anticline region (Fig. 5.8). This is based in part on the obvious basal position of all these units within the Moenkopi, but also on distinct lithologic similarities. The deltaic and marine upper part of the Black Dragon probably corresponds to the Ali Baba Member in the salt anticline region. This is based, more than anything, on well-founded correlations between units that bound them above and below. The overlying Sinbad Limestone Member is pivotal in deciphering Moenkopi relations. *Meekoceras*, an important Lower Triassic index fossil, is widespread in the Sinbad and is also found in the Sewemup Member in the Salt Valley anticline in Arches. The presence of *Meekoceras* in both areas allows for one of the more confident correlations in the Moenkopi. It also provides a crucial tie in the middle of the Moenkopi that strengthens correlations between under- and overlying strata that are based solely on position. The lack of common lithologies or fossils in overlying strata limits the level of detail to correlation between the Sinbad, Torrey, and Moody Canyon members to the west

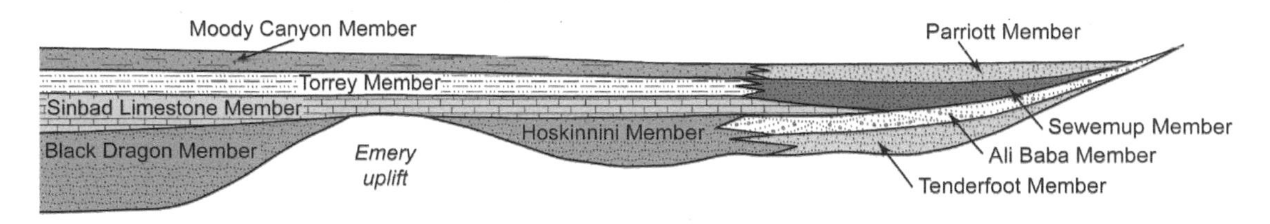

Fig. 5.8. Schematic stratigraphic cross section of the Moenkopi Formation showing interpreted relations between the well-studied western members and the members in the salt anticline region to the east.

with the Sewemup and Parriott members in the salt anticline area. These uncertainties highlight the lack of strong ties between these strata and the problems of correlating marine strata with continental rocks. Essentially, they contain *Meekoceras* (Sinbad and Sewemup members) or are positioned above *Meekoceras*-bearing strata. Another complication, as pointed out earlier, is that the Parriott may be younger than any western Moenkopi strata and therefore would have no equivalent.

Tr-3 Unconformity between Moenkopi and Chinle Formations

The contact between the Lower Triassic Moenkopi and the Upper Triassic Chinle Formation is an unconformity marked by the absence of Middle Triassic strata. This represents an omission of up to 10 million years of geologic history. Although the unconformity, designated Tr-3 by USGS geologists G. N. Pipiringos and R. B. O'Sullivan (1978), is present throughout the Colorado Plateau region, its nature varies over short distances in southeast Utah and western Colorado.

In the westernmost extent of the Chinle, in western Colorado, the formation laps onto the remnants of the once great Uncompahgre highlands. Along this edge of the basin the Chinle rests on deeply weathered Precambrian granite and metamorphic rocks. This contact represents almost 1.5 billion years of missing geologic record! In the salt anticline region a short distance to the west, the Chinle may overlie any number of units, including the Moenkopi, Cutler, or Paradox formations. The relationship between the Chinle and these underlying strata is angular, recounting a history of salt movement immediately preceding Chinle deposition. Over much of the region this angular discordance is less than 7° and so is barely discernible. Locally, however, along the flanks of the salt anticlines, the angle is much higher (Stewart and Wilson 1960).

Moving farther west, onto the relatively stable Moenkopi coastal plain, the Tr-3 unconformity adopts yet another configuration. In this region the Chinle overlies the Moenkopi, and bedding is parallel, but the contact is erosional. This erosion is deeper and more pronounced in parts of the San Rafael Swell, Circle

Cliffs, and Monument upwarp. For example, in the San Rafael Swell, Ron Blakey (1974) traced marker beds in the Moody Canyon Member laterally to discover that 240 feet of the unit was locally removed prior to Chinle deposition. Similarly, in the Circle Cliffs a narrow swath of Moody Canyon has been cut out, and the resulting channel is filled with basal Chinle conglomerate. This equates to the erosion of ~295 feet of Moenkopi, the thickness of the Moody Canyon in adjacent areas (Blakey 1974). Regional incision was initiated by a drop in sea level and the evolution of a high-energy river system that at first excavated channels and only later, during the Late Triassic, deposited sediment in them. This action was enhanced by a climate change from very arid to much wetter conditions. Yet the story is not that simple. The deeper and more obvious channels are centered on modern-day uplifts. This suggests that tectonic forces were also at work at this time, although the activity was weak. Erosion patterns on the Tr-3 unconformity point to mild uplift during the Middle Triassic in the San Rafael Swell, Circle Cliffs, and Monument upwarp as the Chinle drainage network was being established.

Chinle Formation and Equivalent Strata

The Upper Triassic Chinle Formation may be the most heterogeneous and colorful unit to cover the Colorado Plateau. The formation consists of sandstone, siltstone, conglomerate, mudstone, and limestone, all deposited in a continental setting. The variety of lithologies in the Chinle is exceeded in number only by the spectrum of colors it displays. Vivid hues of purple, red, orange, green, and gray color the mounded shale slopes and sandstone ledges. Moreover, the formation has yielded an abundance of fossil plants and animals, ranging from freshwater clams, gastropods, and crayfish to a variety of vertebrates, including amphibians, fish, and reptiles.

Like the Moenkopi, stratigraphic relations in the Chinle throughout southeast Utah and southwest Colorado are extremely variable. These differences are attributed to such familiar influences as the much reduced Uncompahgre highlands, the continued rise of salt along the Utah-Colorado border, and the relative stability west of these areas.

In the San Juan Mountains of southwest Colorado, the Dolores Formation correlates with both the Chinle and the lower part of the overlying Wingate Sandstone to the west. In the salt anticline region a short distance north, the Chinle Formation is recognized, but the numerous members that are defined over most of southeast Utah cannot be distinguished; however, some tentative correlations have been made between these two areas based on similarities in lithology and position within the formation.

Although large-scale tectonic activity had essentially reached a standstill during Chinle deposition, the supercontinent of Pangea had reached its maximum size with the accretion of several large continental fragments far to the east, including Kazakhstan and parts of China and southeast Asia. During this time sea level dropped to an all-time low, exposing the most extensive landmass in Earth history, one that stretched uninterrupted from pole to pole. The final ingredient in this mix was the Late Triassic latitude of the Colorado Plateau, which was between 5° and 15° north of the equator. All of these conditions conspired to form a very different climate from the arid conditions that prevailed both before and after.

Most of the features in the Chinle Formation point to a warm, strongly seasonal monsoonal climate on the Late Triassic Colorado Plateau (Dubiel and others 1991). Moisture came in the form of abundant summer precipitation, with little variation in the warm temperatures over the course of the year. High rainfall came from the summer warming of the vast sea to the west. As the ocean-derived moisture drifted over the continent, rain fell. Although this seasonal inundation has been modeled theoretically and mathematically, the real evidence lies in the rocks and fossils of the Chinle Formation. The wet, monsoonal setting will be apparent in the following east to west survey of the formation across the canyon country of southeast Utah and southwest Colorado.

The Dolores Formation of Southwest Colorado

Chinle equivalent strata in the San Juan Mountains of southwest Colorado are part of the Dolores Formation, named for exposures in the Dolores River valley near Rico, Colorado (Cross and Purrington 1899); however, the top of the Dolores also contains strata suspected to correlate with the Jurassic Wingate Sandstone, which overlies the Chinle over much of the Colorado Plateau. The two are not easily divisible in the San Juan Mountains. Because the conditions of deposition in southwest Colorado were so different from those elsewhere in the Chinle basin, detailed correlations between the Chinle and the isolated occurrences of the Dolores are uncertain and speculative.

The Dolores sits unconformably on the Cutler Formation. The Moenkopi, which intervenes between the Cutler and Chinle formations over much of the Colorado Plateau, did not reach this far east. Over most of its extent this unconformity is simply erosional. A notable exception is around Ouray, Colorado, where the Cutler was tilted prior to Dolores deposition, producing an angular unconformity that suggests mild local tectonic activity in the Early Mesozoic. The Dolores is, in turn, overlain by the Middle Jurassic Entrada Sandstone. This contact is an unconformity that represents the absence of Early Jurassic strata. Dolores thickness varies considerably, from ~130 feet in Ouray to more than 800 feet to the southwest near Stoner, Colorado, in the upper Dolores River valley.

The Late Triassic age of the Dolores, thus its equivalence to the Chinle, is based on the discovery of a phytosaur skull near Telluride. Phytosaurs were widespread, crocodile-like reptiles that are useful for Late Triassic age determination. In fact, a detailed chronology has been established using phytosaur species (Lucas and others 1997).

The Dolores Formation is a mixture of rock types dominated by mudstone, siltstone, and fine-grained sandstone, with minor conglomerate. These rocks represent an assortment of continental environments. According to Robert Blodgett (1984, 1988), one of the few geologists to work on the Dolores in recent years, these environments include fluvial channels and floodplains, low-lying areas occupied by ephemeral ponds, and eolian sand sheet deposits. All these were situated at the foot of the deteriorating Uncompahgre highlands. The proximity to these low hills is one of the influencing factors that makes the Dolores different from equivalent strata to the west.

The base of the Dolores to the north, in Ouray, consists of more than 30 feet of conglomeratic sandstone, which includes clasts of granite and a variety of metamorphic rocks (Stewart and others 1972b). In the southern San Juan Mountains, near Durango and Stoner, basal Dolores strata consist of fine, greenish gray sandstone and minor conglomerate. The sandstone displays an assortment of crossbedding, and conglomerate clasts consist of limestone, chert, granite, and feldspar. These basal beds represent an initial high- to moderate-energy braided stream system emanating from the remnant highlands. It is likely that the dramatic shift to a wetter climate spurred the rejuvenation of this stream system, at least temporarily. Conglomerate clasts may have been recycled from the uppermost Cutler Formation as the Dolores rivers cut down into the coarse underlying strata. It is likely that the continued erosion of the lingering highlands immediately to the east also supplied some gravel.

The middle part of the Dolores consists of red siltstone and very fine-grained sandstone. This monotonous succession is broken sporadically by large lenses of limestone pebble conglomerate; within these lenses are crossbedded sandstone and siltstone. These sparse lenses represent the channel deposits of low-energy meandering streams. The siltstone and sandstone enveloping the lenses display ripple and horizontal laminations. These fine sediments, which dominate the middle Dolores, were deposited on the floodplain that occupied this foothills area. Molluscs, mostly Unionid bivalves that preferred freshwater environments, are found in some siltstone beds, suggesting the presence of lakes and ponds on low-lying parts of the floodplain. This middle part of the Dolores signifies a decrease in energy of the Dolores fluvial system, possibly due to a reduction of relief in the adjacent highlands, or to a decrease in runoff as the climate slowly turned again toward aridity.

The upper Dolores is dominated by red sandstone, with lesser siltstone. The fine-grained sandstone contains horizontal bedding and low-angle cross-stratification, and forms thick, laterally continuous units that weather into smooth, steeply sloping slickrock ledges. The sand accumulated in an eolian sand sheet blown in on the wind from the west, only to be trapped against a backstop of the low, but effective, Uncompahgre hills. The sand sheet delineates the east margin of a monstrous dune field that blanketed most of Utah. To the west these dune deposits form the Wingate Sandstone and mark a return to the very arid climate that endured through the Late Paleozoic and Early Triassic.

Chinle Formation of the Salt Anticline Region, East-central Utah and West-central Colorado

The Chinle Formation in the salt anticline area, like the underlying Moenkopi, fluctuates radically in thickness and lithology. The formation expands to more than 700 feet thick locally, but in other places, such as across the Sinbad Valley anticline, it pinches out completely. Over much of the region the Chinle unconformably overlies the Moenkopi; however, along the flanks of some of the salt anticlines, where older units have been dragged up by the rising salt, it lies on the steeply tilted Cutler or Paradox formations. To the east, well onto what were the Uncompahgre highlands, the Chinle rests on Precambrian igneous and metamorphic rocks. Here the contact marks about 1.5 billion years of missing Earth history.

Three different units have been recognized at the base of the Chinle in the salt anticline region. This diversity is due to the salt movement that created small isolated basins and adjacent anticlines. Although each of these units has distinct characteristics, they are limited to this region and are difficult to trace outside the area.

The oldest Chinle strata in this region are exposed in an isolated outlier along the lower walls of the Colorado River canyon between Castle Valley and Moab, Utah. Rocks in this unit consist of gray sandstone and conglomerate, and mottled gray, purple, and red siltstone. Exposed thickness in this lone locality is ~200 feet, but because the base is covered, its true thickness is unknown. This unit is unconformably overlain by the sandstone and conglomerate that make up the base of the Chinle elsewhere in the region. Prior to deposition of the overlying strata, this basal unit was tilted 10°, creating the first of several local angular unconformities within the Chinle in this area (Fig. 5.9 and Plate 5). Although the unit could conceivably be part of the Moenkopi, the mottled color and the presence of conglomerate are more common with the Chinle. According to Stewart and others (1972a, 1972b), this unit was deposited much earlier than the rest of the Chinle, probably while the Tr-3 unconformity that separates the Moenkopi and Chinle was devel-

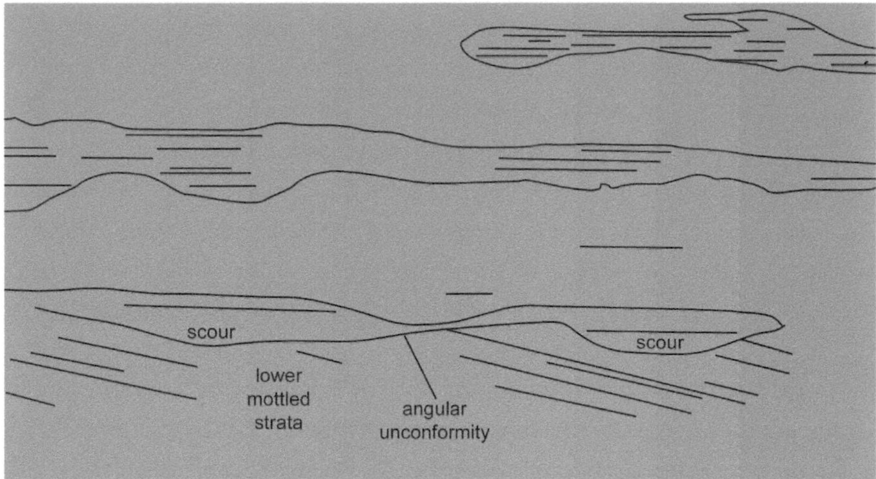

Fig. 5.9. Lower part of the Chinle Formation showing the obvious angular unconformity along the Colorado River between Castle Valley and Moab, Utah.

oping over the rest of the Colorado Plateau. The continued rise and degradation of the Castle Valley and Moab salt anticlines—which bordered this area to the east and west, respectively—probably stimulated this localized deposition. The preservation of this detritus was ensured by the withdrawal of salt from deep beneath this intervening area, driving subsidence.

Elsewhere across the salt anticline region a younger, informal unit known simply as "mottled strata" marks the base of the Chinle. The occurrence of this unit is widespread but discontinuous. The name originates from its distinctive mottled coloration in countless combinations and shades of red, purple, brown, gray, and green. Lithology is dominated by siltstone, and the unit is less than 20 feet thick. Mottled strata in the salt anticline region have an equivalent in the similar colorful rocks that make up the base of the Chinle in the San Rafael Swell to the west, where it is known as the Temple Mountain Member.

FIG. 5.10. Vertical burrows in the basal mottled strata of the Chinle Formation exposed along the Colorado River between Castle Valley and Moab, Utah. The burrows are the lighter-colored rock. The lens cap at the top of the photo for scale is 67 mm in diameter.

The contact between the mottled strata and underlying units is gradational and poorly defined, regardless of the underlying lithology. This is due to the intense mottling, which extends downward through the Tr-3 unconformity and into underlying rock. This is true whether the underlying unit is the Moenkopi or the Cutler, as in the salt anticline region and southwest Colorado. It even affects Precambrian metamorphic rocks in western Colorado, where the Chinle laps onto the weathered core of the fading Uncompahgre highlands.

Intense mottling is attributed to weathering and soil-forming processes during alternating wet and dry seasons. The involvement of underlying rocks indicates a postdepositional process for the coloration. Dubiel and others (1991) have explained the mottling as a reflection of a seasonally fluctuating water table driven by monsoon conditions. As the water table rose and fell with the seasons, alternating reducing and oxidizing conditions affected the sediments. Innumerable cycles of these fluctuations generated an uncommonly irregular distribution of iron oxide in the deposits. Eventually they became inscribed with a gaudy imprint of this extreme climate.

Locally abundant, unique trace fossils are common in the basal mottled strata and mottled strata that occur throughout the Chinle and Dolores formations (Dubiel and others 1987, Hasiotis and Mitchell 1989). These cylindrical burrows are vertical to steeply inclined, but may curve to a horizontal orientation. They range up to 14 cm in diameter and, in some instances, extend downward more than 2 m, although most are shorter (Fig. 5.10). Where present, their density obscures any original sedimentary structures, and the vertical, varicolored stripes of the burrows contribute to the mottled appearance of the host strata. The burrows are confined to fine-grained sediments, mostly mudstone, siltstone, limestone, and fine sandstone.

These burrows originally were interpreted as the dwellings of lungfish (Dubiel and others 1987). This was based on similarities with burrows in other rock units, including Permian rocks in Oklahoma in which well-preserved, complete lungfish skeletons were discovered in some burrows. Based on observations of modern lungfish in Africa, burrowing probably was motivated by an annual dry season, providing an escape from the periodic dessication of the lakes, marshes, and wetlands that hosted these vertebrates. This interpretation strengthens the thesis of Dubiel and others (1991) of a seasonal, monsoon-type climate that was alternately very wet and dry.

As often happens in geology, a later, more-detailed examination of Chinle burrows has led to a different interpretation of their origin. Stephen Hasiotis, one of the few geologists to specialize in trace fossils, and Charles Mitchell (1989) looked at similar Chinle burrows in Canyonlands and unearthed clear-cut evidence linking their origin to crustaceans, specifically crayfish. Within the sediments that fill the burrows, they actually discovered fossil crayfish. Dubiel and his colleagues originally discounted the possibility of such an origin because the previously oldest known crayfish was from Upper Jurassic rocks in Europe, some 60 million years younger. The discovery of crayfish in the Chinle makes them the oldest known. Hasiotis and Mitchell conceded that some of the numerous burrows that they examined might, in fact, be lungfish burrows, as some differ slightly from the unequivocal crayfish burrows. The great vertical length of the crayfish burrows is likely an artifact of the seasonal fluctuation of water tables. It is believed that the crayfish burrowed deeper during the dry season in an attempt to keep their living chambers below the declining water table. Thus, the burrows, whether of lungfish or crayfish origin, reflect extreme changes in the water table tied directly to the seasonality of precipitation.

Over much of the salt anticline region the base of the Chinle is marked by a thin (< 20 feet), discontinuous unit of sandstone and conglomerate. Pebbles are composed of granite, gneiss, limestone, quartzite, and siltstone. The unit weathers to a light tan color, making the contact between the Moenkopi and the Chinle, both of which are red in this area, much easier to pick.

This lower part of the Chinle is difficult to match with the well-established and easily defined members a short distance to the west. However, the diverse strata that comprise the lower Chinle here represent a mosaic of river, floodplain, lake, and swamp deposits (Stewart and Wilson 1960), all of them probably present at any given time. The lowest areas between salt anticlines hosted lakes and swamps that likely captured any streams that bisected them, at least temporarily. Streams

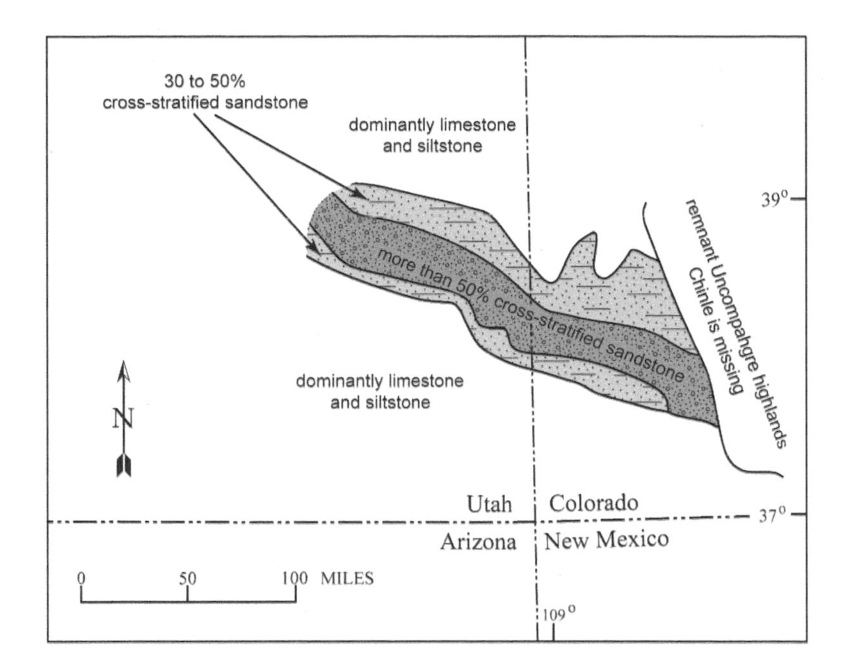

FIG. 5.11. Facies map of upper Chinle Formation showing the northwest-trending belt dominated by sandstone surrounded by finer-grained facies. The sandstone-dominated facies represents a northwest-flowing river system that originated in the ancestral Uncompahgre highlands, whose remnants endured as lower-lying uplands. Limestone and siltstone facies to the north and south represent ponding of water on a low-relief floodplain. Modified after Stewart and Wilson 1960.

initiated on the Uncompahgre hills, but probably picked up sediment along the way as they incised earlier strata that were caught up in the rising anticlines.

The upper Chinle in this region consists of red beds of siltstone, sandstone, and limestone with lesser conglomerate (Stewart and Wilson 1960). These strata constitute the bulk of the formation in the salt anticline area. Interbedded sandstone and siltstone contain cross-stratification that indicates a fluvial channel origin, whereas combinations of siltstone and limestone are suggestive of lakes. Structureless and horizontal stratified siltstone formed on floodplains.

Stewart and Wilson (1960) constructed a map plotting the regional distribution of the various rock types in the upper Chinle. A prominent feature of this map is a northwest-trending belt, ranging more than 20 miles wide, where cross-stratified sandstone and siltstone are dominant (Fig. 5.11). The belt extends from the Uncompahgre highlands northwestward across the salt anticline region and is traceable into central Utah. This sandstone facies outlines the path of a large, northwest-flowing river system that originated in the highlands of southwest Colorado. The rivers apparently cut undisturbed across the salt anticline region. This suggests that the rise of salt, which had affected sedimentation for so long, had slowed or even halted by this time.

As noted above, correlations between Chinle strata in the salt anticline area with its members to the west are difficult, but that has not stopped some geologists who have made some tentative efforts toward solving this problem. However, without an understanding of these western members, such discussion is futile. With this in mind we turn to the west, where stratigraphic relations have been established with more success. The following section surveys the members of the Chinle in southeast and east-central Utah in stratigraphic order—that is, from bottom to top, as they would be encountered walking up a steep hillside of Chinle strata.

Chinle Formation in southeast Utah: Cedar Mesa, Canyonlands, and San Rafael Swell

The **Temple Mountain Member**, the oldest unit in the Chinle to the west, is mostly confined to the San Rafael Swell, where it is widespread. It is named for its type locality at Temple Mountain in the southern San Rafael Swell, a towering butte that was an important uranium mining district in the boom of the 1950s (Robeck 1956). The Temple Mountain forms the lower apron of the butte, where it weathers into purple ledges above the brown red Moenkopi. These colorful sandstone ledges are capped by cliffs of the Moss Back Member, which is the main host of uranium ore and, in this area, is pocked by prospects.

The Temple Mountain Member consists of interfingering siltstone, mudstone, and sandstone. Its defining characteristic is the unique mottling of all the rock types in colors of purple, red, brown, and gray white. Sandstone locally is conglomeratic and contains coalified logs, some with traces of uranium. The member fills a broad erosional swale cut into the top of the Moenkopi over a large part of its exposure in the San Rafael Swell. At Temple Mountain the member is more than 30 feet thick, but it reaches a maximum of 100 feet in the axis of the swale. Toward the edge of this broad channel the Temple Mountain thins and eventually pinches out.

Temple Mountain deposition was initially controlled by incision into the top of the underlying Moenkopi prior to sedimentation. This early incarnation of the Chinle drainage network was cut during the Middle Triassic as large rivers sliced a northwest swath to the sea, bisecting the monotonous coastal plain. This reinvigoration was stimulated by a drop in sea level and probable uplift southeast of the Colorado Plateau.

Following the downcutting and entrenchment of broad channels, sand, gravel, and mud began to fill the hollows as they were dropped from the rivers. Channel orientations and crossbedding point to a northwest-directed flow toward the shoreline that sat somewhere in central Nevada. It likely was a slight rise in the level of this sea that initiated the turnaround from an eroding to a depositing river system. Even a small shift in sea level can be enough to push a fluvial system across this threshold.

The **Shinarump Member** holds the basal position of the Chinle over much of the Colorado Plateau, particularly in northern Arizona and southern Utah. Coarse sandstone and conglomerate dominate the Shinarump, which coarsens to the southeast and fines northwestward. Conglomerate clasts are composed of quartz, quartzite, and chert, although volcanic-derived pebbles locally comprise a significant fraction. The Shinarump is a cliff former, and because it is typically succeeded by less-resistant strata, it weathers back into broad benches that overlook the red Moenkopi slopes.

The history of the Shinarump coincides with the history of geological investigations in western North America. In one of the earliest scientific endeavors in the region, Major John Wesley Powell (1876) identified and named several discrete associations of strata in northern Arizona and southern Utah that could be traced northward across the Colorado Plateau. These packages included, from bottom to top, the Shinarump Group, Vermilion Cliff Group, White Cliff Group, and Flaming Gorge Group. Powell's Shinarump consisted of both the Moenkopi and Chinle formations. It is the only name proposed by Powell that is still in use today. Later, Herbert Gregory (1917), one of the greatest geologists to work the Colorado Plateau region in these early years, revised Powell's nomenclature. Gregory reduced the Shinarump to the strata recognized today and called it the Shinarump Conglomerate. Overlying variegated shale, sandstone, and limestone were named the Chinle Formation for exposures in Chinle Valley on the Navajo reservation of northeast Arizona. Eventually the Shinarump became the basal member of the Chinle Formation (Stewart 1957).

The location and thickness of the Shinarump was governed by the earlier erosion of valleys into underlying rocks. Valleys were etched mostly into the Moenkopi to create great furrows that focused drainage to the northwest and the waiting sea. Ron Blakey and Richard Gubitosa (1983) documented the local absence of 260 feet of underlying Moenkopi, which indicates downcutting to at least this depth. Shinarump thickness varies over a short distance, and the thickest successions fill the deepest parts of the valleys. The maximum thickness is ~330 feet, and the member thins to zero toward the valley edges (Blakey and Gubitosa 1983). Numerous discrete valleys have been identified and named by geologists working in different places throughout its extent. A main trunk valley sliced northwest through central New Mexico, northeast Arizona, and south-central Utah (Fig 5.12). This trunk river was fed by innumerable tributaries that entered from the northeast, southeast, and southwest, producing a dendritic drainage network whose size rivaled anything in existence today. Several of these tributaries have been named for localities where they were recognized (Thaden and others 1964, Blakey and Gubitosa 1983, Dubiel 1994). The Vermilion Cliffs paleovalley tapped a volcanic highland to the southwest; the White Canyon, Moab, and Temple Mountain paleovalleys in southeast Utah had headwaters in the remnant hills of western Colorado and entered the trunk valley from the northeast (Fig. 5.12). Work by Nancy Riggs and others (1996) suggests that the headwaters of the trunk system extended as far east as northern Texas and western Oklahoma.

Shinarump deposits mark the paths of energetic braided rivers. Initially these rivers were confined to the valley bottoms, but as sedimentation progressed, the valleys gradually filled with sand and gravel. Eventually the rivers were liberated

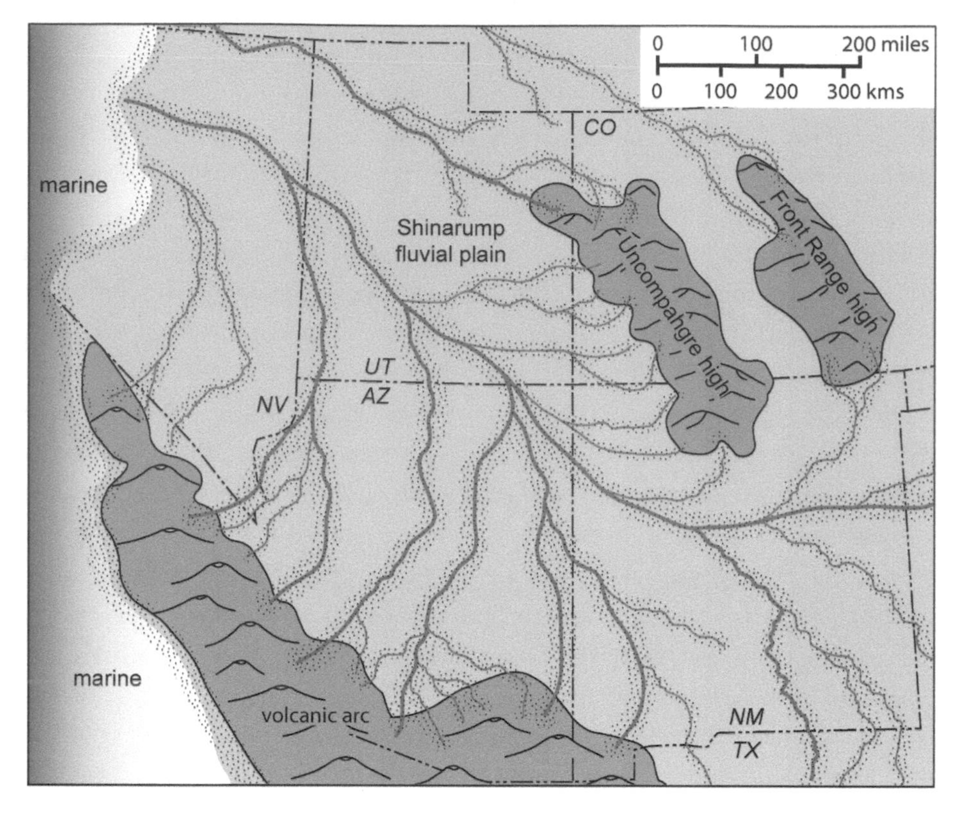

Fig. 5.12. Paleogeography of western North America during deposition of the Upper Triassic Shinarump Member of the Chinle Formation. Note that Shinarump sediments were derived from a variety of sources, including the remnants of the Ancestral Rocky Mountains (the Front Range and Uncompahgre highs), a volcanic arc along the southwest margin of North America, and highlands in Oklahoma and Texas related to the remnant Ouachita-Marathon orogenic belt. Modified after Blakey and Gubitosa 1983.

from the restraints of the valleys and were able to spread their detritus over the smoothed landscape. As a result, the latest Shinarump deposits blanket a wide area.

This river system, on a regional scale, swept sediment to the northwest from the headwaters in the Oklahoma-Texas area, across the Four Corners, and eventually to the shoreline in central Nevada (Fig. 5.12). Along the way, numerous tributaries from the remnant Ancestral Rockies and the chain of volcanic mountains to the west-southwest contributed significant amounts of sediment to the main trunk river.

The **Monitor Butte Member** succeeds the Shinarump over most of southeast Utah, reaching a maximum thickness of more than 300 feet in the Cedar Mesa–San Juan River area. In the San Rafael Swell, to the northwest, it thins to less than 100 feet. Northeast of these areas, including in Canyonlands, the Monitor Butte is absent. It is named for exposures on the remote Monitor Butte, an isolated erosional remnant situated just south of the lower San Juan River, near Piute Farms on the Navajo Reservation.

The fine sandstone, mudstone, and shale of the Monitor Butte are a striking contrast with coarse underlying strata, reflecting an equally dramatic shift in depositional setting. This change was driven by a tremendous influx of fine volcanic sediment, blown in on the westerly winds from the great chain of volcanoes that lined the basin to the west and southwest. This **volcanic arc**, as these volcanic mountain belts are called, belched huge quantities of fine ash into the air that then wafted into the Chinle basin. This inundation clogged the rivers with fine, claylike ash, pushing them from their channels and paralyzing the drainage system. As water spilled across the floodplain, a disorganized string of lakes, marshes, and wetlands evolved.

As tributaries poured slurries of ash into the main stem, dams formed and lakes filled behind them. Large-scale, northwest-sloping sandstone and shale beds mark the locations of small deltas that methodically filled the lakes as rivers dropped their sediment load into the slack water (Blakey and Gubitosa 1983). Black carbonaceous mudstone with fossil fish scales and bones delineates the marshy fringes of the lakes. Thin coal beds represent vegetation-tangled bogs and wetlands (Dubiel 1994). Abundant olive green shale and mudstone are dominated by **bentonite**, a type of clay formed from the alteration of volcanic ash. Bentonite weathers to a puffy, popcorn-like texture, and the mounded green mudstone slopes of the Monitor Butte are mantled with several inches of this "fluffy" clay. Fossil plants in the Monitor Butte include ferns and giant horsetail plants, also attesting to the shallow, fresh-water environment.

The **Moss Back Member** succeeds the Monitor Butte and, in a setting reminiscent of the Shinarump, fills a wide (~50 miles), northwest-trending swale through southeast Utah. This channel, named the Cottonwood paleovalley by Blakey and Gubitosa (1983), can be traced from the White Canyon–Abajo Mountains area in the southeast, across the Orange Cliffs in the Maze District of Canyonlands, and into the San Rafael Swell to the northwest. The southwest margin of the paleovalley is incised into the Monitor Butte Member, whereas the northeast bank is cut into Moenkopi strata or, locally, the basal mottled strata of the Chinle (Fig. 5.13).

The Moss Back consists of brown to gray conglomerate and sandstone. These cliff-forming strata are dominated by quartz sand and pebbles of limestone and siltstone. Pebbles of quartz, quartzite, and chert appear less frequently. Cross-stratification is the most common internal structure and is displayed in a spectrum of styles and dimensions.

All the observed features in the Moss Back add up to deposition in a gravelly, northwest-flowing, braided stream system whose channels were confined to the broad Cottonwood paleovalley. Although this river was never, at any given time, 50 miles wide, the braided channels snaked back and forth across the confined plain, ultimately generating a wide, northwest-trending sheet of sandstone and

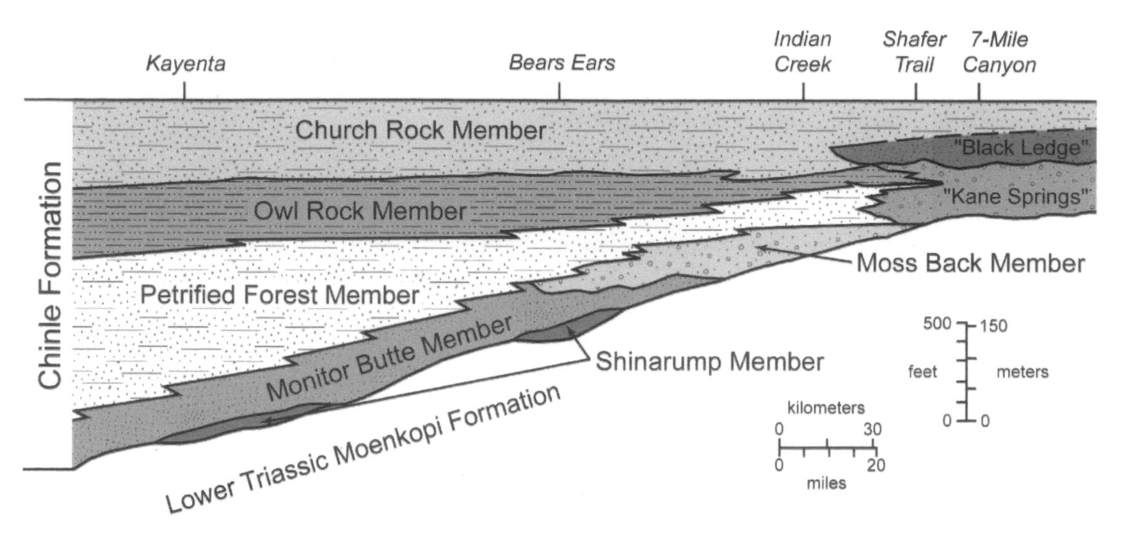

FIG. 5.13. Stratigraphic cross section of the various members of the Chinle Formation across the eastern part of the Colorado Plateau (in northern Arizona and southeast Utah) showing intertonguing relations, correlations, and thinning of many of the members to the northeast in the salt anticline region. Note the informal Kane Springs and Black Ledge units in this region. Modified from Blakey and Gubitosa 1983.

conglomerate. Conglomerate pebbles of limestone and siltstone were cannibalized from adjacent floodplain deposits as the river expanded its borders by cutting new channels into these finer sediments.

The **Petrified Forest Member** and equivalent strata cover a large part of New Mexico, northern Arizona, and southern Utah. The member unconformably overlies the Monitor Butte Member to the south. To the north, however, the member conformably overlies and interfingers with the underlying Moss Back Member (Fig. 5.13). Farther to the northeast, where the Moss Back pinches out against the rising salt structures, the Petrified Forest Member turns into the informally designated "Kane Springs strata" (Blakey and Gubitosa 1983). The Kane Springs strata warrant a different designation because they are a different lithology, but also because they are the lateral equivalent of the upper Moss Back and Petrified Forest members, and part of the overlying Owl Rock Member (Fig. 5.13). For the sake of clarity, they are discussed with the Petrified Forest Member. Because of uplift and erosion on the salt anticlines, Kane Springs strata lie unconformably on the Moenkopi or Cutler formations. These strata are continuously well exposed along canyon walls and on the bounding ramparts of the Island in the Sky District in northern Canyonlands, and to the southeast in Lisbon Valley.

The Petrified Forest Member is dominated by gray, brown, red, and purple mudstone with sparse conglomerate. Mudstone is bentonitic, indicating a continued flood of airborne volcanic ash into the basin from the erupting volcanoes to the southwest. Conglomerate consists of thin lenses of cross-stratified limestone pebbles wrested from floodplain soils with caliche nodules. These strata also contain abundant vertebrate remains, lungfish fragments, freshwater invertebrate fossils such as gastropods and unionid bivalves, and locally abundant plant remains, including the petrified logs for which the member is named.

The setting of the Petrified Forest Member was a reincarnation of the earlier Monitor Butte system. The Four Corners region was a marshy, low-relief floodplain cut by sinuous, meandering rivers and dotted with ponds. Large conifer trees towered over shadowed river banks, and the sky was periodically darkened by rains of volcanic ash. Inhabiting the waterways were lungfish and crocodile-like phytosaurs up to 30 feet long. Essentially, the Petrified Forest plain was a vast tropical lowland, lush with well-watered ferns and trees.

The informally designated **Kane Springs strata** are continuously well exposed immediately southwest of Moab and a short distance farther west, along the Green River and its tributary canyons. These rocks were extensively studied by Northern Arizona University geologist Joe Hazel Jr. (1994); the following discussion is drawn mostly from his work.

Kane Springs strata define a northwest-trending belt of limestone pebble conglomerate and coarse sandstone that contrasts sharply with the time-equivalent Petrified Forest and Owl Rock members immediately to the southwest. The belt of Kane Springs conglomerate is situated just within the salt anticline region, along its southwest margin, where the salt anticlines are more subtle features. Still, localized salt-related uplift and subsidence near this margin was enough to induce Kane Springs deposition and cause the Moss Back, Petrified Forest, and Owl Rock members to pinch out along this boundary (Fig. 5.13).

According to Hazel (1994), the river channel deposits that dominate the Kane Springs range from low-energy, high-sinuosity types to those of high-energy braided streams. Regardless of the variations, features in these deposits indicate an unwavering northwest-directed flow, aligned with the trend of local salt structures. It was these structures that confined the river system to its narrow corridor. The Moab–Spanish Valley salt anticline bounded these rivers to the northeast, while the less conspicuous, salt-cored Cane Creek anticline and Shafer dome hemmed the rivers in along their southwest margins. Separating the Moab and Cane Creek uplifts was the Kings Bottom syncline, its subsidence driven by the withdrawal of salt from beneath it. Individual sequences of fluvial deposits in the Kane Springs clearly thicken in the Kings Bottom syncline and thin or pinch out across the Cane Creek anticline. These relations indicate that the location of the Kane Springs rivers was largely regulated by salt movement which focused their channels into the downwarping syncline.

The surrounding salt uplifts also supplied the limestone pebbles that dominate Kane Springs sediment. As the Moab anticline rose, soils that had developed on its flanks were stripped off, a casualty of the abundant runoff and steepened slopes. The calcareous nodules that formed in these soils were resistant to breakdown, at least over short transport distances, and became a part of the rivers' sediment load. A local source is required because carbonate cannot tolerate for long the turbulence of transport in a river channel. If transported for long distances, these carbonate clasts most certainly would have been destroyed by abrasion.

The **Owl Rock Member** conformably overlies the Petrified Forest Member and covers approximately the same area. It reaches a maximum thickness of 395 feet in southeast Utah and is characterized by pastel-colored mudstone and numerous thin, ledgy limestone beds. Mudstone is much less bentonitic than the underlying Petrified Forest and Monitor Butte members, suggesting that volcanic activity to the southwest was waning.

According to USGS geologist Russell Dubiel (1994), the Owl Rock was deposited in a widespread lake and marsh system linked by the occasional wandering river (Fig. 5.14). Dubiel envisioned an Everglades-like setting in which freshwater trickled northwestward to an eventual rendezvous with the sea in central Nevada. The numerous alternations of mudstone and limestone suggest high seasonal runoff that flushed mud into the system, followed by slack-water periods punctuated by carbonate precipitation. Abundant burrows indicate ideal conditions for crayfish and other invertebrates.

The overlying **Church Rock Member,** which marks the top of the Chinle Formation in southeast Utah, consists of orange red sandstone and siltstone. Based on similar lithologies and stratigraphic position, the Church Rock has been correlated with the widely scattered Rock Point Member in northeast Arizona and western New Mexico. Originally the Rock Point was considered to be the basal member of the overlying Wingate Sandstone (Stewart and others 1972a, 1972b), but it has since been reassigned to the Chinle. The Church Rock is also the equivalent of much of the Dolores Formation in southwest Colorado, again based on similarities in rock type and position. Equivalent strata in northern Canyonlands and the San Rafael Swell include the informally designated Black Ledge sandstone (see Fig. 5.13). Exposures of this resistant unit are covered with a thin veneer of black desert varnish, hence the name. Uppermost Chinle strata are irregularly distributed over the Colorado Plateau, a consequence of the widespread erosion that followed Chinle deposition. The contact between the Chinle and the overlying Jurassic Wingate Sandstone has been designated as the J-0 unconformity.

The Church Rock and its equivalents were deposited in a variety of settings, all signaling an increasingly arid climate. Diagnostic features in sandstone range from large-scale crossbeds indicative of eolian dunes, to wide sandstone lenses floored with pebble conglomerate that record the passage of rivers. Thin interbeds of siltstone and sandstone represent intermittently wet playa mudflats that, during increasingly drawn out dry periods, were blanketed with sheets of fine, windblown sand.

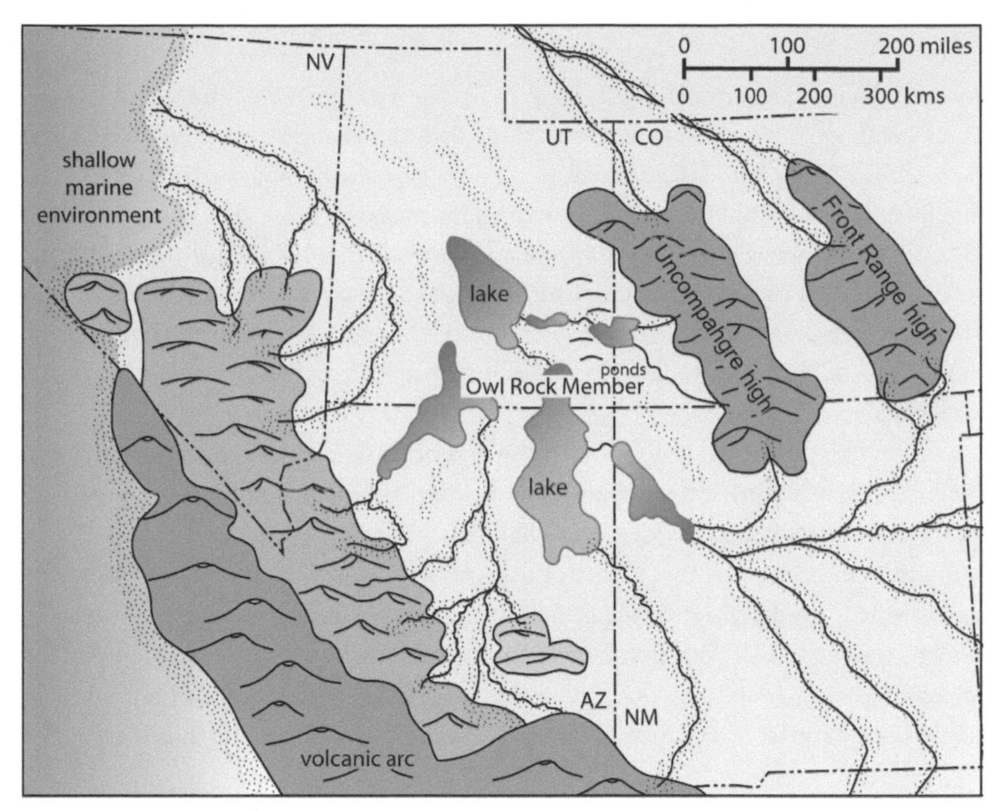

Fig. 5.14. Paleogeographic reconstruction of western North America during deposition of the Owl Rock Member of the Chinle Formation. Note that the earlier drainage pattern that initiated with the basal Shinarump Member is maintained, but the channels have been filled and the regional gradient has dropped, resulting in the development of a series of ponds. It is these low-energy deposits that characterize the Owl Rock Member. After Blakey and Gubitosa 1983.

The sandstone- and conglomerate-dominated **Black Ledge** deposits differ appreciably from their finer-grained equivalents to the south and represent a high-energy fluvial setting. The Black Ledge mimics the underlying Kane Springs strata in lithology and location, indicating similar, if not identical, controls on sedimentation. During Black Ledge deposition, coarse sediment was shed from the southwest flank of the Moab anticline into the subsiding trough that had developed within the adjacent Kings Bottom syncline. The Cane Creek anticline bounded this trough on the southwest, trapping the rivers into a northwest-trending belt. These salt-induced controls are illustrated by a conspicuous thickening of the Black Ledge within the syncline and a pronounced thinning or pinchout as the strata approach the crest of the Cane Creek anticline (Hazel 1994).

Features contained within most of the Chinle Formation point to a much wetter climate than those that preceded or followed its deposition. On a large scale, a surplus of water is told by widespread lake, pond, and marsh deposits. The presence of lush, well-watered forests is indicated by fossil ferns, giant rushes, and horsetails, and by abundant petrified conifer logs up to 200 feet long and 8 feet in diameter. A strongly seasonal, warm monsoonal climate, in which very wet seasons alternated with dry periods, has been proposed by Dubiel and others (1991). Their evidence for such dramatic variations includes the mottled strata that occur throughout the Chinle. These unique multicolored, striped beds formed during frequent, radical fluctuations in the groundwater table driven by alternating wet and dry seasons. The deep vertical burrows that populate these mottled strata, whether excavated by crayfish or lungfish, also suggest such fluctuations. The great depth of these burrows, some up to 2 m deep, are believed to reflect the magnitude of the water table shift as the burrowers struggled to remain below the water table as it dropped.

During the latest Triassic, toward the end of Chinle deposition, the climate in western North America reverted back to aridity, ending a short but unique interval of Colorado Plateau history. This change is indicated by the presence of eolian sandstone in the uppermost Church Rock and Rock Point members. Such deposits are absent in underlying Chinle strata. This climate change was but an introduction to the extremely arid conditions that would dominate the Colorado Plateau in the following Jurassic Period. In fact, as we shall see, the Jurassic Colorado Plateau was the center of the most extensive eolian dune fields in Earth history.

Chapter 6

The Jurassic Period

Sand and More Sand

The great palisades of Jurassic sandstone are one of the characteristic features of the Colorado Plateau and make it one of the most remarkable landscapes in the world. These towering cliff formers are host to some of the most spectacular canyons imaginable, including Zion, the deeply incised slots of Paria and Escalante canyons, the remains of Glen Canyon and its numerous tributaries, and the fins and arches of Arches National Park. Canyonlands National Park is almost completely encircled by these vertical orange walls, including the great northern ramparts of Island in the Sky and Dead Horse Point.

The Jurassic Period spans a period of 62 million years, from 206 to 144 Ma. Rocks deposited on the Colorado Plateau during this time record a variety of depositional settings and climates. The Jurassic dawned with far-reaching eolian sedimentation of the Glen Canyon Group, which includes Wingate Sandstone, the Kayenta Formation, and Navajo Sandstone (Fig. 6.1). By the Middle Jurassic, a shallow, fingerlike seaway extended southward into western Utah. This shallow marine incursion altered the face of the plateau region considerably. While marine sediments blanketed the floor of this narrow sea, its margins and outlying areas continued to record aridity in the form of evaporites and eolian sandstone. These sedimentary rocks make up the San Rafael Group, composed of, in ascending order, the Page, Carmel, Entrada, Curtis, and Summerville formations (Fig. 6.1). The final chapter of the Jurassic Colorado Plateau is recounted by the Upper Jurassic Morrison Formation, world-renowned for its abundant dinosaur fossils. Morrison river deposits record a reversal of drainage patterns in western North America. These rivers traversed the Colorado Plateau, flowing from southwest to northeast. They originated in highlands of Nevada and were connected with the

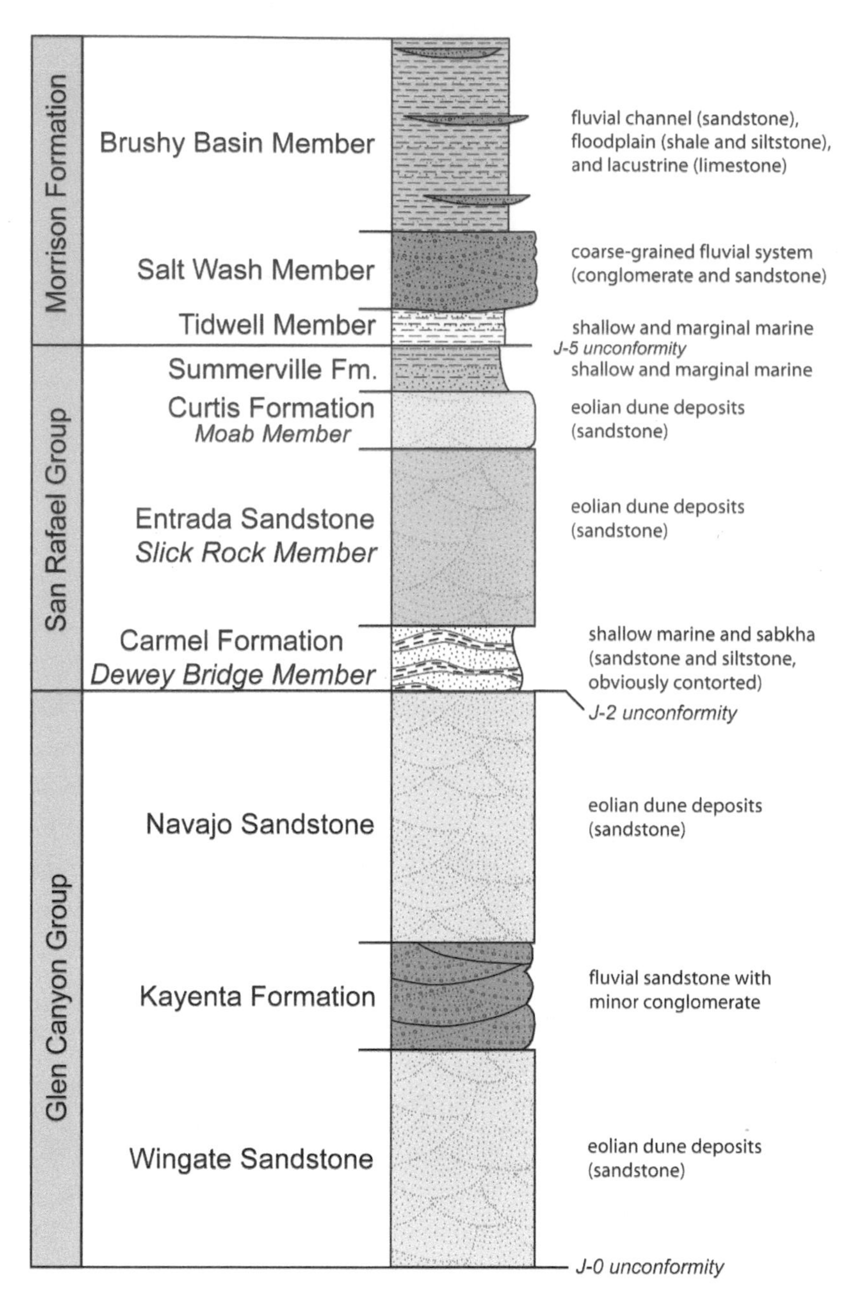

Fig. 6.1. General stratigraphic column for Jurassic rocks in the Moab–Arches National Park area.

receding sea in modern-day Wyoming. The Morrison also marks a return to wetter conditions, although the question of how humid it was remains debatable.

The Tectonic Setting of the Jurassic

The wholesale reversal of drainage directions across the Colorado Plateau was the consequence of a great swell of mountains to the west in modern-day Nevada and eastern California. Although mountains existed in this area during the Triassic, they were separated from the Plateau region by a sea, and their influence was minimal. These Jurassic mountains were of two distinct types and origins: volcanic and compressional. Over most of the Jurassic Period both types were active in western North America, often at the same time. Ultimately both types were generated by the large-scale interactions of the same tectonic plates. To the west an east-moving oceanic plate was thrust beneath the continental plate of North America. As the oceanic plate sank ever deeper beneath the continent, it was heated and began to melt. The magma that was produced then rose toward the surface, eventually to erupt as volcanoes. Additionally, at various times throughout the Jurassic, large islands that had evolved on top of the east-moving oceanic plate were rafted in on it, as if brought in on some gigantic conveyor belt. The thicker crust of these islands, however, was not so easily subducted. Instead they were scraped off as the thinner oceanic crust began its downward journey. The island landmasses were then accreted, or "welded," to the western margin, expanding western North America incrementally and generating huge amounts of compression farther inboard, pushing the mountains upward. It was through this activity that the western margin of North America slowly grew to the dimensions that we recognize today.

Besides reversing earlier drainage patterns, Jurassic tectonism affected the Colorado Plateau by increasing subsidence along its western edge, resulting in a pronounced thickening of individual units in that direction. In compressional mountain belts, the viselike squeeze of compression thickens the crust by folding and faulting, essentially piling masses upon masses of rock. Imagine, for example, a previously undeformed belt that is 50 miles wide. If that belt is then compressed into a zone only 25 miles wide, its thickness doubles, and the load on both the underlying and surrounding lithosphere (the uppermost 100 km of Earth) increases substantially, causing it to subside. Because the lithosphere behaves as a semirigid plate, maximum subsidence takes place directly beneath the mountain belt, and it gradually decreases away from this zone (Fig. 6.2). Strata deposited in this zone of subsidence will be thickest immediately adjacent to the mountain belt and will gradually thin away from it. The strong asymmetric geometry of some Jurassic strata on the Colorado Plateau—that is, the pronounced thickening to the west and thinning to the east—probably reflects the effect of the mountainous load in Nevada.

The relationship between Jurassic mountain building and subsidence on the Colorado Plateau was well documented by Christian Bjerrum and Rebecca Dorsey (1995) while at Northern Arizona University. Their extensive review of the timing and location of Jurassic deformation, coupled with a sound understanding of Jurassic sedimentation on the plateau, enabled them to numerically model subsidence patterns in an effort to find those that best fit the observed thicknesses

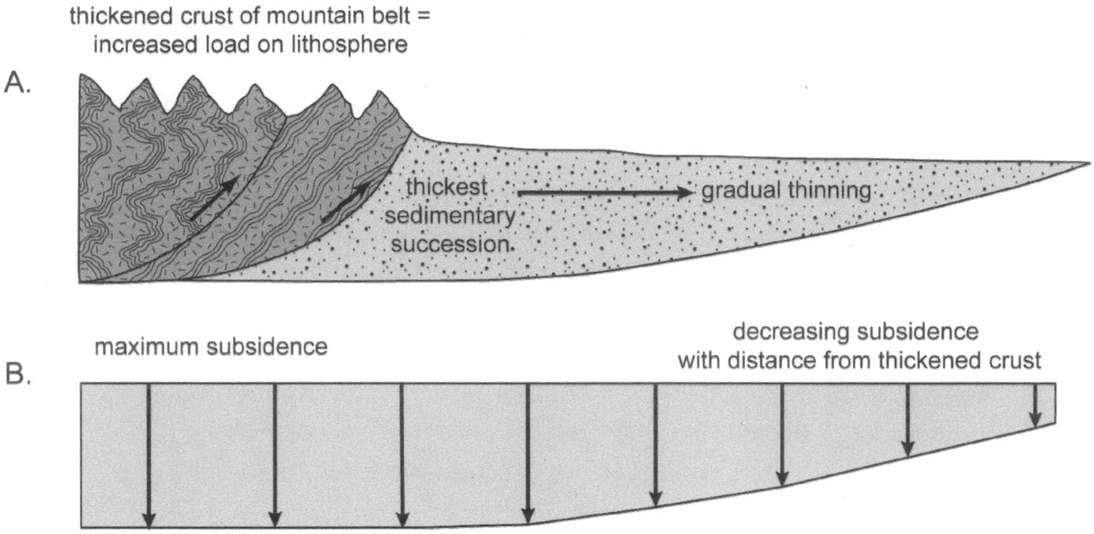

FIG. 6.2. A, schematic cross section through the thickened crust of a mountain belt to an adjacent sedimentary basin showing magnitude of subsidence and the decrease with distance from load of mountain belt; B, diagram illustrating simple changes in the amount of subsidence in Figure 6.2A.

and facies patterns of the Jurassic strata. Such computer modeling allows one to easily consider and adjust the numerous influencing factors until the modeled basin geometry matches the actual sediment fill patterns of a given sequence. While it *is* just a model, the various factors used to generate it can suggest directions for further field studies; that is, the model can be tested. Bjerrum and Dorsey's model provides a convincing argument for the loading-induced subsidence that is typical of foreland basins that develop adjacent to a compressional mountain belt. Significantly, it adds details to a setting that previously was speculative and controversial.

In their modeling, Bjerrum and Dorsey found a best fit using a mountain belt that was 300 km wide with an elevation of 2.5 km (> 8000 feet), depicting a reasonable orogenic belt of moderate relief. Although the Lower Jurassic Glen Canyon Group exhibits an obvious westward thickening, it was difficult to model because of a lack of time constraints in the fossil-poor eolian strata. Modeling of the overlying lower part of the Carmel Formation provided a better fit and suggested the front of the orogenic belt was located in eastern Nevada at this time. This model also fits well with documented deformation in this area. Thickness patterns in the Middle and Upper Jurassic Carmel, Entrada, and Morrison formations fit with the eastward migration of the mountain front into western Utah by this time—again, a reasonable hypothesis. In addition, their modeling identified an interval of relative tectonic quiescence in latest Jurassic time based on an abrupt decrease in subsidence over the Colorado Plateau. This lull continued into the Cretaceous Period, when mountain building resumed with a renewed fury.

The Wingate Sandstone

The Lower Jurassic Wingate Sandstone and equivalent strata comprise the base of the Jurassic System and are spectacularly exposed over most of the Colorado Plateau. This was obvious to the first geologists in the region, including C. E. Dutton, one of a handful of great early geologists in the American West. Dutton (1885) described the Wingate as "a massive bright red sandstone. Out of it have been carved the most striking and typical features of those marvelous plateau landscapes which will be subjects of wonder and delight to all coming generations of men."

The uniform grain size and well-cemented nature of the sandstone, combined with the absence of other rock types, promotes erosion into smooth, vertical walls that extend unbroken for miles. In southeast Utah, where the sandstone ranges 300 to 500 feet thick, these orange and red walls can be seen for great distances. Wingate cliffs loom continuously over the Colorado River from Dewey Bridge westward to Potash, confining this great artery for 45 miles. It also lines the upper walls of Castle Valley and the adjacent Moab Valley. In Castle Valley a narrow rib of Wingate has been hewn into the monolithic Castleton Tower and the nearby Rectory, complete with the Priest and Nuns. The steep, solid cliffs and vertical cracks in these sandstone towers have made them a popular destination for rock climbers. The red promontories and serpentine escarpments of Wingate also elevate the Island in the Sky and nearby Dead Horse Point, in the northern part of Canyonlands. Indian Creek, along the south entrance to Canyonlands, also hosts a world-class climbing area thanks to its smooth, crimson walls of Wingate with a high concentration of vertical fractures. Nearby, North and South Six-shooter Peaks are two isolated fingers of Wingate that aim skyward, stretching far above the surrounding landscape. To the west, across the Colorado River, the Orange Cliffs, named for the impenetrable wall of Wingate that forms them, mark the western boundary of the Maze. Finally, the Wingate Sandstone made up the deep, shaded alcoves and soaring tapestried walls of Glen Canyon before it was inundated beneath the water of Lake Powell. It was from these unparalleled exposures that the Glen Canyon Group received its name. The Wingate makes up the basal unit of the Glen Canyon Group.

The Wingate is separated from the underlying Chinle Formation by the regional J-0 unconformity. This contact is a sharp and obvious line that marks an abrupt change from the slope-forming Chinle shale to the vertical orange walls of Wingate Sandstone. Throughout its extent the Wingate consists of fine sandstone dominated by large- and small-scale cross-bedding with lesser horizontal bedding. Thick successions of these strata are separated by minor thin and wavy bedded sandstone units. Collectively, the features contained within the Wingate indicate eolian deposition in an immense erg, one that covered most of the Colorado Plateau. Thinner, wavy bedded units reflect intervals of sabkha deposition.

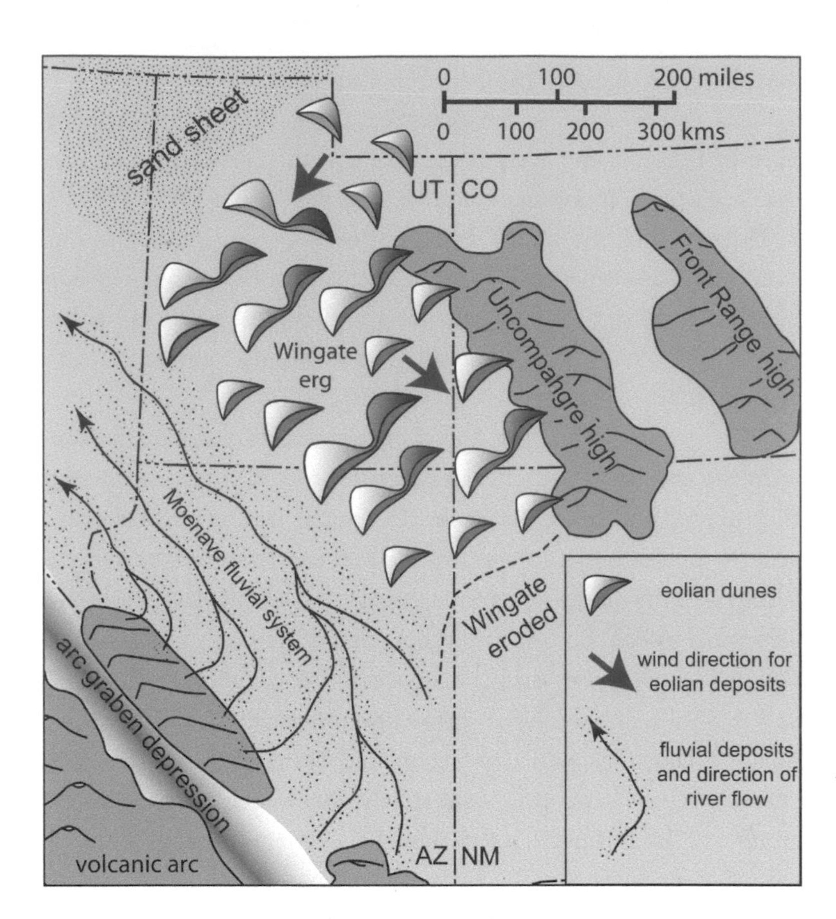

FIG. 6.3. Paleogeography of the Colorado Plateau during deposition of the eolian Wingate Sandstone and fluvial Moenave Formation. After Peterson 1994.

The areal extent of the Wingate is huge; except for northern Arizona, it blankets much of the Colorado Plateau (Fig. 6.3). Its eastern limit was controlled by the slowly wasting Uncompahgre highlands, which lingered as enough of a barrier to impede the eastward flood of sand. In western Colorado the Wingate gradually thins and pinches out against these low hills. To the southeast, in the San Juan basin of New Mexico, the Wingate thins rapidly and is truncated by erosion along a northeast-trending line that cuts through northwest New Mexico. Along its south margin the Wingate intertongues with the lower part of the fluvial Moenave Formation over a 30- to 60-mile-wide zone. This gradational contact trends northwest, passing through Tuba City, Arizona, and Paria, Utah. In the Vermilion Cliffs near Lees Ferry and in Zion National Park, Utah, the Wingate is absent, its cliffs replaced by the sandstone and shale slopes of the Moenave. The Moenave was deposited in a large, ephemeral, northwest-flowing river system (Clemmensen and others 1989). In northwest Utah the Wingate thins and grades into low-relief sand sheet deposits that probably mark the depositional edge of the erg.

Northeast of the zone of intertonguing between the Wingate and Moenave, the Wingate is dominated by eolian erg deposits. Although the Wingate cliffs appear massive and homogeneous from most perspectives, a close examination

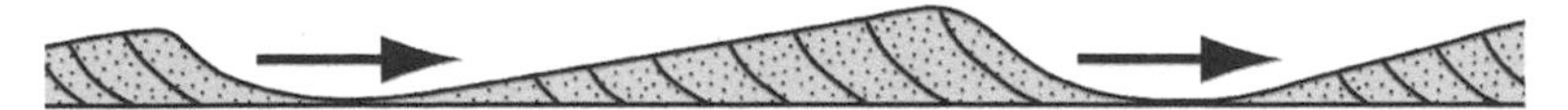

A. Initiation of dune field - low water table and abundant sand

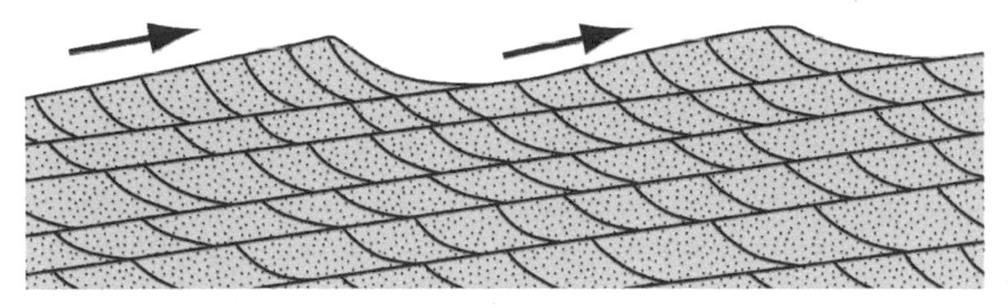

B. Continued development of dune field and vertical accumulation of dune deposits

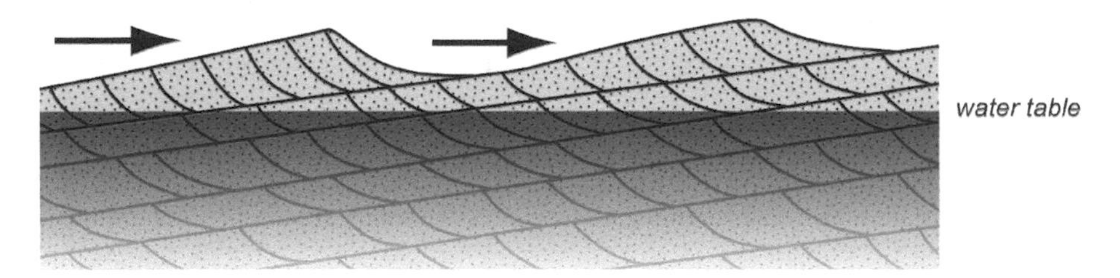

C. Rise in water table cuts off sand supply and erosion of sand above water table begins

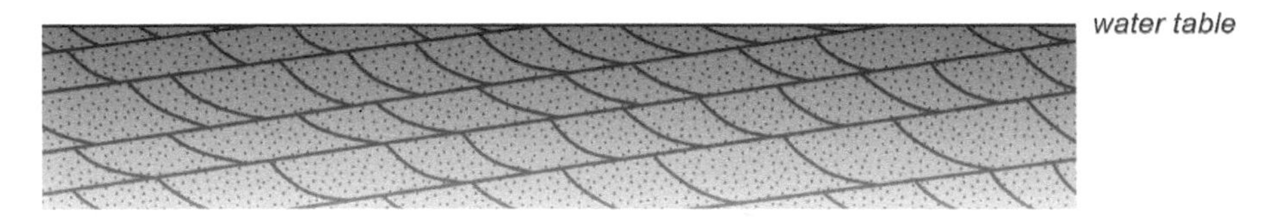

D. Lack of new sand due to cutoff of supply results in erosion down to water table

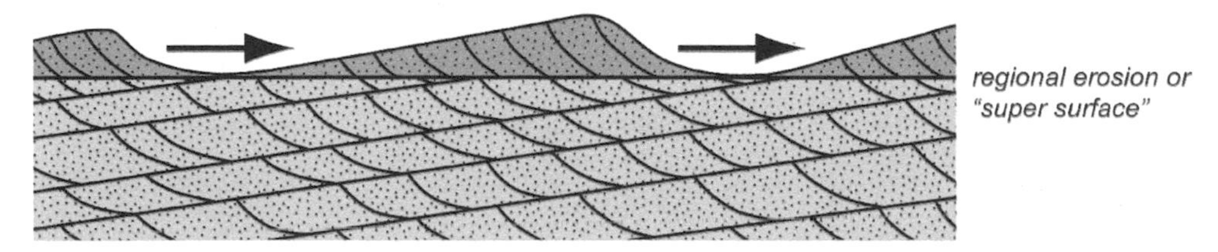

FIG. 6.4. Development of horizontal super surfaces in eolian sequences due to the rise of the groundwater table and cutoff of sand supply, followed by erosion of sand above the water table. Erosion does not continue below the water table because wind cannot dislodge damp sand.

reveals subtle cyclic variations. While at Northern Arizona University, Matthew Nation (1990) conducted the first detailed regional study of the Wingate. Nation recognized several obvious horizontal erosion surfaces within the formation that could be traced throughout his study area, which included southeast Utah, northern Arizona, and southwest Colorado. These "super surfaces," as they are known, separate six discrete sequences documented in the towering Wingate walls. Each sequence overlies one of these regional erosion surfaces. Although there are internal variations, a typical sequence begins with a thin unit of red, wavy-bedded, silty sandstone of probable sabkha origin. This is overlain by small-scale cross-bedded and horizontal bedded sandstone, marking the imminent return of the erg. This, in turn, is overlain by a thick succession of large-scale cross-bedded sandstone, the deposits of large dunes that herald the recovery of the erg. These dune deposits are truncated by another erosion surface, and the sequence begins again.

The individual sequences identified by Nation (1990) are interpreted to record the episodic demise and subsequent rejuvenation of the giant Wingate erg. According to Nation, the extensive erosion surfaces formed while the regional groundwater table was at its highest (Fig. 6.4). Sand below the water table was stabilized; that above the water table was winnowed away. The rise in the water table likely was driven by a sea level rise, which would have inundated the coastal plain to the north and west. Because this coastal plain is believed to have supplied most of the sand, its inundation eliminated this source at the same time the groundwater table was rising. Without an influx of new sediment, the incessant wind cut down to the subhorizontal water table, creating the regional super surface. Low-lying areas that were below the water table became inland sabkhas and ponds in which silty sand and local lenses of limestone were deposited. The subsequent drop in the water table (and sea level) again made available abundant sand. Resurrection of the erg was initially marked by the appearance of small dunes and sand sheets, represented in the succession by small-scale cross-bedding and horizontal, stratified sandstone. The overlying, large-scale cross-bedded sandstone, which dominates the sequences, signifies the full recovery of the erg and its rapid expansion to cover most of the modern Colorado Plateau region.

The cyclic eolian sequences recognized throughout the Wingate are attributed to the rise and fall of the regional water table, which is probably related to sea level fluctuations to the northwest. Similar controls are envisioned for eolian sedimentation in the Permian Cedar Mesa Sandstone, although the Wingate erg was much more extensive.

Like the earlier Cedar Mesa and White Rim eolian deposits, and the subsequent wind-blown sediments of the Navajo and Entrada sandstones, the Wingate sands were molded mostly by strong northwesterly winds in an arid climate. This calls for a practically unlimited sediment source to the northwest. Many geologists have hypothesized the presence of a vast coastal plain that was, at various times, fed huge quantities of sand by various west-draining river systems. The exact origin of this enduring sand source has been a long-standing and controver-

Fig. 6.5. The Kayenta Formation along the Colorado River just north of Moab. Note the ledgy weathering and lens-shaped beds, which represent discrete river channels in the Kayenta fluvial system. Arrows point out the approximate bases of the channels.

sial problem. Recent progress toward its solution is addressed below in this chapter's discussion of Navajo Sandstone.

The large-scale eolian sedimentation in the Wingate erg was finally terminated, at least in south and central Utah, by the development of large, west-flowing rivers. The deposits of this great river system make up the overlying Kayenta Formation.

The Kayenta Formation

The Kayenta Formation consists mostly of red sandstone and siltstone deposited in the wide, shallow channels of west-flowing rivers. Kayenta exposures consist of a series of thin, broken, cliff-forming ledges on top of the uninterrupted vertical Wingate cliffs. A close look at these ledges reveals numerous stacked and overlapping lenses, each representing the sand-filled channel of an ancient river (Fig. 6.5).

The Kayenta covers much of northern Arizona and southern Utah, and extends into western Colorado, where it pinches out eastward against what were the low, lingering hills of the Uncompahgre highlands (Fig. 6.6). In northern Utah the Kayenta grades into the massive eolian deposits of Glen Canyon Sandstone, which is the lateral equivalent of the entire Glen Canyon *Group* (Wingate, Kayenta, and Navajo formations) to the south. Because there was no break in eolian activity to the north, and therefore no fluvial Kayenta Formation, it is impossible to differentiate any units, and the easily divisible Glen Canyon Group becomes a

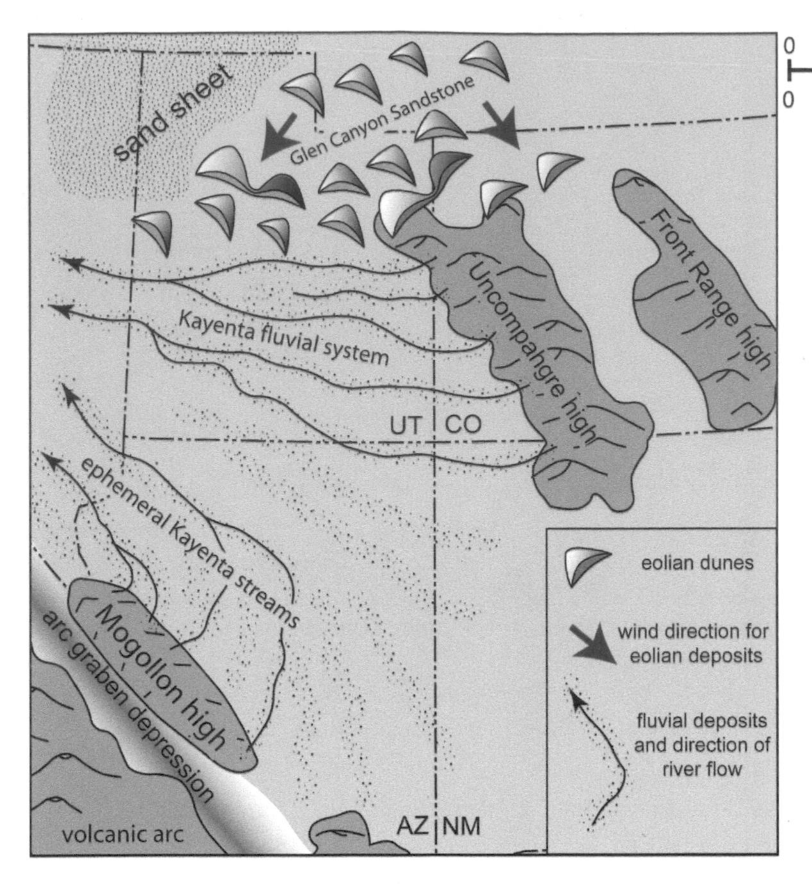

Fig. 6.6. Paleogeography of the Colorado Plateau during deposition of the fluvial Kayenta Formation. Note the perennial streams to the north draining the Ancestral Rocky Mountains (the Uncompahgre high) and the ephemeral streams to the south draining the Mogollon high. After Peterson 1994 and Luttrell 1993a.

single formation: Glen Canyon Sandstone. To the west, Kayenta rivers emptied into a seaway somewhere in eastern Nevada.

Based on a combination of features, the contact between the Kayenta and underlying Wingate has recently been interpreted as an unconformity (Riggs and Blakey 1993). Localized concentrations of gravel at the top of the Wingate are believed to be lag deposits—coarser sediment left behind and concentrated as finer sediment was winnowed away over a long time period. In other places the top of the Wingate contains evidence for soil development, a process that typically requires thousands of years of nondeposition. Wherever it is exposed, the contact is sharp, and in some places, where erosion cut deeply into the Wingate prior to Kayenta deposition, it shows up to 50 feet of relief.

In one of the important early reports on Jurassic strata of the Colorado Plateau, USGS geologist J. W. Harshbarger and his colleagues (1957) recognized a distinct difference between the Kayenta in Arizona and that in Colorado and Utah. Although in both areas it is of fluvial origin, in Arizona the Kayenta is mostly siltstone, whereas in Utah and Colorado it is dominated by sandstone. Geologist Patty Luttrell (1993a, 1993b) of the Museum of Northern Arizona later studied the Kayenta in more detail and discovered further differences in the formation between the south and the north. In the southern part of the basin the Kayenta contains sedimentary structures indicative of deposition in ephemeral streams with only seasonal flow. In contrast, the formation to the north was deposited in large perennial streams. The final difference, and one that likely influenced the previous disparities, was sediment source. Sediment in the southern half of the basin came from a volcanic-dominated source known as the Mogollon highlands, located along the Arizona-Mexico border (Fig. 6.6). Sand-dominated strata in Colorado and Utah sourced the shrinking Uncompahgre highlands. The ancient granitic and metamorphic rocks that made up this long-lived sediment source favored the production of sand over silt, and the perennial flow in these rivers easily wafted any silt-sized sediment that was present farther downstream. The ephemeral versus perennial stream setting within the Kayenta may have been related to the different latitudes of the source areas. The southern location of the Mogollon highlands may have produced less runoff than the slightly more northern latitudes of the Ancestral Rockies. Even today the distance between these two areas produces dramatic differences in climate and precipitation.

The seasonal rivers that traversed Arizona flowed northwest from the Mogollon highlands to the sea in eastern Nevada (Fig. 6.6). Rivers that emitted from the Uncompahgre hills cut a westward path on their way to the same seaway. Among these rivers and their wide floodplains stood small clusters of dunes, reminders of the massive erg that threatened this region from the north.

Uppermost Kayenta strata intertongue with the overlying eolian Navajo Sandstone. As sand blew into the Kayenta basin from the northwest, the rivers struggled to keep their channels clear and maintain their seaward push. Eventually, however, the dry, windblown sand overcame the weakening waters of the rivers, and the region again became buried beneath a thick blanket of fine sand.

The Navajo Sandstone

The eolian Navajo Sandstone consists almost completely of buff-colored sandstone characterized by an incredibly uniform grain size and large-scale cross-bedding (Fig. 6.7 and Plate 6). Although it locally takes on various shades of red or brown, its dominantly yellow-white color makes it easy to recognize and differentiate from the crimson of the underlying Kayenta and Wingate sandstones.

Where it is incised by the Colorado River in southeast Utah, the Navajo rises above the waterway as steep cliffs. In many places these walls are slashed by vertical black streaks, curtains of manganese oxide and lichens that accumulate where water has poured over them, and deeply shaded alcoves have been carved where springs emit from the porous sandstone (Fig. 6.8). In southeast Utah, however, the Navajo has been eroded back from the steep walls of the underlying Kayenta and Wingate. Instead it blankets the extensive plateaus above these cliffs with broad, undulating expanses of beehive domes and mounds, and rounded landscapes

Fig. 6.7. Outcrop of Navajo Sandstone showing its typical large-scale cross-stratification and dome-shaped weathering.

of large parallel fins—all fascinating places to explore the innumerable slots and grottos concealed in these sandstone expanses. Great arrays of aligned fins are especially prominent on the plateau immediately east of Moab, where the Slickrock Trail is located. In fact, the slickrock from which this popular bike trail gets its name is Navajo Sandstone. Massive jointed fins of Navajo also make up the plateau just west of Moab, an area known as Behind the Rocks (Fig. 6.9 and Plate 7). In these exposures, and everywhere else that the Navajo has been cleaved into great fins, it is due to the earlier formation of regional fracture systems. As the Navajo was unearthed by erosion, water widened and deepened the parallel fractures.

The Navajo records the largest erg in Earth history—a vast sand sea that blanketed all of Utah and a large part of western Colorado, where its eastward expansion was impeded by the last vestiges of the Ancestral Rockies (Fig. 6.10). Northward the windblown sand extended into Wyoming and part of Idaho, where it is called the Nugget Sandstone. To the west, fingers of fine sand stretched into Nevada and southeast California, where it again encountered topographic obstacles in the volcanic mountain belt known as the Sierra Nevada magmatic arc. A similar situation existed in southern Arizona, where the dune deposits are interbedded with volcanic rocks (Riggs and Blakey 1993).

› **Fig.** 6.8. Navajo Sandstone along the Colorado River near Moab, Utah, exhibiting dome-weathering and an alcove formed by spring-sapping.

˅ **Fig.** 6.9. The cliffs in the shaded foreground are Wingate Sandstone, and the fins in the middle ground, in the Behind the Rocks area, are Navajo Sandstone. The La Sal Mountains loom in the background.

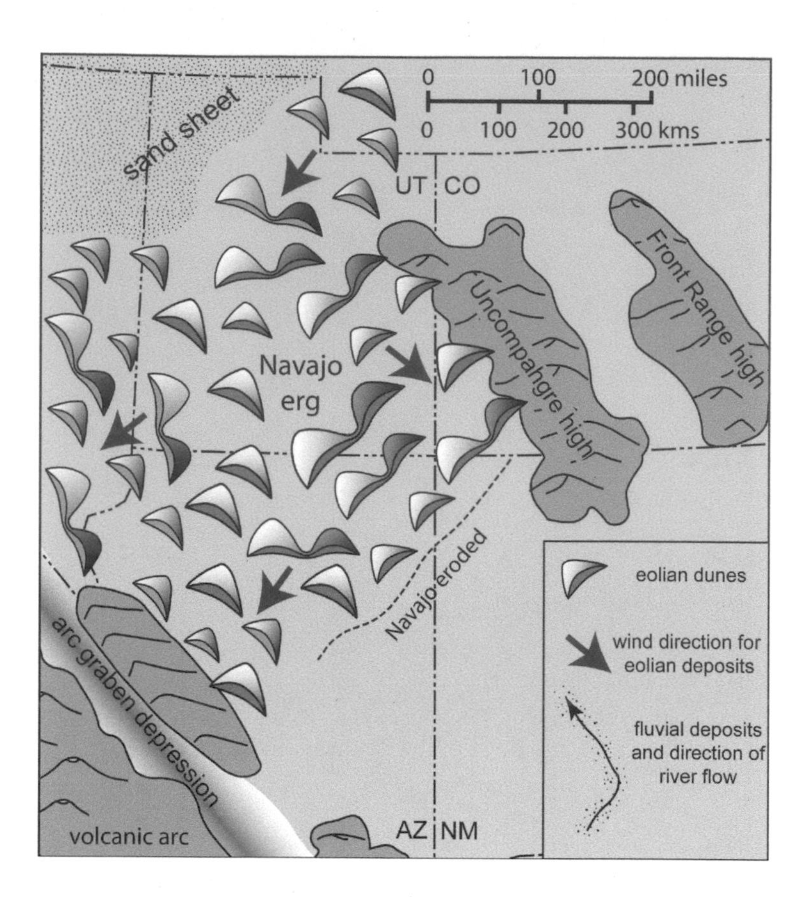

FIG. 6.10. Paleogeography of the Colorado Plateau during deposition of the eolian Navajo Sandstone. After Peterson 1994.

Regionally, the great Navajo sand blanket forms a pronounced wedge as it dramatically thickens westward across southern Utah. The sandstone thins to a feather edge in southwest Colorado but quickly swells to 200 feet along the Colorado-Utah border (Gregory 1950). In Arches National Park, a short distance to the west, it reaches a maximum of 550 feet, and continuing west into central Utah (Capitol Reef), it expands to 1400 feet. The Navajo attains a maximum thickness of 2000+ feet in Zion National Park, southwest Utah, where its unbroken, overhanging white walls completely dominate the scenery.

Although the Navajo erg was a harsh and monotonous expanse, local departures from this severe environment occurred in the form of freshwater ponds that formed in low-lying, wind-scoured areas between large dunes. Evidence for these deviations occurs as rare, widely dispersed carbonate lenses sandwiched between thick beds of cross-stratified sandstone. Geologist James Gilland (1979) of Brigham Young University studied one of these lenses near Moab in detail. Well exposed on outcrops above State Highway 279, along the Colorado River between Moab and Potash, the carbonate lens ranges to ~6 feet thick and stretches laterally for approximately a half mile. This thin unit is darker and more resistant than the surrounding sandstone and consists of a mixture of limestone, dolomite, and sandstone.

Although the bulk of the Navajo is barren of fossils or evidence of life, this pond deposit hosted a variety of organisms. Sandstone that marks the margin of the carbonate lens is extensively churned by horizontal and vertical burrows. This concentration of trace fossils marks the shoreline of the freshwater pond and likely were made by some type of wormlike invertebrate organism. Abundant palynomorphs (fossil pollen and spores) also were identified in the carbonate by Gilland (1979). These indicate that vegetation flourished along the fringes of the pond, taking advantage of the rare well-watered environment. Clear impressions of plants in the pond margin sediments likewise indicate a lush setting. The impressions of these round-stemmed, reedlike plants are very similar to the modern plant *Equisetum*, which today can be found growing on damp, sandy floors throughout southeast Utah's canyons. Large dinosaur tracks are also a common feature in these pond sediments and surrounding deposits. Three-toed tracks up to 37 cm in length are embedded in these sediments throughout the area. In fact, a highway sign on the road below these strata points out these tracks, which are clearly displayed on the bedding surface of a tilted block that can be viewed from the road. Dinosaur fossils are extremely rare in the Navajo, a testimony to the brutal Early Jurassic conditions in the Colorado Plateau region. Where it is present in the Navajo, evidence for life is concentrated around these pond deposits, which represent oases where vegetation and lower invertebrates thrived, and where dinosaurs converged to feed and drink. Despite the popular *Jurassic Park* movies, the Early Jurassic world of western North America was a harsh and unforgiving environment.

Ponds in the Navajo erg were formed by the rise of the groundwater table, which flooded the low-lying hollows in the interdune areas. This must have occurred numerous times because carbonate lenses are sparsely distributed throughout the Navajo over a large region. There is even evidence for pond deposits from upper Navajo strata that have long since been removed by erosion. This is told by a layer of chert and carbonate pebbles at the top of the Navajo that defines the J-2 unconformity.

The J-2 unconformity is one of six Jurassic unconformities documented and traced by USGS geologists George Pipiringos and Robert O'Sullivan (1978) throughout the western interior of North America. Although some of these unconformities represent only a few million years, their great extent marks a significant regional shift from deposition to erosion. These gaps in the rock record impair our attempts to reconstruct parts of Earth's past. Unconformities around the world represent a greater portion of Earth history, by far, than do the actual rocks that *are* preserved. Because of this, the cryptic components of these unconformities become vital clues for interpreting events for which there may be no other record. The J-2 unconformity that separates the Navajo Sandstone from the

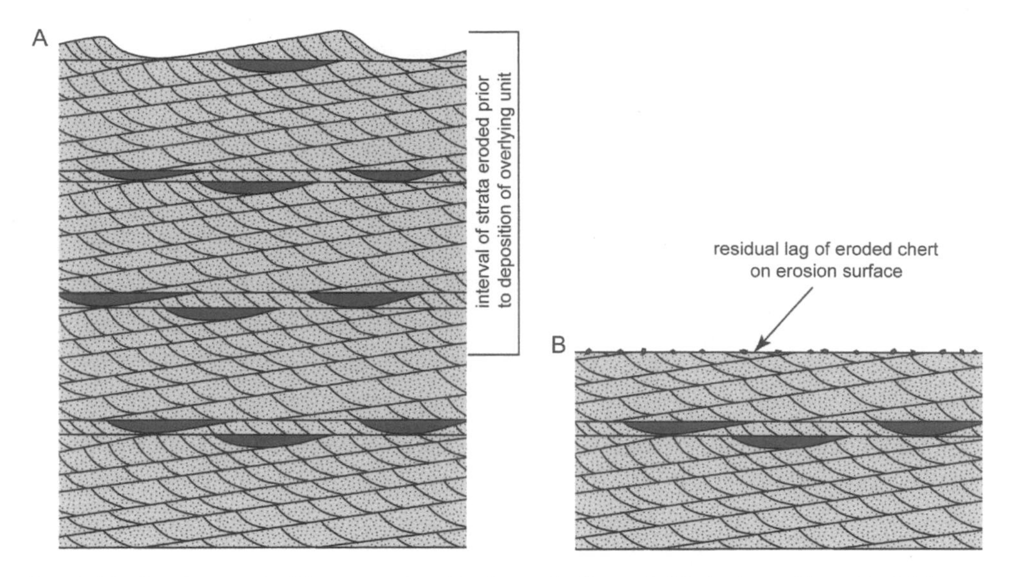

Fig. 6.11. Development of the unconformity shown in Figure 6.12. A, deposition of a thick succession of eolian Navajo Sandstone with numerous interdune pond deposits of limestone and chert. As large-scale erosion ensues, sand is easily winnowed away. B, a concentrated lag of fragmented, resistant chert remains on top of the erosion surface after the thick succession of overlying sand was eroded. The unconformity shown in Figure 6.12, which separates the Navajo from the overlying Carmel Formation, is marked by a lag of angular chert formed by these processes.

overlying Carmel Formation in southeast Utah contains several such clues. Angular chert and carbonate pebbles that litter the J-2 surface are remnants of pond deposits from the upper part of the Navajo (Fig. 6.11). These are interpreted to record an extended period of wind-driven erosion that stripped sand of the upper Navajo from the region. Because carbonate and associated chert in the pond deposits were lithified into rock even as they were laid down, wind did not move this material. Instead, as tens to hundreds of feet of fine sand were blasted from the area, pond deposits broke into fragments and slowly settled into the degrading erosion surface to concentrate along a single horizon as a lag (Fig. 6.12). The very angular shape of all fragments, coupled with their sporadic but widespread distribution, precludes a waterlain origin. The change from deposition to erosion is attributed to a rise in sea level that inundated the coastal plain to the north and west, and cut off the voluminous sand source. Without a sediment load of fine sand, the powerful winds instead eroded, cutting deeply into the earlier erg deposits. Evidence for a rise in sea level as a trigger for unconformity development lies in the overlying marine sandstone, which suggests that the sea was nearby during the erosion.

The Source for Permian and Jurassic Eolian Sand

Upon viewing the numerous Permian and Jurassic eolian sandstones that spread across the Colorado Plateau, the inevitable question is, where did all that

Fig. 6.12. A, the sharp contact of the unconformity separates the large-scale, cross-stratified Navajo Sandstone below from the horizontally stratified Dewey Bridge Member of the Carmel Formation. B, the surface of the unconformity, which is speckled with a lag of angular chert fragments. Photos are from the contact in Arches National Park.

sand come from? Crossbedding in these rocks dip consistently to the southeast, indicating that the formative winds, and sand, came from the northwest (by modern coordinates). Eolian deposition was stimulated by periods of low sea level and the resulting exposure of wide swaths of shallow sea-floor sand to the northwest, where the shoreline lay. As the sea receded, the huge tracts of unprotected sand were scoured by winds, and the inland ergs grew. When sea level rose, the sea floor again was inundated, the sand source was cut off, and the ergs stagnated or disappeared. This, however, just redirects the question: What was the source for this continuous, high-volume supply of shallow marine sand?

Ron Blakey (1994), who has worked extensively on the eolian deposits of the Colorado Plateau, has suggested that much of the Jurassic sand ultimately came from far to the southeast. According to his model, this large-scale transfer began in the Late Triassic with the wide-ranging Chinle river system, which originated in the earlier Ouachita orogenic belt centered in southern Oklahoma and Texas. The main Chinle trunk river, which probably rivaled the modern Mississippi in size and length, ferried huge quantities of sediment to the northwest, dumping it along the shoreline in Nevada. The Ancestral Rockies of Colorado and New Mexico also contributed large amounts of sand to this system. Later, in the Early Jurassic, west-flowing Kayenta rivers continued to dispense sand to the western shoreline, only to be returned eastward by strong Navajo winds. These coastal sands later were mobilized in the Middle Jurassic to form the eolian Page and Entrada sandstones.

More recent, detailed work on this sand source problem has been conducted by William Dickinson and George Gehrels (2002) of the University of Arizona on the scale of individual sand grains. Their study has produced some surprising results and requires a new model for the shuffle of sand to the western North American shoreline. In their study, sparse, sand-sized grains of the mineral zircon were carefully separated from the dominant quartz sand in the Permian and Jurassic eolian sandstones that blanket the Colorado Plateau. Although it is not common, zircon is an ultraresistant mineral that occurs in trace amounts in most igneous rocks. It is much harder than quartz and is fiercely resistant to breakdown by both mechanical (e.g., abrasion) and chemical means (e.g., alteration or dissolution). For these reasons it is relatively common (< 1 percent) in most quartz-rich sands that have, by water or wind, bounced their way across great distances. The most significant aspect of zircons is that they contain small amounts of uranium locked into their crystal structure which, over long periods of time, decay to lead. The time at which these zircons crystallized in their original igneous host rock can be determined using a mass spectrometer to precisely measure the amounts of various uranium and lead isotopes (U-Pb method), all within a single, sand-sized grain of zircon. Armed with the knowledge of the rate at which uranium decays to lead, the ratio of uranium to lead in the grain can be used to calculate the age of the zircon, thus the age of the igneous source of the zircon. While it is certainly a more complex procedure than outlined here, that is the essential basis for the results of Dickinson and Gehrels, who recognized several discrete age populations for the zircons they analyzed, including some that could only have come from the Appalachian Mountains on the east coast of North America.

In their analyses of 468 zircon grains, Dickinson and Gehrels identified three different source terranes based on ages. Half of the zircons fell into a broad but geographically distinct age group ranging from 1315 to 310 million years ago (Ma). There are no recognized bedrock terranes of this age range in western North

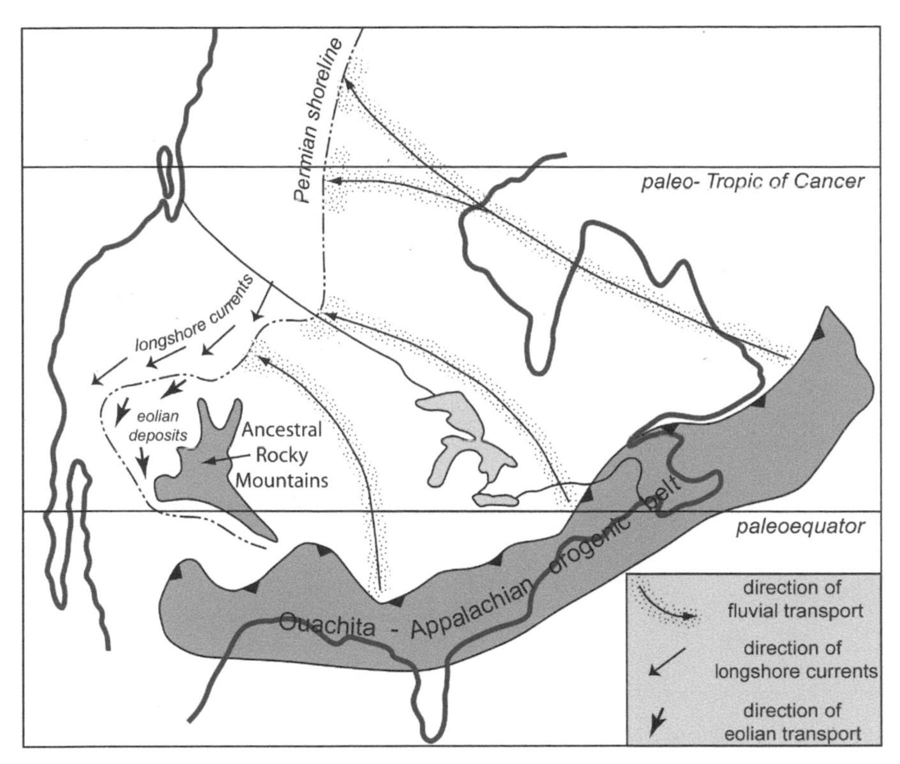

FIG. 6.13. Map showing sediment transport paths during the Permian as hypothesized by Dickinson and Gehrels (2003). The transport of sediment from the Ouachita-Appalachian orogenic belt is based on the ages of zircons found in Permian through Jurassic eolian sediments on the Colorado Plateau. Certain populations of dated zircons from those sediments could only have come from igneous rocks to the east. The extensive Ouachita-Appalachian orogenic belt as a source area for sediment to the west also explains the huge volume of sand available for eolian redistribution after being swept southwestward by longshore currents. Modified after Dickinson and Gehrels 2003.

America, but this age group fits very well with documented ages of igneous rocks in and around the Appalachian Mountains along the *east* coast (Fig. 6.13). As we have seen, these mountains formed during the Pennsylvanian Period, at approximately the same time as the Ancestral Rockies, so they could have supplied some of the sand. According to the model of Dickinson and Gehrels, abundant Appalachian-derived detritus was shed northwestward and swept into a large, transcontinental river system that traversed the central part of the continent and, after a long, abrasive journey, was dumped along a coastline to the northwest (Fig. 6.13). From there, longshore marine currents shuffled the sand southwestward, parallel to the shoreline. Eventually it was positioned west and northwest of the Colorado Plateau region and blown back onto the continent. Here its protracted journey was finished in the great Permian and Jurassic ergs.

A quarter of the remaining analyzed zircons yielded ages consistent with basement rocks in the Ancestral Rockies (1800 to 1335 Ma). The remaining quarter come from the oldest rocks on the North American continent (3015 to 1800 Ma).

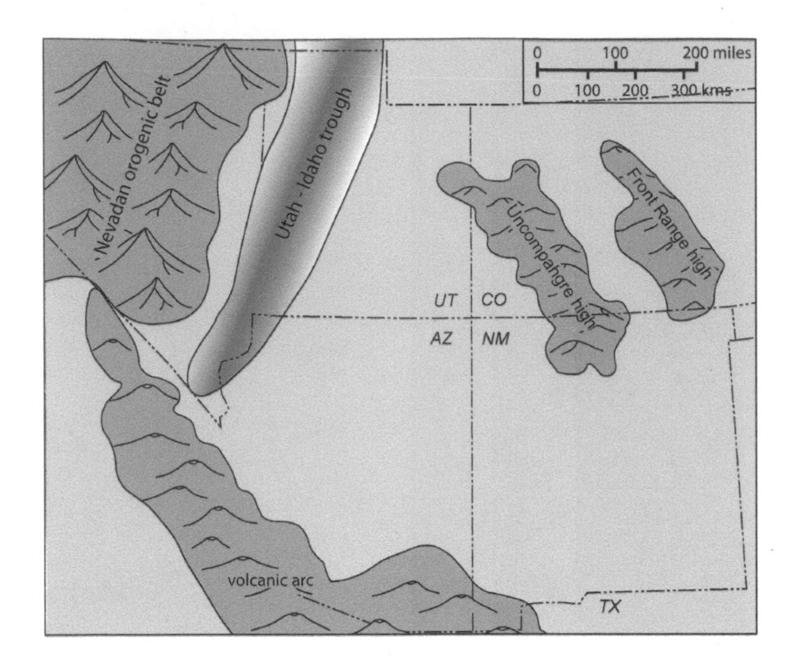

Fig. 6.14. Overview of the structural elements that affected sedimentation during the Middle Jurassic. The Sundance seaway lay to the north, in Wyoming, and periodically extended southward into the Utah-Idaho trough. During times of high sea level the Sundance may have extended eastward onto the west flank of the Uncompahgre highlands.

These rocks are concentrated in the central interior of the continent, and sediments from them were introduced into the transcontinental river system as it traversed the area on its northwestward path (Fig. 6.13). This supplemental sand may have come directly from the weathered basement rocks or may have been recycled sediment that came from earlier sandstones that ultimately sourced these rocks.

The San Rafael Group

Middle Jurassic strata of the Colorado Plateau were named the San Rafael Group by USGS geologists James Gilluly and John Reeside Jr. (1928) for superb exposures in the San Rafael Swell. At this type locality the group originally was defined as interbedded terrestrial and shallow marine deposits of the Carmel, Entrada, Curtis, and Summerville formations (see Fig. 6.1). This definition was modified by Peterson and Pipiringos (1979) upon their identification of the Page Sandstone at the base of the group in the San Rafael Swell. The components of the group vary slightly with geography across the southeast quarter of Utah. For instance, the Summerville Formation is absent east of the Moab area, where it turns into the equivalent Wanakah Formation in part of western Colorado. In southwest Colorado, near the Utah-Colorado border, a tongue of the Todilto Formation extends into the region from the southeast, separating the Entrada and Summerville formations. Although the original definition of the San Rafael Group was published in 1928, it remains a controversial work-in-progress, with several revisions that have yet to be accepted or discarded (e.g., Lucas and Anderson 1997). In fact, a review of the literature on these strata makes it appear as though every geologist that has ever worked on the group has had a different opinion on

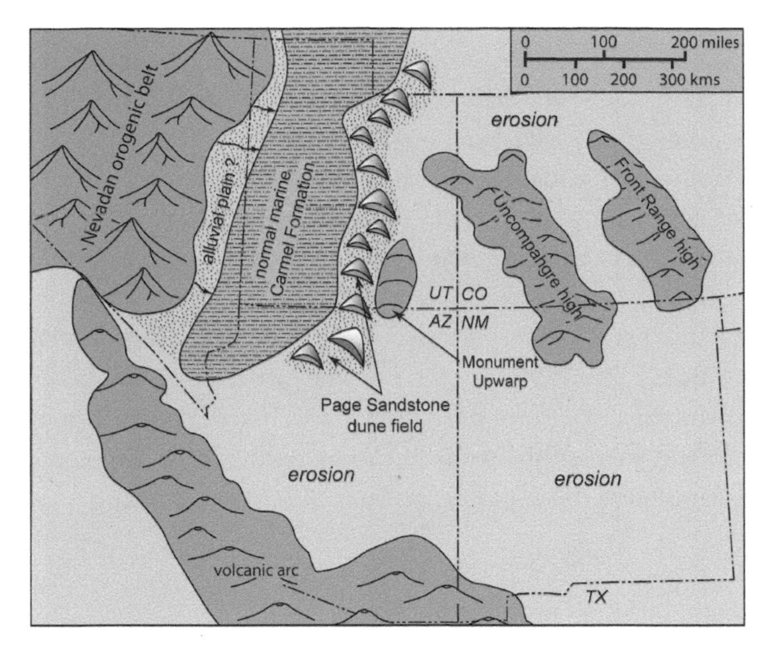

FIG. 6.15. Paleogeography of the Four Corners region showing the coeval environments for the shallow marine Carmel Formation and eolian Page Sandstone of the San Rafael Group. The alluvial plain between the Nevadan orogenic belt and the fingerlike seaway is hypothetical as all strata of this age were eroded from that area during the eastward advance of the mountain front. After Peterson 1994.

its organization! Only through time, continued fieldwork, and debate will these disagreements be settled.

Sedimentation in the San Rafael Group was governed by rapid subsidence in the Utah-Idaho trough, a linear, north-trending downwarp that dominated western Utah. As discussed earlier, subsidence was driven by mountain building immediately to the west, in the Elko orogenic belt. From the viewpoint of depositional setting, subsidence created an inlet for an incursion of the Sundance seaway from the north. This fingerlike seaway invaded from the Wyoming area and stretched southward to the Utah-Arizona border (Fig. 6.14). To the east, in southeast Utah and western Colorado, the San Rafael Group consists of alternating deposits of shallow marine, tidal flat, sabkha, and eolian origin. Sediments varied with the whims of the expanding and contracting seaway, one of several factors that have contributed to the complexity of these strata.

The Page Sandstone

In south-central Utah and the San Rafael Swell to the north, Page Sandstone makes up the base of the San Rafael Group. This eolian sandstone, which overlies the identical Navajo Sandstone, was defined by Fred Peterson and George Pipiringos (1979) for exposures surrounding Page, Arizona, near Glen Canyon Dam. The mesa that the town of Page is perched on is composed of this sandstone. It originally was an unnamed upper part of the Navajo, but the discovery between the units of the chert pebble horizon that marks the J-2 unconformity led Peterson and Pipiringos to separate the two and designate Page Sandstone as the base of the San Rafael Group. At its type locality along the Utah-Arizona border, it is 185 feet thick. It thins northward to ~100 feet in the San Rafael Swell, and east of there it quickly thins and pinches out, to be replaced by the Dewey Bridge Member of the Carmel Formation. The Page Sandstone is absent in the Moab area.

According to paleogeographic reconstructions by Fred Peterson (1994), the Page dune field was confined to a narrow, north-south belt that paralleled the eastern shoreline of the seaway (Fig. 6.15). This is clearly shown in the rock record by interfingering between eolian sandstone and marine deposits of the Carmel Formation to the west. Dune-sculpting winds blew to the south, culling sand from the exposed shoreline of the broader Sundance seaway in Wyoming and redistributing it southward. Eventually, the Page dune field was eliminated by a rise in sea level as the narrow seaway expanded from its confines. This had the dual effect of drowning the dune field and submerging its sand supply to the north. Drowning is indicated by marine deposits of the upper Carmel that overlie the Page throughout its extent.

The Dewey Bridge Member of the Carmel Formation

In southeast Utah, around the Moab area and east to the Colorado border, the deep red siltstone and sandstone of the Dewey Bridge Member of the Carmel Formation overlies the Navajo Sandstone and the J-2 unconformity. The Dewey Bridge correlates with the upper part of the Carmel to the west but has a long history of mistaken identity and improper correlations. In fact, the Dewey Bridge has been a longstanding problem in Colorado Plateau geology, although it is one that was man-made. The first geologist to document the geology of this region, C. H. Dane (1935), correctly placed these strata in the Carmel Formation. Later, McKnight (1940) mapped the area between the Colorado and Green rivers and recognized the transition between the gypsum, limestone, and siltstone that characterizes the Carmel to the west, and the red siltstone and sandstone to the east. Many years later, Wright and others (1962) named it the Dewey Bridge Member for exposures on the north side of the Colorado River near Dewey Bridge, between Moab and Cisco. The problem was that they placed it with the overlying Entrada Sandstone instead of the Carmel! This was based on similarities with the Entrada farther west (Capitol Reef area), where it also consists of interbedded red siltstone and sandstone. Thus, even though it was laterally continuous with Carmel strata, for 38 years the Dewey Bridge remained a member of the Entrada. Finally, in 2000, Utah Geological Survey geologist Hellmut Doelling published a paper that properly placed the Dewey Bridge back into the Carmel, where it belongs. This is its status today.

The Dewey Bridge, as defined since the recognition of the J-2 unconformity (Pipiringos and O'Sullivan 1978), consists of a basal 15 to 20 feet of yellow-white, horizontal-bedded sandstone. This is overlain by red-brown, irregularly bedded sandy siltstone ranging from 60 to 160 feet thick. These red-brown siltstones originally constituted the entire Dewey Bridge and are an obvious sight on the

FIG. 6.16. A view of the Dewey Bridge Member of the Carmel Formation in Arches National Park showing its characteristic large-scale convolutions, the underlying Navajo Sandstone, and the overlying Slick Rock Member of the Entrada Sandstone.

switchbacks of the entrance road into Arches National Park. These strata also are unmistakable for their pervasive large-scale convolutions in which the bedding has been thrown into irregular, undulatory folds 30 feet high and up to 300 feet wide (Fig. 6.16).

If not for the chert pebble horizon of the J-2 unconformity, the basal sandstone of the Dewey Bridge would be indistinguishable from the underlying Navajo Sandstone. The only difference is the horizontal stratification of the basal Dewey Bridge compared to the dominantly high angle crossbedding of the Navajo. Similarities in the sediments across the J-2 suggest the basal Dewey Bridge was derived from reworking of the upper Navajo. Others have suggested that this basal sandstone may be a finger of the Page Sandstone that extended eastward from the coastal dune field (e.g., Doelling 2000). In the Courthouse Rock area, ~12 miles north of Moab, the sandstone contains dinosaur tracks, suggesting deposition in a less-hostile environment than the harsh, windswept setting of the Navajo. The well-preserved nature of these tracks also indicates the sand was damp, pointing to a wetter setting as well.

The red-brown, sandy siltstone that makes up most of the upper Dewey Bridge is difficult to decipher. The lack of diagnostic sedimentary structures and the variable thickness of the beds hinder any simple interpretation. Abundant silt suggests a waterlain setting, and equivalent Carmel strata a short distance west, in the San Rafael Swell, contain fossiliferous limestone of certain shallow marine origin. The depositional environment of the Dewey Bridge has been variously interpreted as simply shallow marine (Molenaar 1981), a low-lying lagoonal setting (Craig and Shawe 1975), or a broad tidal flat (Doelling 2000). The lack of fossils, or even trace fossils, suggests an unfavorable environment for organisms. The absence of sedimentary structures renders a tidal flat setting unlikely as the back and forth currents would surely have generated recognizable ripples or crossbeds. The most likely scenario, although it is one based on the *lack* of diagnostic evidence, is a broad, low-relief sabkha on the margin of the shallow seaway. It is likely that the seaway rarely, if ever, flooded the sabkha. Instead, the sabkha remained moist by the lateral infiltration of seawater through the porous sediments. By this method, the sabkha surface remained wet, but currents that would have generated sedimentary structures rarely swept across the surface.

One telling aspect of the Dewey Bridge is an increase in sand content eastward; it also thins to the east, eventually pinching out in Colorado against the flanks of the Uncompahgre highlands. Together these features indicate that the highlands remained a subtle but still influential factor in sedimentation. As the shrinking highlands continued to send sand west, its flanks were slowly buried in their own detritus. Thus, these once mighty highlands continued their slow journey into oblivion.

A prominent event in the history of the Dewey Bridge that must be addressed is the origin of the large-scale, irregular folds that characterize it over a large part of southeast Utah. These undulatory beds are especially obvious in Arches and other areas north of Moab, where the member is well exposed. Although there is no consensus on their inception, several possibilities have been proposed. One hypothesis is that gypsum originally was interbedded with the siltstone and has since been dissolved by groundwater while in the subsurface. This would have thrown the insoluble siltstone into convolutions as it collapsed into the voids. An alternate scenario invokes dewatering of the Dewey Bridge as the Entrada sediments were laid on top. In this case, residual water in the saturated and unconsolidated Dewey Bridge would have been squeezed out by the crushing force of the Entrada sand load. The upward expulsion of these pore waters would have dragged the silt with it, producing the folds. A more recent interpretation relates the convoluted beds to a controversial meteor impact that may have formed Upheaval dome in northern Canyonlands (Alvarez and others 1998). This possibility is discussed in detail in an earlier section (see Chapter 2) but is recounted briefly here. The idea hinges on a mid-Jurassic meteor impact soon after Dewey Bridge deposition, so that it remained unconsolidated. An impact at Upheaval dome would have generated a shock wave equivalent to a magnitude 8.0 earthquake (Alvarez and others 1998). As these shock waves rippled radially away from the impact site, unconsolidated sediments were deformed by a combination of the passage of waves through the sediments and the resulting expulsion of water.

Regional dewatering and deformation of unlithified sediments is a common by-product of large earthquakes. An interesting aspect of this hypothesis is the concentration of deformed Dewey Bridge strata in the vicinity of Upheaval dome, with the degree of deformation decreasing with distance. As yet, however, there is no independent evidence for the timing of an impact, if that is indeed how Upheaval dome was formed.

During Dewey Bridge deposition the paleogeography of the Colorado Plateau and surrounding areas consisted of several elements that remained important through the Middle Jurassic. From west to east these included the Nevadan orogenic belt, a mountainous region that occupied much of eastern Nevada (see Fig. 6.14). Bordering this to the east was a shallow seaway that submerged the western half of Utah. This elongate seaway was a south-extending finger of a broader sea that covered Wyoming to the north. The invasion of sea into Utah was induced by subsidence in front of the mountains. Marine limestone, gypsum, and mudstone of the Carmel Formation mark the limits of this seaway. Farther east, along the Green River, marine deposits interfinger with Dewey Bridge siltstones. These sabkha deposits stretch unbroken across eastern Utah and into western Colorado, where they pinch out against the mounds that were vestiges of the once towering Uncompahgre highlands.

The Dewey Bridge records a high sea level, an event that began with the drowning of the Page eolian system. The eastward overlap of marine Carmel strata across the Page marks the widening of the formerly narrow seaway as sea level rose. This also enhanced the lateral infiltration of seawater to form the broad, featureless Dewey Bridge sabkha. The ensuing sea level drop, however, shifted environments once again, resurrecting for a final time an extensive eolian system.

The Slick Rock Member of the Entrada Sandstone

In the Canyonlands and Arches region of southeast Utah the Entrada Sandstone, which conformably overlies the Carmel, is made up solely of the Slick Rock Member. The definition of the Slick Rock Member goes back to 1962, when Wright and others grouped the underlying Dewey Bridge, the Slick Rock, and the overlying Moab Member into the Entrada Sandstone; however, with the Dewey Bridge now properly aligned with the Carmel, and the Moab Member correctly placed with the Curtis Formation (Doelling 2000), the Slick Rock becomes the lone and largely unnecessary subdivision of the Entrada.

The Entrada Sandstone in eastern Utah and over its widespread occurrence across Colorado is of eolian origin. The red-brown to orange sandstone weathers into vertical cliffs and steep, rounded bands of slickrock. Internally the Entrada consists of uniform, fine-grained sand and on a larger scale displays graceful, sweeping arcs of crossbedding. These are interrupted regularly by sparse horizontal stratification. In Arches National Park, where the Entrada is easily the

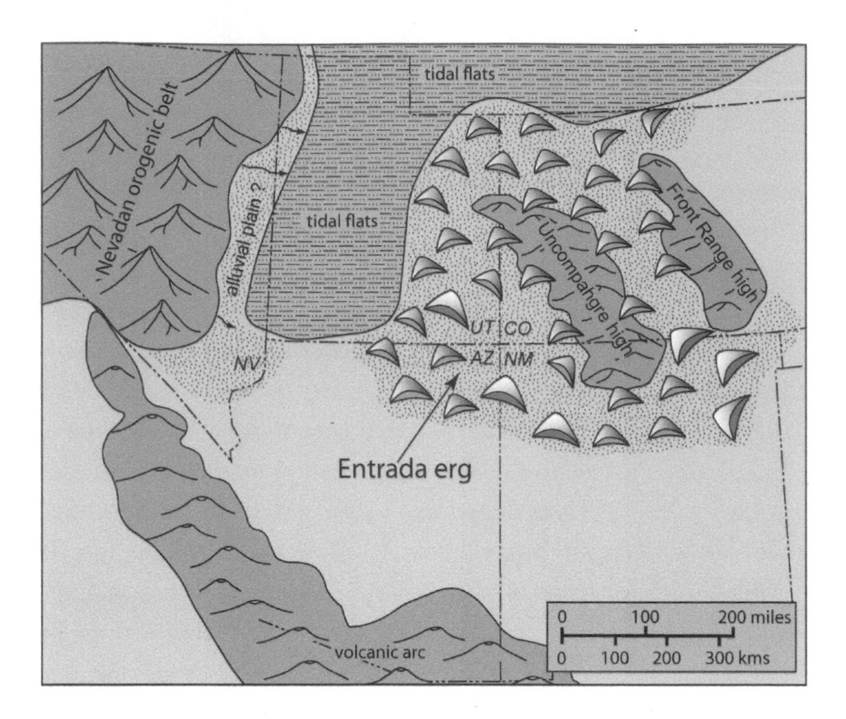

FIG. 6.17. Paleogeography of the Four Corners region showing the coeval environments for the tidal flat and eolian facies of the Entrada Sandstone of the San Rafael Group. The alluvial plain between the Nevadan orogenic belt and the seaway is hypothetical as all strata of this age were eroded from that area during the eastward advance of the mountain front. After Kocurek and Dott 1983.

most prominent rock unit, it reaches 350 feet thick. Within the red-brown, joint-bounded fins of the Slick Rock Member is one of the greatest concentrations of arches in the world. The Entrada forms the bulbous cap of Balanced Rock, as well as the arches of the Window area and the spectacular fins and arches of the Fiery Furnace and the Klondike Bluffs. In short, the Entrada Sandstone makes Arches the uniquely bizarre landscape that it is today.

The Entrada extends over a huge area. The eolian facies is recognized even as far as the southeast corner of Colorado and blankets most of the remnant Ancestral Rockies. It also covers the northern half of New Mexico and a large part of northern Arizona. In Utah the eolian Slick Rock Member is limited to the eastern quarter of the state, stretching northward to the Wyoming border, where it abutted the wide shoreline of the Sundance seaway (Fig. 6.17). In the remainder of Utah, west of the Green River, the Slick Rock Member shifts to interbedded red siltstone and sandstone, reminiscent of the underlying Dewey Bridge to the east. This western Entrada marks the depression of the seaway with sabkha and tidal flat deposits. This zone of active downwarping continued westward to the Utah-Nevada border, where it stopped against the rising Elko highlands.

The colossal Entrada erg stretched farther east than any previous eolian deposit in the region. Crossbedding in the Slick Rock Member and its equivalent strata in Colorado and New Mexico indicate a consistent, south-directed wind, thus a sand source to the north. The unparalleled areal extent of the erg is best explained by the expansion of the Sundance seaway across Wyoming, immediately to the north. Prior to Entrada deposition, the seaway was confined to the western half of Wyoming. During Entrada deposition, however, the seaway expanded as far east as modern-day Nebraska, creating a wide, sandy shoreline. Even small drops in the level of this broad, shallow sea resulted in the exposure of vast areas of sandy sea floor to turbulent blasts of wind, whirling the sand southward. The width of the shoreline accounts for the exceptional extent of the eolian deposits. Because they came from the north, these dunes did not have to overcome the topographic barrier of the Uncompahgre highlands in order to spread eastward. Instead, sand sifted across the Precambrian cores of the Ancestral Rockies, places that had not seen deposition for more than 100 million years.

The Moab Member of the Curtis Formation

Like the Dewey Bridge Member, the Moab Member for many years was considered to be a part of the Entrada Sandstone (e.g., Craig and Shawe 1975). This placement was based on similarities in lithology with the underlying Slick Rock Member. The problem with this designation is that the Moab Member—or Moab Tongue, as it was sometimes called—could be traced laterally west into the Curtis Formation and to the east into the lower Summerville Formation. This problem finally was remedied in 2000, when Hellmut Doelling correctly redefined the Moab Member as a part of the Curtis Formation.

The Moab Member is a resistant white sandstone that forms a caprock cliff above the Entrada. Although the rock types are similar, the white gray Moab Member contrasts with the orange red of the underlying Slick Rock. The Moab Member attains a maximum thickness of more than 100 feet just southeast of Moab and thins in all directions from there (Wright and others 1962). Where thickest it is characterized by large-scale eolian stratification. Laterally, as it thins, it grades into similar sandstone, but with a shift to horizontal stratification and low-angle crossbedding. For instance, in Colorado National Monument near Grand Junction, at the eastern edge of its extent, the Moab consists of 50 feet of horizontal-bedded white sandstone. South of there, in the Paradox Valley area, it grades into the Summerville Formation; to the west, near the San Rafael Swell, it grades into the Curtis Formation. The Curtis is separated from the underlying Entrada by the J-3 unconformity, which is marked by a widespread chert pebble conglomerate layer. The unconformity has yet to be recognized to the east, between the Slick Rock and Moab sandstones.

Sandstone of the Moab Member represents a small but sprawling dune field that evolved along the southeast margin of a shallow seaway. Large-scale crossbedded sandstone in the Moab area marks the center of the dune field and grades radially into the flat-bedded sandstone of the sand flat that skirted the dunes. Locally, along the north and west margins of the dune field, the eolian sand may have been reworked by tidal currents of the bounding seaway.

Regionally, mild uplift in the Elko orogenic belt to the west rippled eastward into what formerly had been the Utah-Idaho trough. This, in turn, caused the narrow seaway to be pushed east into central Utah. Fluvial sand and gravel from these uplands were distributed eastward across a narrow plain and deposited along the west margin of the seaway. Shale and fine sandstone accumulated within the seaway, which bounded the dune field along its north and west margins. These varied strata make up the bulk of the Curtis Formation.

The Summerville Formation

Named for Summerville Point in the northern San Rafael Swell by Gilluly and Reeside (1928), the Summerville Formation comprises the top of the Middle Jurassic San Rafael Group and is bounded at its top by the J-5 unconformity.

In the San Rafael Swell the Summerville overlies the marine facies of the Curtis Formation and reaches its maximum thickness of ~300 feet. In northern Utah, equivalent strata are identical to the Curtis, so the Summerville is not recognized, and the combined succession is lumped into the Curtis. East of the San Rafael Swell, in the Green River and Moab area, the Summerville intertongues with, and overlies, the eolian Moab Member. Finally, in western Colorado, where the Curtis was not deposited, the Summerville overlies the Slick Rock Member of the Entrada.

The Summerville is easily recognized by its thin beds of red and brown alternations of mudstone and sandstone. Interbeds of white gypsum are also common and contribute to its pin-striped appearance. Sandy beds exhibit oscillation ripples, and clay-rich units contain sporadic mudcracks. Collectively, the rock types and features found in the Summerville point to deposition on a tidal-influenced, low-relief coastal plain (Caputo and Pryor 1991). Symmetrical ripples were molded by tidal currents, whereas mudcracks formed during low tide as the recently settled mud baked under the searing Jurassic sun. Similarly, gypsum precipitated in low-lying evaporative pools that were remnants of the receding tide waters.

The Summerville records the final stages of the narrow Middle Jurassic seaway that stretched southward through central Utah. As sea level dropped, the seaway gradually contracted northward into northern Utah and southern Wyoming. Evaporative ponds of a broad sabkha served as vague reminders of the former shallow embayment. Eventually the regression of the sea, coupled with renewed uplift to the west, dictated that erosion take the place of deposition. This regional erosion produced the J-5 unconformity, the last of the widespread Jurassic unconformities in western North America. Thus, over its extent on the Colorado Plateau, the San Rafael Group is separated from the overlying Late Jurassic Morrison Formation by the J-5.

The formations that make up the San Rafael Group hold in common the influence of the seaway which, for most of the Middle Jurassic, reached no farther south than the modern Utah-Arizona border. The alternating expansion and contraction of the seaway, and its eastward shift, which was driven by tectonism, exerted a continuous control on sedimentation. The shallow seaway hosted a spectrum of deposits, including carbonates, clastics, and evaporates. Periodic regressions of the seaway provided a voluminous source for landlocked eolian deposits in the form of exposed shoreline sediments. The tie was broken in the Late Jurassic as the seaway receded northward and tectonism was reactivated with a renewed vigor. The dominantly fluvial Upper Jurassic Morrison Formation records this change as most of the Colorado Plateau became a vast fluvial plain.

The Morrison Formation

The Upper Jurassic Morrison Formation is separated from the underlying San Rafael Group by the regional J-5 unconformity (see Fig. 6.1). In eastern Utah and western Colorado the Morrison is dominated by three distinctive members that, in ascending order, include the Tidwell, Salt Wash, and Brushy Basin members. Other important but less extensive members in southeast Utah include the Bluff/ Junction Creek Sandstone in the Four Corners area and the Recapture Member in the Blanding–Abajo Mountains area.

The Morrison is the youngest unit in the thick blanket of Jurassic rocks on the Colorado Plateau and is very well exposed throughout the eastern half of the province. It has been eroded in the western part of the Plateau region due to post-Jurassic uplift. The colorful fluvial and lacustrine deposits mark the end of the seaway that invaded the region during the Middle Jurassic, and are world-renowned for their abundant dinosaur fossils. In fact, Dinosaur National Monument, along the northern Utah-Colorado border, is named for the great concentration of dinosaur bones recovered from the Morrison.

The Morrison attracted much attention after World War II for its abundant uranium deposits. Innumerable rough, rocky dirt tracks on the Colorado Plateau dead-end at pits excavated in the Morrison, remnants of Cold War prospectors on the hunt for the elusive uranium motherlode. According to the U.S. Department of Energy, half of the uranium reserves in the United States are contained in the Morrison Formation. It is for this reason that the formation was so intensely studied in the 1950s and 60s in a collaborative effort by the USGS and the U.S. Department of Energy.

The J-5 Unconformity

The J-5 unconformity of Pipiringos and O'Sullivan (1978), which underlies the Morrison over its entire extent, was generated by a combination of erosion and subtle tectonic activity. Erosion likely was initiated by the final northward retreat of the Sundance seaway. Mild uplift within the stable interior of the Colorado Plateau at this time is well illustrated by local stratigraphic relations (Peterson 1986). For instance, across the Emery uplift, an earlier incarnation of the San Rafael

Swell, the lower Morrison (Tidwell and Salt Wash members) locally thins and pinches out. Here the uppermost Brushy Basin Member overlies the Summerville. Thus, during the early part of the Late Jurassic, the structure was high enough to hinder deposition, but not so high as to promote erosion of the Summerville. Fred Peterson (1986) documented similar relations in northern Arizona along the modern Black Mesa uplift. In the Monument upwarp region (modern Cedar Mesa–lower San Juan River canyon) the upwarping appears to have occurred a bit later, as the Brushy Basin Member thins over its top.

Three different members of the Morrison Formation make up the base and overlie the J-5 unconformity across southeast Utah and southwest Colorado. The eolian Bluff Sandstone makes up the basal Morrison to the south, along the San Juan River around its namesake town, Bluff, Utah. This eolian unit extends eastward into southwest Colorado, where it is called the Junction Creek Sandstone. Throughout most of southern Utah, the Tidwell Member blankets the unconformity. Locally, where the Tidwell and Bluff/Junction Creek are absent, the Salt Wash Member rests on the J-5 erosion surface. In the following discussion the Bluff/Junction Creek will be addressed first, followed by the Tidwell.

The Bluff Sandstone Member and Junction Creek Sandstone

The Bluff Sandstone Member has been placed in and out of the Morrison Formation throughout the history of regional geologic studies. In one of the classic early works on the region, *The San Juan Country* (1938), geologist Herbert E. Gregory originally placed the Bluff in the Morrison Formation. A later study, however, put it in the underlying San Rafael Group (Harshbarger and others 1957). Still later, O'Sullivan (1980) again placed it with the Morrison. This was based on detailed work that established intertonguing relations between the Bluff and what are now known as the Tidwell and Salt Wash members. In addition, O'Sullivan documented the presence of the J-5 unconformity beneath the Bluff. Since it was recognized, the J-5 has been used to define the top of the San Rafael Group. More recently, however, geologists Spencer Lucas and Orin Anderson (1997) have argued that the Bluff should again be included in the San Rafael Group and, in fact, lay *below* the J-5 unconformity. This newly proposed alignment is similar to one that earlier was found to be erroneous and was discarded (O'Sullivan 1980), so it is unlikely to be embraced by the geological community. Considering all the available evidence, the Bluff Sandstone is here considered to be a member of the Morrison Formation.

As if the correlation problem of the Bluff was not enough, it becomes the Junction Creek Sandstone in southwest Colorado (Goldman and Spencer 1941). The unit was named for exposures of the white eolian sandstone along Junction Creek, which flows through the town of Durango. This naming occurred before its continuity with the Bluff to the west had been demonstrated by subsurface data.

It is another human-induced stratigraphic problem that has yet to be cleaned up. In the meantime, it is side-stepped by referring to the entire sandstone body as the Bluff/Junction Creek. The Junction Creek extends northward from its type locality to the town of Gunnison, Colorado, where it sits on Precambrian igneous and metamorphic rocks. Here it was the first unit to be deposited on the ancient and long-lived erosion surface on the eastern flank of the Ancestral Uncompahgre highlands.

The Bluff Sandstone is continuously well exposed in the immediate area surrounding the town of Bluff. Here the member reaches 340 feet thick and forms a prominent orange red cliff band cut by sparse vertical fractures. The top weathers to rounded domes. From Bluff the sandstone thins irregularly in all directions. To the west, toward Comb Ridge, it pinches out over a short distance. Eastward, in Durango, the equivalent Junction Creek Sandstone is 265 feet thick. It thins to the north, and near Gunnison it is only ~60 feet thick. A few miles north of Gunnison it pinches out completely.

The fine-grained sandstone dominated by large-scale crossbedding indicates an eolian origin for the bulk of the Bluff/Junction Creek. Lesser horizontal stratification represents interdune environments or localized sabkhas. The lower half of the Junction Creek around Gunnison, however, is horizontally bedded, with some bedding surfaces exhibiting high-relief, symmetrical ripples of obvious aqueous origin. It is possible that these northernmost deposits record the margin of the receding seaway during this time. Alternatively, they may represent a lake margin in which the clean eolian sand was reworked into ripples as small waves lapped against the shoreline. The upper half of this succession consists of eolian strata that are typical of the unit elsewhere.

Despite minor, localized variations in depositional setting, the Bluff/Junction Creek strata represent a widespread eolian dune field (Fig. 6.18). Eolian sand extended eastward from Comb Ridge well into southwest Colorado. From there it stretched northward across the trace of the earlier Uncompahgre highlands before wedging out along their eastern margin. Dip directions from crossbeds indicate that the wind was directed east to northeast, pointing to a sediment source to the west. The Bluff disappears to the west, across the Monument upwarp and its modern-day expression of Comb Ridge, suggesting that this feature had a subtle but influential presence during Bluff deposition. According to USGS geologist Fred Peterson, who has worked on the Morrison over several decades, the eolian sand likely was winnowed from the vast alluvial plain to the west and blown across the upwarp to pile up on the downwind side; however, it is not believed that the Monument upwarp at this time possessed enough relief to act as a source for sediment.

Interfingering relations between the Bluff/Junction Creek eolian system and the Tidwell and Salt Wash strata indicate these depositional environments existed

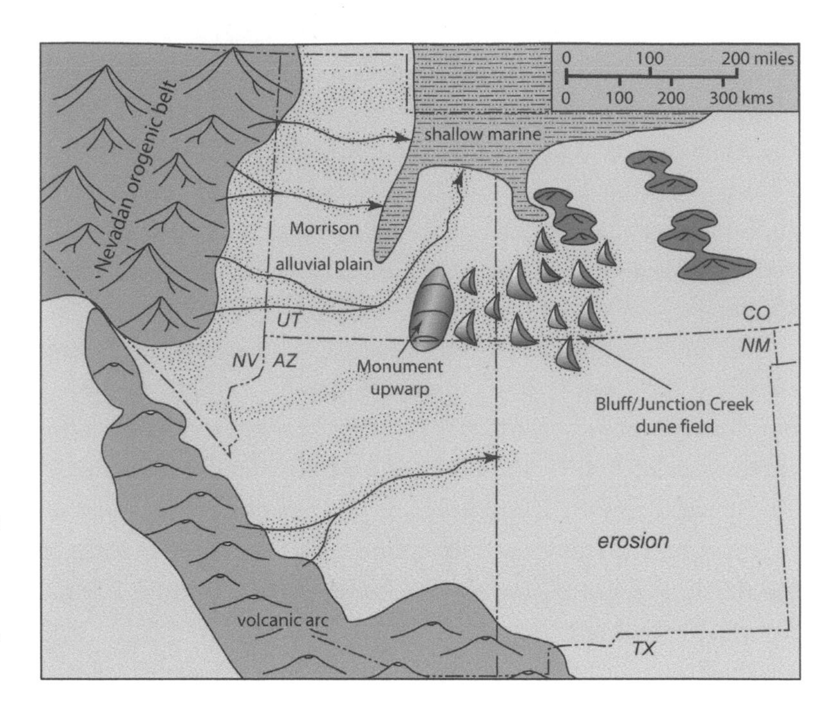

Fig. 6.18. Late Jurassic paleogeography of the Four Corners region showing the environments for the eolian Bluff and Junction Creek sandstones in the Four Corners area, the shallow Tidwell sea, and fluvial sediments coming off the Nevadan orogenic belt. After Turner and Peterson 2004.

side by side. Initially, the eolian system coexisted with the diverse Tidwell environments; it continued to be active into the following early phase of Salt Wash sedimentation. The Tidwell Member, probably the most enigmatic and widespread to occupy the basal Morrison Formation, is discussed next.

The Tidwell Member

The basal Tidwell Member of the Morrison overlies the J-5 unconformity and interfingers with the Bluff and the overlying Salt Wash Member across the eastern half of the Colorado Plateau. The green and red mudstone that dominates the Tidwell is easily differentiated from the sandstone-dominated Bluff and Salt Wash members. It is, however, very similar to the underlying Summerville and, in fact, originally was part of that formation. Later, Gilluly and Reeside (1928) recognized an unconformity separating the red Summerville mudstone from the overlying green and red mudstone sequence and lumped the sequence with the Salt Wash. Subsequent workers in the region placed this lower contact in different parts of the mudstone-dominated succession, resulting in years of misleading correlations and unnecessary confusion. Finally, Fred Peterson (1988) formally designated the green and red mudstone beds above the J-5 unconformity as the Tidwell Member of the Morrison. Over much of its extent the lower contact with the J-5 unconformity is marked by coarse sand grains and pebbles in an otherwise fine-grained sandstone.

Although it is dominated by colorful, thinly bedded mudstone, the Tidwell regionally hosts an assortment of rock types. In the San Rafael Swell, near its type locality, the Tidwell is ~100 feet thick and consists mostly of gray green mudstone. The base is marked by a prominent, thick white gypsum bed. In the Henry basin to the south, where red mudstone dominates, white gypsum also defines the base. To the east, in the Moab region and western Colorado, the base is composed of fine sandstone with coarse sand grains and pebbles interspersed along its base. The remainder is mostly red and gray mudstone. Across the region the Tidwell contains scattered lenses of dark gray, organic-rich limestone.

The variety of rock types in the Tidwell clearly portrays the diverse environments coexisting at any given time. Basal gypsum in the Henry Mountains and San Rafael areas represent low-lying evaporative pans filled with ponded seawater. The water was stranded as the earlier, fingerlike seaway that endured through the Middle Jurassic receded northward into present-day Wyoming. Evidence for derivation from seawater is found in carbon isotopes from the gypsum with a fingerprint of Jurassic seawater (Peterson and Turner-Peterson 1987). Small lenses of pebbly sandstone scattered throughout the Tidwell suggest the presence of small streams that bisected a broad, muddy floodplain. Interbedded limestone lenses and thin sheets of colorful mudstone depict a floodplain pocked by ponds in which limestone formed. Mudstone accumulated as streams periodically spilled from the constraints of their limited channels. Black fragments of carbonized plant material in mudstone indicate a vegetated floodplain. Widely dispersed bodies of fine sandstone contain evidence for an eolian origin, whereas others are interpreted as remnant sandbars in moderate-sized lakes (Peterson and Turner-Peterson 1987). Uppermost Tidwell mudstones interfinger with coarser sandstone and conglomerate of the overlying Salt Wash Member, suggesting that they are the laterally equivalent floodplain deposits of the high-energy Salt Wash river system that succeeded the Tidwell.

The Salt Wash Member

The coarse-grained Salt Wash Member overlies the Tidwell and Bluff Sandstone over most of its extent. In rare instances where these members are absent, it rests on the J-5 unconformity. Sandstone and conglomerate that dominate the Salt Wash weather into brown cliffs slashed by sparse recesses that form in less-resistant, thinner mudstone beds. The Salt Wash is present across the Colorado Plateau and ranges up to 400 feet thick, but mostly is much thinner. It was deposited by rivers whose detritus tells of renewed uplift to the west and southwest. The tectonic unrest that began at this time has continued across western North America to the present day, a span of more than 150 million years.

The Salt Wash to the west, in south-central Utah, is dominated by pebble conglomerate, and mudstone intervals are rare and thin. Moving eastward, however, to the Arches/Canyonlands area and western Colorado, sediment size diminishes, and sandstone replaces conglomerate. Additionally, the thick, continuous cliffs that characterize the Salt Wash in Capitol Reef convert to steep mudstone slopes

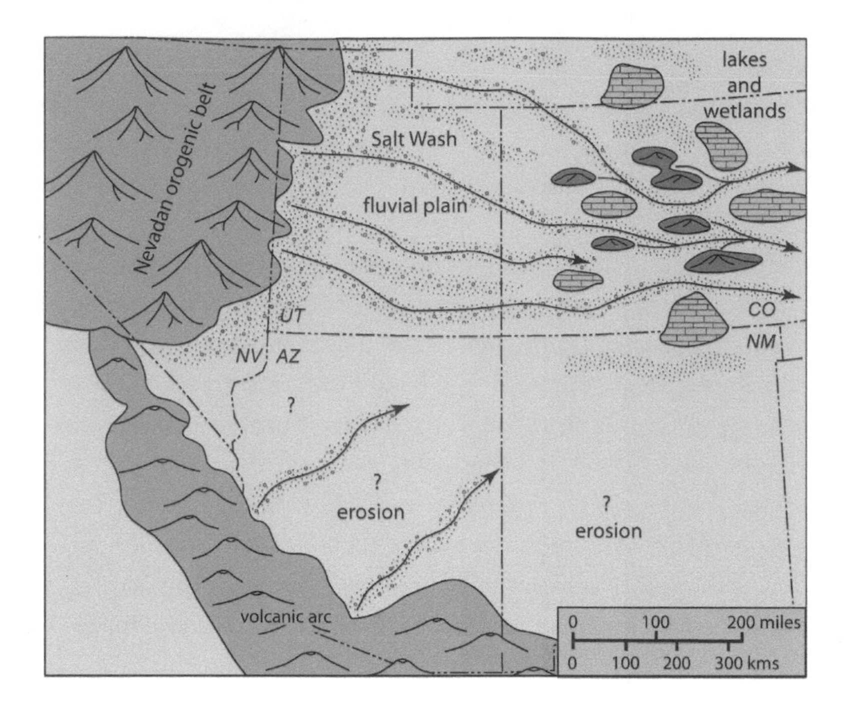

Fig. 6.19. Late Jurassic paleogeography of southwestern North America showing the depositional setting for the Salt Wash Member of the Morrison Formation. Note the low, remnant hills in the vicinity of the Ancestral Rocky Mountains, and the ponds and wetlands in which lacustrine limestone was deposited. After Turner and Peterson 2004.

supported by scattered sandstone lenses. In fact, the Salt Wash to the east is predominantly red, green, and purple mudstone (Doelling 2000). In this area thin sandstone ribbons trace the paths of Late Jurassic river channels, whereas varicolored mudstone marks the more voluminous floodplain deposits.

Crossbeds in sandstone and conglomerate throughout the Salt Wash consistently slope to the east, indicating a large-scale west-to-east drainage for the first time since the initiation of the now-buried Ancestral Rockies. After a lengthy eastward traverse across the Plateau region, the tired rivers arced northeast to feed the shoreline of the shrinking seaway (Fig. 6.19).

The rejuvenation of the western highlands and the onset of Salt Wash deposition drove the wholesale reversal of the regional drainage network. Uplift generated a renewed influx of sediment that exceeded subsidence in the earlier Utah-Idaho trough and filled it. This led to the development of a vast fluvial plain across which the rivers flowed, unobstructed by the now-buried Ancestral Rockies. As they sliced across this plain, the gradients of the rivers were reduced with increasing distance from their headwaters. This decrease corresponds to a decline in the rivers' energy and their ability to drag sediment downstream. The eastward shift from conglomerate to sandstone, as well as the concurrent increase in mudstone content, clearly illustrates the relationship between stream gradient and energy, and sediment size.

The rise of the source highlands in western Utah and Arizona marks the initiation of the Sevier orogeny, an important mountain-building episode that pulsed through the West, lasting until the end of the following Cretaceous Period, when it was succeeded by yet another orogenic event. It has also been proposed that the Mogollon highlands in southern Arizona were an important contributor of sediment (Fig. 6.19). Mudstone beds in the Salt Wash and the overlying Brushy Basin Member contain abundant altered volcanic ash in the form of the clay mineral bentonite. This particular clay mineral is easy to identify because it swells when wet and upon drying produces a puffy, popcorn-like veneer that crunches underfoot. During Morrison deposition airborne volcanic ash blew east on westerly winds that swept across the Sierra Nevada volcanic arc. This extensive chain of volcanic mountains sprawled across eastern California and arced (hence the name) southeastward through southern Arizona to form part of the Mogollon highlands. Today the bowels of this monstrous volcanic arc are spectacularly exposed as the serrated granite peaks of the Sierra Nevada.

The Brushy Basin Member

The Brushy Basin Member conformably overlies the Salt Wash across most of the Colorado Plateau and makes up the top of the Morrison Formation. The exception to this simple relationship occurs to the southeast, in the San Juan basin of northwest New Mexico. Here the stratigraphy of the Morrison changes with the addition of the Recapture (lower) and Westwater Canyon (upper) members between, and partially equivalent to, the Salt Wash and the Brushy Basin. The Recapture Member is partially equivalent to the uppermost Salt Wash and the lower part of the Brushy Basin Member. The overlying Westwater Canyon Member correlates with the middle part of the Brushy Basin.

The Brushy Basin Member ranges up to 540 feet thick and is dominated by mudstone that weathers into colorful, rounded slopes. These multihued mounds emanate various shades and combinations of green, purple, white, and red. The gracefully curving slopes are interrupted by sparse ledges, mostly lenses of sandstone and thin, discontinuous beds of gray limestone.

Brushy Basin mudstones contain numerous beds of tuff, rock formed from volcanic ash blown into the basin during explosive eruptions in the adjacent Sierra Nevada arc. Moreover, Brushy Basin sediments in an area encompassing the northwest part of the San Juan basin in New Mexico, and much of the earlier Paradox basin in western Colorado and southeast Utah, contain a unique assemblage of alkaline minerals. According to USGS geologists Christine Turner and Neil Fishman (1991), the regional bulls-eye map pattern of smectite (a clay mineral), clinoptilolite, a mixture of analcime and potassium feldspar, and in the center, sodium feldspar, marks the location of a colossal, shallow alkaline lake (Fig. 6.20). Turner and Fishman named this Late Jurassic body of chemical-laden water Lake T'oo'dichi', a Navajo term meaning "bitter water." This uncommon mineral assemblage is intimately associated with the ubiquitous volcanic tuff beds. The alkaline elements and silicon dioxide that combined to form this assemblage came from the unstable volcanic glass that makes up the ash as the lake and groundwater

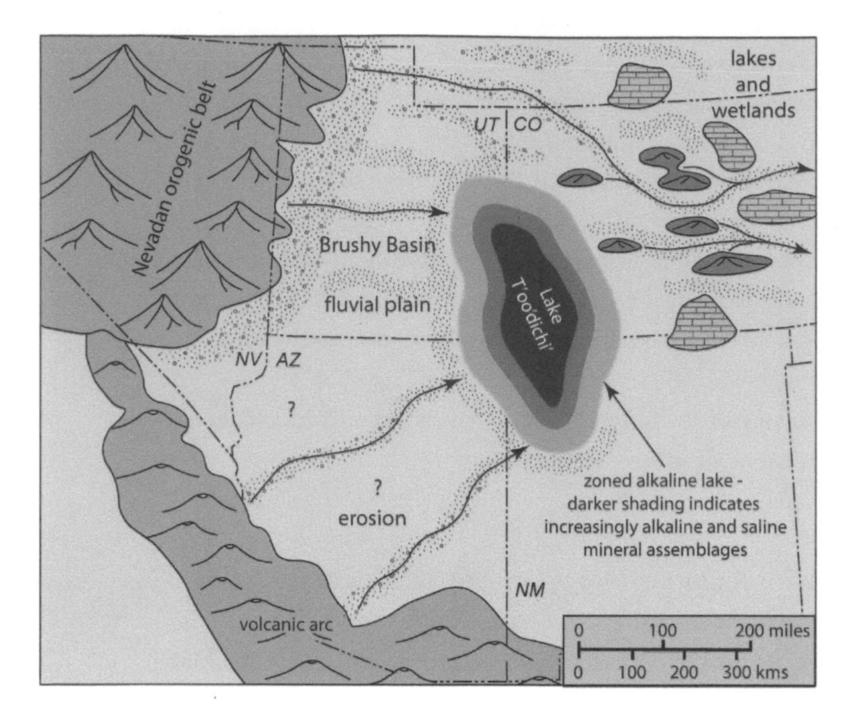

FIG. 6.20. Late Jurassic paleogeography of southwestern North America showing the depositional setting for the Brushy Basin Member of the Morrison Formation. The basin occupied by the ephemeral Lake T'oo'dichi' was bounded by the subtle Monument uplift along its west margin and by the last remnants of the ancestral Uncompahgre uplift to the northeast. After Turner and Fishman 1991 and Turner and Peterson 2004.

leached them from the tuff. Some of the minerals formed from the alteration of the volcanic sediments, whereas others precipitated in the pore spaces of the sediments directly from the alkaline waters.

The various mineral types in Lake T'oo'dichi' sediments, as well as features within these sediments, suggest an arid climate. This interpretation clashes with earlier interpretations of a humid, subtropical climate. These earlier calls for a wet setting were based on the discovery of abundant large herbivore dinosaur fossils in the Morrison. Lake T'oo'dichi', however, is interpreted as a giant playa lake, one that periodically dried up. Such a complete, intermittent dessication requires an arid climate with only brief wet intervals. When full, the lake stretched east to west for ~200 miles, and north-south for ~300 miles, its shores reaching from Grand Junction, Colorado, to Albuquerque, New Mexico. Mudcracks, root traces of plants, and river channel deposits in lake center sediments attest to extended dry periods. Regardless of the evaporation of surface water, Turner and Fishman (1991) propose that a chemical soup of groundwater remained in the shallow subsurface to continually react with the tuffaceous sediments.

The shallow lake basin was surrounded by mud flats, and the entire region must have been a vast lowland. The playa lake was bounded to the west by a fluvial plain slashed by the shimmering ribbons of energetic streams. Smaller streams may have flowed only sporadically, during the short-lived wet seasons; however, there likely were larger, permanent rivers that issued from the rising Sevier mountains to traverse the arid plain. Grain by grain, these rivers methodically ferried the Sevier highlands seaward. It was along these rivers, with their abundant, jungle-like riparian zones, that the dinosaurs congregated to feed. The herbivorous dinosaurs were attracted to the verdant fringes of the river, while the predatory carnivores were drawn by the herbivores on which they preyed.

Although they are absent from the Canyonlands/Arches area, the Recapture and Westwater Canyon members stretch northward into southernmost Utah and Colorado. Moreover, they are laterally equivalent to the Brushy Basin and figure prominently in the paleogeography during its deposition. Because of this, they will be briefly discussed here. The lower of the two, the Recapture Member, interfingers with both the upper Salt Wash and the lower part of the Brushy Basin. It is a slope-forming, heterogeneous assemblage of mudstone, sandstone, and rare limestone beds. The Recapture records a northeast-flowing river system, although some sandstone bodies represent moderately large and occasionally continuous dune fields, providing further evidence of aridity. According to Peterson and Turner-Peterson (1987) the best way to differentiate the Recapture from similar sandstones in the adjacent Salt Wash and Brushy Basin members is the presence of abundant feldspar in the form of sand grains and pebbles. Although it requires close inspection and the ability to recognize the mineral feldspar (it is commonly pink), it reveals a different drainage system from the rest of the Morrison. These sediments come from a unique source highlands in which granite was exposed. In contrast, most of the Morrison sediments (sand and pebbles) are derived from a sedimentary source terrane. These granitic highlands have been traced back to probable Jurassic uplifts in south-central Arizona and southeast California and are responsible for the discrete Recapture drainage system.

South of the Four Corners region the Westwater Canyon Member succeeds the Recapture. The member is dominated by crossbedded sandstone and conglomerate that form red, ledgy cliffs. These high-energy fluvial deposits were laid down in the shallow, braided channels of northeast-flowing rivers. Like the underlying Recapture, the Westwater Canyon Member contains pebbles of feldspar and granite, but quartzite, chert, and quartz pebbles are also common. Besides similarities in geographic extent, it is likely that these two members shared a source terrane to the southwest as well. The rivers of the Recapture and Westwater Canyon provided a link between the rising source highlands in south-central Arizona and the giant, if ephemeral, Lake T'oo'dichi', into which they emptied. Thus these members record the proximal, high-energy fluvial facies of a regional drainage network, while the playa lake deposits of the Brushy Basin depict the more distal aspect of the same system.

Uranium in the Morrison Formation and the Colorado Plateau

The Morrison Formation has been the main source for uranium on the Colorado Plateau, which itself is a world-class uranium province. The yellow ore was first noted in 1881 by prospector Tom Talbert in the Roc Creek area of Montrose County, Colorado, although he didn't know what it was at the time (Cohenour 1967). Tolbert was looking for gold and silver, and at that time nobody knew or cared what the yellow, clayey mineral was. It wasn't until the 1940s, after the development of the atomic bomb, that uranium became a valued resource. In 1946 President Truman established the Atomic Energy Commission (AEC) in an effort to identify and exploit domestic sources for the valuable ore. After preliminary surveys, in 1948 sixteen ore-buying stations were established across the country, twelve located on or around the Colorado Plateau, including at White Canyon, Monticello, Moab, and Thompson, just north of Moab. Once ore began to be produced in sufficient quantities throughout the region, mills were built in Green River, Mexican Hat, and Moab. The boom was on, and during the 1950s more than 500 people, mostly geologists and technicians, were employed by the AEC across the Colorado Plateau. It was recognized early on that the main uranium ore–bearing units were the Triassic Chinle Formation and the Morrison Formation, providing a target for the numerous independent prospectors. Most of our basic understanding of these formations comes from AEC-funded studies conducted by the USGS during this time.

A brief survey of the uranium districts in southeast Utah emphasizes the importance of the Morrison Formation to the occurrence of uranium. The region has been divided into discrete districts by the AEC and the USGS based on geographic and geologic considerations (Fig. 6.21; Stokes 1967). The following review begins to the north with the San Rafael district, and works southeast to the Four Corners area to end with the White Canyon district. Many important uranium districts occur south of the Four Corners, in New Mexico and Arizona, but because they are not part of the region that is the focus of this book, they are not included in this discussion.

The San Rafael district, which encompasses the main uplift of the San Rafael Swell, includes the unique and poorly understood collapse feature of the Temple Mountain cluster of uranium mines. Like the other occurrences in this district, the main ore body at Temple Mountain is situated in the basal sandstone of the Chinle Formation. The Morrison is not present over most of this district due to erosion, and thus is not a significant factor; however, along the west flank of the San Rafael Swell lies the Cedar Mountain district, which is characterized by small, scattered uranium deposits in the Salt Wash and Brushy Basin members of the Morrison (Stokes 1967).

Farther east, the Thompson district includes numerous small mines on the northeast flank of the Salt Valley anticline, along the eastern boundary of Arches National Park. All the uranium in this district comes from lenses of sandstone, ancient river channels, in the Salt Wash Member of the Morrison. Numerous old mines and small waste piles mark Yellow Cat Flat, the main focus of the Thompson district. Immediately to the south is the Moab district, marked mostly by small deposits at the south end of Spanish Valley. Most of the mines are in the Salt Wash Member in the Pack Creek syncline, a downwarp that extends westward from the La Sal Mountains into Spanish Valley.

In the Gateway and Paradox districts, which straddle the Colorado-Utah border, scattered but common uranium deposits also are concentrated in sandstone lenses in the Salt Wash. This area includes the earliest Roc Creek discovery by Tom Talbot.

The Monticello district was the richest for uranium production and included the biggest discovery on the Colorado Plateau, Charlie Steen's surprisingly rich Mi Vida mine. The Mi Vida was situated on the southwest flank of the Lisbon Valley salt anticline with the high-quality ore coming from the basal Chinle Formation. Some smaller deposits occur in the Permian Cutler Formation. Elsewhere in the district, hundreds of discrete, moderate-sized prospects have been identified in the Salt Wash Member exposed in canyons that dissect the Great Sage Plain, an extensive tableland that fringes the Abajo Mountains to the south and east.

The White Canyon district lies west of the Monticello district, north of the San Juan River, and east of the Colorado River (Fig. 6.21). As in the San Rafael district, the Morrison Formation here has been stripped away by erosion. Consequently, all the significant uranium deposits are in the Triassic Moenkopi or Chinle formations. The largest known deposit in this district is the Happy Jack mine, located in the Shinarump Member of the Chinle high on the south rim of lower White Canyon. The Happy Jack is a unique deposit because it contains uraninite in association with various copper ore minerals. After the Monticello district, the White Canyon is the richest district on the Colorado Plateau.

Although it is difficult to determine with certainty the origin of uranium, or the controls on its ultimate accumulation, several conditions common to the Colorado Plateau deposits are considered. First, the ores are confined to porous, fluvial sandstone with good permeability. The uranium was probably introduced by chemical-laden groundwater that flushed through the permeable sandstone conduits. The uranium-bearing sandstone bodies are underlain by shale or some other impermeable rock type forming a barrier to the downward migration of fluids. As the migrating groundwater paused or encountered new chemical conditions (e.g., reducing conditions), uranium ore and associated minerals precipitated in the pore spaces. In some instances they replaced clumps of carbonaceous plant material or petrified logs that were deposited at the bottom of the ancient river channels. Uraniferous logs are common in both the Chinle and Morrison formations.

Uranium, in general, originates in plutonic and volcanic rocks. Because it tends to be widely disseminated through these igneous rocks, it is usually of no economic value. Uranium must somehow be concentrated into an economically viable ore deposit before it is worth extracting. There are thousands of small deposits

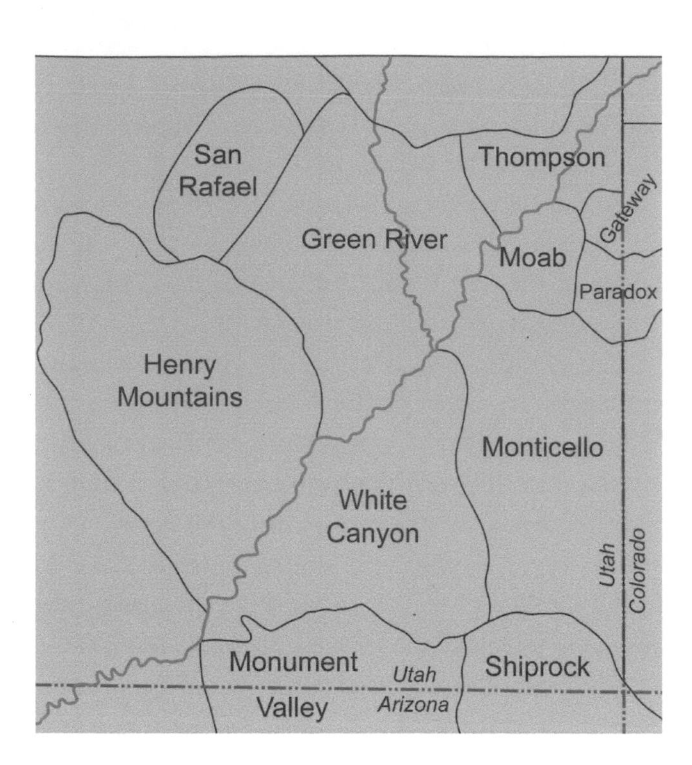

Fig. 6.21. Uranium districts of southeastern Utah and adjacent areas. After Stokes 1967.

scattered throughout the southeast Colorado Plateau that simply are not worth the money and effort that would be required to mine them. The uranium ore in the basal sandstones of the Chinle and Morrison formations likely came from the tuffaceous shale and siltstone beds that make up the upper parts of both formations. The abundant volcanic ash that blew in from the Sierra Nevada arc to accumulate across the floodplains contained small amounts of uranium. The elements that made up the unstable volcanic glass were easily released and mobilized by slowly percolating fluids. As radioactive elements were leached from these source beds, they were slowly transferred to the permeable sandstone conduits that lay below. Here the chemical-rich fluids moved laterally through the sandstone channels until encountering the proper conditions for precipitation as uranium ore and associated minerals. Slowly, over millions of years, the originally dispersed uranium evolved into highly concentrated ore in the basal sandstones.

The last factor to be considered here is structural control, particularly along the salt anticlines that characterize much of the region. Although some uranium deposits have no relationship to these large folds, many important deposits do. For instance, in the Thompson district the uranium deposits contained in the Yellow Cat Flat area occur along the northeast flank of the Salt Valley salt anticline. In addition, a large number of prospects in the Paradox district are clustered high on the northeast flank of the large Paradox Valley salt anticline. Similarly, the rich deposits of the Mi Vida mine and its numerous associated mines and prospects are situated along the southwest flank of the Lisbon Valley salt anticline. The scenario for the emplacement of these deposits is one of a rising groundwater table migrating upward along the flanks of the salt anticlines. As the water table in the porous, folded sandstone units slowly rose toward the axis of the large folds, the uranium-charged water at some point encountered chemical conditions that triggered the precipitation of uranium minerals. In this way the salt anticlines probably acted to concentrate the uranium ore.

The great Colorado Plateau uranium boom came to an end in 1958 when the AEC announced a gradual slowdown in their buying program. This immediately eliminated prospecting across the region, as ore was to be bought only from already producing mines. Eventually even that was phased out. Gradually the numerous mills in the region closed. By 1971 the price of U_3O_8 dropped to $6.20/lb, and only two of the early mills endured, one in Uravan, Colorado, and the Atlas Minerals mill in Moab (Chenoweth 1996). Surprisingly, however, by 1976 the price of U_3O_8 had rebounded to more than $40/lb, and a second, but short-lived, boom ensued. This price increase resulted from the development of nuclear energy, which required large amounts of uranium. This new boom pushed the construction of a new mill in Blanding, which was completed in 1980. Unfortunately, the price plummeted to $27/lb that same year, and by 1983 the boom was over. By this time only the Blanding and Moab mills remained open, scraping by on the small amounts of ore being extracted from the few active mines scattered across the Colorado Plateau. After a struggle, the Moab mill shut down in 1989, and in 1990 the Blanding mill also closed. Today the price of a pound of U_3O_8 is less than $47, and the legacy of uranium on the Colorado Plateau consists of memories and piles of toxic and radioactive waste. In many cases the mills have been dismantled, and the waste piles have been removed or stabilized and buried. Some, however, are still exposed. These piles of radioactive waste remain a threat to the land and water of the region and await U.S. government funding for cleanup.

The Jurassic-Cretaceous Boundary

As in many aspects of geology, there is some controversy over depositional activity across the Jurassic-Cretaceous boundary (144 Ma) on the Colorado Plateau. For example, some have suggested that the Brushy Basin Member extends across this boundary into the earliest part of the Cretaceous Period. These interpretations come from several dating methods applied to minerals in the volcanic ash beds contained in the Brushy Basin. Fission-track ages (Kowallis and Heaton 1987) and K-Ar radiometric dates (Bowman and others 1986) indicate the uppermost Morrison is younger than 144 Ma and so extends into the Cretaceous. However, subsequent $^{40}Ar/^{39}Ar$ analyses by Kowallis and others (1991) from the same locality that earlier was dated by Bowman and his colleagues (1986) in Dinosaur National Monument yielded an age of 153 Ma, which falls well within the Jurassic Period. Most geologists today accept the $^{40}Ar/^{39}Ar$ method as the best dating system available, as it overcomes many of the problems and errors inherent to fission-track and K-Ar methods. Considering all the methods that have been used to determine the minimum age of the Morrison, it appears that its age lies mostly, if not wholly, within the Jurassic Period. It remains a distinct possibility, however, that continuing investigations could change this determination.

An additional complication to this dilemma is geologist William Aubrey's (1996) contention that the basal part of the Cedar Mountain/Burro Canyon Formation (hereafter referred to as the Burro Canyon Formation) should be part of the underlying Brushy Basin Member of the Morrison. Aubrey considers the entire Morrison Formation to be of Late Jurassic age and asserts that the basal Burro Canyon conglomerate interfingers with shale of the uppermost Brushy Basin, forming a gradational contact, and thus the two must be approximately the same age.

The Burro Canyon Formation consists of two discrete units, the previously mentioned basal conglomerate and an overlying thick, shale-dominated unit. Previous workers considered the basal conglomerate to be separated from the underlying Morrison Formation by an unconformity. Evidence from fossil pollen or palynomorphs from the upper, shale-rich part of the Burro Canyon demonstrates an Early Cretaceous age (Tschudy and others 1984). Aubrey's interpretation is that the undated basal conglomerate should be part of the Morrison, whereas the overlying shale unit is undoubtedly younger. According to Aubry, the unconformity lies between these two units. The evidence, however, is sparse, and these relations have not been accepted by most geologists. They are mentioned here to help illustrate the full range of problems associated with the Jurassic-Cretaceous boundary on the Colorado Plateau. The Early Cretaceous Burro Canyon Formation and its equivalent to the west, the Cedar Mountain Formation, are discussed in detail in the following chapter.

The Cretaceous Period

The Sea Appears

The Cretaceous Period (144 to 66 Ma) in western North America was a time of dramatic sea level rise, accelerated tectonic activity, and climate change, all of which are clearly recorded in the rocks of the Colorado Plateau. Cretaceous sedimentary rocks in southeast Utah are continuously exposed along the sinuous escarpment of the Book Cliffs, which loom north of Interstate 70 and parallel the highway from Grand Junction, Colorado, to Green River, Utah. As viewed from the interstate, the otherworldly vista is capped by a corrugated rim of brown sandstone and floored by fluted gray shale slopes. These extensive exposures have been the subject of numerous recent studies as they preserve a world-famous record of the interplay between fluvial and marine systems. Elsewhere in southeast Utah the soft, shale-dominated Cretaceous strata have been stripped away by erosion. Locally, however, basal Cretaceous sandstones endure as resistant caps to the colorful Jurassic rocks.

Although nowhere near as continuous as the Book Cliffs, other extensive exposures of Cretaceous strata in surrounding regions form similar sandstone-capped mesas. The Mesa Verde region of southwest Colorado is one such area. The rocks of Mesa Verde stretch southward into the San Juan basin of New Mexico, where Cretaceous strata constitute one of the largest sources for natural gas in North America. Another large, isolated tract of Cretaceous rocks is Black Mesa in northeast Arizona. Here thick deposits of Cretaceous coal are being mined to supply power plants that provide a large part of the electric supply to the southwest United States. Although the Cretaceous sedimentary record has largely been eroded from the Colorado Plateau, the few remaining outliers provide important energy resources for the United States and allow for an accurate reconstruction of the Cretaceous Colorado Plateau.

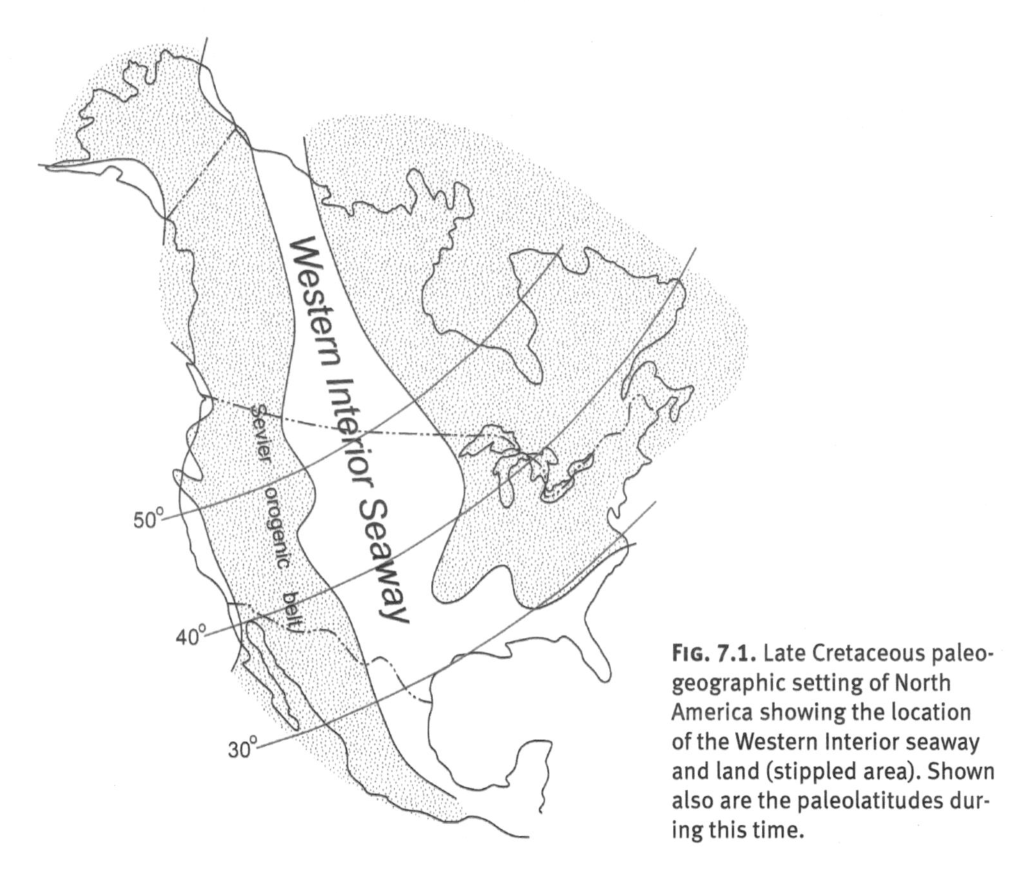

Fig. 7.1. Late Cretaceous paleogeographic setting of North America showing the location of the Western Interior seaway and land (stippled area). Shown also are the paleolatitudes during this time.

Geologic Setting

The Western Interior seaway drowned much of west-central North America throughout the later part of the Cretaceous Period. This extensive, north-south oriented seaway stretched north to the Arctic and southward to the Gulf of Mexico (Fig. 7.1). Although the sea was rising and expanding laterally through most of the Late Cretaceous, it grew in an oscillatory fashion. As thick deposits of black, organic-rich mud blanketed the deeper parts of the sea, the shoreline, which periodically occupied parts of Colorado and Utah, was marked by great swamps in which abundant vegetation flourished and died, creating vast tracts of coal. As the shoreline migrated west or east with the whims of sea level, so did the coal swamps. The shoreline deltas in which the swamps formed were fed sediment by large, east-flowing rivers that originated farther west, in the mountains of the Sevier orogenic belt along the Nevada-Utah border.

The Sevier orogenic belt in western North America was the principal factor in Cretaceous sedimentation, as the great mountain chain not only supplied the sediments that record its presence, but also drove the high rate of subsidence that promoted the preservation of up to 20,000 feet of sediment. This north-south-trending mountain range extended northward well into modern-day Alaska and southward to Mexico. The range was bounded on the east by an alluvial plain that sloped east to the Western Interior seaway. At the latitude of the Colorado Plateau these mountains initially developed in eastern Nevada. As time progressed, however, the mountain front was gradually pushed into western Utah. Huge volumes of detritus from these mountains were fed eastward into the rapidly subsiding basin. The location of the mountain front at any given time was marked by a fringe of coarse conglomerate. Farther to the east the sediments become finer-grained due to a consequent decrease in the gradient and energy of the turbulent Cretaceous rivers. Eventually the fine clay fraction was flushed into the Western Interior seaway to settle to the deep sea floor as black, organic-rich ooze.

Through most of the earlier part of the Mesozoic (the Triassic and Jurassic periods) the climate was marked by aridity. This is clearly displayed in the rock record by abundant eolian sandstone and evaporites in the shallow marine deposits. During this time most of North America was positioned between the paleo-equator and the latitude of 30° N, putting it well within the arid latitudes that straddle either side of the equator. But this wasn't the only factor. North America was situated in the interior of the Pangean landmass, isolated from the immense Panthalassan sea that could potentially have supplied moisture to the dessicated land. Although the western half of North America lay along the western edge of Pangea, and so was adjacent to the sea, several generations of mountains swelled along the west coast during the Mesozoic Era. The various incarnations of these coastal mountains created a high barrier to any moisture that may have wafted eastward. As the clouds rose into the cooler atmosphere to traverse these mountains, they released their watery load and dissipated.

During the Cretaceous Period the climate of western North America became much wetter as the result of several contributing factors. First, as the continents that formed Pangea continued to separate, North America drifted north, out of the arid lower latitudes to a location between 30° and 60° N. Although it was located in a humid climate zone, the rising mountains of the Sevier orogenic belt had the potential for creating a precipitation barrier similar to that of the earlier Mesozoic. The Cretaceous, however, saw an unprecedented rise in sea level that inundated much of the North American landmass. Despite the presence of mountains that may have blocked incoming moisture from the Pacific Ocean, the Western Interior seaway that covered much of the land east of these mountains supplied more than enough moisture to offset this loss. By Mid-Cretaceous time approximately one-third of the continental landmass of the world known today was submerged beneath this sea, creating vast tracts of water for evaporation into the atmosphere. Finally, oxygen isotope analyses ($^{18}O/^{16}O$) of Cretaceous fossil shells suggest a seawater temperature that was 15°C warmer than that of similar latitudes today. This would have greatly enhanced the evaporation rate. The Cretaceous of North America is believed to have been dominated by a humid, subtropical climate.

Plate Tectonics

The global plate tectonic setting played a monumental role in Cretaceous deposition on the Colorado Plateau and surrounding regions. The dominant factor worldwide was an acceleration in the production of new oceanic crust at mid-ocean ridges. In the Cretaceous version of the Pacific Ocean, the revitalization of the mid-ocean ridge resulted in an increased rate of subduction along the west margin of North America. One consequence was the continued eruption and evolution of the Sierra Nevada volcanic arc, as well as a later eastward migration of igneous activity. The consequence most relevant to this discussion was an increase in compressional stress along the western North American margin, which generated the Sevier orogenic belt along the Utah-Nevada border. As time progressed, however, the mountains pushed eastward, well into western Utah. Although mountain building in the region had initiated much earlier, the scale and eastward push of the mountain front reflected a renewed intensity.

Along the great length of the Sevier belt, which stretched from Mexico to Alaska, the style and composition of the mountains remained amazingly consistent. The fold and thrust belt, as this style of uplift is known, consists of thick wedges of folded Precambrian and Paleozoic sedimentary rocks that were thrust tens of kilometers eastward, up and over younger rocks. This lateral movement of kilometer-thick slabs occurred along "sled-runner"-shaped thrust faults, initially moving horizontally along the flat part of the low-angle fault and finally curving upward to its termination. Near its terminus the fault steepens and ramps upwards, so as the moving sheet of rock broke the surface, it rose vertically. This fault geometry created a formidably steep mountain front that towered high above the adjacent foreland basin.

The mountains of the Sevier fold and thrust belt were composed of numerous massive thrust sheets stacked like gargantuan shingles that sloped westward, toward the origin of the compression. As the mountain belt grew, it widened by expanding eastward, toward the stable continental interior. This migration occurred by breaking out a new thrust in front of the former frontal thrust. Overall, geologists estimate that the Sevier orogenic belt shortened the width of western North America in an east-west direction by ~100 kilometers.

The enormous sedimentary basin that bounded the Sevier mountain belt to the east was a direct result of the fold and thrust style of mountain building. **Foreland basins**, as these types are called, form adjacent to the fold and thrust belts when the increased load of the thrust sheets depresses the crust over a broad area centered beneath the belt. Maximum subsidence is beneath the thickest buildup in the mountain belt. Although this subsidence is not enough to cancel the topographic heights of the belt, it does reduce their overall elevation as their roots sink into the crust (Fig. 7.2). The degree of subsidence gradually decreases away from the mountainous load. The rate of this decrease depends

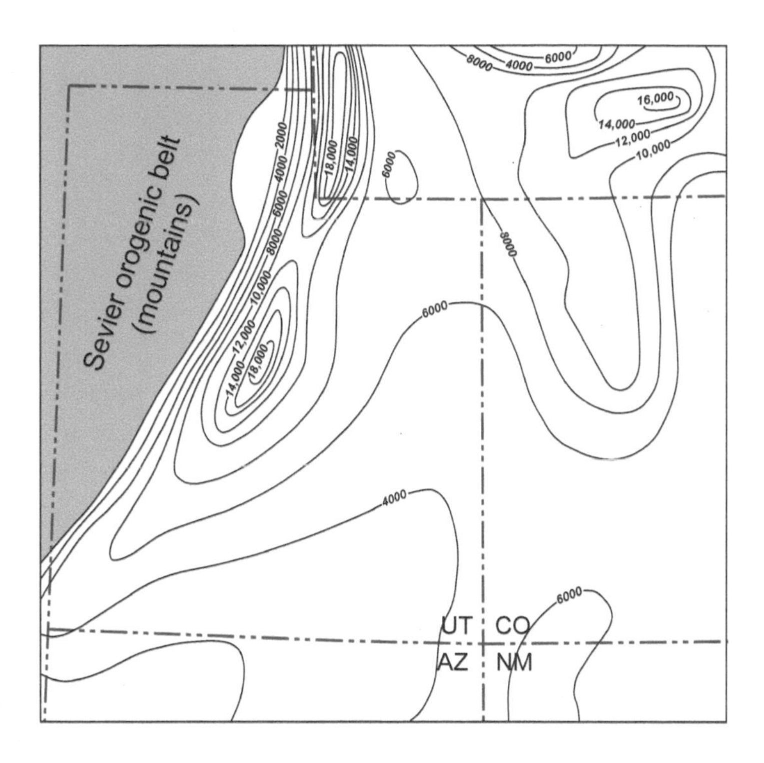

Fig. 7.2. Isopach map of the Upper Cretaceous strata on the Colorado Plateau. Numbers represent thickness in feet. Isopach lines represent the thickness of Upper Cretaceous strata in a particular place. The preserved thickness of strata is an indicator of how much subsidence took place, and the map clearly shows patterns of subsidence. Upper Cretaceous strata patterns show a dramatic increase in subsidence immediately adjacent to the Sevier orogenic belt. The strongly asymmetric subsidence pattern shown here typifies foreland basins such as this. Note also the decrease in subsidence away from the mountain belt. From McGookey and others 1972.

on the strength of the crust and the magnitude of the load. Excluding the area beneath the actual mountain belt, maximum subsidence occurs directly adjacent to the mountain front, in the proximal part of the foreland basin—in this case, in western Utah. Traversing east across the foreland basin, subsidence diminishes regularly across Utah and Colorado. Thus, the Western Interior foreland basin had a pronounced asymmetric geometry in a west-to-east transect, with a very thick (~19,000 feet) sediment fill against the mountain front and a gradual thinning eastward until Cretaceous sediments disappear in eastern Kansas.

Another major effect of increased mid-ocean ridge activity was the worldwide rise of sea level and expansion of the Western Interior seaway. As the production of oceanic crust at mid-ocean ridges accelerated, the volume of new, hot crust increased substantially in the ocean basins. The hot crust was more buoyant, so it sat higher relative to the older, colder crust. The increased production of oceanic crust caused larger parts of the sea floor to sit higher than normal, effectively displacing huge volumes of water worldwide so that it spilled across the continental landmasses.

Cretaceous Strata

In contrast to the vibrant polychromatic hues of the underlying strata, Cretaceous rocks of the Colorado Plateau tend toward shades of brown and gray. These drab colors are due to the subtropical climate of the time that generated abundant organic carbon, and the various environments that enhanced its preservation. The deposits span the realm of clastic sediments, from boulders at the mountain front to black, organic-rich mud in the quiet depths of the Western Interior seaway. Continental environments ranged from alluvial fans to various fluvial types, delta and coal swamps, and beach settings. Marine deposits varied from tidally influenced shorelines and shallow shelves to very deep offshore environments. Far to the east, in modern-day Kansas, away from the influence of the Sevier-derived clastics, the basin gradually shallowed, and fine-grained, chalky limestone was deposited.

Cretaceous sedimentation dawned with deposition of the Cedar Mountain and Burro Canyon formations in rivers that tumbled northeastward from the rising Sevier mountains. This was followed by an interval of nondeposition and erosion. The Dakota Sandstone overlies this unconformity and forms the base of an immense succession of sandstone and shale. Like the earlier Cedar Mountain/Burro Canyon, the Dakota initiates with fluvial deposits but gradually shifts upward into shallow marine sandstone, and eventually grades up into the thick, deep marine deposits of the Mancos Shale. As sea level oscillated and the shoreline shifted back and forth in an east-west direction, fluvial/deltaic sandstone and marine shale competed for space, and eventually the deltaic Mesa Verde Group was laid down. It is this interplay that is so clearly etched into the ramparts of the Book Cliffs. A detailed account of Cretaceous sedimentation in the Book Cliffs of Utah and Colorado follows.

Lower Cretaceous Cedar Mountain/Burro Canyon Formation

The Lower Cretaceous Cedar Mountain and Burro Canyon formations are correlative, fluvial-dominated units of Early Cretaceous age, yet they differ enough that separate names arguably are warranted. The Cedar Mountain Formation, as defined by W. L. Stokes (1944, 1952), covers the western part of Utah's Colorado Plateau, from west-central Utah adjacent to the Sevier thrust belt from which it was derived, eastward to an arbitrary northeast-trending boundary that follows the Colorado River through Canyonlands National Park (Fig. 7.3). East of this line, equivalent strata become the Burro Canyon Formation, which was defined by W. L. Stokes and D. A. Phoenix in 1948. The Burro Canyon reaches its eastern limit in western Colorado, where it pinches out. Both formations rest unconformably on the Brushy Basin Member of the Morrison. Additionally, both consist of fluvial conglomerate and sandstone, and colorful floodplain mudstone. The differences, however, lie in the details of the two formations.

Thickness trends in Lower Cretaceous rocks across the Colorado Plateau detail the effect of the Sevier thrust belt on sedimentation (Fig. 7.4). In a west-to-east transect across the Plateau, the Pigeon Creek Formation is the westernmost equivalent to the Burro Canyon/Cedar Mountain succession. It is confined to a narrow strip of west-central Utah and abuts what was the front of the Sevier

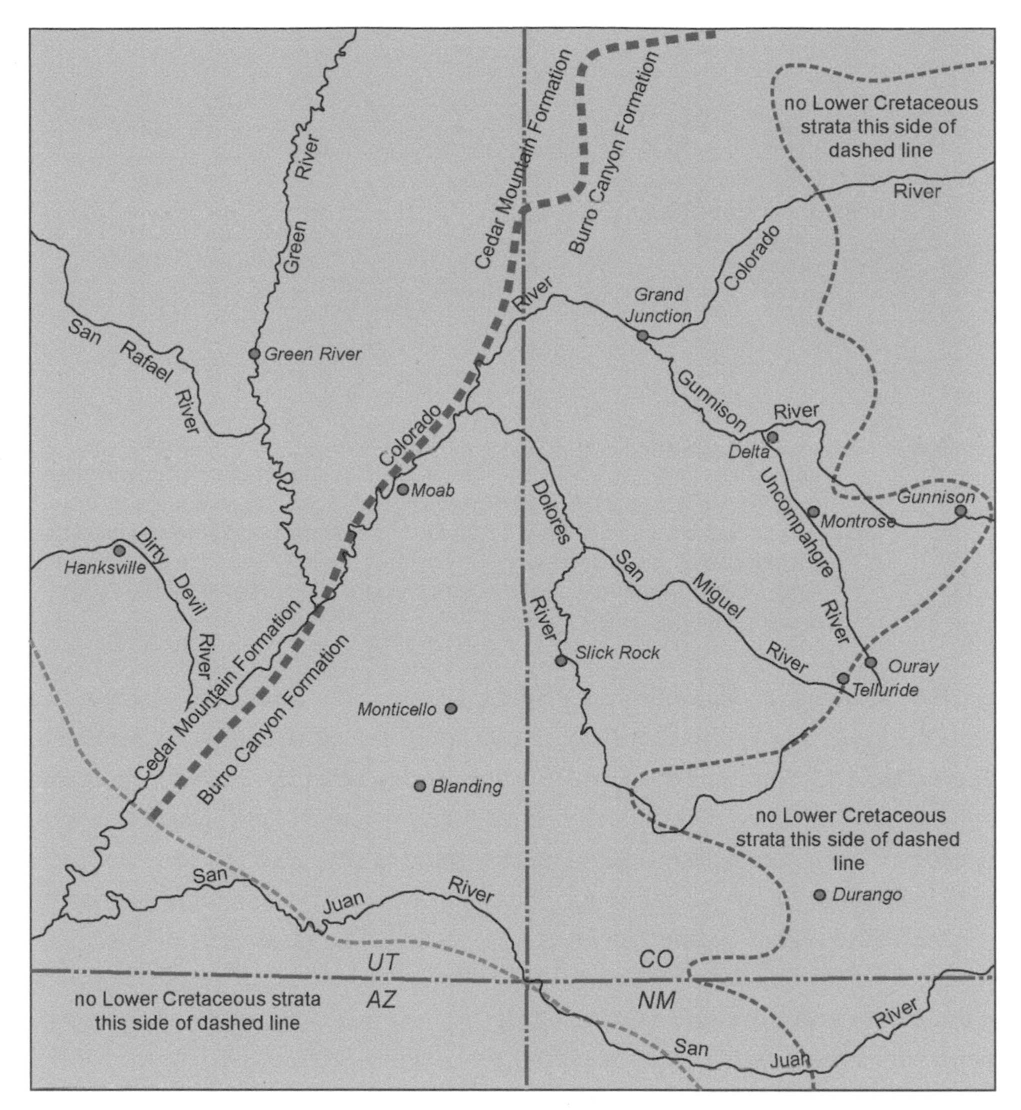

FIG. 7.3. Map showing the extent of Lower Cretaceous strata on the southeast part of the Colorado Plateau and the confusing nomenclature for these strata. The northeast-trending dashed line that follows approximately the trace of the Colorado River shows the delineation of the Cedar Mountain Formation to the west of the line and the Burro Canyon Formation to the east. From Tschudy and others 1984.

belt during the Early Cretaceous. According to Peter Schwans (1988), who recognized and defined the formation, the Pigeon Creek consists of variegated shale, sandstone, and conglomerate of fluvial origin, and limestone that represents shallow lakes. The formation attains a maximum thickness of 3300 feet in its westernmost exposures but thins rapidly to less than 500 feet approximately 15 miles to the east. Moving east, the lithologically similar Cedar Mountain Formation of east-central Utah reaches a greatest thickness of 560 feet just west of the San Rafael Swell, near the town of Castle Dale (Craig 1981). The formation thins gradually to less than 130 feet near its eastern limit along the Colorado River. East of the Colorado River, where it becomes the Burro Canyon Formation, it continues

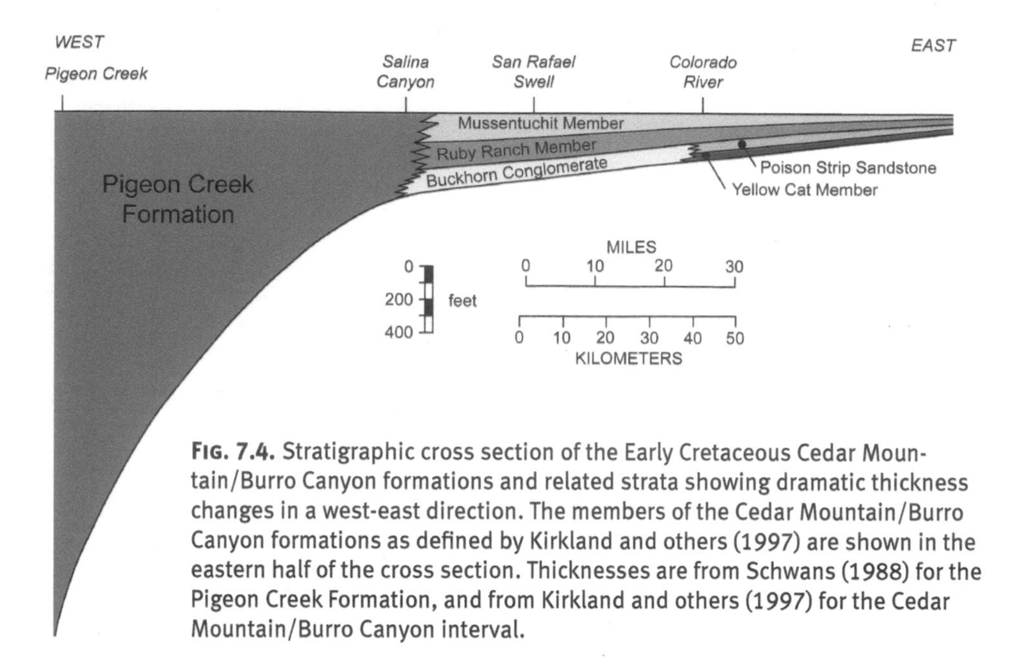

FIG. 7.4. Stratigraphic cross section of the Early Cretaceous Cedar Mountain/Burro Canyon formations and related strata showing dramatic thickness changes in a west-east direction. The members of the Cedar Mountain/Burro Canyon formations as defined by Kirkland and others (1997) are shown in the eastern half of the cross section. Thicknesses are from Schwans (1988) for the Pigeon Creek Formation, and from Kirkland and others (1997) for the Cedar Mountain/Burro Canyon interval.

to thin, although it does so in an irregular pattern. The formation pinches to a zero thickness in western Colorado, immediately east of the town of Gunnison. The variable thickness of the Burro Canyon is illustrated by its range of 100 to 250 feet within the relatively small area of Arches National Park (Doelling 2000). Geologist W. G. Aubrey (1996) suggests that some of the seemingly random thickness changes in this area may be explained by recurrent movement of salt in the salt anticline region during the Early Cretaceous. Overall, these thickness patterns across the Colorado Plateau exhibit a classic foreland basin asymmetry, with a maximum preserved thickness along the zone immediately adjacent and parallel to the thrust front. Thickness patterns of Lower Cretaceous strata mirror the amount of subsidence during that time interval. Thus, maximum subsidence was confined to a trough-like zone just east of the mountain belt and gradually diminished to the east, away from the rising mountains, until Lower Cretaceous strata disappear in western Colorado. The apparent absence of Lower Cretaceous rocks farther east may reflect subsequent erosion of the rocks rather than nondeposition. The unconformity at the top of the Lower Cretaceous succession contains widespread evidence for such erosion.

The Cedar Mountain Formation occupies only a small western part of the area that is considered in detail here. Its western counterpart, the Pigeon Creek Formation, is located on the extreme western margin of the Colorado Plateau and is far from the area under consideration; however, Lower Cretaceous strata across the Colorado Plateau are vital to understanding the Early Cretaceous evolution of the region, so they will be briefly addressed below. The following discussion begins with the Pigeon Creek, which is the most proximal to the Sevier thrust belt; this is followed by the Cedar Mountain and, finally, the easternmost Burro Canyon Formation, completing the west-to-east transect (Fig. 7.4).

Exposures of the Pigeon Creek Formation are confined to the Gunnison Plateau and the west side of the adjacent Wasatch Plateau, which defines the west edge of the Colorado Plateau at this latitude. At its type locality, on the west side of the Gunnison Plateau, the formation reaches more than 3000 feet thick and can be divided into two distinct, informal members (Schwans 1988). The lower, mudstone-dominated member is ~600 feet thick and consists of red mudstone and interbedded pebbly sandstone. The monotony of these fluvial floodplain and channel deposits is broken by sparse freshwater limestone deposits that identify the presence of lakes on the muddy floodplain. The thick upper member, at 2600 feet, makes up the bulk of the formation at Pigeon Creek and is dominated by sheetlike conglomerate deposits; interbedded red mudstone is only a minor component. The upper member records the development of alluvial fans that fringed the mountain front looming immediately to the west. All indicators of flow direction in the Pigeon Creek Formation show eastward sediment transport. As the formation thins basinward to the east, it does so with the thinning of the upper conglomeratic member to less than 500 feet in that direction (Schwans 1988).

To the east, the base of the Cedar Mountain Formation is defined by the Buckhorn Conglomerate Member, composed of up to 80 feet of yellow gray, ledge-forming, coarse sandstone and conglomerate. These coarse sediments record the continued rise and eastward push of the Sevier mountain belt in the form of turbulent rivers that plunged eastward from the mountainous heights. The upper, thicker part of the Cedar Mountain is dominated by purple, red, and green mudstone, with fewer lenses of coarse sandstone. Mudstone was deposited on a vast, low-relief floodplain across which placid rivers flowed, their passage marked by the sandstone lenses. Cross-stratification in Cedar Mountain sandstone indicates an eastward flow for its rivers, suggesting a simple continuation of the east-directed Pigeon Creek fluvial system.

Farther east, in southeast Utah and western Colorado, the Burro Canyon Formation makes up the Lower Cretaceous succession. It is named for a type locality near Slick Rock, Colorado, where the formation is typified by channel-shaped bodies of pebbly sandstone encased in green mudstone (Stokes and Phoenix 1948). The basal contact is defined by irregular scours that locally cut deeply into the underlying Brushy Basin shales. The pairing of sandstone and mudstone formed from the fluvial channel/floodplain couplet that characterizes Lower Cretaceous strata throughout the region. Sparse limestone lenses accumulated in small, ephemeral freshwater lakes that dotted the floodplain. Although the Burro Canyon contains no basal conglomerate that correlates with the basal Buckhorn Member of the Cedar Mountain, the pebbly sandstone lenses dominate the lower

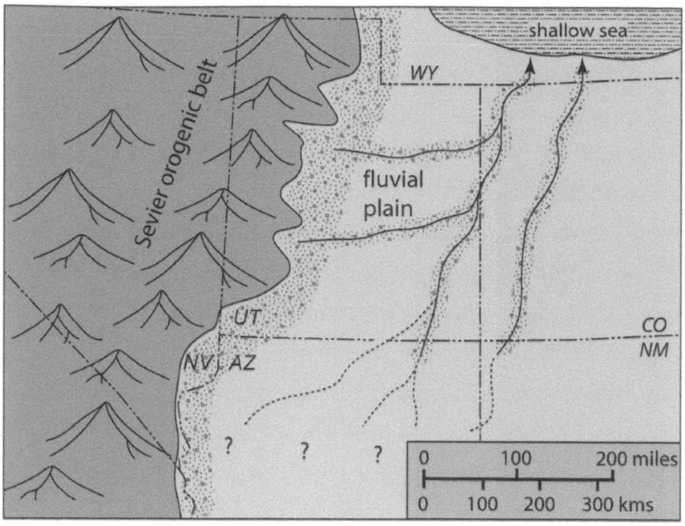

Fig. 7.5. Paleogeography of the Lower Cretaceous Cedar Mountain/Burro Canyon formations showing two source areas: one to the south that fed the Burro Canyon system, and one to the west, in the Sevier orogenic belt, that provided sediment to the Pigeon Creek and Cedar Mountain formations.

part. In contrast to the Cedar Mountain, paleoflow indicators from sandstone in the Burro Canyon show a northern flow direction for the river system.

Disparities in flow direction between the Cedar Mountain and Burro Canyon formations are likely a consequence of each originating in a different source area (Craig 1981), suggesting that the two represent distinctly different drainage networks that eventually merged. The Cedar Mountain system had a source in the Sevier thrust belt to the west, in western Utah, and flowed east across Utah. In contrast, Burro Canyon rivers ran north, emanating from highlands somewhere to the south in Arizona (Fig. 7.5). These two systems met in northwest Colorado, where the combined rivers continued northeast to the sea that lay in Wyoming. It is in northwest Colorado, along the arbitrary line between the two formations, that they appear to interfinger, supporting the confluence of these drainages in this area.

The Early Cretaceous on a global scale was a time of remarkable diversification of terrestrial life, probably an effect of the warm, tropical climate. A major evolutionary milestone during this time was the appearance of **angiosperms**, or flowering plants. This development corresponds to a broad diversification of insects and herbivorous dinosaurs, as well as a parallel expansion of carnivorous species. In fact, the first appearance of angiosperms coincides with an evolutionary change throughout the entire terrestrial ecosystem.

Prior to the Early Cretaceous the dominant plant type was the gymnosperm, mostly woody plants whose seeds came from cones or formed on leaves. They relied on wind for pollination and dispersal. Angiosperms had the advantage of flowers, which attracted insects, and later birds, for pollination. Additionally, their seed-bearing fruits were eaten by animals, and the undigested seeds were efficiently spread over wide areas in their feces. Angiosperms also matured more rapidly and so produced seeds over a shorter cycle. Because of this rapid growth, they quickly recovered from browsing by the herbivorous dinosaurs, providing an increased and widespread food supply. This benefited the herbivores as they became more diverse and abundant, and they in turn provided more prey for the carnivores. In this way the angiosperms boosted the entire food chain. The evolution of insects similarly parallels angiosperm development as told by the earliest moth and bee fossils that are found in Cretaceous rocks. Cross-pollination by insects likely broadened the diversity of angiosperms by creating new species. Finally, the best evidence so far discovered for the age of the Cedar Mountain and Burro Canyon formations comes from primitive angiosperm pollen recovered from carbonaceous layers in these rocks (Tschudy and others 1984). The rapid evolution of angiosperms through the Cretaceous Period and the unique morphology of the pollen from individual species make these fossils an important tool in determining the age of nonmarine Cretaceous rocks.

Dinosaurs were experiencing a rapid phase of evolution during this time and constitute an important faunal component of the Cedar Mountain/Burro Canyon formations. Although these strata lie directly above the dinosaur-rich Morrison Formation, they were for many years considered to be barren of such remains. It was not until the discovery of numerous bones in the San Rafael Swell that paleontologists realized the abundance and diversity of the Early Cretaceous dinosaur fauna (DeCourten 1998). Since then two important localities have been established in the region. One is in the northern part of the San Rafael Swell. The second is north of Moab, along the flanks of the Salt Valley salt anticline, which makes up much of Arches National Park.

The discovery and analysis of the Cedar Mountain/Burro Canyon dinosaur fauna has resulted in a detailed revision of the stratigraphy, age, depositional setting, and climate of the unit, as well as insight into dinosaur evolution (Kirkland and others 1997, Kirkland and others 1999). In the pursuit of dinosaurs these workers have identified four additional members adjacent to or overlying the basal Buckhorn Conglomerate (see Fig. 7.4). These new interpretations are based predominantly on lithologic criteria, but are reinforced by similarly positioned changes in plant and dinosaur fossils. The new subdivisions have yet to stand the test of acceptance by the geologic community, but they appear to be based on valid criteria. The members and their characteristics are briefly recounted below.

The Yellow Cat Member overlies the basal Buckhorn Conglomerate, or, where the Buckhorn is absent, it sits on the Morrison Formation. The base of the Yellow Cat consists mostly of a thick limestone bed interpreted by many as a long-lived soil zone (e.g., Aubrey 1996), although Kirkland and others (1999) have considered the possibility of lake deposits. Where both the Buckhorn and the limestone are absent, the base is marked by a scattered distribution of smoothly polished, well-rounded pebbles thought by some to be **gastroliths**, pebbles ingested by dinosaurs and stored in their stomachs to aid in digestion. Most of the member is composed of drab green mudstone that is readily differentiated from the more colorful mudstone of the underlying Morrison Formation. Winding ribbons of sandstone and limestone lenses trace the paths of rivers and the locations of waterholes, respectively. The mudstone that encases these lenses represents the alluvial plain that was sliced by the rivers and on which a varied population of dinosaurs lived and died.

The rich dinosaur fauna recovered from the Yellow Cat Member records a diverse ecosystem (Kirkland and others 1999). Dinosaur fossils discovered so far include a new genus of polcanthid ankylosaur—a large, slow, and heavily armored creature with bony plates and protruding spikes—and a great, clublike tail that augmented its defenses. The gigantic herbivorous sauropods are well represented by the titanosaurids and camarasaurids. The bipedal herbivore *Iguanodon ottingeri* has also been identified. The carnivorous predators are represented by the bones of a therapod that resembles the Jurassic-age *Ornitholestes* found in the underlying Morrison Formation, and the large dromaeosaurid predator *Utahraptor ostrommaysorum* (Kirkland and others 1993). *Utahraptor* was designed for killing and was the largest of the terrifying dromaeosaur predators, measuring 20 feet in length and probably weighing in at around 1000 pounds (DeCourten 1998). The large biped had powerful hind legs for running, and curving, sickle-like claws up to 9 inches long. These claws undoubtedly were used to dispatch and feed on its unfortunate prey. These vicious carnivores clearly sat at the pinnacle of the food chain in this extensive Early Cretaceous basin.

The Poison Strip Sandstone Member of the Cedar Mountain overlies the Yellow Cat Member in eastern Utah and consists of a continuous ribbon of conglomeratic sandstone traceable from the vicinity of Green River, Utah, eastward to the Utah-Colorado border area. Like the underlying Yellow Cat, the Poison Strip Sandstone may be the lateral equivalent of the Buckhorn Conglomerate in the San Rafael Swell area to the west. This meandering stream channel deposit has so far yielded specimens of the large ankylosaur *Sauropelta* and the ornithopod *Tentosaurus* (Kirkland and others 1999). In addition, large conifer logs and the stalk-like cycads *Cycadeoidea* and *Monathasia* have been recovered from this member in exposures around Arches National Park.

The Ruby Ranch Member overlies the Poison Strip Sandstone in the east, from the Utah-Colorado border to the east side of the San Rafael Swell. Unlike the two underlying members, the Ruby Ranch continues west across the Swell to blanket the Buckhorn Conglomerate farther west. The member consists mostly of slope-forming, pale purple mudstone with limited lenses of sandstone and limestone. A prominent feature of this member, and one that has been noted by every worker since Stokes (1944), is the overwhelming abundance of carbonate nodules in the mudstones. These resistant, fist-size limestone nodules litter the mudstone slopes throughout their exposure. The irregular nodules formed in shallow horizons of the floodplain soils during the alternating, very wet monsoonal periods and dry seasons. Dissolved calcium carbonate was washed down into the soil during the wet season and precipitated out as a solid during the dry periods as the soil moisture evaporated. Carbonate precipitated by this process requires extended periods of time to develop, which suggests that the 100-foot-thick mudstone-dominated unit accumulated slowly over a prolonged period. The sparse sandstone bodies in the Ruby Ranch form sinuous, shoestring-like ridges that may be traceable for over a mile. These are the product of low-energy rivers that wound lazily across the muddy floodplain.

Although it is the least known of the Cedar Mountain assemblages, the Ruby Ranch fauna include an assortment of Early Cretaceous dinosaurs. Several of these also occur in the underlying members, including the iguanodontid *Tenontosaurus* and the large ankylosaur *Sauropelta*. A collection of bones separated from a single lens of limestone is believed to represent a combination of adult and juvenile sauropods similar to *Pleurocoelus*. DeCourten (1998) has suggested that these small sauropods (adults less than 20 feet long) congregated and perished at a water hole on the floodplain. The presence of carnivores is told by the preservation of the claws and teeth of these predators (Kirkland and others 1999). One large, curving claw may represent the fierce carnivore *Deinonychus*, which has been recovered from correlative strata in Montana, although its true origin awaits the recovery of more diagnostic bones. The teeth have tentatively been assigned to *Acrocanthosaurus*, although their extraordinary size may eventually result in the recognition of a new, huge predator. Angiosperm pollen recovered from the Ruby Ranch Member is both diverse and abundant, indicating that these plants were in full bloom at this time. The abundant calcareous nodules that litter the mudstone slopes point to a semiarid, monsoonal climate that seasonally was very wet, but at other times was very dry.

The Mussentuchit Member succeeds the Ruby Ranch Member across its extent and represents a pronounced change from underlying strata. The lower contact is marked by an abrupt shift to the gray, smectitic mudstone of the Mussentuchit, easily recognized because it weathers into a popcorn-like surface. This puffy weathering veneer is a characteristic of volcanic-derived smectite clays wherever they are exposed. Besides the obvious color change, these mudstones contain none of the carbonate nodules that are so abundant in the underlying member. The Mussentuchit also hosts sparse sandstone lenses of fluvial channel origin and thin, black lignite horizons. This form of low-grade coal indicates abundant vegetation and a corresponding shift to a much wetter climate than that inferred for underlying units. In fact, lithologically the Mussentuchit has more in common with the overlying Dakota Sandstone, from which it is supposedly separated by an unconformity, than the underlying members of the Cedar Mountain/Burro Canyon succession with which it has been aligned.

The flora and fauna of the Mussentuchit Member also suggest a wetter climate. The numerous plant fossils recovered so far consist of conifers, cycads, abundant flowering plants, and ferns, including the common treelike fern *Tempskya* (DeCourten 1998). The somber gray shades of the mudstone derive from the widely dispersed black, carbonaceous fragments of these various plants. It was by the rise and fall of dense thickets of these plants that the thin lignite horizons were formed. The vertebrate fauna of the member contains elements of both Early and Late Cretaceous faunas, and represents a significant transitional assemblage that is yet to be fully documented and understood. The fauna include an assortment of crocodiles, turtles, amphibians, lizards, and dinosaurs. Primitive mammals recovered from the member include bones of the world's oldest marsupial (DeCourten 1998).

Dinosaurs identified from the Mussentuchit are an interesting and varied lot. Herbivore bones recovered so far include the nodosaurid *Animantarx ramaljonesi*, a primitive iguanodontid, and *Telmatosaurus*, a hadrosaurid which is popularly known as a duck-billed dinosaur. Numerous teeth have also been discovered in these sediments. Fortunately dinosaur teeth are unique, at least at a family level, allowing for the identification of a number of dinosaurs that would otherwise remain unknown. Herbivore teeth include those of ceratopsians, pachycephalosaurs, and small sauropods. Carnivores are represented by the teeth of dromaeosaurids and small tyrannosaurids, the earliest evidence of this family in North America.

Dinosaur paleontologist James Kirkland and his colleagues (1999) have noted that the nearest ancestors of the hadrosaurs, ceratopsians, and tyrannosaurids in the Mussentuchit are part of an Early Cretaceous Asian fauna. Although these dinosaurs or their close relatives later became common elements of the Late Cretaceous North American fauna, they were unknown on this continent prior to this time. Their discovery in the upper part of the Cedar Mountain Formation changed this and marks their first appearance on the North American continent. Kirkland and others (1999) proposed that during the Early Cretaceous, while sea level was at a low stand, a land bridge emerged to connect eastern Asia to Alaska and the vast North American continent. This land bridge would have provided the very different eastern Asian dinosaurs the opportunity to migrate to North America, eventually spreading across the landmass. Although speculative at this time, it is a plausible scenario and undoubtedly will be the subject of continued research.

Regardless of the ancestry of some of the dinosaurs in the Mussentuchit, the discovery and analysis of the Cedar Mountain fauna over the last twenty years has provided a significant evolutionary bridge between the better-known Jurassic and Late Cretaceous faunas of North America. While there will always be gaps in the fossil record due to a lack of preservation or erosion of the host sediments, the Cedar Mountain Formation holds the answers to many important questions concerning dinosaur evolution. The dinosaurs from the formation provide several "missing links" in the rise of these fascinating animals.

The contact between the Cedar Mountain/Burro Canyon Formation and the overlying Dakota Sandstone has long been considered an unconformity" a gap in the rock record that straddles the Early and Late Cretaceous boundary. Recent work on the Mussentuchit Member, however, may suggest otherwise. The recognition of a volcanic ash layer immediately above a dinosaur-bearing horizon provided a convenient means for radiometrically dating the upper Cedar Mountain. High precision $^{40}Ar/^{39}Ar$ analyses from minerals in this ash have yielded an age of 98.39 +/- 0.07 Ma (Cifelli and others 1997). In comparison, the absolute age of the Early/Late Cretaceous boundary is 98.5 Ma, essentially placing the ash layer at the boundary. The ash bed is overlain by several tens of feet of gray shale, pushing the Cedar Mountain unequivocally across the boundary and into the Late Cretaceous.

Besides a revision in the age of the Cedar Mountain Formation, the work of Kirkland and his colleagues (1997, 1999) has shown that in the southwestern San Rafael Swell the upper part of the Mussentuchit grades into the overlying Dakota Sandstone. In this area discontinuous sandstone beds at the top of the member grade up into the obvious sheetlike sandstone body of the Late Cretaceous–age Dakota. Moreover, gray carbonaceous mudstones with lignite horizons in the Mussentuchit appear to have a closer affinity with the similarly carbonaceous Dakota Sandstone than with the underlying Cedar Mountain strata. These factors led Kirkland and coworkers to speculate that the Mussentuchit may fit better with the overlying Dakota; however, they could come to no consensus on the problem, and the Mussentuchit remains a member of the Cedar Mountain Formation. Still, over most of its exposure throughout the region the contact between the Cedar Mountain/Burro Canyon strata and the overlying Dakota Sandstone is sharp and erosional, suggesting that the contact is unconformable.

Upper Cretaceous Dakota Sandstone

The Upper Cretaceous Dakota Sandstone forms a regional, sheetlike deposit of sandstone with lesser amounts of conglomerate, siltstone, and coal. This great blanket of sediment stretches from the plains of eastern Colorado to western Utah. Lithologically the Dakota succession is surprisingly consistent throughout the region; three distinct parts, each representing a different environment, are recognized. These predictable shifts in rock type record a transition from basal fluvial sandstone and conglomerate upward into interbedded sandstone, siltstone, and coal deposited in a large delta at the land-sea interface. The uppermost Dakota consists of sandstone and siltstone with numerous marine fossils and invertebrate trails and burrows, all indicative of a shallow marine environment. The Dakota grades vertically into the thick, deep marine Mancos Shale. This succession as a whole chronicles a rise in sea level that can be recognized worldwide.

Late Cretaceous sedimentation in western North America largely was controlled by two competing factors. The first was an acceleration of mountain building in the Sevier orogenic belt with the continued rise and eastward push of this great range. This chain of mountains, which stretched from northern Canada southward to Arizona, flushed huge amounts of sediment eastward into the adjacent foreland basin. The second major influence was the slow but dramatic rise and fall of the Western Interior seaway and the resulting oscillations of the shoreline. These fluctuations controlled the nature and distribution of the sediments that poured from the Sevier mountains.

The Dakota lies unconformably on the Cedar Mountain/Burro Canyon succession or, where that is absent, on the Morrison Formation. In most exposures across the eastern Colorado Plateau the lower third consists of sandstone and conglomerate. Basal conglomeratic sandstone grades upward into sandstone as conglomerate diminishes. To the west, however, conglomerate increases and becomes coarser grained. Pebbles throughout its extent are dominated by chert and quartzite. These strata are arranged into stacked and overlapping cross-stratified packages that scour into each other, indicating alternating episodes of erosion and deposition. Consistently east-dipping cross-stratification, coupled with the composition of these sediments, confirms a highland sediment source in the older Paleozoic sedimentary rocks that were uplifted in the thrust sheets of the Sevier orogenic belt. The rivers that routed these sediments across Utah and Colorado flowed eastward toward their eventual rendezvous with the Western Interior seaway, which at this time lay in east-central Colorado.

The middle third of the Dakota is characterized by gray, carbonaceous sandstone, siltstone, mudstone, and coal. Impressions of plant stems and twigs are common, as are black clots of carbonaceous plant remains. Coal beds are low grade and generally less than a foot thick, so are of no economic value. All these attest to abundant vegetation, a consequence of the warm, tropical Late Cretaceous climate. Although the coal beds are thin, it should be pointed out that the generation of a 1-foot-thick coal bed requires more than 100 feet of intensely compressed and concentrated plant matter. The middle Dakota preserves the record of a swampy shoreline cut by winding, slack-water rivers. These rivers, which fed large deltas, coursed though a lush, shadowed landscape of swampy forests. The main conduits of the river and delta channels were fringed by stagnant, coffee-colored swamp waters.

The upper Dakota consists of clean, fine-grained sandstone that weathers into a thin but continuous cliff-forming bench. This tan sandstone mostly is cross-stratified, horizontally stratified, and ripple laminated. Locally the cross-stratification dips in opposing directions in successive beds, a feature known as herringbone crossbedding. Ripples are symmetric. Both features document the back and forth currents of rising and falling tides that molded the shallow sea floor sands as they dragged first shoreward, then seaward. Another diagnostic feature in deciphering these sandstone beds is the abundant trace fossils made by invertebrate organisms. These include locomotion tracks on the surfaces of some beds, dwelling burrows excavated into the sediments, and feeding traces of organisms on the surface and within the sediments. Although it is difficult to determine which organisms made these particular traces, it has been shown that many in the upper Dakota were confined to sandy, shallow marine environments (Fig. 7.6).

Trace fossils, in general, are valuable for deciphering the detailed aspects of marine environments. Their numerous and diverse patterns make them simple to identify and furnish information on conditions such as water depth, especially in the absence of shelly fossils. Specific assemblages of trace fossils have been used to recognize ancient depths as shallow as the intertidal zone down to depths of more than 6500 feet.

The top of the Dakota shifts gradually into the dark gray, organic-rich shale of the Mancos Shale. As the Dakota sands become finer-grained upwards, the thickness of the interleaved shale beds increases until they take over. Collectively, the Dakota and its gradational upper contact with the Mancos form a classic transgressive sequence that documents a widespread rise in sea level. The vertical shift from high-energy river deposits to those of low-energy coastal rivers, deltas, and swamps marks the westward migration of the shoreline as sea level rose. As the sea continued to expand west, the shoreline was drowned and transformed into a shallow, sandy shelf environment. Eventually the continued rise led to the burial of the Dakota by the deep marine muds of the Mancos Shale.

In a far-reaching transgressive sequence such as the Dakota, the age of a particular facies varies with location. For example, the swampy delta deposits become younger to the west. At no place along an east-west transect is this deposit the same age. Rather, it records an instant of geologic time that the delta was active at that particular locality. The rise in sea level, which controlled the westward migration of the shoreline delta, is in human terms an imperceptibly slow process. It took several million years for the shoreline to shift from central Colorado to central Utah, and although the delta/shoreline deposits form a continuous blanket across the region, it is everywhere a different age. This holds true for any facies in the Dakota—and for the formation as a whole. The corollary to this is that for any given instant of time, all the environments represented by the Dakota succession were present side by side (Fig. 7.6). Wherever the delta sands and muds were dropped, high-energy rivers could be found to the west, while burrowed offshore sands were accumulating immediately to the east. Still farther east the deep sea muds were simultaneously accumulating in the quiet depths of the Western Interior seaway.

One of the most difficult things to resolve in the geologic record is the rate of ancient processes such as transgressions. Detailed work by Fouch and oth-

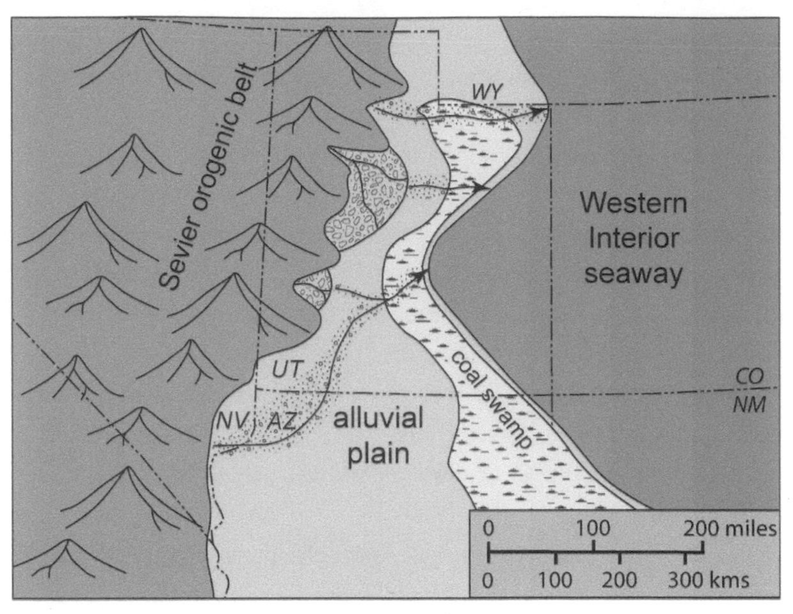

FIG. 7.6. Paleogeography of the Colorado Plateau region during the Late Cretaceous showing the various environments that were active during Dakota Sandstone and Mancos Shale deposition. The Dakota is represented by (from bottom to top) fluvial conglomerate and sandstone; deltaic and swamp deposits of sandstone, mudstone, and coal; and shallow marine sandstone. The Mancos Shale consists of deep-water shale and sparse sandstone. As sea level rose, the shoreline shifted westward, depositing rocks of what were adjacent environments on top of each other. Thus, as the shoreline migrated west, shallow marine sandstone was deposited on top of earlier coal deposits. Similarly, coal deposits were laid down on earlier fluvial sandstone and conglomerate.

ers (1983), and Rex Cole (1987), however, has provided for just such an opportunity. In examining marine fossil assemblages along the Dakota-Mancos contact in the Book Cliffs, Fouch and his colleagues identified two distinctly different fossil zones: one along the Utah-Colorado border, and another 90 miles to the west, near Price, Utah. Predictably these fossil zones documented the westward younging of this contact, but it also permitted an evaluation of the actual rate of transgression. According to Rex Cole (1987) the fossil zone at the Utah-Colorado border represents an absolute age of ~92.5 million years, while that from Price carries an age of ~91.5 million years, establishing an age difference of 1 million years. This 90-mile westward migration over a 1-million-year period translates to an average rate of 0.5 feet/year. It should be noted that this rate is averaged over a 1-million-year period; in reality, it probably was irregular, with more rapid and slower rates, and even some small-scale sea level drops over this time.

Mancos Shale

The gradual vertical shift from the shallow marine Dakota Sandstone into the deep-water Mancos Shale signifies a continuation of the greatest sea level rise of the Cretaceous Period—a time distinguished by an extraordinarily high sea level and numerous fluctuations. This transgressive and subsequent regressive sequence records what is known as the Greenhorn cycle, one of six transgressive-regressive events that have been deciphered from Cretaceous rocks in western North America (Fig. 7.7; Kauffman 1977). The first of these cycles (Skull Creek) actually triggered deposition of the earlier Cedar Mountain/Burro Canyon sequence. Also preceding the Greenhorn transgression was the Mowry cycle, which is well documented in the Dakota Group of Colorado's Front Range area. The Mowry could easily have been part of a continuous transgression that reached a zenith with the Greenhorn except for a minor drop in sea level that separates the two (Fig. 7.7). The Greenhorn cycle was followed by three more cycles whose signatures are clearly stamped into the Book Cliffs.

The Mancos Shale is dominated by dark gray, organic-rich, laminated shale with common, thin sandstone and siltstone layers and thin bentonite beds. The shale erodes into slopes and mounds, and typically accommodates wide valleys wherever it is exposed due to its nonresistant nature. It may also form steep slopes where it is protected by a sandstone cap, as at Mesa Verde in southwest Colorado and along the Book Cliffs in east-central Utah and western Colorado. The namesake of the formation is the Mancos River valley, situated along the north escarpment of Mesa Verde (Cross and Purington 1899). At this type locality the Mancos is 2250 feet thick and is overlain by sandstone of the Mesa Verde Group, also named for the area. To the north, the next complete exposure lies 150 miles away in the Book Cliffs, which loom just north of Interstate 70. Here, north of Arches National Park, near Crescent Junction, the Mancos reaches ~3500 feet thick. Farther east, near Grand Junction, Colorado, it thickens to 5600 feet. In the Book Cliffs, with some minor exceptions, the shale rises uninterrupted into the Mesa Verde sandstones. One exception is a 2-foot-thick sandstone bed believed to be a distal representative of the Ferron Sandstone Member that thickens to more than 1000 feet only 75 miles to the west.

Although the Ferron Sandstone Member of the Mancos is very thin in eastern Utah and indiscernible in western Colorado, it is briefly discussed here because it records the regressive phase of the Greenhorn cycle and to the west is a regionally important coal-bearing unit. Near Arches and along the I-70 corridor in that area the Ferron is an indistinct sequence of interbedded siltstone and shale capped by a thin bed of resistant, fossiliferous sandstone. Throughout this area it forms a low-lying ridge amidst the gray, rounded shale mounds. These beds are the distal eastern expression of a 1000-foot sequence of fluvial-deltaic sandstone, siltstone, and coal in Castle Valley, along the western flank of the San Rafael Swell. As the sea receded in the Castle Valley area, the shoreline was trailed by sediment-laden rivers draining eastward out of the nearby Sevier highlands. Wherever these rivers met the sea, large deltas built seaward.

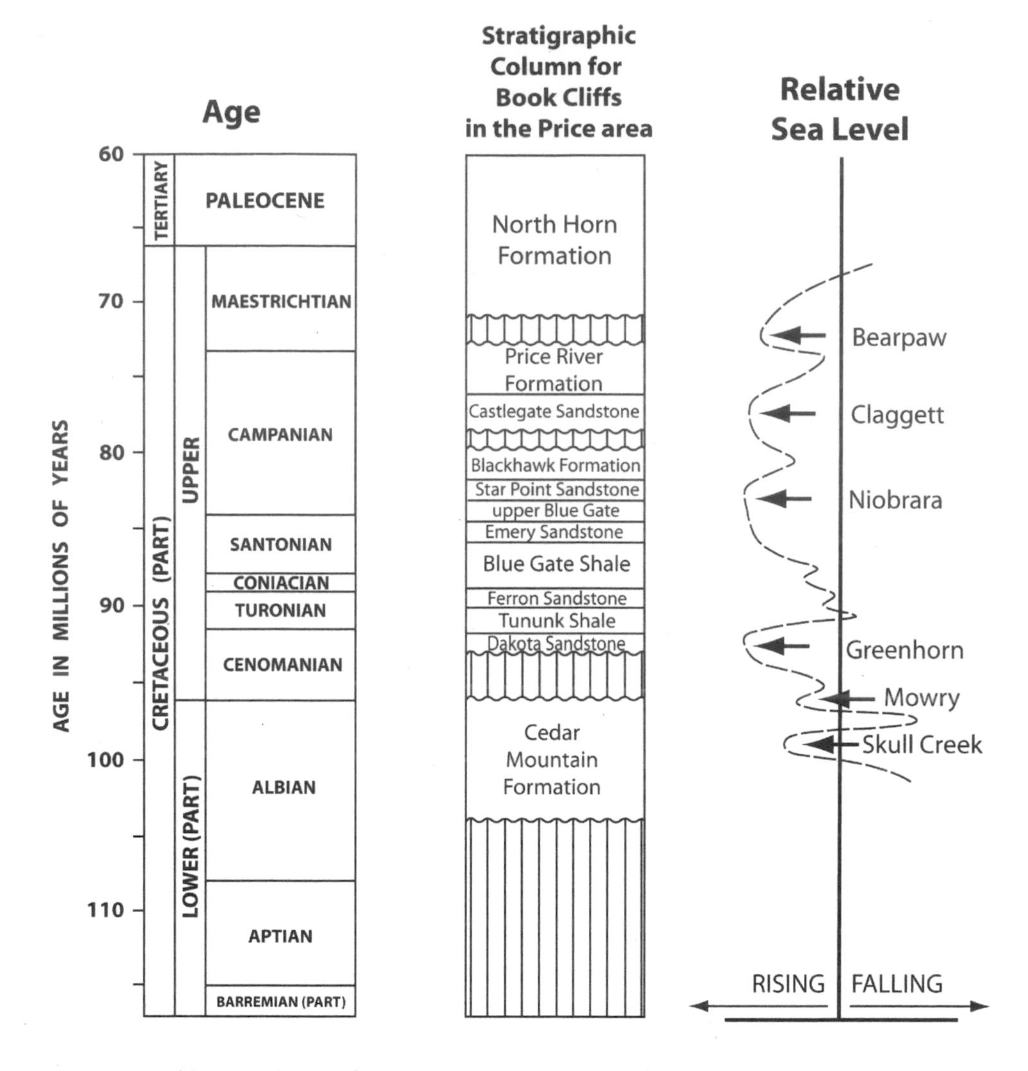

FIG. 7.7. Combination time scale for the Cretaceous and Early Tertiary, stratigraphy of the Book Cliffs, and relative sea level and cycles recognized in the sediments of the Western Interior seaway. After Cole 1987.

In the past 25 years the Ferron has been studied closely due in large part to its numerous coal deposits and their potential for producing coal bed methane; it also has similarities with important subsurface oil and gas reservoirs on the North Slope of Alaska and in the coastal area of the Gulf of Mexico (e.g., Ryer 1981, Ryer and McPhillips 1983, Anderson and others 1997, Garrison and others 1997). The continuous sandstone cliffs of the Ferron form gargantuan murals that allow geologists a unique opportunity to study in detail the characteristics that are prohibitively expensive, if not impossible, to glean from sparse and widely scattered subsurface data. Luckily for geologists it is cheaper and ultimately more informative to send them to the spectacular cliffs of Castle Valley than to haphazardly drill holes in Alaska and the Gulf of Mexico. These intensive studies have identified eight distinct deltaic sequences in the Ferron, each marked by

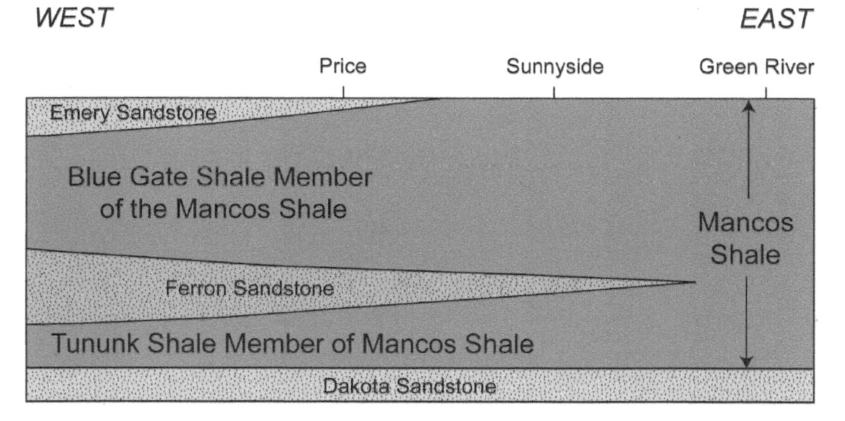

Fig. 7.8. Schematic stratigraphic cross section showing the split of the lower Mancos Shale into two members by the eastward progradation of the Ferron Sandstone into the Western Interior seaway.

a coal bed at the top. Although the Ferron ultimately documents the progradation of deltas during the Greenhorn regression, which separates it from the subsequent Niobrara cycle, the eight stacked and overlapping delta systems record the smaller-scale oscillations of the sea. Each shift, whether a rise or a fall, altered the delta-building process. Finally, where it is present, the Ferron sandstone body splits the shale facies of the Mancos into two discrete members. From the Henry Mountains in the south to Castle Valley and the Uinta basin in the north, the Ferron is underlain by the Tununk Shale and overlain by the Blue Gate Shale (Fig. 7.8). Farther east, however, in eastern Utah and western Colorado, the Mancos is an unbroken succession of shale with only a thin, sandy reminder of the great deltas that lay to the west.

Deposition of the Ferron continued unabated for a million years, but as the sea continued its eastward retreat, deposition ceased and erosion took over. The top of the Ferron over much of its extent is an unconformity that in the Henry Mountains represents an erosional interval of 2 million years (Peterson and others 1980). Farther to the north, in the western Book Cliffs and Castle Valley, the duration of erosion may have been shorter; however, duration increases to the west. The unconformity represents a decreasing period of time to the east until it disappears somewhere east of the San Rafael Swell. While erosion was taking place in a terrestrial setting to the west, sedimentation in the offshore marine environment of eastern Utah and western Colorado continued uninterrupted.

The subsequent transgression, which drowned the erosion surface at the top of the Ferron, opened the Niobrara cycle and spurred deposition of the Blue Gate Shale Member of the Mancos (see Fig. 7.7). As the rising sea pushed west, back over the delta sands, the region was blanketed by a thick layer of black mud. The depositional setting for the Blue Gate Shale was identical to that of the Tununk, across which the Ferron deltas had advanced earlier.

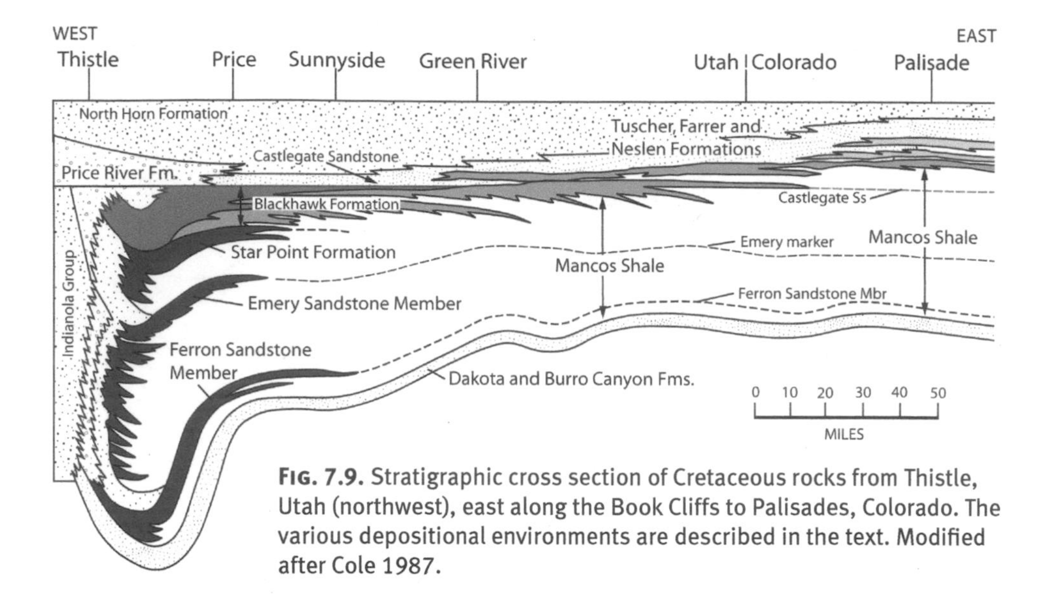

Fig. 7.9. Stratigraphic cross section of Cretaceous rocks from Thistle, Utah (northwest), east along the Book Cliffs to Palisades, Colorado. The various depositional environments are described in the text. Modified after Cole 1987.

Deposition of the Blue Gate Shale in east-central Utah continued for about 4 million years, but it was not without interruption. In fact, at the same position that the earlier Ferron deltas built to such a great thickness, another thick wedge of sand accumulated, spreading out across the thick mud on the sea floor. This body, the Emery Sandstone Member of the Mancos, reaches 800 feet thick at Castle Valley and divides the Blue Gate Shale into an upper and lower part (see Fig. 7.7). A short distance to the northeast, near the town of Price, and to the east across the San Rafael Swell, the Emery thins considerably, and the thick sandstone units of Castle Valley give way to alternating thin sandstone and black shale beds. The Emery thins to less than 50 feet along the Utah-Colorado border, where it is dominated by shale with thin, silty sandstone beds (Fig. 7.9).

According to geologists Paul Matheny and Dane Picard (1985), most of the Emery in the extensive Castle Valley exposures represents a tide-dominated, shallow marine setting that shifted to subtidal and offshore marine environments a short distance to the east and northeast. Although the Emery attains a maximum thickness in the same area as the deltaic Ferron Sandstone, it was by very different processes. Matheny and Picard suggest that rather than being fed from the delta to the shoreline, most of the Emery sand was diverted by longshore currents from a sand-rich area to the northeast. **Longshore currents** are shallow marine currents that run parallel to the shoreline, or "along the shore." The pattern of interbedded sandstone and shale, even in the thick, sand-dominated Castle Valley exposures, records several smaller-scale sea level fluctuations over shorter intervals than, for instance, the Niobrara cycle, of which the Emery was a small part. East of the San Rafael Swell the much thinner sandstone beds are deeper shelf deposits that formed during the small-scale regressive phases. These minor sea level drops permitted tidal currents and storm surges to drag coarser sand and silt into the deeper reaches of the seaway.

Through what likely was a combination of subsidence, continued sea level rise, and a shutdown of the sediment supply, Emery deposition came to a close. As the sea deepened and the shoreline was displaced to the west, the Emery was blanketed by a thick layer of black mud. The westward shift of the shoreline caused the once broad plain that separated the mountains from the sea to shrink to a narrow strip of coarse sediment. Farther to the east, while both the Emery and the upper Blue Gate Shale were being deposited, fine mud was raining to the sea floor without pause.

Star Point Sandstone

The regressive stage of the Niobrara cycle produced a slow retreat of the sea that began with the uneventful continuation of shale deposition across eastern Utah and most of Colorado. As the drop in sea level persisted, however, the first of a long line of increasingly wide sand bodies was laid down. This initial response to the eastward retreat was limited and produced the Star Point Sandstone. Exposures of the Star Point are clearly seen along the western escarpment of Castle Valley and extend almost unbroken northeastward into the Book Cliffs near Price, where it reaches a maximum thickness of 520 feet (Cole 1987). Farther east the Star Point sand bodies thin and disappear into the interminable mass of gray marine shale (Fig. 7.9). It is absent in the Book Cliffs above the town of Green River, Utah.

The Star Point Sandstone grades upward from the underlying Blue Gate Shale and consists of three discrete units separated by thick tongues (50–60 feet) of gray shale. Sandstone beds represent various types of coastal and shallow offshore sand complexes, including some delta deposits (Marley and others 1979, Newman and Chan 1991). Trace fossil and fossil assemblages, as well as sedimentary structures, indicate a shallow marine setting affected by normal wave activity and storm surges. Intervening shale beds indicate a periodic deepening and show that the withdrawal of the sea was not a smooth, continuous event, but was the cumulative effect of several smaller-scale oscillations.

In some places the top of the Star Point Sandstone is marked by a ledge of bleached white sandstone with abundant root traces, interpreted as an ancient soil zone or paleosol. This suggests that by the end of Star Point deposition, sea level had dropped enough to locally expose these upper sands, allowing them to host plants. Where the bleached zone is present, the contact with dark shale and mudstone of the overlying Blackhawk Formation is sharp. Where the bleaching is absent, however, the Star Point sandstones interfinger with the Blackhawk Formation in a gradational relationship (Marley and others 1979).

The Blackhawk Formation

The Upper Cretaceous Blackhawk Formation is a combination of sandstone, shale, siltstone, and coal deposited in a wide range of fluvial-deltaic and shallow marine environments. The Blackhawk is more extensive than the underlying Star Point, stretching from the Wasatch Plateau to the west, and pinching out eastward near the Utah-Colorado border (Fig. 7.9). A maximum thickness of ~1200 feet is attained in the Book Cliffs around Price. Western exposures of the formation conformably overlie the Star Point Sandstone. To the east each successive sandstone sequence penetrates farther into the mass of Mancos Shale (Fig. 7.9). These six discrete sand sequences, each designated as a formal member of the formation, record the stepwise retreat of the seaway that makes up the regressive phase of the Niobrara cycle. Coal is an important economic component of the formation, particularly to the west, in the Wasatch Plateau and Castle Valley, where it has been mined intermittently since 1898 (Johnson 1978).

Sandstone comprises ~60 percent of the Blackhawk in Castle Valley and occurs in three types of bodies, each defined by geometry, lateral extent, internal features, and associated rock types (Marley and others 1979). Comprising the first type are thin sheets of fine-grained sandstone that are bounded above and below by coal. Top and bottom contacts typically are sharp. In most sandstone beds the upper contact is cut by root traces, remnants of the tangle of vegetation that evolved into the overlying coal bed. These sandstone sheets were spread across the low-lying, swampy areas between the active distributary channels of the delta (Fig. 7.10). When sand-laden floodwaters careened through the sinuous delta channels, they burst through the leveed banks, blanketing the interdistributary swamps with a thin sheet of rippled sand. The wetlands flora were instantly buried by the sand, interrupting their growth; however, the overlying coal suggests that recovery of these plants was rapid.

The other two types of sandstone bodies in the Blackhawk are distinctly lens-shaped in cross-section, but they differ in dimensions. The larger of the two ranges up to 24 feet thick and 300 feet wide (Marley and others 1979). These lenses contain fine-grained sandstone with various types of cross-stratification and are the distributary channels of deltas (Fig. 7.10). These great conduits shuttled water and sediment eastward from the Sevier mountains to the receding shoreline of the Western Interior sea. Through the continued influx of sediment, the Blackhawk delta system methodically prograded eastward, filling in the moderately deep sea and burying the black Mancos mud with a thick blanket of sand.

Smaller lenses are similarly filled with crossbedded sandstone and trace the paths of minor channels that branched off the larger distributary channels (Fig. 7.10). Laterally these sandstone ribbons grade into coal, carbon-rich shale, or thin sandstone sheets, all representing various types of swamp deposits. These small

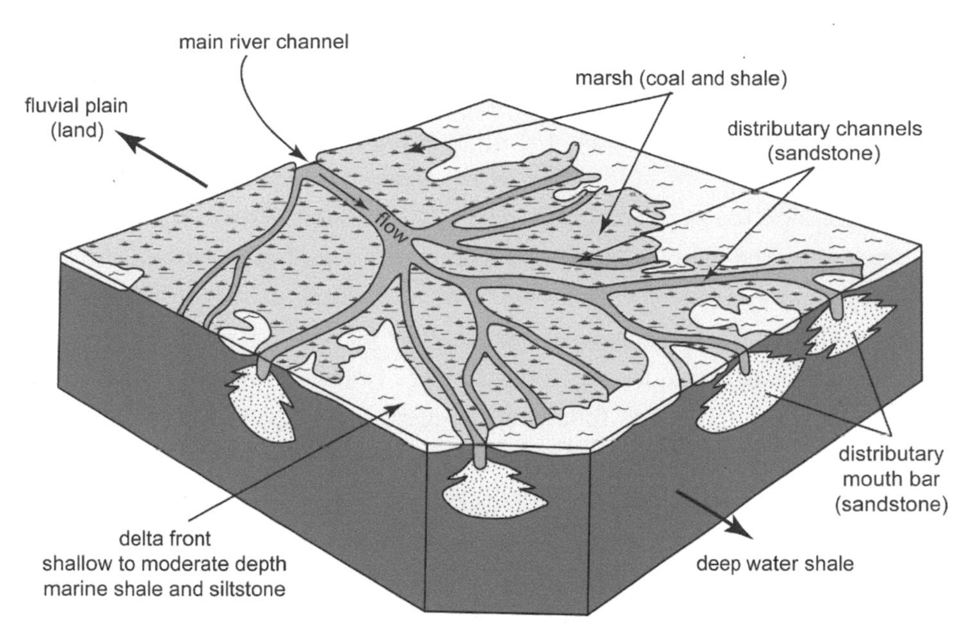

Fig. 7.10. Block diagram of the delta setting of the Blackhawk Sandstone and most river-dominated deltas in general. Branching channels are distributary channels that transport sandstone from the upstream river system. Marshes between the distributary channels are dominated by coal and shale.

watercourses siphoned flow from the main channels during floods, diverting water into the steamy marshlands.

Coal beds are common in the Blackhawk, especially in the lower reaches, where individual beds range up to 14 feet thick, although the thickness varies greatly. Coal swamps developed along the coast in protected lagoons and in the interdistributary lowlands farther inland. A detailed analysis of plant material from the Blackhawk by geologist Lee Parker (1976) revealed a plant community dominated by angiosperm trees and evergreen conifers, with ferns crowding the understory closer to the ground. The flora documented by Parker indicate a warm, temperate to subtropical climate along the Blackhawk coastal plain, a large-scale depositional setting closely resembling the modern Louisiana coast.

For each of the six members that make up the Blackhawk Formation there is, from west to east, a landward fluvial plain facies, delta and shoreline deposits, and a finger of shallow marine sandstone that extended seaward into the deep water of the Mancos Shale (Fig. 7.10). Each successive member stretched farther east, eventually producing, in a west-east direction, a 150-mile-wide sandstone deposit across the Mancos. The tonguelike sandstone of the basal Spring Canyon Member spread only as far east as the town of Price. This was followed by a temporary sea level rise that pushed the shoreline back to the west. Deposition of the overlying Aberdeen Member ensued when the sea resumed its eastward retreat. The Aberdeen sands pushed about 15 miles farther east than the underlying Spring

Canyon. Following this same zig-zag migration pattern, the overlying members steadily pushed eastward. Each successive member extended a wedge of sand another 15 to 45 miles east, until the uppermost Desert Member pinched down to a zero thickness a few miles short of the Utah-Colorado border. Through this erratic regression the sea began its gradual abandonment of the Colorado Plateau.

Etched into the winding escarpment and reentrants of the Book Cliffs, the Blackhawk is one of the world's best and most continuously exposed records of the regional advance of a fluvial/deltaic/shallow marine system during a regression. The Blackhawk was deposited during the sea level lowstand that separates the Niobrara cycle from the following Claggett cycle (see Fig. 7.7). It has also been suggested by some researchers that the upper Blackhawk, particularly to the east, may have been influenced by incipient tectonic unrest associated with the Laramide orogeny, a regional mountain-building event that later completely altered the face of the Colorado Plateau and generated the uplift of the Southern Rocky Mountains (e.g., Cole 1987, O'Byrne and Flint 1995, Yoshida 2000).

To the west the contact between the Blackhawk and the overlying Castlegate Sandstone is an erosional unconformity. Traveling east from the east rim of the Wasatch Plateau to the town of Green River, the unconformity represents a decreasing interval of time that is unrecorded. Farther east, at Westwater Canyon, near the Utah-Colorado border, the unconformity turns into a conformable contact, indicating that deposition across the Blackhawk-Castlegate boundary continued without pause. The Cretaceous strata of the Western Interior have been studied in such great detail that a very high resolution biostratigraphic framework has been established for both terrestrial and marine fossils (e.g., Kauffman 1977). Moreover, these detailed fossil zones have been correlated with numerous radiometric ages (real numbers), allowing geologists to establish an absolute age chronology for these successions. The combination of detailed biostratigraphy and absolute ages has also allowed the interval of missing time represented by the unconformity between the Blackhawk and the Castlegate to be determined in unparalleled detail. In its westernmost exposures along the east walls of the Wasatch Plateau the thin line of the unconformity represents a 3.5 million year gap in the rock record. Eastward, in Price Canyon, the missing time interval shrinks to ~1.5 million years. The decrease continues east to Green River, where it is reduced to approximately half a million years. Finally, along the Utah-Colorado border and beyond into Colorado, the unconformity disappears into a mass of marine sediments in which continuous deposition prevailed.

The Castlegate Sandstone, the Buck Tongue of the Mancos Shale, the Sego Sandstone, and the Neslen Formation

The Castlegate Sandstone is exposed without break along the east facade of the Wasatch Plateau for some 60 linear miles. This cliff band terminates north at the Price River canyon, but picks up again on the north side as part of the Book Cliffs, where it extends continuously for another 160 miles, first to the southeast, then east until the Castlegate thins to a pinchout near the Utah-Colorado border (Fig. 7.9). Once again the detailed biostratigraphic framework that has been established for the Cretaceous Period of western North America allows this formation's age to be tightly bracketed between 79 and 74 Ma (Fouch and others 1983). Its type locality is tucked into the upper reaches of Price River canyon, just above the town of Castlegate, where three informal members have been recognized based on lithology and depositional setting (Chan and Pfaff 1991). The three divisions of the 636-foot-thick formation include a lower member composed of cross-stratified sandstone and pebble conglomerate that marks the passage of an energetic, east-flowing, braided river system; a middle member dominated by large lenses of sandstone encased in carbonaceous mudstone, interpreted as the product of slower, meandering rivers; and an upper member, also known as the Bluecastle Tongue, that signifies a shift back to a coarse-grained, braided river system. Equivalent strata a short distance to the west coarsen into a massive cobble and boulder conglomerate that accumulated on a bajada of coalescing alluvial fans at the foot of the Sevier mountains. Because of their different nature these strata are part of the Sixmile Canyon Formation of the Indianola Group (Lawton 1986).

Braided streams of the lower member endured as far east as Green River, where a decrease in gradient provoked a change to more sedate, meandering rivers. This transition is told by large lenses of sandstone (river channels) embedded in sheetlike bodies of silty mudstone (floodplain). These winding ribbons of sluggish water met the delta/shoreline environment ~15 miles east of the Green River. A short distance farther east, in modern-day Colorado, the sandy offshore deposits were reduced to a feather edge and vanished into a bottomless mass of black Mancos Shale (Fig. 7.9). A subsequent sea level rise again allowed the sea to invade the region, blanketing the delta and meandering stream deposits with a finger of dark marine mud as far west as Green River.

The transgression that followed deposition of the lower Castlegate triggered a westward shift in facies. At the type locality in the Price River canyon this change stacked low-energy river deposits of the middle member on the braided stream facies. Farther east, in the vicinity of Green River, a series of correlative marginal marine and offshore marine deposits record this sea level rise and its subsequent fall. Here the deeper marine Buck Tongue of the Mancos Shale abruptly overlies the fluvial/deltaic lower Castlegate and extends a sliver of gray silty shale 65 miles westward, far into the variable mass of Mesa Verde Group sandstones (Fig. 7.9). The Buck Tongue grades upward into the fine-grained sandstone and siltstone of the Sego Sandstone. Fossil oysters, trace fossils, and various types of cross-stratification and ripples indicate deposition in a tidal flat to shallow marine setting (Lawton 1983). The Sego is, in turn, succeeded by the Neslen Formation, which represents the eastward advance of a low-gradient coastal plain and coal swamps that grade east into tidal flats. The Neslen is blanketed by the coarse sandstone of the Bluecastle Tongue of the Castlegate, which heralds the rejuvenation of braided rivers across the region. The marine strata sandwiched between the braided river deposits of the Castlegate thus track a complete transgressive-regressive sequence within the 5-million-year duration of its deposition.

Although clearly affected by sea level fluctuations, deposition of the Castlegate and its eastern equivalents is believed to have been influenced mostly by tectonic unrest in the adjacent Sevier mountain belt (van de Graff 1972, Fouch and others 1983, Willis 2000). In a detailed analysis of Late Cretaceous and Early Tertiary river systems across Utah, geologist Tim Lawton (1983) proposed that the initial progradation of the lower Castlegate was driven by the rise of discrete thrust sheets in the Sevier fold and thrust belt. The following transgression, recorded by the Buck Tongue, is attributed to increased subsidence driven by the thrust load to the west, or a temporary reduction of incoming sediment while subsidence continued at a normal pace. Either scenario could have produced an "apparent" sea level rise. Deposition of the overlying Sego and Neslen formations marks the gradual filling of this new space, pushing the sea back to the Utah-Colorado border area. According to Lawton, the renewal of energy in this system records a resurgence of uplift to the west, initiating the eastward spread of coarse sand across the expanding coastal plain.

The Price River, Farrer, and Tuscher Formations

The latest Cretaceous Price River Formation conformably overlies the Castlegate along the upper escarpments of the Price River Canyon. To the east, in the Book Cliffs around Green River, the formation becomes the Farrer (lower) and the Tuscher (upper) formations. The contact between the Farrer and the underlying Bluecastle Tongue is sharp, marked by the bleached upper surface of the Bluecastle. This oxidized zone suggests prolonged exposure and an unconformable contact. In a study of palynomorphs (fossil pollen) across this line, however, Fouch and others (1983) discovered that very little to no time was missing. It is likely that the duration of weathering was shorter than the resolution available with the fossil record.

The Price River Formation consists of 600 feet of ledge-forming, pebbly sandstone that was deposited by energetic, sediment-laden, braided rivers. These lively rivers were a continuation of the earlier systems that redistributed the detrital remnants of the mountains eastward. The eastern equivalent of the lower Price River, the Farrer Formation, reaches up to 1000 feet thick just east of Green River but thins westward to less than 300 feet before disappearing completely across the crest of the northern San Rafael Swell. This pinchout is attributed to later erosion rather than a lack of deposition. The Farrer is made up of large, lens-shaped bodies of sandstone contained within more continuous siltstone and mudstone deposits; these represent the in-channel deposits of meandering rivers and their adjacent floodplains, respectively (Lawton 1983). The Price River–Farrer transition marks the change from proximal, braided rivers to a more sedate, meandering system with distance from the mountainous source and consequent decrease in gradient.

The Tuscher Formation succeeds the Farrer and is the eastern equivalent of the upper Price River Formation. At its namesake Tuscher Canyon, east of Green River, the formation consists of 600 feet of sandstone and siltstone that coarsens upward into conglomeratic sandstone. Like its predecessor, the Tuscher thins and pinches out to the west along the flanks of the San Rafael Swell. Lower sandstone and siltstone beds recount the passage of large, east-flowing, meandering rivers. In contrast, coarse-grained, upper pebbly sandstones record the northeast-flowing paths of braided rivers. The package as a whole represents the rejuvenation of the drainage system with a conversion from relatively low energy to high energy rivers. In addition, this transition marks a reorganization of the drainage network on the broadening alluvial plain.

The change in drainage patterns that occurs with the Price River Formation in the west and the Farrer/Tuscher succession to the east coincides with a pronounced change in the composition of the sand that makes up these units (Lawton 1983, 1986). It is common for geologists to look in detail at the minerals that make up the individual grains in a particular sandstone in an effort to decipher the makeup of the source highlands. In many cases the sandstone composition, coupled with a reconstruction of drainage patterns obtained from crossbedding, can point back to the source terrane, or the mountains that supplied the sand, even if those mountains have long since been flattened by erosion. This is a vital step in reconstructing the paleogeography of the region, and one that geologist Tim Lawton (1983, 1986) has used effectively for the latest Cretaceous of east-central Utah. In his study Lawton documents an abrupt change in sandstone composition from quartz-rich sand (more than 90 percent quartz grains) in underlying Mesa Verde Group sandstones, to a sand composition in the Price River–Farrer/Tuscher formations with up to 50 percent of the grains made of lithic rock fragments. The remainder of the grains are quartz.

In general, the quartz-rich sandstones that make up the lower Mesa Verde Group are typical of most sandstones from around the world and throughout geologic time. Several factors contribute to this abundance. First, the mineral quartz is by far the most common mineral in the rocks that make up the continental landmasses, and therefore the rocks most likely to decompose to sand. Second, of all the common rock-forming minerals in the continental crust, quartz is the hardest, so the most resistant to breakdown and destruction. Thus, the disintegration of almost any rock into sediment and its introduction into the natural grinding mill of rivers generate an overwhelming quantity of quartz sand. This is reflected in the composition of the sandstone that it eventually forms.

In contrast to the single-mineral sand grains of quartz, lithic rock fragments typically are multimineral sand grains that are pieces of preexisting rock—for example, sedimentary rock fragments such as chert or limestone, or volcanic rock fragments such as basalt or andesite. Any rock type can degrade to sand-sized lithic rock fragments. These fragments, however, are not as tough as quartz and tend to be destroyed in the abrasive turbulence of rivers and wave-battered shorelines. So while quartz sandstone is common, lithic-rich sandstone is significant in its sudden presence in the Price River Formation and its lateral equivalents. This change represents the uplift and exposure of new rock types in the nearby Sevier orogenic belt, but it is not quite that simple.

Although lithic fragments appear in both the Price River Formation to the west and its eastern equivalents, the Farrer and Tuscher formations, these geographically separated successions differ substantially in the types of lithic grains that they contain (Lawton 1983, 1986). The Price River Formation contains a large component of sedimentary-derived lithic fragments, including sand-sized grains of shale, sandstone, limestone, and chert. In contrast, the eastern facies (Farrer/Tuscher) shows an abundance of volcanic lithic fragments, a unique addition to a sand grain population that previously appeared to hail exclusively from the voluminous sedimentary rocks of the Sevier mountain belt. In evaluating these compositional differences in conjunction with paleocurrent data, Lawton (1983, 1986) reconstructed a setting in which the volcanic-bearing Farrer/Tuscher sands were derived from some not-yet-recognized volcanic terrane to the southwest. This sand was then transported by northeast-flowing rivers that ran parallel to the mountain front and merged with the Price River drainage a short distance north of Green River, in east-central Utah (Fig. 7.11). The drainage of the northeast-flowing Price River Formation emitted more directly from the east-advancing Sevier mountain belt, which remained dominated by uplifted Paleozoic and Mesozoic sedimentary rocks. The renewed uplift and eastward push of the mountain front decreased the travel distance of the resulting detritus destined for the Price River Formation, increasing the abundance of sedimentary lithic fragments.

The interpretation that the Farrer/Tuscher succession is the distal part of a continuous drainage system that originated to the southwest fits well with the age-equivalent uppermost Cretaceous strata to the south. In the vast Kaiparowits Plateau of southern Utah, the Kaiparowits Formation and the overlying Canaan Peak Formation also record the introduction of volcanic lithic fragments into the foreland sands (Goldstrand 1992, Lawton and others 2003). Crossbedding in these fluvial deposits similarly shows northeast-directed paleoflow. Thus it appears that the Farrer/Tuscher succession is the distal segment of a continuous mountain front–parallel system that, to the south, is represented by the Kaiparowits and

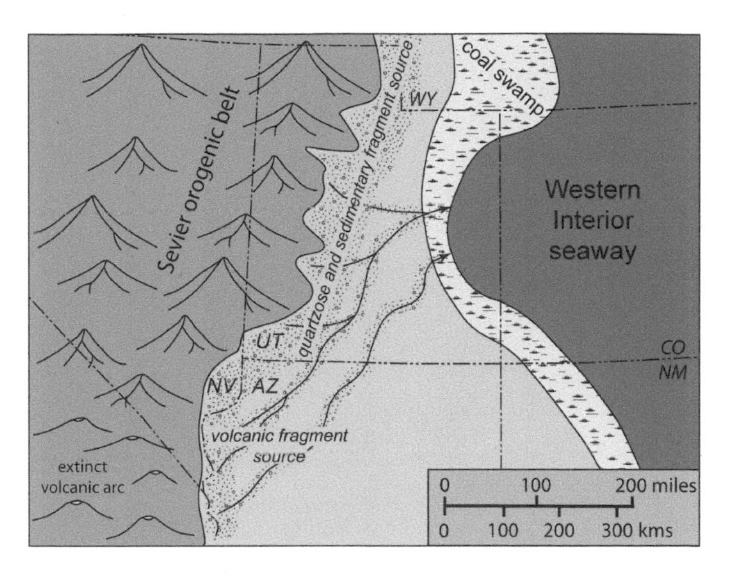

Fig. 7.11. Paleogeography and sediment source areas for the Late Cretaceous Price River and Farrer-Tuscher River systems. Volcanic fragments in the Kaiparowits Formation in southern Utah and the equivalent Farrer and Tuscher sandstones to the north were derived from the volcanic terrane that likely resided in the southern California area. At the same time, quartz and sedimentary rock fragments continued to be shed eastward from the Sevier orogenic belt to merge with the volcaniclastic Kaiparowits system, diluting its volcanic content in the Farrer-Tuscher sediments. Modified from Lawton 1986 and Lawton and others 2003.

Canaan Peak deposits (Fig. 7.11). This regional drainage network likely was born in highlands in the California–southern Nevada region, where it cut headward to the west to tap the extensive volcanic arc that marked the Cretaceous west coast. From there it entered the foreland and flowed northeast, just east of the Sevier mountains, which were in their final spasms of uplift. Along this course the main trunk river was fed by various tributaries that debouched from the imposing highlands, each adding their load of quartzose sediment to the system. The Price River Formation represents one such tributary.

The final pulse of Cretaceous deposition and tectonism on the Colorado Plateau is recorded in the uppermost, conglomeratic part of the Tuscher Formation. Associated with these terminal Cretaceous strata are previously unrecognized beds of localized conglomerate that rest above the Tuscher but lie beneath documented Tertiary sediments. These "pebbly beds," as they are informally called by Lawton (1986), are also thought to be of Cretaceous age. The recurrence of conglomerate in the Tuscher and pebbly beds marks a rejuvenation of these northeast-flowing rivers. This change is coupled with a shift in composition from the earlier, volcanic-rich sediments to a more quartzose composition. Lawton has suggested that these changes herald the incipient uplift of the San Rafael Swell and possibly other monoclinal upwarps that today dominate the structure of the Colorado Plateau.

The rise of monoclinal upwarps across the Colorado Plateau disrupted the simple, earlier configuration in which large rivers flowed east or northeast from the Sevier orogenic belt. These upwarps generated new sediment sources as earlier Mesozoic strata were exhumed to provide a recycled supply of quartz-rich sediment. The prior northeast-flowing rivers were deflected around these upwarps, causing the once-continuous foreland basin to be segmented into smaller, but connected basins that received sediment from multiple sources. According to

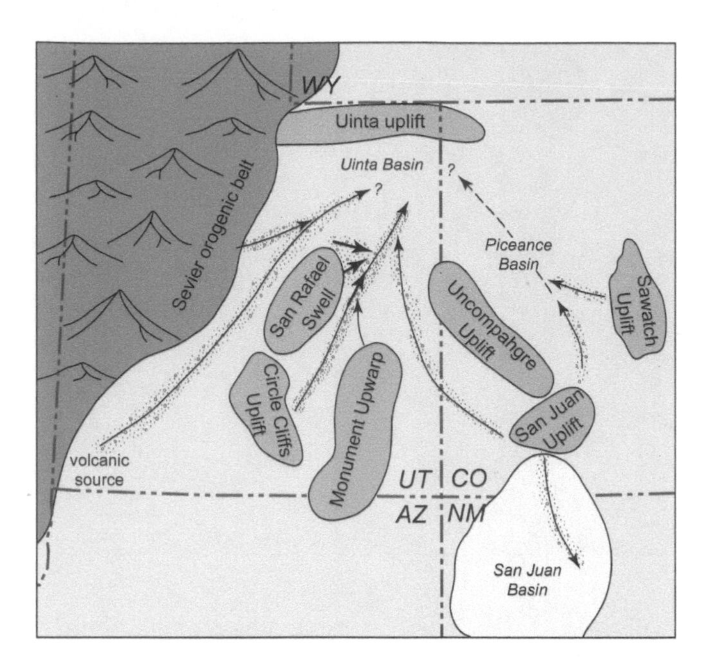

FIG. 7.12. Paleogeography during latest Cretaceous deposition of the upper part of the Price River and Tuscher formations showing various Laramide uplifts that segmented the earlier continuous foreland basin and disrupted the drainage patterns. Arrows show paths of rivers that drained from these highlands. After Lawton 1986.

reconstructions by Tim Lawton (1986) these highlands included the San Rafael Swell, the Circle Cliffs uplift, the Monument upwarp, and the Uinta uplift, which bounds the Colorado Plateau to the north (Fig. 7.12).

This new stage of uplift marks the opening of the Laramide orogeny in western North America, the most recent of a continuous series of mountain-building events that began in the Jurassic. Laramide deformation spread eastward across the Colorado Plateau and into modern-day Colorado and New Mexico to form the modern Southern Rocky Mountains. As told by the sediments, this event initiated near the end of the Cretaceous (65.5 Ma) and continued into the Tertiary to ~50 Ma. In contrast to the stacked, shingle-like uplifts of the thrust-faulted sedimentary strata of the previous Sevier orogeny, the Laramide shoved deep-level basement rocks upward to produce a broad zone of upwarps and high mountains separated by deep sedimentary basins. As Laramide deformation rippled eastward, uplift in the Sevier belt ceased. Through the Early Tertiary (the Paleocene and Eocene epochs), the compressional stress of the orogeny was transferred eastward, across the previously undisturbed foreland basin that had endured unscathed throughout the Cretaceous. The details of the Laramide orogeny will be addressed in further detail in the following chapter on the Tertiary Period, as it was during that time that the mountain-building event reached its zenith.

Terminal Cretaceous Extinction—Gateway to the Tertiary

The end of the Cretaceous Period at 65.5 Ma is marked by one of the greatest extinction events in Earth history. This event marks not only the Cretaceous-Tertiary boundary, but also draws the important line between the Mesozoic and

Cenozoic eras. At this time approximately half of all life on Earth was killed, including land animals and plants, and marine organisms both large and small. Possible causes for this worldwide extermination have, for the past twenty-five years, been the subject of thousands of articles and books in scientific and popular literature alike, and have stimulated a vigorous and sometimes bitter debate. This surge of interest was triggered by the proposal of Alvarez and others (1980) that this extinction was the result of an asteroid impact with Earth. The sudden demise of all dinosaurs after the collision of Earth with an extraterrestrial object captured the public imagination as no other event in Earth history could. Predictably, the hypothesis raised the hackles of many geologists, who had been trained to discount any geologic explanation that called on catastrophic processes. Thus, the bar was raised to a considerable height for the proponents of the impact hypothesis.

Although the disappearance of the dinosaurs is most commonly cited in discussions of the terminal Cretaceous extinction (herein referred to as the "K-T" extinction, using the geologic symbols for the Cretaceous and Tertiary periods), it affected both land and marine dwellers, from gigantic to microscopic. On land the dinosaurs and pterosaurs (flying reptiles) were wiped out completely. Land plants also were hit hard, severely damaging the bottom of the terrestrial food chain. Mammals, which rebounded to take over the Cenozoic land, were nonetheless affected. Although they survived, there was a swing from a marsupial-dominated Cretaceous population to one of mostly placental mammals in the Tertiary. In the seas microorganisms whose populations plummeted at the K-T boundary include the planktonic (free-floating) foraminifera and coccoliths. Molluscs also were decimated, although some recovered to repopulate and dominate the floors of the shallow Cenozoic seas. Ammonites, swimming predators that had hunted the seas since the Devonian Period, went extinct. Finally, the giant reptiles that had adapted to the sea died out completely.

Although it has not been unanimously embraced by Earth scientists, most believe the K-T extinction event was triggered by the collision of an asteroid with Earth. The overwhelming burden of evidence for such a catastrophic scenario appears to have been overcome by the hundreds of scientists who have worked on pieces of this global puzzle. A great body of diverse evidence from around the world has been generated by collaborations between scientists of all kinds, including geologists, oceanographers, biologists, physicists, and chemists. As growing volumes of supporting evidence have poured in, however, skeptics maintained that without the discovery of the actual impact crater (the "smoking" crater, if you will), it was meaningless. Finally, after a decade of increasingly detailed analyses of K-T boundary sediments, much speculation, and an eventual focus on the Caribbean region, Hildebrand and others (1991) identified the Chicxulub crater in the northern part of Yucatán, Mexico, as the impact structure. Subsequent studies have confirmed this. This report spurred a flurry of activity that continues to this day.

The first line of evidence, and the one that led to the original impact hypothesis, was accidentally discovered in Gubbio, Italy, by the father-son team of Luis and Walter Alvarez. In a thin clay layer previously identified as the K-T boundary based on fossil evidence, they recognized an anomalously high concentration of the element iridium. Although it is extremely rare in the Earth's crust (upper ~50 km), iridium is slightly more abundant in the mantle and in meteorites. Alvarez and others (1980) proposed that such a concentration could only have come from an extraterrestrial source, and they suggested an asteroid. While many waved the iridium off as a local phenomenon, subsequent work identified an anomalous iridium spike in K-T boundary sediments at 105 sites around the world, from both terrestrial and marine deposits.

Another convincing line of evidence is shock lamellae in grains of quartz, feldspar, and zircon found in K-T boundary sediments at more than 50 sites. Shock lamellae are regular, wavy fractures that permeate relatively hard minerals when submitted to instantaneous intense pressure. Such features previously have been identified only from known meteor-impact sites and are generally accepted as evidence of an impact event.

Another crucial piece of the K-T puzzle was the discovery of microscopic tektites at more than 60 boundary sediment localities worldwide. These microtektites originated from microscopic blebs of impact-generated molten rock, formed during the exceptionally high temperature of a meteor impact. **Tektites** are defined as pieces of nonvolcanic glass, typically with an aerodynamic shape that forms as the molten material solidifies while hurtling through the air at hypersonic velocities. During a large meteor impact, bits of molten rock spew into the atmosphere as the ejecta of the crater-forming process. The widespread distribution of these solidified microdroplets may be enhanced by the atmospheric transport of the ejecta before their eventual drop to Earth.

In general, the compositional term *glass* refers to a solid, noncrystalline substance formed from the rapid cooling of molten rock. Solidification is so rapid that atoms of the various elements that make up the melt are unable to come together and bond to form minerals. Instead, the ions fuse into an unordered mass to form an amorphous solid. Regardless of its origin, however, whether it is from magma or an impact-generated melt, glass has a characteristic chemical composition that can be analyzed for comparison with the bulk chemistry of various rock types.

The ability to chemically fingerprint these far-flung glassy fragments from the K-T boundary provides a means to trace their source. Two distinctive types of glass have been recognized in this layer. These include a more-abundant black glass and a less-common yellow glass. Yellow microtektites are rich in CaO and SO_3, suggesting derivation from limestone ($CaCO_3$) and anhydrite ($CaSO_4$), both present in abundance in the Yucatán region and as breccia fragments in the Chicxulub crater. The black glass has a bulk chemical composition of the volcanic rock andesite. These match deep-level rocks in the Chicxulub crater that were recognized earlier during the drilling of exploratory petroleum wells. Within the deep confines of the crater is a thick (< 1 km) layer of intermixed andesite and andesitic glass, believed to have formed during impact melting of the sedimentary rocks. Inclusions of shocked quartz in the andesite and the absence of these rocks in drill holes immediately outside the crater point to an impact melt. In addition to correlations in tektite chemistry, the size and abundance of these spherules increase markedly towards Chicxulub crater. Finally, Swisher and others (1992) used the $^{40}Ar/^{39}Ar$ isotope system to obtain dates from both the melt layer from Chicxulub crater and tektites from the boundary layer in Haiti. This precise analytical method produced essentially identical ages of 65 Ma. Taken collectively, data from K-T boundary tektites strongly support their origin as ejecta of the crater-forming meteor impact at Chicxulub at the Cretaceous-Tertiary time boundary.

The thickness of the K-T boundary layer decreases away from the Chicxulub impact site in the southern Gulf of Mexico. Thick proximal sections include Coxquihui, Mexico, where the layer reaches 90 cm, and on the island of Hispaniola, where it ranges from 12 to 70 cm. In addition, Haitian tektites reach up to 2 cm in diameter. In western North America, in Colorado's Raton basin, ~2200 km from the Chicxulub site, the layer thins to 1–2 cm. At distances of more than 7000 km it thins down to only 2–3 mm (Smit 1999).

The well-preserved Chicxulub crater mostly lies beneath the waters of the southern Gulf of Mexico; however, a small, arclike segment is "exposed" along the heavily vegetated northern tip of the Yucatán Peninsula (Fig. 7.13). Most of the current knowledge of the crater's geometry comes from seismic reflection experiments that have criss-crossed the crater (Morgan and Warner 1999). According to interpretations based on this seismic survey, the crater diameter from rim to rim is 145 km (90 miles). Moreover, the deformation of the impact actually penetrates the crust and extends down into the uppermost mantle.

It is apparent that a large meteor impact occurred at Chicxulub crater at the K-T boundary, and that many organisms perished at about the same time. The question that arises is, how did the impact affect so many organisms around the world? On a local level the devastation would have taken on many faces (Kring 2000). In the region within a ~1000 km (620 miles) radius of the impact site the air blast alone would have toppled all trees. Reconstructions from the sedimentary record suggest that tsunamis, or gigantic ocean swells, triggered by the impact may have reached 300 m (1000 feet) high as they rippled across the Cretaceous version of the Gulf of Mexico. These disruptions would have been intensified by the magnitude 10 earthquakes that must have accompanied the impact. Finally, a thick ejecta blanket would have buried any surviving life within a few

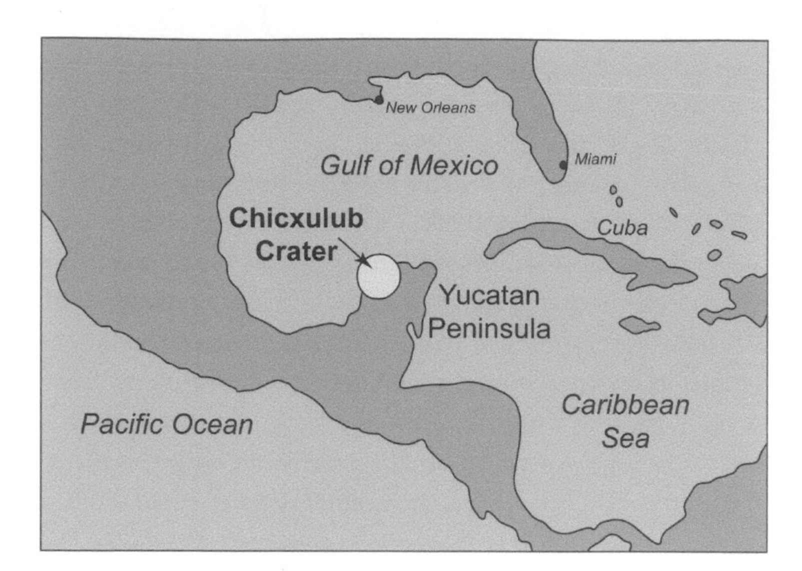

FIG. 7.13. Map of the Gulf of Mexico region showing the location of the Chicxulub Crater, the site of meteor impact.

hundred kilometers of the crater. Yet all of these devastating effects would have been confined to the Gulf of Mexico and its immediate surroundings. What killed off the organisms in the rest of the world?

When Alvarez and others (1980) first proposed their impact hypothesis, they attributed the extinctions to the secondary effect of debris ejected high into the atmosphere. Although this theory has not changed much in the past few decades, it has been refined to the effect of dust and specific gases (Kring 2000). Dust and vaporized seawater generated by the impact were launched through the atmosphere and into space. Upon exiting the atmosphere, the ejecta plume spread around the globe before eventually settling back to Earth. This would have blocked sunlight, darkening the skies for up to a year and killing organisms, mostly plants, that relied on photosynthesis for survival. This effectively truncated both continental and marine food chains at their base. The impedance of sunlight would also have rapidly decreased continental temperatures over the short term.

The vaporization of limestone, anhydrite, and silica-rich basement rocks likely produced several gases easily capable of inducing sudden climate changes (Kring 2000). These include SO_2 and SO_3 from anhydrite, CO_2 and H_2O from limestone and seawater, and Cl (chlorine) and Br (bromine) from a combination of impacted materials. Sulfates would have generated sulfuric acid rain, contributing to the demise of terrestrial and aquatic plants, as well as further reducing sunlight. Elevated levels of CO_2 and H_2O likely produced a greenhouse effect, causing a long-term increase in global temperature. The volume of Cl and Br estimated to have been blown into the atmosphere would have been more than enough to destroy the ozone layer, removing a vital protective barrier from the atmosphere. Taken collectively, these results would have dramatically affected climate over different time scales and in different, and sometimes contradictory, ways. However, considering the unprecedented scale of the Chicxulub impact and these predicted effects, it is not unreasonable, or even difficult, to envision such catastrophic results on ecosystems worldwide.

During investigations of the K-T boundary layer over the past few decades, the only complete section found in southwest North America was in the Raton basin, which straddles the Colorado-New Mexico border. The apparent absence of K-T boundary sediments in Utah and across the Colorado Plateau has been attributed to a combination of the Late Cretaceous regression of the sea and the onset of the Laramide orogeny. Both promoted erosion rather than sedimentation, and where deposition was occurring, it was dominated by high-energy rivers with a slim chance of preserving such a thin, inconspicuous clay layer. Recently, however, detailed investigation into a Cretaceous-Tertiary succession in southern Utah has pinpointed a fully preserved section of K-T boundary sediments.

Geologist Rose Difley and her colleagues (2004) from Brigham Young University recently identified a thin K-T boundary succession in the North Horn Formation of Emery County, Utah. The strata are preserved in low-energy lacustrine deposits ~685 feet (208 m) above the base of the formation. The boundary had previously been approximated based on fossils and is ~15 feet (4.4 m) above the last dinosaur fossils, and 33 feet (10 m) below the first Paleocene (Early Tertiary) turtle fossils. The succession rests on fossiliferous limestone and a thin coal bed. The kaolinite-rich boundary clay unit above the coal is 1 cm thick and is overlain by 22 cm of mudstone containing shocked quartz, altered microtektites, and charcoal. The fortuitous preservation of these thin layers of fine sediment is attributed to the local development of ponds or lakes on the North Horn floodplain. The discovery of this succession is a testament to the careful analysis and detailed study of the North Horn sequence by Difley and colleagues. Recognition of this thin but important interval will undoubtedly aid subsequent workers in identifying with precision the K-T boundary elsewhere on the Colorado Plateau.

Chapter 8

The Tertiary Period

The Rise of the Colorado Plateau

The Tertiary Period (66 to 1.6 Ma) marks the rise of mammals and the molding of the Earth, both biologically and physically, that we see today. The global extinction of the dinosaurs and numerous other organisms at the close of the Cretaceous opened the door for a different population of organisms that eventually became dominated by mammals.

When discussing the strata and events of the Tertiary Period, geologists tend to use its more refined subdivision, called epochs (Fig. 8.1). Refinement of the Tertiary time scale is possible due to the greater preservation of Tertiary sediments worldwide, which provides a more detailed record. The fossil organisms used to define these epochs are better preserved than their Paleozoic and Mesozoic counterparts, and life was more diverse. Tertiary sedimentary basins, in large part, are still basins, and thus their original context has not been complicated by later deformation. In addition, these strata have not been buried at deep levels in the crust, where they would likely be altered, possibly making fossil material unrecognizable. In short, the more complete preservation of Tertiary strata allows for a finer and more detailed time scale for these younger rocks. In the following discussions, a combination of epochs and absolute ages are used to describe the timing of events and the age of rock units.

During the Early Tertiary, mountain building in western North America continued over an expanded area. This reversed the long-lived trend of deposition with a shift to regional erosion, the process that dominates across the region to this day. Erosion was also stimulated toward the end of the Cretaceous by the withdrawal of the sea. On the Colorado Plateau the main result of this widespread erosion was the later evolution of the Colorado River drainage system, which ultimately led to the sculpting of the canyons, mesas, and buttes that make the modern Colorado Plateau landscape so unique.

The Tertiary dawned with the Laramide orogeny, a continuation of Cretaceous mountain building, but with a pronounced eastward shift into what was previously an extensive sedimentary basin. As uplift proceeded, the Western Interior basin was segmented into high-relief uplifts separated by relatively small intervening basins (Fig. 8.2). It was during the Laramide orogeny that the modern Southern Rocky Mountains of Wyoming, Colorado, and New Mexico formed. On the Colorado Plateau this great mountain-building event resulted in monoclines—the great, ramplike folds that are so boldly displayed across its expanse. Prominent monoclines in southeast Utah include Comb Ridge, the San Rafael Swell, and the Waterpocket Fold in Capitol Reef National Park. The huge, single-limbed fold of Colorado National Monument—which drops beneath the city of Grand Junction, Colorado—is also a structure of this generation. Although sediments likely were deposited in basins on the downwarped margins of these monoclines, subsequent erosion has stripped these Tertiary deposits from most of the Plateau. As regional erosion began in the later Tertiary, it was the uppermost succession of earlier Tertiary sediments that were the first to go. Tertiary sediments are preserved, however, to the north above the Book Cliffs, where they cap the well-preserved Cretaceous succession (discussed in the previous chapter). Elsewhere on the Plateau, Tertiary sediments are preserved in Bryce Canyon National Park, where they make up the colorful red and white strata that form its undulating fins and hoodoos.

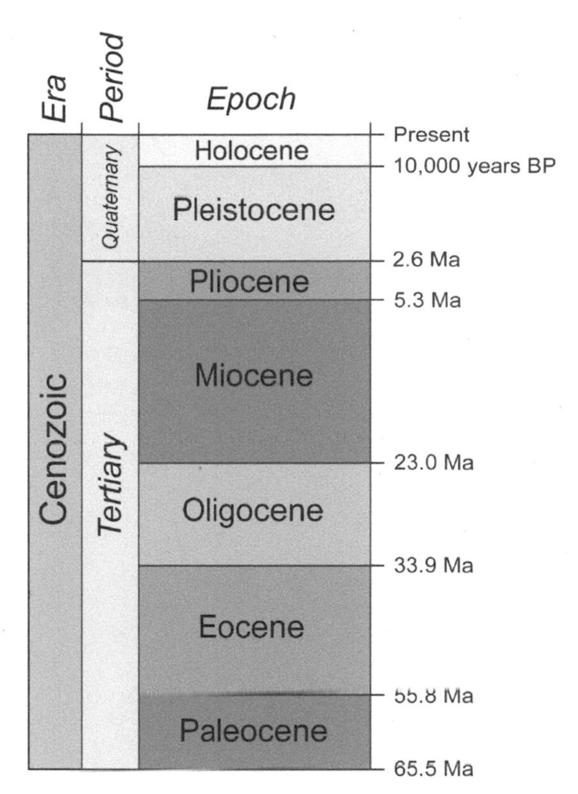

Fig. 8.1. Geologic time scale showing the epochs and numerical ages of the boundaries for the Tertiary and Quaternary periods.

The worldwide extinction event that defined the end of the Cretaceous Period led to a rapid diversification of life as survivors evolved and adapted to voids in the food chain. During the Cretaceous, mammals were small, shrewlike animals that mostly scurried through the forest understory, trying to avoid the dinosaurs. With the demise of the dinosaurs, however, the terrestrial ecosystem was thrown wide open, and mammals rapidly expanded to fill the niches. Within 15 million years the tiny mammals had evolved into a varied population that had taken to

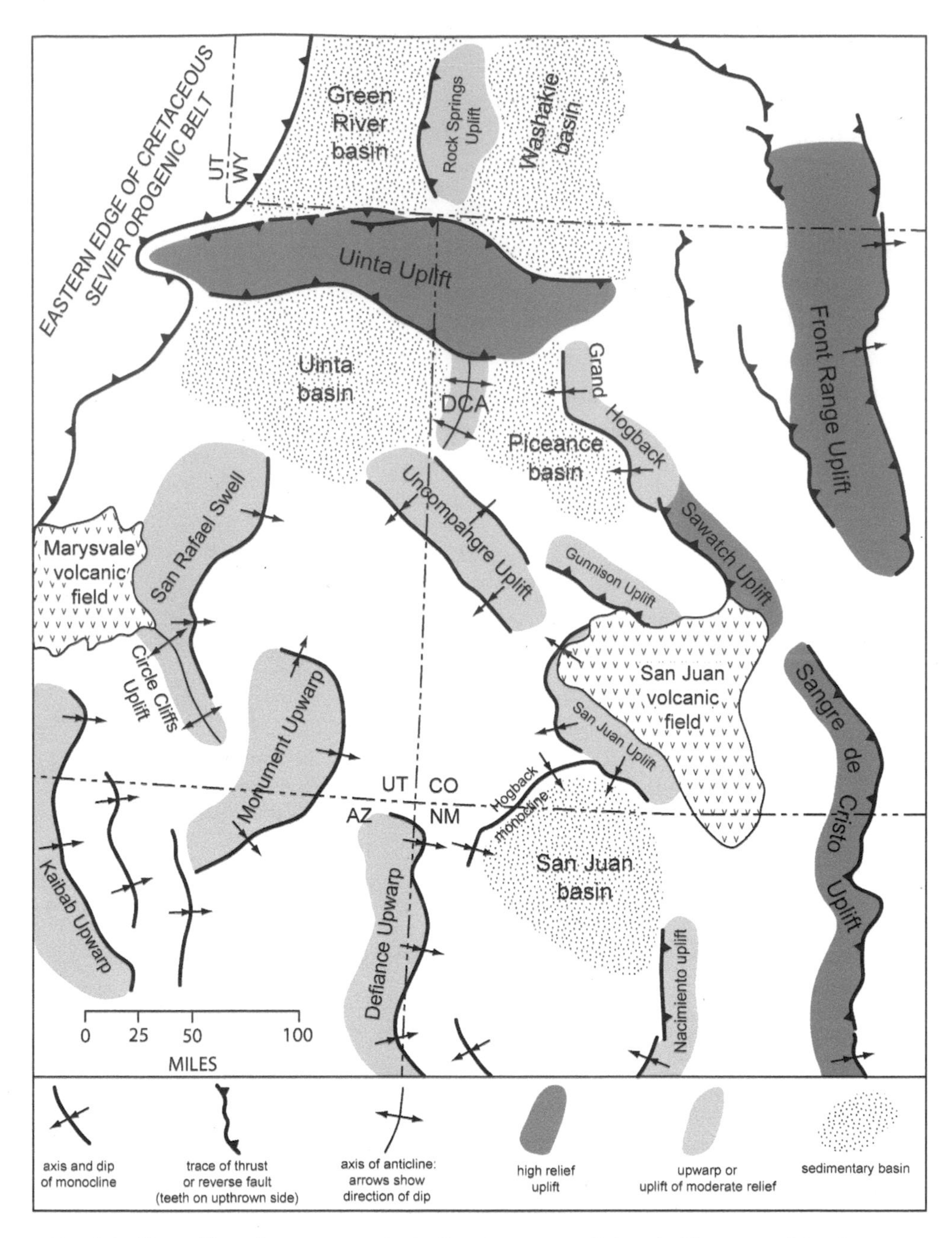

FIG. 8.2. Map of Four Corners area, including the eastern Colorado Plateau and Southern Rocky Mountains, showing various Laramide tectonic elements. Between the Uinta and Piceance basins is the low-relief Douglas Creek Arch (DCA on map). After Dickinson and others 1988.

the air with bats, to the sea with whales, and on land with a variety of animals, including some hoofed mammals that grew to the size of elephants. Just as Early Tertiary mountain building in North America controlled the processes that gave us our modern landscape, the rise of mammals eventually delivered our modern animal kingdom.

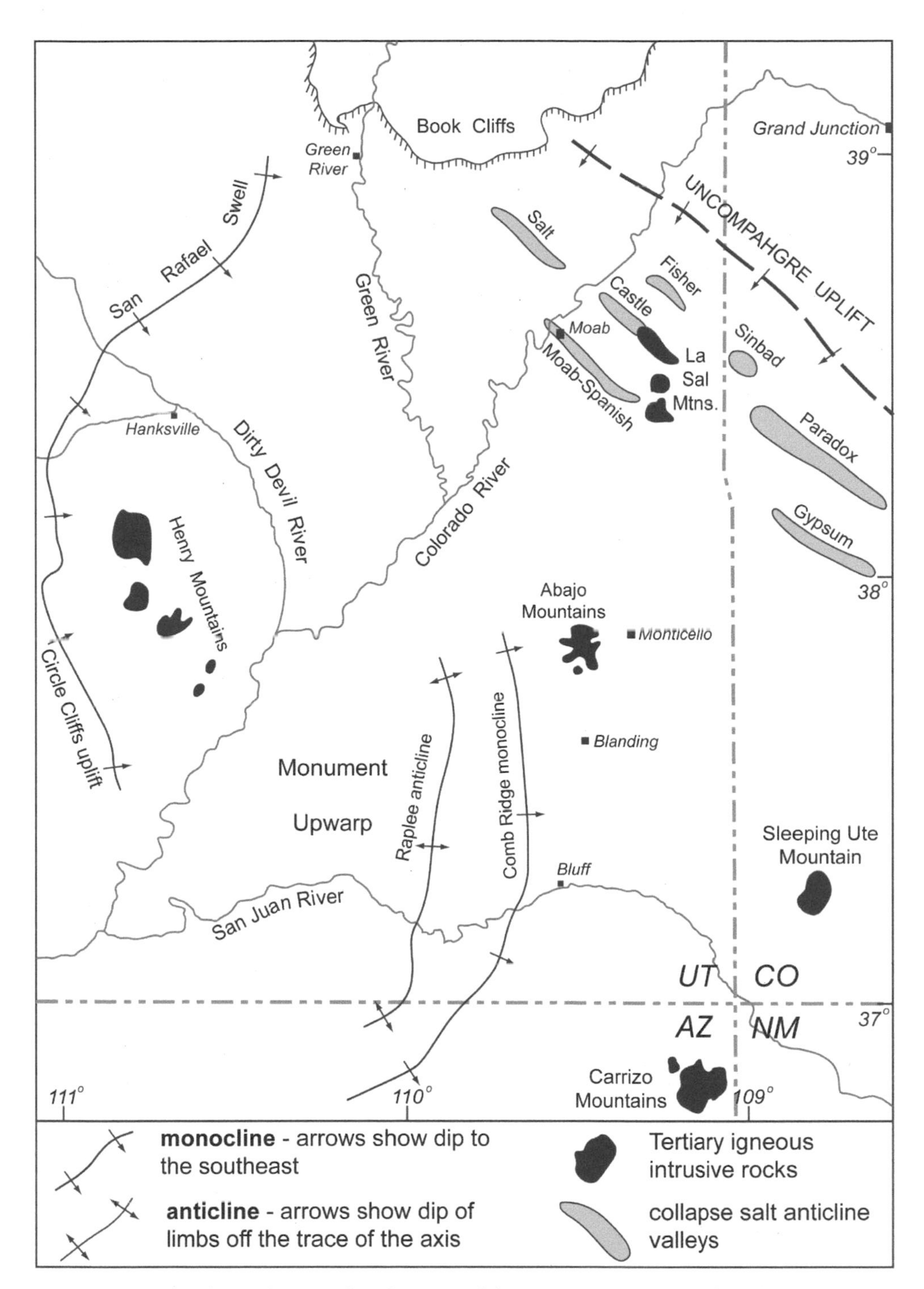

FIG. 8.3. Map showing various Tertiary features of the Four Corners region, focusing on southeast Utah, including major Laramide folds, prominent salt anticline valleys, and igneous intrusive rocks. Modified after Molenaar 1987.

A unique component of Tertiary landscape development was the localized intrusion of magma into shallow levels of the crust that pushed up the La Sal, Abajo, and Henry mountains, as well as Sleeping Ute Mountain (Fig. 8.3). These great alpine islands loom high above the canyons of southern Utah and stretch upward to more than 12,000 feet. Each of these isolated ranges formed as a cluster of intrusions, mostly in the form of mushroom-shaped laccoliths. All these great laccolith clusters of the Colorado Plateau are of similar composition and overlap in age, ranging from 32 to 23 Ma—similarities that signal a shared origin. Since their intrusion, the incessant movement of water and, in the La Sal Mountains, glaciers have bared the granitic cores of these intrusive centers.

The Laramide Orogeny

Deformation attributed to the Laramide orogeny began toward the end of the Cretaceous and continued to ~50 Ma, a period of about 20 million years. Deformation stretched eastward 1500 km, from the inactive front of the Sevier highlands in western Arizona and Utah into Nebraska and Kansas, where the effects are barely discernible (Tikoff and Maxson 2001). In a north-south direction the most severe zone of deformation forms the backbone of the modern Rocky Mountains from Montana south through western Wyoming and Colorado, and into New Mexico. On the Colorado Plateau, sandwiched between this lofty spine and the inactive Sevier highlands, the strong, stable crust limited deformation to large, archlike folds and spectacular monoclinal ramps. East of the Rocky Mountains, Laramide deformation was still more subtle, expressed mostly as low-relief arches and shallow basins that are hidden beneath younger sediments (Tikoff and Maxson 2001). Like the Colorado Plateau, these low-relief structures reflect the stable crust of the Midcontinent.

Today it is generally agreed that the driving force behind the Laramide orogeny was east-directed compression caused by the subduction of oceanic crust (the Farallon plate) beneath the west margin of North America. Because this subduction zone had been active since at least the Late Triassic, a profound change must have occurred in the Late Cretaceous to suddenly push the mountain front so far eastward, from western Utah into central Colorado. The change that triggered the Laramide is somewhat controversial and has been the subject of healthy debate. Any plausible model for this important event must address several major differences between the Laramide and the earlier Mesozoic orogenies. The most obvious change was geographic, with a pronounced eastward shift of mountain building across what previously was an enormous and actively subsiding sedimentary basin. Another major difference was the structural style of the deformation. Deformation during the previous Cretaceous Sevier orogeny was confined to the thick succession of sedimentary rocks of the upper crust; basement rocks were not involved, at least in the eastern limits of the orogenic front. In contrast, the Laramide uplifts—whether the sharp peaks of the Rockies or the less-precipitous, monoclinal upwarps of the Colorado Plateau—are characterized by

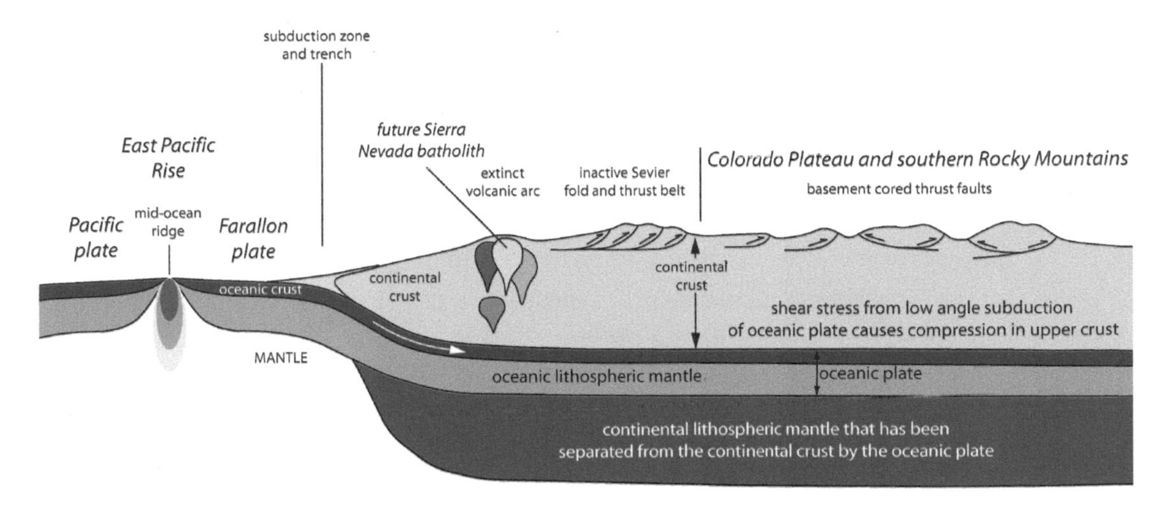

FIG. 8.4. The generally accepted plate tectonic model for western North America during the Early Tertiary development of the Laramide orogeny. As the mid-ocean ridge in the Pacific Ocean drifted closer to the subduction zone, the hot, buoyant oceanic crust no longer easily subducted and instead remained at a shallow level during subduction. This low-angle subduction separated the continental crust from its lithosphere, generating large shear stresses at the base of the continental crust as far inland as Utah and Colorado.

massive reverse faults that cut deeply downward through basement rocks into the middle and lower crust. The last major transformation was the suspension of igneous activity across western North America during the Laramide, even though regional magmatism preceded and followed the orogenic event. Even the long-lived, subduction-generated Sierra Nevada volcanic arc was extinguished with the onset of the Laramide. Since the theory of plate tectonics was embraced in the 1960s, several reasonable scenarios for the Laramide orogeny have been proposed. The most plausible of these are briefly recounted below.

One of the first models of the Laramide orogeny, and the one most widely accepted today, was proposed in 1978 by William Dickinson and Walter Snyder. This model calls on a change in the angle at which the Farallon oceanic plate was subducted beneath North America. The relatively steep angle of Mesozoic subduction was reduced during the Late Cretaceous to a low angle so that the oceanic plate, after being pulled beneath North America, flattened as it moved eastward (Fig. 8.4). As this oceanic plate scraped its way eastward along the base of the continent, huge shear stresses were transmitted upward through the continental crust to form compressional reverse faults. The flat geometry of the subduction allowed these stresses to affect the crust much farther inboard, well east of the earlier limits of uplift. Many of the faults along which uplift took place were reactivated from earlier faults and crustal weaknesses. Prominent examples include the bounding faults of the Pennsylvanian Ancestral Rockies, and the monoclines that bound the Monument upwarp and San Rafael Swell, which were sporadically active from the Late Paleozoic through the Mesozoic.

The flatter path of the subducting oceanic plate in the Late Cretaceous explains the far inboard effect of compression and the dramatic eastward push of the Rocky Mountain front. The production of compression as the Farallon oceanic plate sheared eastward across the base of the overlying North American crust accounts for the change in structural style of uplift to a deep-seated thrusting of basement rocks. With the rise of compression from the deep plate boundary, huge stresses were concentrated in the complex of plutonic and metamorphic basement rock. Overlying sedimentary rocks were also affected, but for the most part they passively "went along for the ride."

The shutdown of igneous activity also is conveniently explained by a change in subduction angle. The extensive Sierra Nevada volcanic arc that lined the west edge of North America during much of the preceding Mesozoic was generated by partial melting of the subducting oceanic plate and overlying continental crust. This was the result of the steep subduction angle of the oceanic plate as it plunged to a deep level in the upper mantle, where higher temperature and pressure combined to begin melting it. This magma then rose through the crust to feed the volcanoes, and also to crystallize in deeper levels of the crust to form the gigantic granite batholith of the modern Sierra Nevada. During the Laramide, however, the decrease in the subduction angle and flat trajectory of the oceanic plate stalled magma production. The subducting oceanic plate never reached the depths required for partial melting, and thus the flattened angle of subduction that drove the Laramide orogeny effectively shut down magmatism across western North America.

The missing piece of this model is the ultimate cause of the reduced angle of subduction. The oceanic Farallon plate is gone, consumed long ago beneath the North American plate. But the Farallon crust formed at the mid-ocean ridge (the East Pacific Rise) that split the Pacific Ocean floor, so its mirror image *is* preserved in the modern Pacific plate, formed on the west side of the East Pacific Rise. Although no data will ever be obtained from the Farallon, its western counterpart preserves an identical record of mid-ocean ridge activity. The record salvaged from the Pacific plate shows that the rate of oceanic crust production during the Laramide, beginning in the Late Cretaceous, increased from 7–8 cm/year to ~15 cm/year. This translates to a profound acceleration—in fact, a doubling of the rate at which the Farallon must have been subducted beneath North America.

An equally important factor was the increasing proximity of the mid-ocean ridge to the subduction zone. At this point a brief lesson on subduction zone tectonics may be helpful. A basic premise of plate tectonics is that young oceanic crust, situated near the mid-ocean ridge, is hot and buoyant. This young crust sits high on the sea floor, hence the name "East Pacific Rise." As that crust slowly moves away from the mid-ocean ridge toward the subduction zone in a conveyor belt fashion, it becomes older, cooler, and denser. Because of this increase in

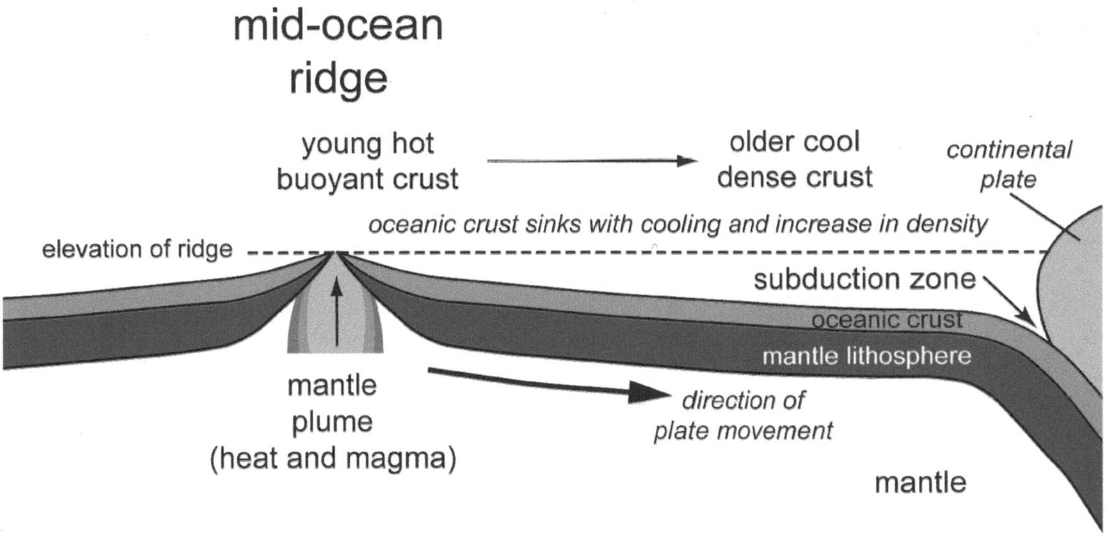

Fig. 8.5. Schematic cross section through an oceanic plate showing the origin of new crust at the mid-ocean ridge and its cooling and sinking as it moves away from the heat of the ridge. Note the changes in relative elevation as the oceanic crust moves away from the mid-ocean ridge. The subduction zone is shown to emphasize the ultimate fate of cool, dense oceanic crust as it pushes against the less dense continental crust.

density, it sits gradually lower with distance from the mid-ocean ridge (Fig. 8.5). When it reaches the subduction zone at the plate boundary, the thin but dense oceanic crust is readily subducted beneath the continental plate. However, if the mid-ocean ridge edges close to the subduction zone, as it did during the Late Cretaceous–Early Tertiary of western North America, the oceanic slab is hot and buoyant, even as it is being subducted. In this setting the oceanic plate would not have been easily drawn to deep levels beneath the continent and may have resisted subduction. Instead, as the East Pacific Rise slowly drifted eastward, the buoyant oceanic plate would have been pushed beneath North America at a subhorizontal angle. The extra horizontal push to this hot, buoyant crust was provided by the increased rate of crust production at the mid-ocean ridge. Thus, the combination of an increase in the rate of subduction and an eastward shift of the East Pacific Rise collaborated to generate uplifts far inland from the subduction zone.

A modern analogue for the flat-slab subduction model for the Laramide orogeny is found on the South American continent in the Sierra Pampeanas of Argentina (Jordan and Allmendinger 1986, Jordan and Alonso 1987). The entire length of western South America is marked by an offshore subduction zone in which the Nazca oceanic plate is plunging eastward beneath the South American continent. This has produced, from west to east, the modern Andes volcanic arc, the sub-Andean fold and thrust belt, and the broad foreland basin. These modern South American features correspond to North America's Cretaceous-age Sierra Nevada volcanic arc, the Sevier fold and thrust belt, and the Western Interior

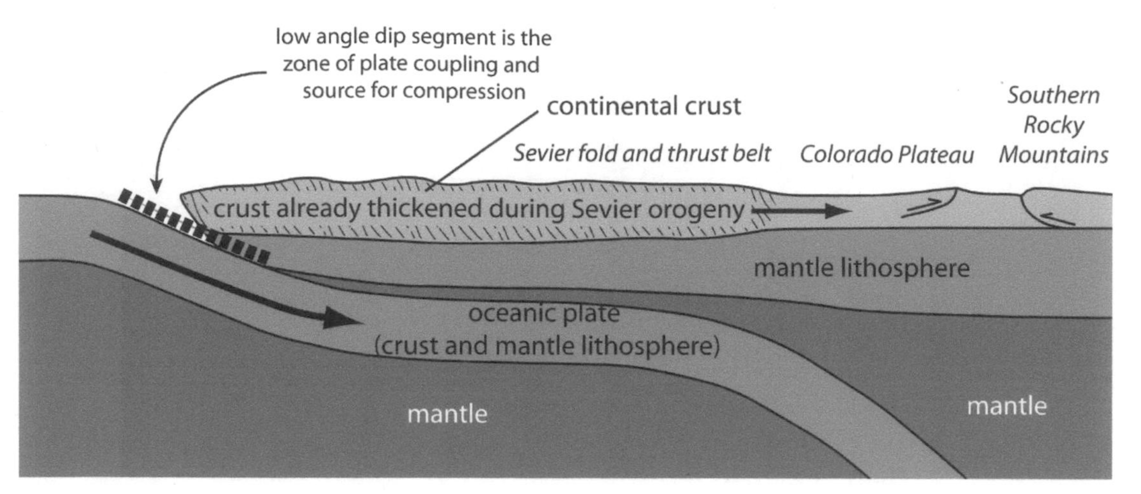

Fig. 8.6. Livaccari and Perry's (1993) model for an alternate source for the compression that drove the Laramide orogeny. In this model the compression originates at the subduction zone, where the oceanic and continental plate come into contact with each other, producing intense compression along the continental margin. Because the area immediately landward had already been compressed during the previous Sevier orogeny, the compression was transmitted farther inland to the Colorado Plateau and the Sevier foreland.

basin, respectively; however, across a west-east segment in Argentina between the latitudes of 26° and 33° S, the broad continuity of the foreland is disrupted by basement-cored uplifts and intervening basins that have been active over the last ~10 million years. Moreover, magmatism has been dormant across this transect over the same time interval. A flat geometry for the subducting Nazca plate along this segment has been documented by tracing subduction-generated earthquakes across this region (Jordan and others 1983). The basement-cored, reverse-faulted Sierra Pampeanas, the magmatic lull, and the recognition of a flat subduction trajectory present some intriguing parallels with the Laramide of North America; the similarities also help to explain why the model of Dickinson and Snyder (1978) has been favored over the years.

The main problem with the Dickinson and Snyder model is that in order for the oceanic Farallon plate to transmit shear stress along the base of the crust, it would have had to split the North American plate into an upper continental, crustal part of the lithosphere and an underlying, upper mantle lithosphere, effectively severing the lower part of the plate (the mantle lithosphere) from the upper continental crust (see Fig. 8.4). The chemistry of post-Laramide igneous rocks across western North America, however, carries the same mantle lithosphere signature that it has recorded since the construction of southwestern North America ~1.7 billion years ago. These disparities have stimulated the development of two alternative models for the origin of the Laramide orogeny (e.g., Livaccari and Perry 1993, Tikoff and Maxson 2001).

In their alternative model, Rick Livaccari and Frank Perry (1993) agree that a shift to flat subduction occurred in the Late Cretaceous, but they find no evidence for the removal of the lower-mantle lithosphere of the overriding North American plate. They argue that the consistent chemical signature throughout geologic history of mantle-derived igneous rocks shows no change in the makeup of the lower part of the North American plate. Instead, Livaccari and Perry suggest that the Farallon plate plunged beneath the entire thickness of the North American lithosphere before assuming a flat eastward path, thus preserving the North American mantle lithosphere. This scenario, however, precludes far-reaching, shear-induced compression at the base of the continental crust as a mechanism for Laramide uplift.

The compression that drove Laramide mountain building in this model was provided by *horizontal end loading*, essentially a stress that was generated far to the west, where the two plates met. Where the Farallon plate began its low-angle descent beneath North America, its buoyancy and accelerated rate of subduction combined to couple the two plates along this downward-diving segment (Fig. 8.6). It was along this zone of coupling that the east-directed compression was generated. But how was this compression transmitted so far inland? As the compression advanced eastward, it first encountered the already overthickened crust of the Sevier orogenic belt. Earlier deformation had brought this region to its compressional limit, causing it to act as a rigid block that transferred the stress eastward to the adjacent Colorado Plateau. Here the thick, stable crust yielded slightly along long-lived weaknesses, pushing up the monoclines along the sinuous traces of deeply buried basement faults. Again, however, the bulk of the stress passed eastward as the rigid Plateau pushed against the crust that bounded it in that direction. There the great pulses of compression encountered numerous, large-scale crustal weaknesses in what today constitutes Wyoming, Colorado, and New Mexico. The faults that accommodated the Southern Rocky Mountains formed during the Precambrian and have been periodically reactivated, most dramatically in the uplift of the Late Paleozoic Ancestral Rocky Mountains and the more recent Laramide mountains. As the subduction-driven compression pulsed from west to east, colossal blocks of rock again grated past each other, squeezing skyward into high-relief mountains. In this way, according to Livaccari and Perry, the monoclines of the Colorado Plateau and the adjacent Southern Rocky Mountains were born.

In an alternative explanation advanced by Basil Tikoff and Julie Maxson (2001), end loading from the west is again called on as the driving force for the Laramide, but Tikoff and Maxson offer a different source for compression: a collision along the west margin with a continental terrane that rafted in on the subducting Farallon plate. This landmass is called the "Baja BC" terrane. According to Tikoff and Maxson, Baja BC collided with North America along the present-day latitudes of northern Mexico and the Mojave Desert of southern California. Following this impact ~95 million years ago, right-lateral shearing driven by con-

FIG. 8.7. Maxson and Tikoff's (1996) model for the plate tectonic setting of the Laramide orogeny. A, the pre-Laramide Cretaceous setting in which the Farallon oceanic plate was subducted, generating the Sierra Nevada volcanic arc and the compression to drive the Sevier orogeny. B, the cessation of subduction as the Farallon plate and the associated mid-ocean ridge are subducted. The collision of the Baja BC terrane generated compression that was transmitted through the already thickened crust of the Sevier belt into the foreland, where the Laramide uplifts rose along preexisting weaknesses. C, the northward translation of the Baja BC terrane creating shear stresses that are similarly transmitted into the foreland, further driving Laramide deformation. Modified after Maxson and Tikoff 1996.

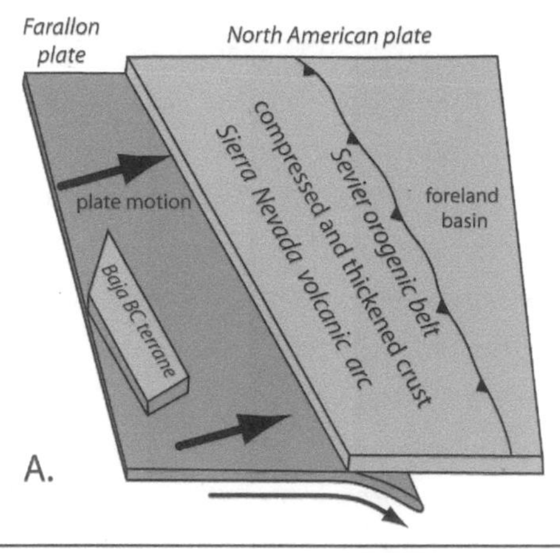

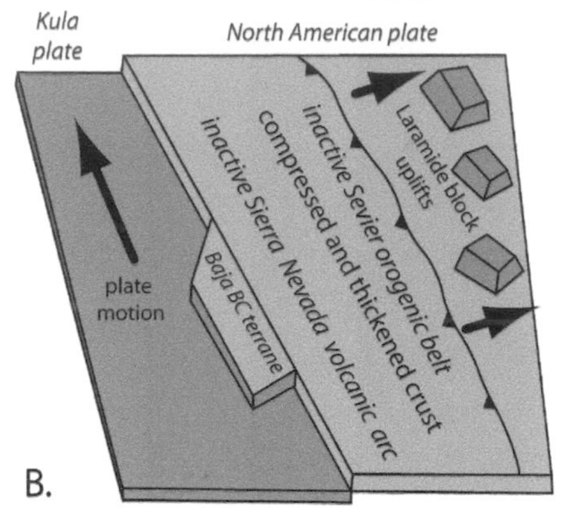

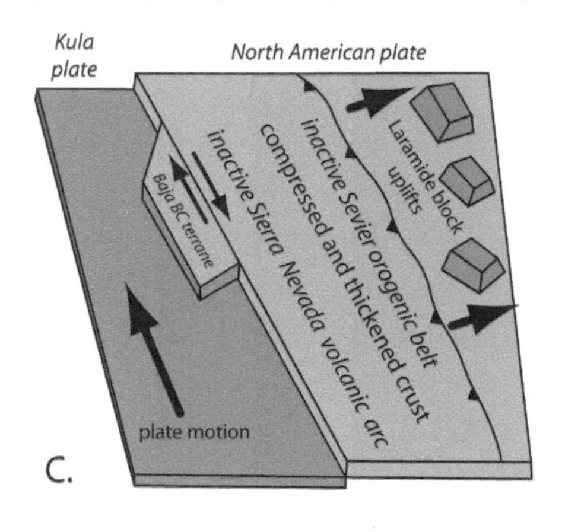

tinued oblique subduction scooted the new terrane northward along a strike slip fault (Fig. 8.7). Today this landmass resides in western British Columbia and southern Alaska, hence the name "Baja BC." The net result of the end loading driven by the docking of Baja BC was the eastward transmittal of compression and a large-scale buckling of the lithosphere of western North America. The large, wavelength folds that mirror this buckling stretch across the Colorado Plateau, Southern Rocky Mountains, and farther east into the High Plains region to eastern Kansas, where the fading deformation finally dies out. The preexisting weaknesses across this region broke into the large reverse faults that bound the dramatic Rocky Mountain ranges and which core the monoclinal ramps of the Colorado Plateau.

The Laramide Orogeny on the Colorado Plateau

Regardless of the ultimate source of compression, the Laramide orogeny across the Colorado Plateau generated a number of northwest-to-northeast-trending monoclines and many smaller-scale folds. It is the unique expression

FIG. 8.8. Aerial view to the north of the north end of Comb Ridge monocline. The light-colored strata on the left (west) side are the Permian Cedar Mesa Sandstone. Note how the strata flatten to horizontal on the west side to form Cedar Mesa, the type locality for this sandstone unit. The western edge of the Abajo Mountains can be seen in the background to the east.

of these large monoclines that punctuate the extraordinary scenery of the Colorado Plateau. These regional-scale, single-limbed folds in southeast Utah include the San Rafael Swell to the northwest and Comb Ridge, a sinuous monocline that forms a jagged escarpment between the towns of Bluff and Mexican Hat (Fig. 8.8 and Plate 8). The spectacular Raplee anticline, which takes on the appearance of a monocline, also is in this area. The steep west limb of this giant, asymmetric fold is spectacularly exposed along the San Juan River immediately east of Mexican Hat. All the monoclines in this region flank the steep margins of broad upwarps. Studies across the Plateau have shown that most of these folds are cored at depth by large, basement-involved reverse faults, just like the Laramide uplifts of the Southern Rocky Mountains. The extremely thick, overlying sedimentary succession has cushioned some of the offset in these deep faults, so that their surface expression is one of an abrupt, single-limbed fold (Fig. 8.9).

Comb Ridge and the Monument Upwarp

The Comb Ridge monocline stretches 110 miles in a north-south direction from the Abajo Mountains of Utah to Kayenta, Arizona, where it arcs southwest to merge with the Organ Rock monocline (see Fig. 8.3). Along most of its length, the monocline's undulating crest is a fin of east-dipping, resistant Jurassic Navajo Sandstone (Fig. 8.10 and Plate 9). This colossal, east-sloping ramp defines the eastern boundary of the long-lived Monument upwarp. Strata on the upwarp, on the west side of Comb Ridge, lie ~3000 feet higher than the same strata just 3 miles to the east (Kelley 1955a). Erosion of this most recent version (Laramide) of the Monument upwarp has exhumed a vast area of Pennsylvanian and Permian

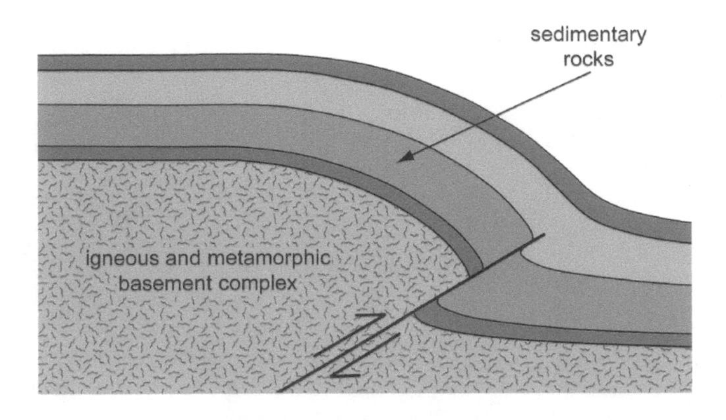

‹ **Fig.** 8.9. Cross section through a typical monocline showing the inferred reverse fault at depth that at the surface is shown as a simple ramp-type fold. After Bump 2003.

⌄ **Fig.** 8.10. Aerial view to the north of Comb Ridge. The main spine of Comb Ridge is Navajo Sandstone, overlain by the Carmel Formation and Entrada Sandstone. The beds in this photo are dipping toward the lower right (east). Note also the parallel fractures in the upper surface of the Entrada Sandstone.

rocks. This includes the deeply incised canyons of Cedar Mesa, which is capped by the Permian Cedar Mesa Sandstone, and the meandering lower reaches of the San Juan River, which slices deeply into the upwarp to expose Pennsylvanian limestone in its depths.

Detailed studies of the strata exposed on and around the Monument upwarp provide abundant evidence for uplift of this same area in the geologic past. Pennsylvanian and Permian rocks thin across the upwarp, suggesting it was a mild topographic high during this time. Pennsylvanian limestones indicate that it was a shallow marine shelf, compared to a rapidly subsiding evaporate basin to the east. It appears that the compression that drove the Ancestral Rockies upward at this time also affected the fault deep beneath Comb Ridge. Similarly, facies changes in the Upper Jurassic strata across this boundary suggest that the upwarp was again a mild uplift that supplied sand to the eolian Bluff Sandstone. The Bluff was deposited on the downwarped side, east of modern Comb Ridge. This episode of uplift may have been related to development of the Nevadan orogenic belt to the west during this time. It thus appears that any time that compression rippled across the region, movement on the ancient fault that underlies Comb Ridge nudged the Monument block upward. Its striking Laramide expression, so dramatically exposed today, is only its most recent incarnation.

The San Rafael Swell

The San Rafael Swell is a large upwarp in central Utah that is bounded on its steep east margin by a monocline. The trace of this great flexure is curved in map view, but trends mostly northeast along its 60-mile length (see Fig. 8.3). Bedding in the Jurassic rocks that line its flank is locally tilted up to 85°. Within the deep canyons that knife through the uplift, strata as old as Permian age have been exhumed. Stratigraphic studies across the spectacular east-diving structure show that the west side has risen ~3500 feet relative to the downdropped east side (Kelley 1955a, 1955b). To the south the fold merges with the Waterpocket Fold, a northwest-trending monocline that defines Capitol Reef National Park. The downwarped eastern sides of these two large folds have been instrumental in preserving vast tracts of easily eroded Cretaceous rocks in the Book Cliffs to the north and the Henry basin to the south.

Like Comb Ridge and most of the larger monoclines across the Plateau, the San Rafael Swell is cored at depth by a regional-scale reverse fault in which the western block of basement rock is thrust up and eastward. It is paired with a downdropped region to the east. Most geologists who have worked on these structures agree that the Laramide faults initially formed during Late Precambrian (~900 to 700 Ma) rifting along the western margin of North America (e.g., Davis 1978, Huntoon 1993, Marshak and others 2000, Bump 2003). Although these earlier faults and fractures initiated during a prolonged period of extension, Laramide reactivation was driven by a very different stress regime, one of intense compression. Thus the most recent movement along these long-lived crustal weaknesses is the opposite of movement they accommodated during their Precambrian genesis. Such a sequence of movements has been well-documented by Peter Huntoon (1993) on monocline/fault pairs in the Grand Canyon, one of the few places on the Colorado Plateau that has been cut deeply enough by erosion to expose these faults.

The reactivation of older weaknesses has occurred repeatedly across the Colorado Plateau and is also evident along the length of the Southern Rocky Mountains to the east. In the Rockies, this younger Laramide version rose vertically along the same weaknesses that guided the Ancestral Rocky Mountains 300 million years ago. It is from this recurrent activity that the Pennsylvanian version derives its name. It should be emphasized that recurrent movement on ancient faults is not unique to this region and, in fact, has been documented in countless examples around the world. There are, however, few geologic regions around the

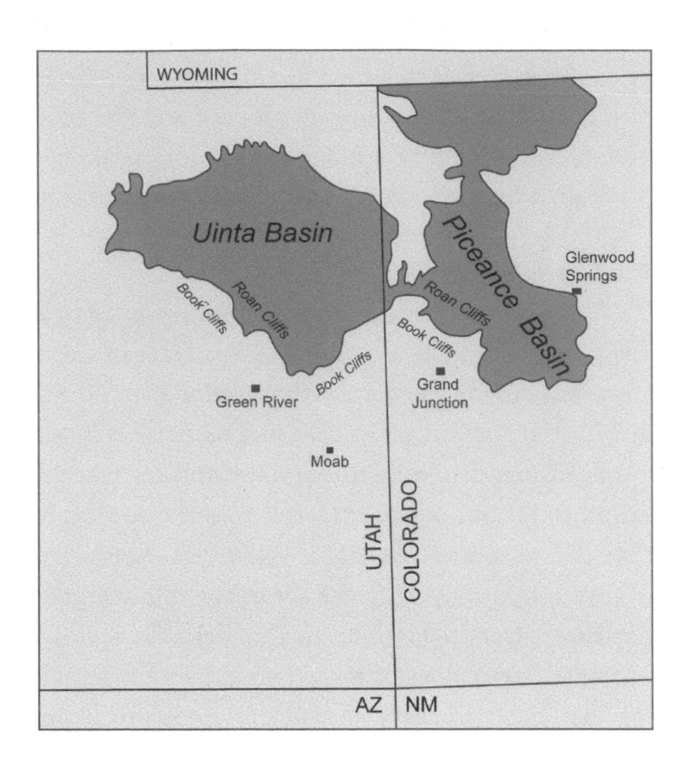

FIG. 8.11. Map showing the extent of Tertiary sedimentary rocks (dark gray) in eastern Utah and western Colorado.

world that offer the combination of continuous lateral exposures of strata that record these sometimes subtle movements and the locally deep levels of incision. These factors provide numerous, well-documented examples of long-lived, intermittently active faults across the region.

EARLY TERTIARY STRATIGRAPHY AND DEPOSITIONAL HISTORY

Tertiary sedimentary rocks associated with Laramide deformation are sparsely distributed across the Colorado Plateau due to a wholesale shift from deposition to erosion approximately 5 million years ago. Across the expanse of southeast Utah, south of the Book Cliffs, the Tertiary sedimentary record has been erased by erosion. Although they almost certainly blanketed much of this region, Tertiary sediments were subsequently stripped away during an erosional interval that began with the initiation of the Colorado River drainage system and continues to this day.

The record is not entirely lost, however, as an extensive tract of Paleocene and Eocene sedimentary rocks are preserved across the vast area north of the east-west palisade of the Book Cliffs (Fig. 8.11). This dissected tableland of Tertiary strata, known as the Roan Plateau, includes the Uinta basin of northeast Utah and the Piceance basin of northwest Colorado (see Fig. 8.2). These rocks are dominated by a 3000-foot sequence of the lacustrine Green River Formation and provide a welcome sedimentary record of Laramide events in the region.

From the latest Cretaceous to Late Eocene times, deposition in northern Utah and northwest Colorado was guided by the basement-cored Laramide uplifts that disrupted the once continuous foreland. These various uplifts segmented the region into a series of intermontane basins surrounded by highlands, which acted as both partitions and sediment sources. These highlands ranged from mild upwarps (Douglas Creek arch), to monoclines (San Rafael Swell), to the abrupt, fault-bounded mountain ranges of the Southern Rockies (e.g., the Uinta Mountains and Sawatch Range, Colorado). The basins at the foot of these uplifts acted as receptacles for the detritus that they shed. The seas had receded, and the basins were occupied by complex systems of high- and low-energy rivers, and lakes of various sizes. For approximately 10 million years the two largest basins, the Uinta and Piceance, were filled by giant lakes.

These two large basins covered the northern Colorado Plateau during the Paleocene and Eocene. The Uinta basin lay to the west, covering much of northeast Utah, and was bordered on the west by the Sevier orogenic belt, which remained a formidable barrier and a sediment source. Its northern boundary was the rising Uinta Mountains, an anomalous, east-west-trending Laramide mountain range. These high mountains today make up the northern boundary of the Colorado Plateau. Along the basin's southeast border was the renewed Laramide version of the Uncompahgre uplift, and to the southwest lay the San Rafael Swell (see Fig. 8.2). The eastern counterpart to the Uinta basin was the similarly extensive Piceance basin, which occupied part of northwest Colorado. This basin was bordered on the north by a more subtle eastward extension of the Uinta Mountains called the Axial arch. To the northeast lay the White River uplift and the lengthy monocline of the Grand Hogback, which graded southward into the higher Elk Mountains. At the southern margin of the Piceance basin was the mountainous Sawatch uplift and the more humble Gunnison and Uncompahgre uplifts (see Fig. 8.2). The Uinta and Piceance basins were separated by the Douglas Creek arch, another subtle Laramide structure that initially acted as low barrier separating the basins, but later provided a connection between the two. These two basins have long and complex histories, and although they were not always connected, they share these histories.

Laramide-related sedimentation began in the latest Cretaceous as the new highlands began to rise. These highlands acted as sediment sources but also were the catalyst for a wide-ranging reorganization of the drainage. Late Cretaceous to Early Paleocene depositional systems ranged from narrow zones of high-energy rivers that fringed the highlands to a more extensive system of sluggish rivers with swampy floodplains. Lakes were present during this early phase, and as the highlands continued to rise and internalize the drainage, the lakes expanded across the basins. From latest Paleocene to Late Eocene ~10,000 feet of lacustrine

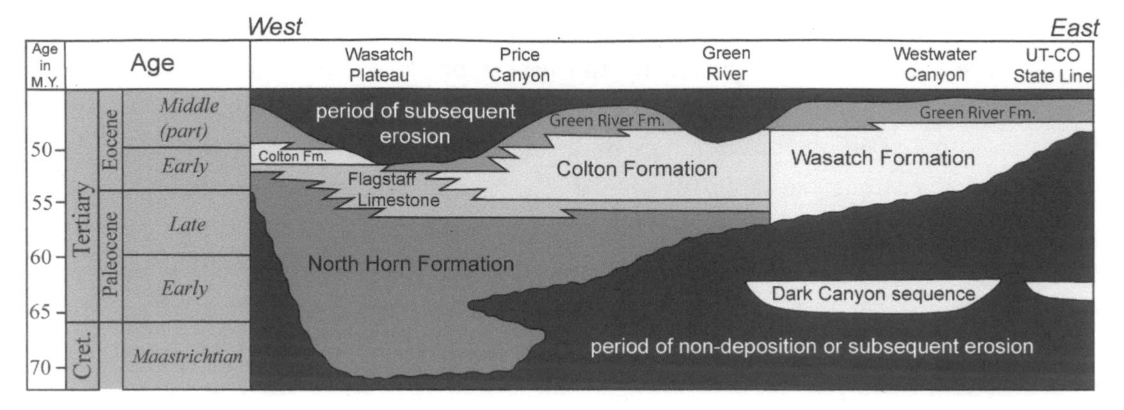

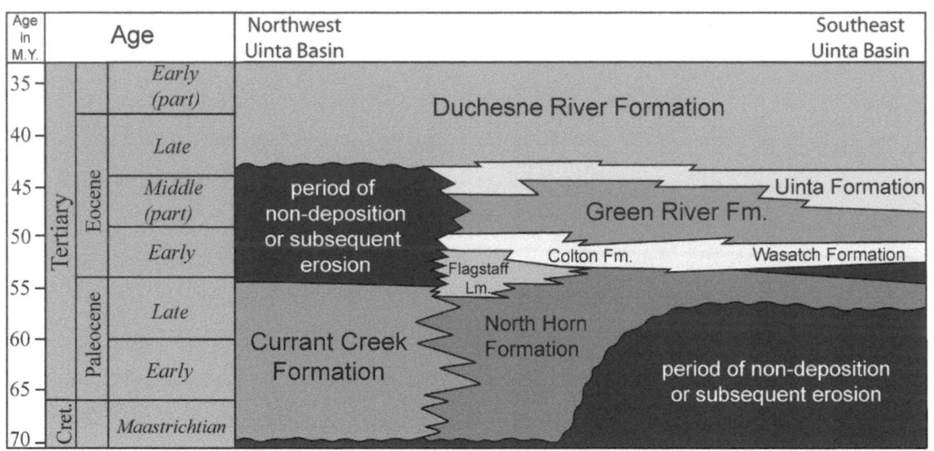

FIG. 8.12. Stratigraphic relations of various Tertiary stratigraphic units in the Uinta basin and Roan Cliffs area of northern Utah. Modified after Isby and Picard 1985 and Franczyk and others 1990.

sediments settled into the basin centers. This succession includes a thick sequence of organic-rich oil shale. Lake sediments also host a particularly rich and varied fossil assemblage that includes abundant fish, plants, reptiles, amphibians, and insects. The following section details the complex history of these two basins, whose sediments make up the richest and most revealing Tertiary deposits on the Colorado Plateau.

The Paleocene Epoch

Laramide-driven deposition in the Uinta basin began in the latest Cretaceous (Maastrichtian) along its western margin, adjacent to the Sevier orogenic belt. The Upper Cretaceous–Lower Paleocene Currant Creek and North Horn formations in this region record the waning activity of the Sevier orogenic belt and the rise of the San Rafael and Uinta uplifts. While these formations were being deposited in the western part of the basin, the central and eastern parts hold no record for this time (Fig. 8.12). It is likely that, although activity was waning in that area, the load of the Sevier thrust sheets forced the continuation of subsidence in the western part of the basin, improving the probability of preserving these sediments.

The Currant Creek Formation

The Currant Creek Formation is exposed in a small area in the northwest corner of the Uinta basin along several drainages that come from the modern Uinta Mountains (Fig. 8.12). Though exposures today are limited, the formation originally extended across a larger area. The Currant Creek Formation rests unconformably on Late Cretaceous Mesa Verde Group sandstones and is unconformably overlain by the Upper Eocene Duchesne Formation. To the south it grades into the North Horn Formation. The Currant Creek consists of 1500 feet of sandstone, conglomerate, and mudstone of fluvial origin. A prominent boulder conglomerate in the upper part indicates an abrupt increase in energy in this river system through time. Collectively, the lithologies and sedimentary structures in the formation indicate a south-flowing, braided river system with a source to the north. Detailed study by Isby and Picard (1983, 1985) identifies two discrete sources for the formation, each active at different times. The lower, finer-grained part originated in the Sevier orogenic belt to the northwest. The river at this time flowed southward, parallel to the mountain front, after emitting from the mountain belt. The coarse, boulder-dominated upper part signals the rise of the Uinta uplift immediately to the north and the emergence of a major new sediment source. Thus, the Currant Creek Formation, though of limited extent, provides an important record of the transition from the long-lived, Sevier-dominated system to a rapidly evolving Laramide-sourced drainage.

The North Horn Formation

The North Horn Formation is a more extensive succession of sandstone, siltstone, mudstone, and thin limestone beds deposited in a spectrum of environments on a widespread, low-relief alluvial plain (Lawton 1985). The formation blankets much of the western Uinta basin, ranging as far west as the front of the Sevier highlands and reaching east to the vicinity of Green River. Its age in the west ranges from latest Cretaceous (Maastrichtian) to Early Eocene. In contrast, the formation in the Green River area ranges from Early to Late Paleocene. Thus the North Horn becomes younger to the east. Throughout its extent the North Horn unconformably overlies various Late Cretaceous Mesa Verde Group sandstones. The time gap represented by this unconformity ranges from less than a million years in the west to ~15 million years in the Green River area (Fig. 8.12) (Fouch and others 1983).

The North Horn grades to the northwest into the coarser-grained Currant Creek Formation. To the northeast the upper part interfingers with lacustrine deposits of Flagstaff Limestone. Throughout its extent the North Horn is succeeded by the lake deposits of the Flagstaff. This upper contact is both conformable and gradational.

A maximum thickness of ~1600 feet for the formation occurs in the west, on the Wasatch Plateau. In this area, where the North Horn also reaches its maximum age, deposition was in an actively subsiding trough bounded to the west by the Sevier mountain front and on the east by the rising San Rafael Swell. The formation thins to the northeast with distance from the trough (Fig. 8.12). Near its eastern limit along the Green River, the thin, ~130-foot succession of sandstone and shale grades upward into the lacustrine Flagstaff limestones.

North Horn sandstones were deposited in a system of east-flowing, low-energy, meandering rivers that looped across a level plain. Siltstone, mudstone, and thin coal beds mark the floodplain that received sediment when the rivers breached their banks. Thin limestone lenses indicate low-lying hollows where slack water and a high water table conspired to form ponds and small lakes. The lazy rivers drifted across the plain, coursing eastward to collect in the subsiding basin center. As the upwarps encircling the broad basin continued to rise, the initially limited lake expanded into a large one that spread across the alluvial plain. The initiation and subsequent growth of this lake is recorded by the Flagstaff Limestone. As it grew in area, the lake continued to be fed by runoff from the surrounding highlands via a fringe of river systems whose continued presence is told by the intertonguing relations between the uppermost North Horn and Flagstaff lake deposits.

The most complete and accurate record of the rise (and fall) of past mountain belts has come not from the mountains themselves, which tend to be dismantled by erosion, but from the enduring fragments of these mountains fortuitously preserved as sediments in adjacent basins. In the previous chapter we saw that the sudden appearance of conglomerate in the Late Cretaceous (Campanian) Tuscher Formation signaled the first pulse of uplift on the San Rafael Swell (Lawton 1986). The upper North Horn and Currant Creek formations also record several tectonic events. Conglomerate in the upper Currant Creek heralds the incipient rise of the Uinta uplift and the formation of the Uinta basin. In contrast, the North Horn records the end of tectonic activity along other margins of the Uinta basin, along the Sevier belt to the west, and on the San Rafael Swell, along the southwest margin (Lawton 1986). This is seen in the relations between the North Horn and underlying Late Cretaceous strata, which have been folded, faulted, and eroded as they were pushed upwards along the margins of these highlands. The upper North Horn, however, is undisturbed and laps onto both of these tectonic features, indicating that no deformation has occurred since its deposition. The North Horn Formation indicates that tectonic activity on these two large tectonic elements was over by Late Paleocene time.

The Dark Canyon sequence

During the early Paleocene, while the North Horn was blanketing the western part of the Uinta basin, the eastern part was being supplied with coarse sediment from highlands to the south. These conglomerate beds, informally referred to as the "Dark Canyon sequence," are the first part of a thick Tertiary succession that overlies the Late Cretaceous unconformity in the eastern Uinta basin (Franczyk and Pitman 1987). The sequence is unconformably overlain by the Paleocene-Eocene Wasatch Formation (Fig. 8.12), a contact that represents a gap of ~6 million years (Franczyk and others 1990). Earlier workers placed these conglomerate beds at some localities with the top of the underlying Late Cretaceous strata, whereas elsewhere they were included with the overlying Wasatch Formation (Fisher and others 1960). For many years these designations stalled efforts to recognize the significance of the Dark Canyon sequence and place in the Tertiary succession. Today the recognition of unconformities above and below—coupled with its Paleocene age, which is based on fossil pollen—has resulted in its current placement at the base of the Wasatch Formation.

Detailed study of the Dark Canyon sequence by USGS geologists Karen Franczyk and Janet Pitman (1987) has revealed many aspects of these localized deposits. The sequence is exposed only north of the Book Cliffs, where deep canyons cut northward into the Roan Plateau. Its western limit is just east of the Green River, and its eastern extent is immediately east of the Utah-Colorado border, near Douglas Pass. Pebble and cobble conglomerate makes up most of the sequence, although sandstone is common in the upper part. Conglomerate clasts are mostly quartzite and chert. A minor but notable component of this population is clasts of silicified limestone that contain Pennsylvanian and Permian marine fossils. According to Franczyk and Pitman, the clast assemblage reflects recycling of clasts from Mesozoic and Late Paleozoic conglomerate that was stripped from rising Laramide uplifts.

The Dark Canyon sequence was deposited in a coarse-grained, braided river system. Paleocurrent indicators record a dominant northwest flow for these rivers, although easternmost outcrops show a northeast-directed paleoflow (Franczyk and Pitman 1987). Reconstruction of this system suggests that rivers issued from the west side of the renewed Uncompahgre highlands to merge with a larger, northwest-flowing river that eventually fed into the Uinta basin center. It is possible that other Laramide uplifts— such as the Monument upwarp, and the Needles uplift in the present-day San Juan Mountains of southwest Colorado—also contributed to this drainage network, although this is speculative. Erosion has removed any trace of evidence for this across southeast Utah. Franczyk and Pitman interpreted the anomalous northeast-directed flow recorded in outcrops along the Utah-Colorado border as rivers that drained off the northern end of the Uncompahgre highlands to flow into the Piceance basin to the northeast. The

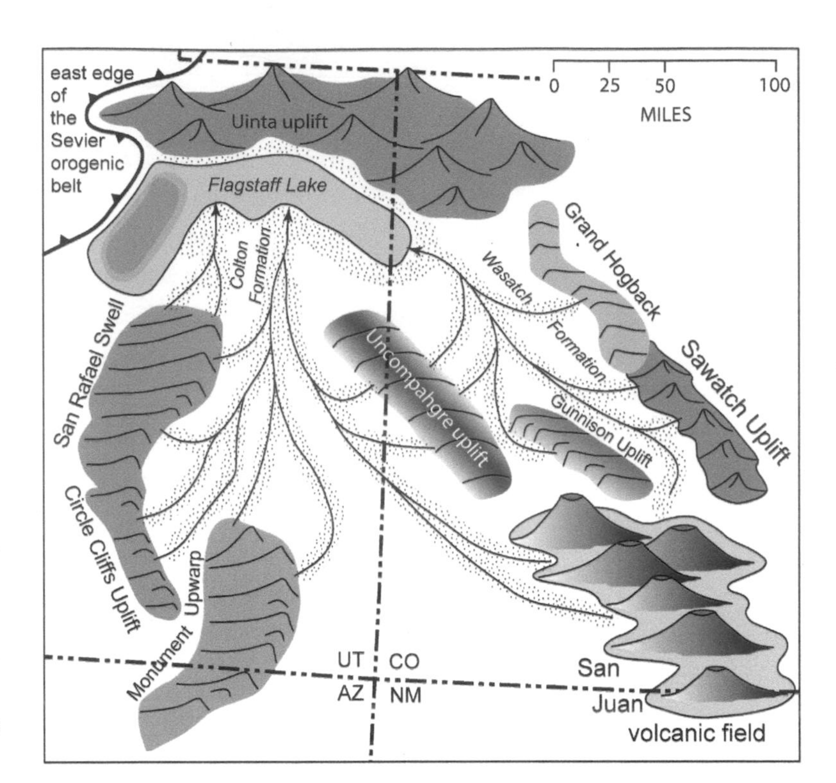

Fig. 8.13. Paleogeographic reconstruction of the Early Eocene Colton (to west) and Wasatch (to east) fluvial systems and the Flagstaff Lake system into which they drained. Note the deeper part of Flagstaff Lake to the west, where the Sevier orogenic belt continued to drive subsidence.

drainage patterns of these rivers show a path across the Douglas Creek arch suggesting that it was not yet a topographic high.

Flagstaff Limestone

The Paleocene-Eocene Flagstaff Limestone records the development of the first extensive lake to occupy the Uinta basin. Through the expansion of the small ponds and lakes of the North Horn Formation, a single large lake evolved. Its initial stage formed in the rapidly subsiding area between the Sevier belt and the San Rafael Swell. By latest Paleocene time, however, the lake in this southern finger of the basin had grown north and eastward, spreading to the west flank of the Douglas Creek arch (Johnson 1985). Stretching the length of the basin, Lake Flagstaff covered an east-west distance of 140 miles. The margins of this gigantic lake were bordered by narrow alluvial plains, as indicated by the interfingering relations between alluvial sediments in the upper North Horn and the lacustrine Flagstaff Limestone. Rivers threading across these plains were conduits for water and sediments from surrounding highlands. Where large rivers poured into the standing water, deltas formed. This is illustrated by the Paleocene-Eocene Colton Formation, created by rivers that drained highlands to the south.

Relations between the Flagstaff Limestone and surrounding strata are varied due to the presence of several sediment sources and fluctuations in lake level. Along its west margin the Flagstaff grades upward from the underlying North

Horn Formation. Even after the development of Lake Flagstaff the continued stream of clastic sediment from the diminishing Sevier belt to the west produced lateral intertonguing between these two units along the west lake margin. Along its southern margin the formation thins, and the upper part grades up into and intertongues with the fluvial/deltaic clastic rocks of the Colton Formation (Fig. 8.13). Finally, in the Price River canyon area, in what would have been the lake center, the Flagstaff grades up into the overlying Eocene Green River Formation. In this area some workers have considered the Flagstaff Limestone to be a member of the Green River Formation (e.g., Fouch 1976, Ryder and others 1976), although this designation cannot be applied throughout the basin.

The Flagstaff Limestone is thickest and best exposed on the Wasatch Plateau, located between the Sevier belt and San Rafael Swell. Here the formation attains a maximum thickness of 980 feet and is subdivided into three members based on changes in lithology that correspond to variations in lake level (Stanley and Collinson 1979). These members, named for well-exposed localities on the Wasatch Plateau, are from base to top the Ferron Mountain, Cove Mountain, and Musinia Peak members. Although the Flagstaff is dominated by lacustrine limestone, variations in lithology, fossil content, and internal features in the rock warrant this subdivision on the Wasatch Plateau. To the east, however, across the main part of the Uinta basin, the formation thins, and these members are not recognizable.

The basal Ferron Mountain Member grades up from the mudstone and sandstone of the North Horn Formation. It is composed of interbedded fossiliferous limestone and mudstone containing abundant freshwater fossils, including molluscs, ostracodes, and charophyte algae. Local, cross-stratified oolitic limestone indicates high energy in some parts of this extensive lake. The Ferron Mountain Member marks a highstand in the water level of Lake Flagstaff. The numerous fossil organisms suggest an oxygenated, shallow freshwater lake during this time (Stanley and Collinson 1979).

The overlying Cove Mountain Member records an environmental shift in Lake Flagstaff. It is dominated by dolomitic mudstone with common mudcracks and scoured channels filled with intraclasts composed of the surrounding mudstone. A 100-foot-thick interval of gypsum also occurs in this member and can be traced across the Wasatch Plateau over an area of 500 km^2. These strata and their internal features suggest a drawdown in lake level and a resulting shift to more saline waters. Additionally, fossils are sparse, but blue-green algae are present locally in the form of stromatolites. These laminated, cabbage-like heads of algae suggest an environment that was adverse to most organisms. Since their initial appearance on Earth more than 3 billion years ago, stromatolites have adapted to a variety of extreme environments in response to predatory grazing. All the components of the Cove Mountain Member suggest a reduced lake level, the contraction of its margins, and a temporary change from a freshwater lake to a large, hypersaline evaporative pan.

The uppermost Musinia Peak Member consists largely of gray fossiliferous micrite and mudstone deposited in the rising water of Lake Flagstaff as it expanded to twice its previous maximum area (Stanley and Collinson 1979). To the southeast the Musinia intertongues with and is succeeded by the fluvial-deltaic Colton Formation. The Musinia Peak Member marks the final highstand in Lake Flagstaff and a return to freshwater conditions; however, by the end of its deposition, the lake had again contracted.

The Colton Formation

The Paleocene-Eocene Colton Formation is exposed in the northwest part of the Book/Roan Cliffs, Price River canyon, and to the southwest on the Wasatch and Gunnison plateaus. Red and gray sandstone and mudstone dominate the formation and range up to 1300 feet (400 m) thick. The Colton intertongues with and overlies the Flagstaff Limestone and is overlain by lacustrine deposits of the Eocene Green River Formation (see Fig. 8.12). This upper contact is recognized by an obvious color change from the red clastic rocks of the Colton to the olive green and buff-colored limestone of the Green River Formation (Morris and others 1991). The Colton was laid down by rivers and deltas flowing northward into the shallow Lake Flagstaff. Sand and mud in the formation originated in Laramide uplifts and volcanic centers to the south and southeast (Zawiskie and others 1982).

Colton lithologies depict a variety of continental environments and have enabled the reconstruction of the Paleocene-Eocene Uinta basin. Sandstone in the Colton forms lenses up to 100 feet (30 m) thick that are characterized by basal scours lined with intraclasts of the same mudstone into which they cut. Above the intraclasts the lenses are filled with trough, cross-stratified sandstone. Large lenses are interpreted as the channel deposits of meandering rivers, whereas smaller lenses represent the transition to distributary channels of deltas that dispersed sediment and water into Lake Flagstaff. Intervening mudstone that encases the channel deposits ranges from red to gray green and black. It is commonly churned up by burrowing organisms and plant roots to the extent that no original sedimentary structures are recognizable. Mudcracks are also common. These floodplain deposits indicate abundant vegetation and occasional dessication following flood events. Scattered through the mudstone are muddy limestone deposits with fossils of turtles, fish, crocodiles, shore birds, and a variety of molluscs. These strata mark the position of small lakes on the poorly drained, low-relief floodplain. Taken together, the various components of the Colton record the processes that occurred on the southern margin of Lake Flagstaff as north-flowing rivers traversed a flat floodplain to empty into the extensive lake (Zawiskie and others 1982).

Sandstone composition in the Colton paints a more detailed picture of regional drainage patterns and paleogeography. Paleocurrent trends from cross-stratification show northwest- to northeast-directed flow for Colton rivers (Zawiskie and others 1982). Thus the quest for the source highlands looks to the south. Besides quartz grains, which dominate all far-traveled sands, the Colton contains sand-sized grains of plutonic, sedimentary, and volcanic origins, as well as feldspar, biotite, and muscovite. Plutonic rock fragments and associated feldspar, biotite, and muscovite probably came from the high-relief Laramide uplifts in western Colorado. Precambrian-age crystalline basement rocks exposed in the nearby Gunnison and Sawatch uplifts are the most likely sources (Fig. 8.13). Sedimentary-derived fragments are less unique, and because all the Laramide uplifts initially were capped by Mesozoic strata, these grains could have come from any or all of them as these sediments were recycled into the Colton. Probable sources near or adjacent to the Uinta basin include the monoclinal upwarps of the Uncompahgre highlands, Monument upwarp, Circle Cliffs uplift, and the San Rafael Swell, as well as the early phases of the higher-relief Gunnison and Sawatch uplifts. In contrast, volcanic rock fragments in the Colton *are* unique, with possible sources limited to Late Cretaceous–Early Tertiary volcanic centers that were situated in the present-day San Juan Mountains of southwest Colorado (Dickinson and others 1968). Cretaceous–Early Tertiary volcanism was rare across the region, and the Colton volcanic fragments almost certainly came from this area.

Rivers that drained the San Juan and Sawatch highlands swept northward, guided by the flanks of the northwest-trending Uncompahgre upwarp. Although the Laramide incarnation of the Uncompahgre was more subtle than its high and steep Pennsylvanian version, it was an important element in molding the drainage patterns and funneling water and sediment from southwest Colorado into the large lake basins to the north. Rivers that paralleled the southwest flank of the upwarp emptied into the Uinta basin along its southeast margin. Drainages that flowed along the northeast side of the Uncompahgre likely fed into the Piceance basin in northwest Colorado, although it is possible that these rivers also fed the Uinta basin (Fig. 8.13). To the west, on the Colorado Plateau, it has been proposed that north-flowing rivers drained the Monument and Circle Cliffs upwarps and merged with tributaries that came off the San Rafael Swell as they entered the Uinta basin (Zawiskie and others 1982).

By the end of Flagstaff/Colton deposition the various highlands that defined the boundaries of the Uinta basin had emerged. Though many would continue to evolve throughout deposition of the overlying Eocene Green River Formation, the basin margins had been established and would control sedimentation until the initiation of the modern Colorado River drainage ~5 million years ago. Many of these Laramide uplifts endure today as dramatic highlands. The Sevier highlands that formed the west basin margin were rendered inactive during the Paleocene but remained a formidable barrier as well as a continued sediment source. To the north lay the east-trending Uinta uplift, which extended across northern Utah into northern Colorado. Today the crest of this uniquely oriented range rises into jagged peaks more than 13,000 feet in elevation. The south margin of the basin is defined by a series of huge, regional-scale monoclines that include, from west to east, the San Rafael Swell, Monument upwarp, and the Uncompahgre uplift. The eastern margin was formed by the subtle but significant Douglas arch, which separated the Uinta basin from the Piceance basin to the east. While this north-trending ridge acted as an important divide through the Paleocene, it was over-topped by the rising Eocene lake waters to combine the Uinta and Piceance basins into a single, colossal lake. Although these large lakes were connected, each basin maintained its identity with a deeply subsiding depocenter marked by organic-rich lake sediments.

The Eocene Epoch

The Green River Formation

The Green River Formation is superbly exposed across an extensive area north of Interstate Highway 70 in northeast Utah and northwest Colorado. The formation has eroded back to the north from the persistent line of Cretaceous sandstone that forms the Book Cliffs. Above this sinuous escarpment, the Green River has been carved into another higher rampart known as the Roan Cliffs. Although the formation does rise abruptly above the flat bench that marks the top of the Cretaceous succession, these "cliffs" are in reality a series of steep ledges and slopes, which is characteristic of all Green River exposures.

The Green River Formation consists dominantly of lacustrine (lake) sediments deposited in a regional lake system that developed fully during the Eocene epoch. The name "Green River Formation" is widely applied to extensive Eocene lake deposits laid down in three different sedimentary basins, all bounded by Laramide uplifts. Although these basins were mostly isolated from each other by various uplifts, when water levels rose in the lakes, the basins were periodically connected. These basins include the greater Green River basin, which occupied a large tract of southern Wyoming; the Piceance basin in northwest Colorado; and the Uinta basin of northeast Utah, which is the focus of the ensuing discussion (see Fig. 8.2). Because the name Green River Formation has been so widely applied, many discussions can be confusing and misleading. Moreover, the geologic history of any of these individual sedimentary basins is complex, and contradictory interpretations abound in the literature. For these reasons, the following discussion is limited to the formation in the Uinta basin, but also discusses the Piceance basin where its history is intertwined with that of the Uinta basin. The greater Green River basin of southern Wyoming is largely ignored.

In general, the Green River Formation consists of two facies that delineate different environments within the giant lake. The shallow lake margin environment to the south and east is marked by sandstone and siltstone deposited in deltas where rivers merged with the lake. Where rivers were small or absent, carbonate shorelines developed with the deposition of oolitic, ostracodal, and stromatolitic limestone. In the deep-water lake center, organic-rich claystone and carbonate mudstone dominated. These quiet-, deep-water deposits include the valuable oil shales for which the Green River is well known, and which have spurred numerous detailed investigations. The shallow lake margin and deep, lake-center deposits interfinger in a complex pattern. When lake levels dropped, margin sediments stretched into the lake center. Conversely, as waters rose, deep-water deposits extended shoreward. These tongues of sediment are important for interpreting lake-level fluctuations and recognizing major changes through the lake's history.

Deposition of the Green River Formation began with the expansion of lakes across both the Uinta and Piceance basins. According to Bradley (1931) and Johnson (1985), lake levels may have risen to the point that they overtopped Douglas Creek arch for a short period, connecting the two basins. Transgressive lake margin deposits in the Uinta basin are dominated by oolitic and ostracodal limestone interbedded with rare sandstone. Oncolites and stromatolites are also common. Blue-green algae that comprise these components attest to clear, shallow water conditions. Ostracodes, a major component of these limestones, are millimeter-sized bivalved organisms. These diminutive shells were concentrated along the high-energy shelf as framework grains. Crossbedding in these and oolitic deposits point to a high-energy lake margin in which waves swept the shallow shelf area. Sandy shorelines were confined to parts of the south and north lake margins, where a supply of clastic sediment was readily available. To the south sand was flushed into the lake by a large, north-flowing river system that met the lake between the northern tips of the San Rafael Swell and the Uncompahgre highlands. The north lake margin was fringed by a narrow band of sand where the high Uintas bounded the basin.

While shallow water sediments accumulated on the shallow shelf, the deeper lake center was blanketed by organic-rich clay that today is the basal part of a thick succession of oil shale. For many years the oil shale of the Green River Formation has been carefully studied for possible oil extraction. Analyses of these basal oil shales show an average oil content of 10 gallons per ton of shale (Johnson 1985). The resource potential and economic aspects of the Green River oil shale are discussed in a subsequent section.

Stage 2 in the Uinta basin is marked by a renewed influx of clastic sediment and the progradation of thick wedges of sandstone and siltstone toward the lake center (Johnson 1985). It has been proposed that oil shale deposition in the deeper lake center was inhibited by this incursion, although it is difficult to document because of the scarcity of subsurface data on Uinta basin center deposits. Exposures along the southeast basin margin indicate that much of the clastic sediment came from the south, as it had earlier.

The third stage in the Uinta and Piceance basins records a shift to carbonate deposition in the shallow shelf and offshore parts of the lakes. Where rocks of this stage are best exposed, in the southeast part of the Uinta basin, stromatolitic limestone and carbonate-rich mudstone blanket the earlier clastic rocks. Clastic deposition endured locally at reduced rates and only where rivers continued to pour into slack lake waters. During this time the Douglas Creek arch was inundated by shallow water. Thus, what was earlier a low divide that separated the Uinta and Piceance basins became a connecting passageway (Johnson 1985). This is well-illustrated by discrete beds of stromatolitic limestone that can be traced across the arch and into the shallow margins of both basins. The shallow water margins of the Uinta and Piceance graded away from the arch into deep water depocenters, where organic-rich carbonate mud was slowly settling to the quiet lake floor. The organic content of this interval increased appreciably, as the oil shale has been assayed at ~40 gallons of oil per ton of rock.

Stage 4 began with a small transgression and the continued hydrologic connection between the Uinta and Piceance basins. The rise in lake level caused carbonate-rich oil shale to expand over the previous shallow basin margins, including the Douglas Creek arch. Interbedded sandstone and siltstone, however, indicate that clastic sediment continued to feed into the lakes from peripheral drainages. An increase in chemical sedimentation is shown by beds of nahcolite, a crystalline sodium carbonate mineral ($NaCO_3OH$) that precipitated from the water column. Nahcolite is more abundant in the Piceance basin, where one bed attains a thickness of 30 feet (Dyni 1974). In the Uinta basin nahcolite occurs mostly as thin, discontinuous beds. The overall thickness of stage 4 deposits reaches a maximum of 1400 feet in the center of the Uinta basin and ~1000 feet in the Piceance. Across the connection of the Douglas Creek arch the interval thins to 200 feet (Johnson 1985).

Stage 5, the final stage of Green River deposition, commenced with another transgression and deposition of the Mahogany zone, a regional oil shale marker interval that blanketed both the Uinta and Piceance basins. The Mahogany zone consists of up to 200 feet of exceptionally rich oil shale, which locally contains up to 60 gallons of oil per ton. Its name comes from the resemblance of its shiny brown to black color to that of polished mahogany wood (Bradley 1931). A bed of volcanic airfall ash within this interval has been radiometrically dated at 46.5 ± 0.6 Ma, putting the upper part of the Green River in the Middle Eocene (Johnson 1985).

Infilling of both basins continued after Mahogany zone deposition with oil shale accumulation to the south. This was cut short along the north basin margins by a massive influx of volcanic-derived sediments from the Absaroka volcanic field in northwest Wyoming. Initially the rivers that were clogged with volcanic detritus drained southward into Lake Gosiute, a large, shallow lake basin in southern Wyoming. Upon filling Lake Gosiute, the rivers topped the low divide and built southward into the Uinta and Piceance basins. Tongues of volcaniclastic sediment extended southward into both basins to interfinger with oil shale that was being deposited in the basin center. Eventually these fluvial sediments, known today as the Uinta Formation, blanketed both basins, ending deposition of the Green River Formation and the long-lived lakes.

Fossils in the Green River Formation

The Green River Formation is famous worldwide for its varied and unique fossil content. The preservation of delicate plants, fish, and insects, to name a few, is attributed to several aspects of Green River deposition. The quiet water of the deep lakes allowed fine clay to settle slowly but constantly to the lake bottom. As a result, any organism that died and drifted to the floor was quickly and gently buried. Rapid burial is vital in preserving any fossil organism, but it was especially important in the preservation of the fragile fossils found in the Green River shales. Changes and variations in lake chemistry contributed to both death and preservation. Massive kill-offs in the ecosystem resulted from changes in water chemistry as the lakes became increasingly saline. As the abundant fish fell to the lake bottom, an anoxic lake floor prevented scavengers from disturbing the death bed. The lack of oxygen on the lake bottom also enhanced the preservation of plant materials such as leaves blown into the lake. The Green River is well known for abundant fossil fish; almost every rock shop in the western United States sells the well-preserved specimens collected from these beds.

Although fossil fish are abundant, it is the smaller, more delicate insect and plant fossils that reinforce the interpretation of a quiet, deep water lake setting. The exceptional preservation of insects and plant leaves provides a glimpse of the lake shore ecosystem as well. Fossil insects and arachnids recovered so far include grasshoppers, crickets, cockroaches, stoneflies, bees, gnats, mosquitoes, dragonflies, several types of beetles, and several species of spiders (Dayvault and others 1995). The formation has also yielded one of the best-preserved fossil scorpions found in the world (Perry 1995); this new species comes from the Green River Formation in Colorado. Identifiable plant fossils include ancestors to modern water lilies, legumes, and leaves of now extinct species related to modern poplar, willow, sumac, and sycamore trees (Johnson and Plumb 1995).

A number of large, terrestrial fossil vertebrates have been reported from a single locality of the Green River Formation by J. Leroy Kay (1957) of the Carnegie Museum. In a preliminary report Kay outlined the discovery of several thousand vertebrate specimens from sandy, deltaic deposits along the east margin of the Uinta basin in Utah. Fossils include the bones of lizards, shore birds, and numerous mammals, including a marsupial, several insectivores and carnivores, three different rodents, and three identifiable primates. These diverse terrestrial vertebrate fauna, in concert with the variety of plant material and insects found in the lake deposits, showcase an environment that was favorable to an assortment of plants and animals. It is rare that all levels of an ecosystem are so well represented in the fossil record. The Green River Formation contains such a record in abundance, providing an exceptional snapshot of the Eocene Colorado Plateau.

The climate of the Colorado Plateau during the Eocene has been reconstructed based on several aspects of the Green River Formation. An analysis of its flora has led to a consensus that deposition occurred under a warm, temperate climate analogous to that of the modern Gulf Coast region of the United States (Knowlton 1923, Brown 1929). Paleontologist E. W. Berry (1925) made it clear, stating, "I know of no member of the Green River or so-called Bridger floras that would not be perfectly at home somewhere in the region between South Carolina and Louisiana . . . at the present time." Another more quantitative analysis based on laminations in Green River sediments and the reconstructed dimensions of the lakes was conducted by USGS geologist Wilmot H. Bradley (1929). By drawing comparisons with modern lakes of similar dimensions and similar types of sediment from around the world, Bradley concluded that during the Eocene the region experienced cool, moist winters and relatively long, warm summers. Temperatures fluctuated widely around a mean of 65°F. Bradley also estimated that 30 to 43 inches of rain fell annually into the lake basins. Although these initial investigations took place in the 1920s, each is based on different lines of evidence, and the results appear to validate each other. Moreover, no subsequent fossil discoveries or interpretations have contradicted these early findings. They have, in fact, strengthened them.

The Uinta Formation

The Eocene Uinta Formation consists of up to 2500 feet of brown volcaniclastic sandstone and siltstone, and conformably overlies the Green River Formation (see Fig. 8.12) (Dane 1954). The Green River/Uinta contact is gradational, with the fluvial Uinta intertonguing with lacustrine shale of the Green River Formation. Stratigraphic relations indicate volcanic-derived sediments entered the Piceance and Uinta basins from the north (Surdam and Stanley 1980). The south-flowing rivers that introduced this sediment originated in the Absaroka volcanic field in northwest Wyoming. In order to reach the Piceance and Uinta basins, however, the rivers first had to fill Lake Gosiute, which occupied the extensive Green River basin to the north (see Fig. 8.2). After Lake Gosiute filled with gravel, sand, and silt, the rivers spilled over a low spot at the east end of the Uinta Mountains and began to dump volcanic sediment into the northern reaches of the Piceance basin. Intertonguing relations between the volcaniclastic sandstone to the north and lacustrine shale to the south clearly show the subsequent southward progradation of river sediments into the Piceance basin. Eventually the rivers turned east and crossed the Douglas Creek arch to extend their volcanic load into the Uinta basin (Johnson 1985). These rivers gradually spread westward but stopped short of the basin center, where volcaniclastic sandstones are absent. Here and farther west, a shallow lake environment endured to the end of the Uinta Formation deposition.

The complex path of sediment from the Absaroka volcanic field across various Laramide uplifts and finally into Lake Uinta has been traced geographically and chronologically through several detailed studies. Richard Mauger (1977) obtained radiometric dates on a number of volcanic ashfall tuff beds that occur throughout Eocene deposits in Utah, Colorado, and Wyoming. By combining these dated units and their locations, both geographically and stratigraphically relative to volcaniclastic sediments, a time frame for the methodical southward shift of these sediments has been established. These dates, along with sediment composition and paleocurrent indicators, were used by Surdam and Stanley (1980) to reconstruct the path of the south-flowing river system that originated in the Absaroka volcanic field. Initially, volcanic-laden streams reached the southeast margin of the Wind River basin at ~49 Ma; these rivers continued to deposit sediment in the region to at least 45 Ma. After filling the Wind River basin, the rivers overtopped the low divide between the southeast flank of the Wind River uplift and the west end of the Granite Mountains uplift and poured into the north end of the huge Green River basin. A tuff bed near the base of volcanic sediments in the Green River basin yielded an age of 45.5 Ma, providing the time at which these rivers extended into the basin center. After filling the Green River basin to capacity, the waters overflowed the east end of the Uinta uplift to discharge into Lake Uinta at the northern end of the Piceance basin. During this time Lake Uinta occupied both the Piceance and Uinta basins. Absaroka-derived fluvial sandstone initially prograded south into the Piceance but eventually turned westward into the Uinta basin. Near its western limit in the east-central part of the basin, a tuff bed within the Uinta Formation was dated at 41.9 Ma. Although the tuff does not occur at the base, it does provide a vital estimate for its arrival into the basin.

By the end of Uinta Formation deposition, most of the large lakes across the region had been filled in with sediment and the lakes replaced by rivers that wound through the broad basins. The exception was the central and western part of the Uinta basin. The Absaroka-fed rivers stopped short of this region and a much-reduced Lake Uinta endured to the end of the Eocene (Johnson 1985). In this shrunken lake up to 2500 feet of lacustrine shale, sandstone, and siltstone was deposited. Lake center deposits contain up to 900 feet of saline deposits that include nahcolite, suggesting that the Uinta basin had no outlet. Fluvial deposits of the Eocene-Oligocene Duchesne River Formation succeed the Uinta Formation.

Eocene-Oligocene(?) Duchesne River Formation

The Duchesne River Formation is preserved and exposed across the northern half of the Uinta basin, where it overlies the Uinta Formation. The contact is defined by a shift from the variegated, fine-grained sediments of the Uinta Formation to the red-brown sandstone of the Duchesne River. Variations in the nature of the contact are the rule; it ranges from intertonguing between the formations to an erosional unconformity in which the Duchesne River cuts deeply into the Uinta. Locally the contact is an angular unconformity (Anderson and Picard 1972).

In an effort to clarify the stratigraphy of the Duchesne River Formation, David Anderson and longtime Uinta basin geologist M. Dane Picard (1972) examined the formation in detail and defined four members. From oldest to youngest these are the Brennan Basin, Dry Gulch Creek, Lapoint, and Starr Flat members. This study was in part motivated by the designation of the formation as the standard section for the latest Eocene Duchesnean Stage. Its fossils provide the standard reference for vertebrate biostratigraphy throughout North America. Intense interest in these strata for their fossil content began with one of the earliest paleontologists to explore the West, the famed O. C. Marsh (1871). His paper was followed by a more detailed account of the Tertiary fossil mammals of the region by Princeton professor H. F. Osborn (1895). Although its fossil mammals continued to receive attention, a formal stratigraphic framework for the formation was not established until Anderson and Picard (1972).

The lowermost Brennan Basin Member is a resistant, cliff-forming succession dominated by sandstone and conglomerate with minor fine-grained mudstone and siltstone. Locally boulders up to 6 feet in diameter make up crude sheets of coarse conglomerate. The member reaches a maximum of ~2000 feet thick at its type locality along the Green River, but it thins to less than 1000 feet east and west of there (Anderson and Picard 1972). The member intertongues with the overlying Dry Gulch Creek Member, which is one reason for the great variation in thickness. Another is likely the simple variation in original depositional thickness. The Brennan Basin Member was deposited by powerful, high-gradient, braided rivers that descended southward into the basin from the renewed heights of the Uinta uplift.

The Dry Gulch Creek Member consists mostly of siltstone and claystone with lesser lenses of sandstone. This red-brown, slope-forming succession attains a maximum thickness of 659 feet at its type locality (Anderson and Picard 1972). The top of the member is precisely defined as the base of the lowest continuous bentonite bed in the overlying Lapoint Member. Because this and the other bentonite beds in the Lapoint Member are individual volcanic ash deposits that settled rapidly from the atmosphere, probably over a week or less, this bed represents a single geologic instant. Moreover, this bentonite bed contains biotite crystals that have been radiometrically dated and yield an age of 39.3 ± 0.8 Ma, placing the Dry Gulch Creek Member and at least the lower part of the overlying Lapoint Member well within the Eocene Epoch (McDowell and others 1973). The Dry Gulch Creek Member records a decrease in river gradient and energy, with a shift to a sluggish, meandering system. Sandstone lenses mark the position of the sinuous river channels, and the abundant finer-grained strata that encase these channel forms represent the surrounding floodplain.

The Lapoint Member consists of red-brown claystone, siltstone, and mudstone, with lesser sandstone and rare conglomerate. The dominance of fine-grained strata has resulted in weathering to poorly exposed slopes and valleys. Fine-grained strata host many laterally continuous, green-gray bentonite beds, the product of altered volcanic ash that repeatedly blanketed the region during Lapoint deposition. It is the previously mentioned lowermost bentonite that conspicuously marks the base of the member and provides an absolute radiometric age of 39.3 ± 0.8 Ma (McDowell and others 1973). The upper contact with the Starr Flat Member is placed at the base of the lowest red-brown sandstone or conglomerate immediately above the highest green-gray bentonite bed. Various fossil vertebrate remains have been recovered from these low-energy deposits, including fossil rodents; large, carnivorous, hyena-like mammals; and hoofed animals such as the even-toed artiodactyls and odd-toed perissodactyles. These hoofed mammals are the extinct relatives of pigs, camels, horses, and rhinoceroses. Lapoint Member strata represent a slow-moving, meandering river system. It is this low-energy setting that contributed to the preservation of the bentonite beds and the remains of the large, strange mammals that characterize the Eocene Epoch.

The Starr Flat Member forms the top of the Duchesne River Formation and consists dominantly of sandstone and conglomerate that weathers to steep slopes and cliffs. Preserved remnants of the member are limited to the southern flanks of the Uinta Mountains, where it lies unconformably on the uplifted edges of older strata (Anderson and Picard 1972). The member extends into these mountains up to an elevation of 8000 feet and represents the remains of a once-thick blanket of coarse sediment. As the uppermost member of the Duchesne River Formation, the Starr Flat has been more susceptible to erosion, and its extent and thickness have been largely controlled by this process. It reaches a maximum thickness of 769 feet at its type locality but rarely reaches 200 feet elsewhere. Overlying sediments consist of younger, unconsolidated gravel in an unconformable contact. No fossils have yet been recovered from the member, so its age is uncertain, but Anderson and Picard (1972) speculate that it likely extends into the Oligocene Epoch. The sheetlike geometry of the strata and their coarse-grained nature indicate deposition in a high-energy, braided river system comparable to that of the basal Brennan Basin Member. It also points to renewed energy, possibly due to resurgent uplift in the Uinta Mountains.

The Duchesne River Formation is composed of fluvial deposits laid down in both braided and meandering rivers. Regardless of the river type embodied by a particular member, all drained southward off the mountainous Uinta uplift, which for the previous 25 million years bounded the basin along its northern margin. But what spurred the shift from the previous west-flowing system to a south-flowing drainage orientation? What drove the sudden evolution of the Uinta highlands as the dominant source area for the region? The answer lies in the renewed uplift of the Uinta highlands and the timing of this activity. According to Ritzma (1974), in the Asphalt Ridge area near Vernal, Utah, exposed reverse faults along the south flank of the Uinta Mountains cut through and deform the underlying Uinta Formation but do not affect the overlying Duchesne Formation. Apparently this pulse of uplift was enough to drive an influx of detritus southward off the reactivated highlands. Thus, the last recognizable pulse of uplift in the Uinta Mountains occurred immediately after deposition of the Uinta Formation, but just prior to the major southward flow of sediment into the northern part of the basin. The Duchesne River Formation is the sedimentary signature of this uplift.

Oligocene-Miocene Igneous Activity

Tertiary magmatism on the Colorado Plateau began in the Oligocene and continued into the Early Miocene in the form of the intrusive laccolith complexes of the La Sal, Abajo, and Henry mountain ranges (see Fig. 8.3). Today these isolated clusters of intrusive rock rise to more than 12,000 feet, towering high above the deeply incised artery of the Colorado River. From any vantage point on the eastern Colorado Plateau, at least one, and often all three, of these snow-laced mountain islands may be seen through the shimmering waves of heat. In the summer the green aspen- and pine-clad slopes and gray, rock-strewn summits offer a cool respite from the intense heat of the redrock desert.

The three major laccolith clusters of the Plateau fall within the same age range and composition, leading most geologists to conclude that they have a similar origin. In general, these intrusive centers can be discussed as a group. The La Sal and Abajo mountains, however, will be treated individually and in more detail due to their prominence in southeast Utah, which is the primary focus of this book. Because they lie outside this region, the Henry Mountains will be discussed only in a general sense, where the information is applicable to all the Colorado Plateau laccoliths. For a more detailed account of the Henrys the interested reader is referred to *Geology of the Parks, Monuments, and Wildlands of Southern Utah*, also by Robert Fillmore (2000).

Laccoliths were first described and defined by pioneering geologist G. K. Gilbert in 1877 from the well-exposed examples in the Henry Mountains of southern Utah. Laccoliths are composed of intrusive igneous rock that rose as magma to shallow levels in the crust, where it infiltrated the veneer of sedimentary rock. Most intrusive bodies invade country rock of various types as large, shapeless blobs of molten rock with fingers of magma that extend irregularly into the surrounding rock. Laccoliths, however, are rigidly defined by their specific and unique relations with surrounding sedimentary layers. In cross-section they are mushroom-shaped, with a flat base that parallels bedding in the underlying sedimentary rocks (Fig. 8.14b). The dome-shaped top of the laccolith formed as the magma squeezed laterally between layers. The upward force of the magma at these shallow levels caused upward inflation of the intruded zone, doming overlying strata upward so that they draped across the top of the laccolith. The "stem" of the mushroom-shaped intrusion is formed by the conduit, or pipe, by which the magma was fed to the shallowly buried strata. Millions of years later erosion stripped away the overlying strata to expose the resistant granitic cores of the laccoliths, which today loom high above the surrounding canyons (Fig. 8.14c).

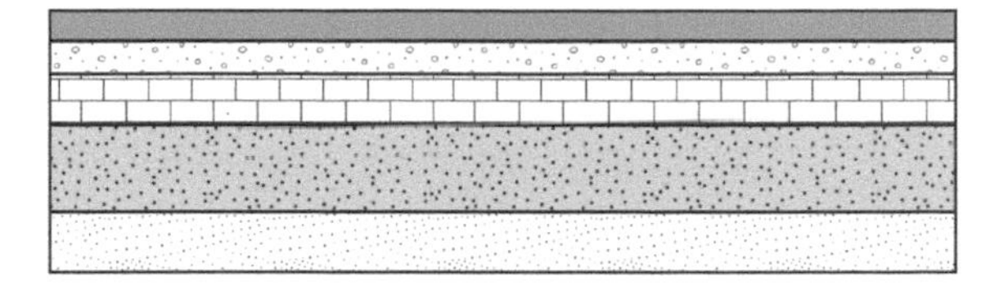

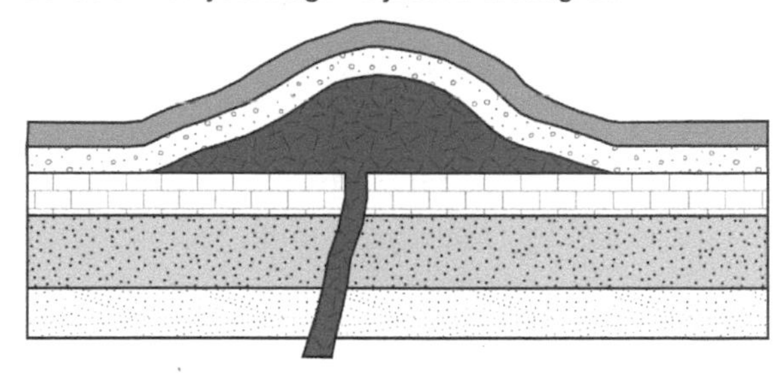

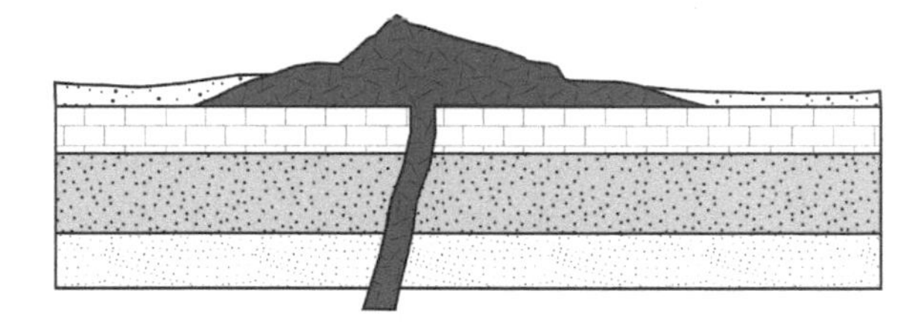

Fig. 8.14. The steplike formation of laccoliths on the Colorado Plateau and in western Colorado, beginning with undeformed sedimentary host rocks (A) and ending with modern exposures of the intrusion's granitic core (C).

The La Sal Mountains Intrusive Complex

The La Sal Mountains igneous intrusive complex is composed of three distinct intrusive centers that collectively form a broad dome with ~2000 feet of relief. Although the larger, more prominent features are laccoliths, the complex also includes associated dikes, sills, and plugs of plutonic rock. The north-south alignment of the three centers has led to their designation as the north, middle, and south intrusive centers (see Fig. 8.3). The northern center was emplaced into the Castle Valley–Paradox salt anticline and is elongated in a northwest-southeast direction, parallel to the orientation of the host anticline. Similarly, the south intrusive center pierces the Moab–Spanish Valley salt anticline and is elongate to the northwest. It is apparent that the geometries of these intrusions were influenced by the presence of the salt anticlines, probably by allowing the magma to initially rise through the same basement faults that controlled anticline development (Ross 1998). The lateral spread of magma at shallower crustal levels may also have been enhanced by the distribution of the thick salt that was concentrated in the anticlines. In contrast, the middle intrusive center invades the syncline that separates the two large anticlines, and in map view it has a circular geometry. The middle mountains contain the highest peaks in the range, including Mt. Peale (elevation 12,721'), Mt. Mellenthin (12,646'), and Mt. Tukuhnikivatz (12,493').

Analysis of intrusive rocks in the La Sal Mountains has produced $^{40}Ar/^{39}Ar$ radiometric dates from 27.9 to 25.1 Ma, suggesting that intrusive activity spanned a period of several million years (Nelson and others 1992, Nelson 1998). The earliest intrusive episode emplaced diorite, which makes up most of the La Sal intrusive rock (Hunt 1958). This was followed by dikes and sills of monzonite. The final phase of igneous activity was the development of breccia, either by the injection of magma along the margins of the igneous bodies, which would have broken up the host sedimentary rocks, or by the forceful release of gas-rich fluids to the surface, resulting in the explosive expansion of gases in the subsurface (Hunt 1958, Ross 1998). The implications for the latter origin are that there would have been a surficial expression of the magmatic activity in the form of volcanic features. Subsequent erosion, however, has obliterated any evidence for such activity, if it ever existed.

In the earliest study of the laccoliths, conducted by G. K. Gilbert (1877) in the Henry Mountains, it was proposed that laccolith complexes initiate as a series of sills. A sill is a tabular-shaped intrusion that extends as a tongue between beds of sedimentary rock and, except for being igneous in origin, has the geometry of a sedimentary layer. At shallow depths the sills are able to expand upward into the mushroom geometry of a laccolith by lifting and bending the overlying strata. In the Henrys, Gilbert concluded that the complexes were cored by a central laccolith overlapped by various satellite laccoliths that had grown out of the surrounding sills. Thus, according to Gilbert, the dominant features of the Henry Mountains are sills and laccoliths.

The next major study of Colorado Plateau laccoliths was undertaken in the La Sal and Henry mountains by USGS geologist Charles B. Hunt (1953, 1958). In his detailed studies of these igneous complexes Hunt concluded that the intrusive centers of both ranges were cored by a central, pluglike stock that fed magma to the laccoliths extending from it. Hunt envisioned the cylinder-shaped stock as a rising magma body that, upon reaching shallow crustal levels, sent lateral tongues of magma radially outward to squeeze between the layers of upper Paleozoic and Mesozoic strata. Some sills inflated upward to form flat-bottomed laccoliths, while others remained sills.

FIG. 8.15. View to the south across the Castle Valley salt anticline of Round Mountain, an igneous plug in the middle of the valley, and the north cluster of the La Sal Mountains laccoliths.

More recent work in the La Sal (Ross 1998) and Henry mountains (Jackson and Pollard 1988, Jackson 1998) has cast doubt on Hunt's central stock hypothesis. Instead, these recent studies cast new light on Gilbert's earlier interpretation of both central and satellite laccoliths. These studies have analyzed deformation in the surrounding folded and faulted sedimentary rocks in an effort to determine the geometry of the intrusive bodies in the subsurface. In both ranges the results point to the presence of an irregular laccolith rather than a central stock. Uncertainties remain due to the lack of exposure in the lower parts of these intrusions; where unearthed by erosion, they have subsequently been covered by talus and thick mantles of loose surficial rock and soil, and so cannot be directly observed. Both research groups admit that in the absence of actual exposures, the presence of a central stock remains a possibility.

Round Mountain is a prominent knob of igneous rock that juts from the center of flat-bottomed Castle Valley at its southern end, isolated from the nearby northern intrusive center of the La Sal Mountains (Fig. 8.15 and Plate 10). This small mountain, located in the shadow of the La Sals, is one of several small plutons that occur adjacent to, but outside of, the main bulk of the northern intrusive center. Round Mountain is especially obvious due to its position in the center of the collapsed Castle Valley salt anticline. Like most of the La Sal intrusives, the pluton is composed of hornblende diorite. Contorted sedimentary rocks of the

Paradox Formation surround the base of the pluton and form small patches of caprock on its summit. The edges of the pluton around its base are engraved by fault-induced grooves, or **slickensides**. These led Ross (1998) to conclude that the pluton is surrounded by a fault. According to his interpretation, sometime after emplacement of the Round Mountain pluton, the Paradox salts were partially dissolved, and the less soluble part of the formation collapsed around the igneous body. Because the igneous rock was deeply rooted, it remained stable as the Paradox dropped around it, encircling it with a fault. While Round Mountain is certainly related to the La Sal Mountains intrusive complex, it remains unknown whether it was fed by the northern intrusive center or had some deeper level of connection to the larger magma source that fed the three intrusive centers of the main La Sals.

The depths of the La Sal plutons at the time of emplacement are estimated to have ranged from 1.9 to 6.0 km, based on restoration of the sedimentary rocks that blanketed the region during the Oligocene (Ross 1998). Shallower levels (~1.9 km) represent the laccolith intrusions of the high mountains, some of which squeezed into upper Jurassic or Cretaceous strata. Deeper-level plutons (~6.0 km) are known from drill hole data from the central part of the middle intrusive center. In a well drilled by Exxon Corporation, multiple levels of intrusive rock were encountered to a depth of ~3600 feet below sea level. The deepest intrusions were hosted by the lower Paradox Formation. Upward from this level, intrusions occur in multiple horizons up to and including the Mancos Shale, the youngest strata to be intruded. These multiple horizons of plutons, likely spreading laterally from a feeder conduit, suggest a Christmas tree–like cross-section geometry for the middle intrusive complex (Ross 1998).

The Oligocene rise of the La Sal Mountains abruptly increased local relief, accelerating erosion. Immediately following the rise of these mountains, the action of water and ice, coupled with gravity, worked to reduce their stature. To date these processes have peeled back the overlying Tertiary and Mesozoic sedimentary rock, exposing the gray igneous cores of the higher laccoliths. The surviving rock of the La Sals continues to be ground and gouged by the frost-shattering action of ice, the flow of water, and the incessant pull of gravity. Until these lofty peaks are reduced to minor knobs or beveled completely flat and their detritus scattered into rivers and seas, the erosional processes will continue unabated.

The Abajo Mountains Intrusive Complex

The intrusive complex of the Abajo Mountains is located ~40 miles south of the La Sals (see Fig. 8.3). The range overlooks the town of Monticello, Utah, which is situated immediately to the east on the forested lower flanks of the Abajos; the town of Blanding lies 10 miles south of the range. The elevation of these igneous mountains exceeds 11,000 feet, and their rounded, gray summits are in constant

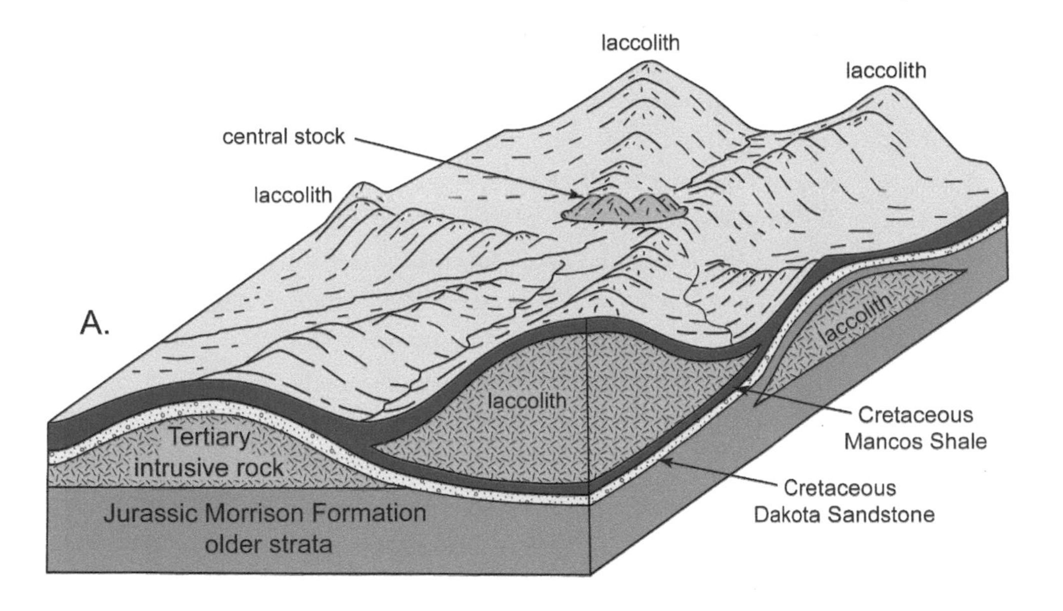

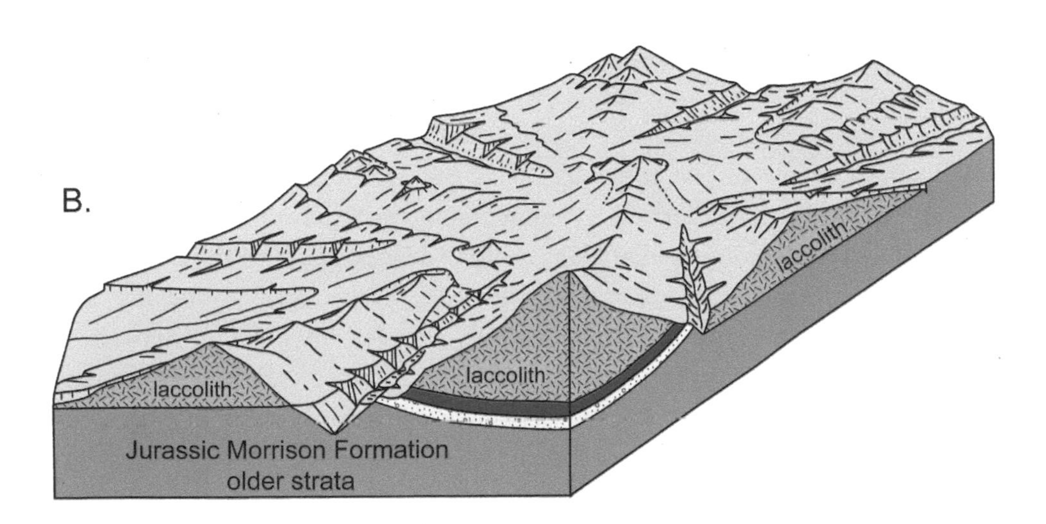

Fig. 8.16. Diagram of laccolith intrusion in the Abajo Mountains. A, the setting soon after intrusion of magma and development of topography driven by formation of laccoliths. Note the central stock that apparently fed magma in a radial pattern to form the surrounding laccoliths. B, the geologic setting after ~30 million years of erosion. The figure shows the drainage development and incision that leave the exposed granitic core of the laccoliths as discrete mountains. Incision continues to occur between the laccoliths. Modified after Witkind 1964.

view from the Needles area of Canyonlands, which bounds the isolated range on the north.

Although the Abajo Mountains are not as well studied and have not recently been reexamined as the La Sal and Henry mountains have, the geology of the range was mapped in detail by Irving Witkind (1964) of the USGS. This study was initiated by the U.S. Atomic Energy Commission to identify and evaluate uranium deposits in the Abajos and surrounding areas. In his mapping, Witkind identified four distinct topographic masses in the Abajos; following the lead of the earlier work of Hunt (1953, 1958) in the Henry and La Sal mountains, he interpreted each as a discrete igneous center cored by a central stock. Satellite laccoliths and sills reached fingerlike from the feeder stock out into the shallow sedimentary rocks.

Documentation of the age of the Abajo intrusions has similarly lagged behind that of their sister ranges. The most recent age determinations come from the use of the fission track method on the minerals zircon and sphene. Using this method, Kim Sullivan (1998) of Brigham Young University obtained ages that range from 28.6 ± 3.4 Ma to 22.6 ± 2.2 Ma. The error range for this method is greater than that for the more precise $^{40}Ar/^{39}Ar$ radiometric system, which was used for the La Sals and Henrys. Still, these ages agree with the age obtained by R. L. Armstrong (1969) using the older and less precise K-Ar radiometric system. By recalculating Armstrong's results using the more recently refined decay constants for the K-Ar system, Sullivan reports an updated age of 28.1 Ma for Armstrong's results for the Abajos. This agrees closely with both her fission track ages and the ages of the La Sal and Henry mountain laccoliths.

Rock in the Abajos, both the intrusive and the sedimentary host rocks, are strikingly similar to those in the La Sals. Plutonic rock of the Abajo Mountains grades from diorite to quartz diorite, as in the nearby La Sals. The lateral infiltration of the magma to form the laccoliths appears to have been guided by Late Jurassic and Cretaceous sandstone at shallow depths. Most of the laccoliths are sandwiched between the Jurassic Morrison Formation (below) and the Cretaceous Burro Canyon and Dakota sandstones (above; Fig. 8.16), or between the Dakota and overlying Mancos Shale. Some deeper laccoliths, however, split units as old as the Permian Cutler and the Triassic Chinle formations.

The four topographic domes of the Abajos identified by Witkind (1964, 1975) include the large West and East mountains; the smaller Johnson Creek dome, located between West and East mountains; and Shay Mountain, which lies north of the other domes (see fig. 8.3). Witkind interprets each of these as a discrete igneous intrusive center with a central parent stock and multiple associated, overlapping satellite laccoliths (Fig. 8.16). The West and East centers host the highest peaks in the range and have been eroded to the deepest levels, providing the clearest views of the range's intrusive features, including the base of some laccoliths. West Mountain center consists of a small central stock less than a half mile in diameter surrounded by five areally extensive satellite laccoliths. The much larger East Mountain center also is cored by a small central stock from which up to eighteen recognizable laccoliths radiate. Six of these laccoliths can be tied directly to the stock, while other laccoliths appear to have been born of these six laccoliths. Johnson Creek center, nestled between the West and East centers, contains five discernible laccoliths. The isolated Shay Mountain center, 3 miles north of the others, is limited in extent and hosts only two identifiable laccoliths. This may be the result of limited erosion in this area; only the roofs of the intrusions have been exhumed, and others may lie beneath. Although exposures are limited, Witkind (1964, 1975) assumed that the smaller centers of Johnson Creek dome and Shay Mountain, like their larger and better exposed neighbors, are cored by a small stock that served as the magma source for the laccoliths.

The Origin of Oligocene Magmatism

In deciphering the origin of the Oligocene laccolith centers of the Colorado Plateau, one must also consider their relationship with coeval igneous activity on the west and east margins of the Plateau. Along the western margin of the Colorado Plateau, in western Utah, lies the voluminous Marysvale volcanic field (see Fig. 8.2). This volcanic center straddles the transition zone between the Colorado Plateau on the east and the Basin and Range geologic province to the west. An eruption age of 34 to 21 Ma places the Marysvale volcanics comfortably within the 32.3 to 22.6 time frame established from the collective ages of the laccoliths (Rowley and others 1998). Along the east margin of the Plateau, in southwest Colorado, lies the larger San Juan volcanic field. Multiple large collapsed volcanoes, known as **calderas,** dominate this massive volcanic field, which ranges in age from 32 to 23.1 Ma (Lipman 1983). Immediately north of the San Juan volcanic field is the West Elk volcanic center, a single stratovolcano that was active from 29 to 30 Ma (Coven and others 1999, Murphy and others 2000). Still farther north, and buttressing this volcano, is the laccolithic cluster of the West Elk Mountains, which consists of at least fourteen discrete laccoliths. These laccoliths share the same composition, and those that have been dated fall within the 30 Ma age range. This north-south oriented line of contemporaneous plutonic and volcanic features in western Colorado defines the east margin of the Colorado Plateau.

The extensive volcanic fields along the margins of the Colorado Plateau are a stark contrast with the isolated, small-volume laccolith centers that represent this interval of magmatism within the Plateau boundaries. This difference is best explained by the thick, largely undeformed crust of the Colorado Plateau. Thickness of Plateau crust ranges 45 to 50 km, whereas crust to the west, in the Basin and Range province, may be only 30 km thick and riddled with faults. In addition, the mafic rock that comprises most of the Plateau crust is stronger and more resistant to deformation than the more silicic crust that dominates to the west. Crust on the east margin, in western Colorado, is thick but has a long history of faulting, which contributes to its weakness. Faults on both sides of the Plateau likely were conduits that facilitated the rise of magma to the surface and probably played a significant role in the location of volcanic activity. The paucity of igne-

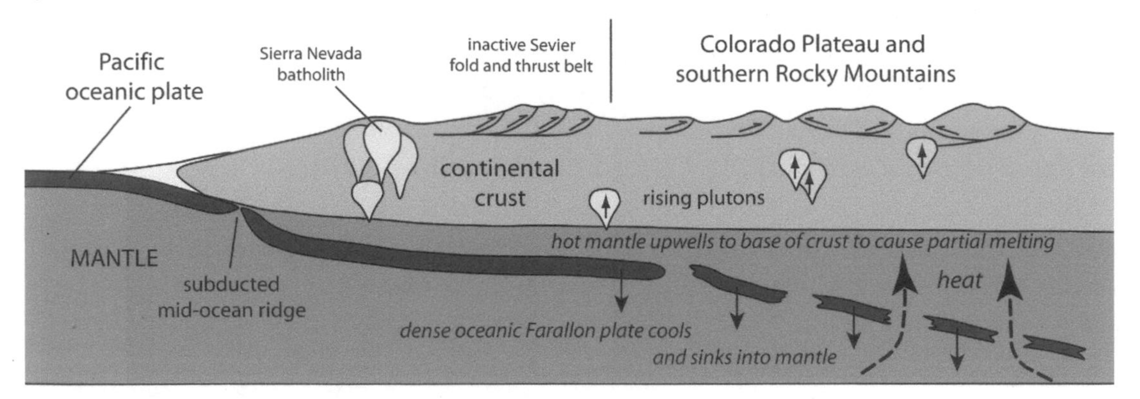

FIG. 8.17. Schematic cross section across western North America after subduction of the mid-ocean ridge in the Pacific Ocean (~25 Ma). At this time subduction ceased and the Farallon oceanic plate is interpreted to have foundered and broken up, sinking into the mantle. The result of this is that hot, deeper mantle rose to replace the foundering plate, heating up the base of the crust. This increase in temperature caused partial melting of the lower crust, producing the magma that fed, at the surface, the San Juan and Marysvale volcanic fields. Plutonic features formed during this event include the Henry, Abajo, and La Sal mountains laccolith clusters, as well as laccoliths in western Colorado located immediately north of the San Juan volcanic field.

ous activity on the Plateau, even while immense volumes of magma were leaking around its margins, is attributed to the thick, long-stable crust underlying the mostly undeformed block.

How does a melt "suddenly" (~30 Ma) form in the mantle across a large region following a period of magmatic quiescence? The answer lies in the geochemistry of the resulting igneous rocks across the region. These rocks bear a geochemical signature of a magmatic arc, meaning that the magma was derived from partial melting of subducted oceanic crust in the mantle. The vital link in this model is the preceding flat trajectory of the subducting oceanic plate beneath the region that presumably drove the Laramide orogeny. The enduring presence of this oceanic plate provided a voluminous and widespread source for arc-related magmatism.

The rapid convergence between the Farallon oceanic plate and the overriding North American plate that drove low-angle subduction began to slow by 50 Ma (Best and Christiansen 1991). As the eastward motion of the oceanic plate stalled, its greater density dragged it into a slow descent into the mantle. As the Farallon oceanic plate foundered, mantle material from beneath the slab rose around it, replacing the cooler oceanic crust with a wedge of hot mantle material (Fig. 8.17). The collapse of the Farallon slab into the mantle caused it to emit fluids that then ascended through the intervening mantle, generating magma along the way (Nelson and Davidson 1998).

The most recent and detailed model for the origin of Colorado Plateau laccoliths comes from the chemical analysis and interpretation of igneous rocks in the Henry Mountains by Stephen Nelson of Brigham Young University and Jon Davidson of UCLA (1998). Because of similarities in composition, this model can also be applied, in a general sense, to the intrusive rocks of the La Sal and Abajo mountains. Certain chemical signatures suggest that the magma originated in the upper mantle with a mafic composition. As the less-dense melt slowly ascended through the mantle, it pooled at the base of the crust, probing for a weakness to exploit and assist in its continued rise. Upon infiltrating the silicic crust, the greater density of the mafic magma, due to its high Fe and Mg content, likely caused it to stall. The pooling of magma in the lower crust allowed for the melting and assimilation of more silicic crust into the melt, and the simultaneous crystallization of mafic minerals, removing this component from the batch. By these processes the original mafic composition of the mantle-derived melt shifted to an intermediate composition. This also decreased its density, promoting a continued, slow vertical journey through the crust, where after reaching shallow levels on the Plateau, it spread laterally to form the laccolith centers and associated sills. The geochemistry of the final rock that made its way into the shallow crust lies between the end member compositions of a mantle source and a crustal source for the melt, recording the various modifications the magma experienced on its ascent (Nelson and Davidson 1998). Despite the cessation of Laramide mountain building, the flat-slab subduction that drove it continued to influence the landscape of western North America with the intrusion of laccoliths and other plutonic bodies, and large-scale, explosive volcanism.

Uplift of the Colorado Plateau—Early or Late Tertiary?

The earliest geologists to explore the Colorado Plateau—including John Wesley Powell, Clarence Dutton, and Grove Karl Gilbert—recognized that the deep incision of the spectacular canyons had been driven by relatively recent uplift. It has long been apparent to geologists that sometime during the last 65 million years, after the Sevier orogeny, the Colorado Plateau and the Southern Rocky Mountains were pushed upward. Although this uplift has long been recognized, no consensus has been reached on exactly when uplift took place, and the cause of uplift is even more controversial (e.g., Pederson and others 2002, Morgan 2003). Recently, based on multiple lines of evidence, researchers have focused on activity in the past 5 million years in attempts to solve this puzzle. So while there is general agreement that uplift occurred, recent work has invigorated the debate on how and when it took place. Some of these hypotheses are outlined below.

The most difficult problem in understanding the uplift history of the Colorado Plateau is that uplift induces erosion; consequently, the sedimentary record normally used to read geologic history has mostly been erased for this time. The clearest evidence for uplift sometime during the Tertiary is the present elevation of Late Cretaceous shoreline deposits of 6990 feet (2117 m) (Pederson and others

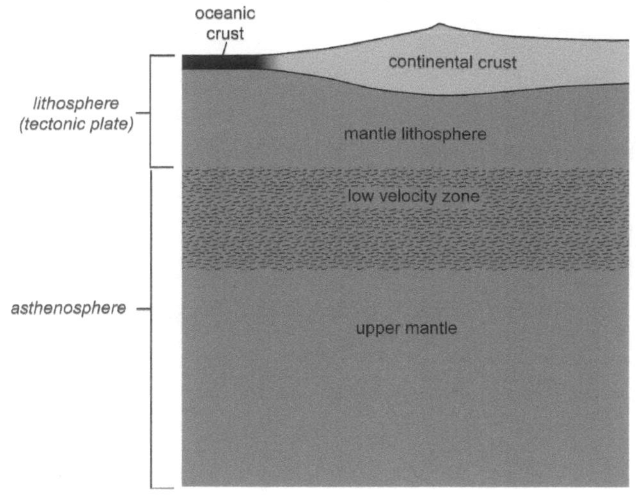

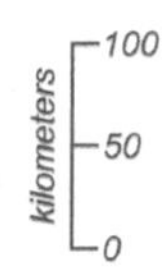

FIG. 8.18. A cross section through the crust and part of the upper mantle showing the various components, including the low-velocity zone in the mantle. This marks the boundary between the lithosphere and the asthenosphere, and is the zone of partial melting along which the tectonic plates move. Note that the complete upper mantle is not shown here.

2002). These were deposited at sea level during the Late Cretaceous, the last time, besides the present, that elevation on the Plateau could be unequivocally documented. Thus uplift took place, but when?

The Nature of the Mantle Lithosphere and Asthenosphere

A basic appreciation of the lithosphere, which is a combination of the crust and uppermost mantle, is vital to understanding the various hypotheses that have been proposed for Colorado Plateau uplift (Fig. 8.18). All the models for uplift in some way involve the lithosphere.

The continental crust that forms the outermost layer of the continents has an average chemical composition of granite, which, relative to the underlying mantle, is low-density material. Continental crust in general ranges from 30 to 70 km thick, and the crust of the Colorado Plateau averages 40 to 45 km thick. The mantle lies beneath the crust, and the boundary between the two is defined by a change from less dense granitic rock to very dense ultramafic rock known as peridotite.

The solid lithosphere (crust and uppermost mantle) is underlain by the asthenosphere, which is hotter and behaves like a plastic. The boundary between the lithosphere and underlying asthenosphere lies within the upper mantle (Fig. 8.18). Although the mantle is mostly composed of solid, ultramafic rock, the dividing line between the mantle lithosphere and the asthenosphere is a zone of partially molten rock. This also defines the base of the tectonic plates.

Thickness of the continental lithosphere varies widely, even across western North America. Present lithospheric thickness depends on recent and past tectonic history, and magmatism, among other factors. As stated before, the crust of the Colorado Plateau averages ~45 km thick, with an equally thick underlying mantle lithosphere, giving an average lithosphere thickness of 80 km (Thompson and Zoback 1979). For comparison, the lithosphere of the long-stable south-

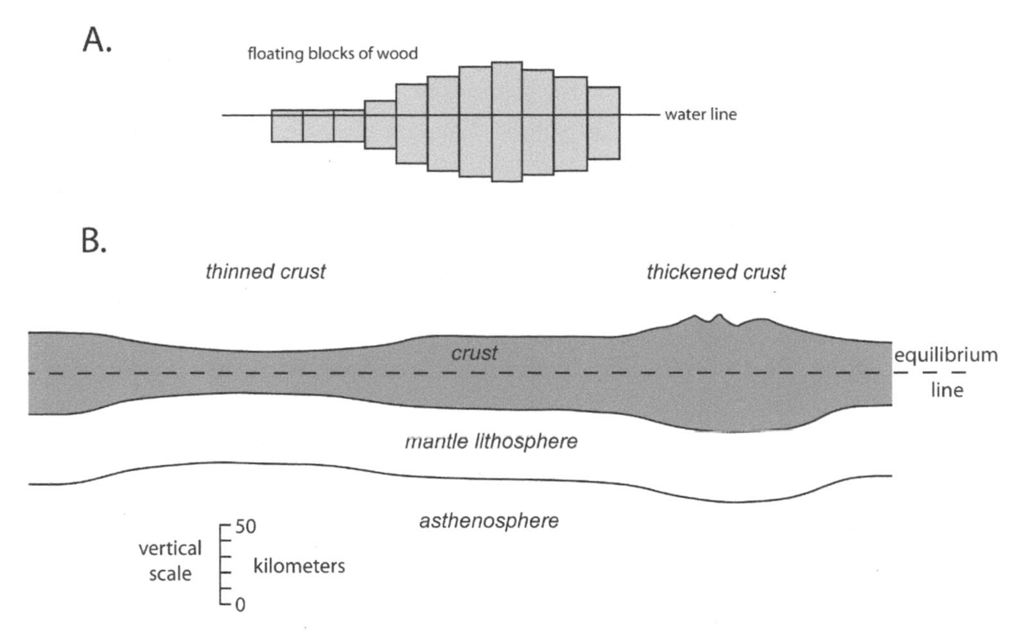

Fig. 8.19. The concept of isostacy. Figure 8.19A illustrates the concept with a series of wooden blocks of equal density floating in water. Although the blocks float, part of each block sinks into the water. The percentage of the block that is below the water level is proportional to its total mass. Thicker blocks sink more than thin blocks. Figure 8.19B shows the same concept with continental crust of constant density rather than blocks of wood. Thicker crust, in areas of mountain building, sinks deeper into the asthenosphere than thinner zones. Again, the volume of crust below the imaginary equilibrium zone (analogous to the water line in Fig. 7.19A) is a function of the total volume of crust there. The sinking and rising action of the lithosphere is accommodated by the slow, viscous flow of the asthenosphere.

ern Great Plains to the east is ~120 km thick. Lithosphere in the Basin and Range province to the west and south, where the crust is being pulled apart and thinned, is only ~60 km thick.

In a simplified view, the solid lithosphere can be considered to float on a relatively viscous asthenosphere, just as blocks of wood float on water. Following this analogy, a thick block of wood sinks deeper in the water than a relatively thin one, given the same density (Fig 8.19a). The water surface defines an equilibrium line: part of the block sits below the line, and some smaller part is above it. Similarly, differences in thickness of the Earth's crust, as altered by tectonic activity, control the degree to which the lithosphere sinks into the underlying asthenosphere, as well as the elevation that it reaches at Earth's surface. If, for instance, crust is thickened during mountain building, its elevation obviously increases. What is not so obvious is that the thickened crust causes the lithosphere to sink deeper into the asthenosphere (Fig. 8.19b). Essentially, the mountainous relief at the surface is mirrored in the subsurface. Mountains have deep roots that develop as they are formed. As the thickened lithosphere sinks deeper, the asthenosphere accommodates by flowing very slowly in a viscous manner away from the thickened area.

Conversely, thinned crust in zones of crustal extension decreases the load on the asthenosphere. This allows the lithosphere to rise buoyantly, so that the crust remains centered on the imaginary equilibrium line (Fig. 8.19b). As with all crust, the lower part mirrors the upper part. So while crustal thinning reduces surface elevation overall, the total reduction is negated by the resultant rise of the lithosphere.

Vertical adjustments in the position of the crust across the imaginary equilibrium line in response to thickness or density changes gives rise to the concept of **isostacy**. When the lithosphere load increases, by whatever means, it sinks more deeply into the asthenosphere. When the load is reduced, the lithosphere rises buoyantly, all the while maintaining a balance at the equilibrium line. The rise and fall of crust by these means at the surface are called **isostatic adjustments**.

Slower, smaller-scale changes in crustal thickness, driven by processes at the surface, may also produce limited changes in elevation by isostatic adjustment. The long-term, large-scale erosion of a mountain belt, for instance, incrementally reduces crust thickness. This should stimulate recurrent mild uplift in the mountain belt to compensate for the decrease in mass.

Just as erosion may drive isostatic uplift, the addition of a large volume of sediment to a basin, typically derived from erosion of an adjacent mountain belt, will contribute to crustal subsidence. The added load of a thick succession of sediments is answered by downwarping, which increases the probability of its preservation. The erosion of mountain belts and the deposition of sediments in adjacent basins are not isolated situations; in fact, the two are closely related. The tectonic setting of a mountain belt dictates that it is always bordered by a sedimentary basin.

Density changes within the lithosphere, due either to a shift in temperature or composition, may also alter the elevation of a region. If the lithosphere is heated by a deeper mantle source or by the intrusion of a large volume of magma, the density decreases and the lithosphere becomes more buoyant. In contrast, the cooling of hot lithosphere increases its density, causing it to sink.

Changes in composition involve a solid-state change to a new rock type without any modification of its original chemical composition. If the lithosphere is depressed deeply enough, through crustal thickening for instance, ultramafic rock in the mantle lithosphere may be submitted to much higher pressure—so much that the original rock is altered to a different, more dense rock type. In this case, the elements contained in the original minerals are forced by pressure to recrystallize into different, more dense minerals with tighter crystal structures. Alternatively, uplift could cause high-pressure rocks to rise into the realm of lower pressure (relatively speaking), causing it to revert to less dense rock. This would have the opposite effect of reducing the mass, augmenting the earlier uplift.

All the possibilities for lithosphere-controlled changes as outlined above have been invoked to explain the rise of the Colorado Plateau. Geologists have yet to reach a consensus, and because many of their hypotheses call on unobservable changes deep below the surface, the problem may never be completely resolved. Still, work on the uplift mechanism continues and has seen a recent resurgence of interest.

Mechanisms for Colorado Plateau uplift fall into two categories based on the inferred timing. The first argues for Early Tertiary uplift during the Paleocene and Eocene epochs. Explanations for this timing are linked to the Laramide orogeny, either by the direct effect of compression or though modifications of the lithosphere associated with Laramide tectonics. The second category involves more recent uplift of the Plateau beginning during the Miocene epoch—more specifically, over the past 10 million years. Although evidence for this later uplift fits better with the more recent erosion on the Plateau, there is no direct evidence for tectonism as the mechanism. Instead, Miocene uplift has been attributed to thermal and compositional changes in the mantle lithosphere and asthenosphere. The inability to directly observe these depths has so far prohibited the testing of the various hypotheses. As a result, models are employed extensively in testing the possibilities.

Evidence and Models for Early Tertiary Uplift

A frequently cited piece of evidence for the early rise of the Colorado Plateau comes from the work of Wolfe and others (1998) on fossil leaves from the Eocene Green River Formation of northern Utah and northwest Colorado. By comparing modern leaf morphology, which is controlled by temperature and climate, and thus their elevation ranges, to the leaves of the well-preserved Green River flora, they concluded that the lake sediments of the Green River Formation were deposited at an altitude of ~9,500 feet. This is near the elevation at which the fossil leaves are found today, implying that little change in elevation has occurred on this part of the Plateau since ~50 Ma, after Laramide uplift. Thus the research of Wolfe and others suggests that most of the rise of the Colorado Plateau is the result of the Laramide orogeny.

In an effort to evaluate the various components of Colorado Plateau uplift, Arizona Geological Survey geologist Jon Spencer (1996) isolated the effect of Laramide crustal deformation, among other things. Spencer analyzed the monoclines, the primary expression of the Laramide orogeny, in several transects across the Plateau to calculate the amount of uplift caused by crustal thickening. Vertical displacement on the monoclines along these transects averaged ~6500 feet (2 km). Of particular note, however, is the amount of actual crustal thickening by this deformation, which translates to an increased load on the lithosphere. This is significant because of previously discussed isostatic compensation. Spencer calculates an average increase in crustal thickness of 1,640 feet across the Plateau, which, because of local isostatic compensation, ultimately produces only a 160-

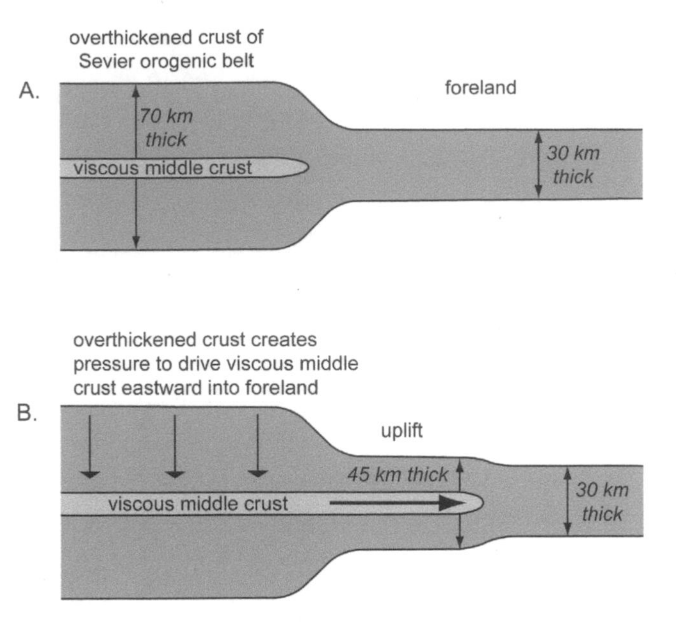

Fig. 8.20. McQuarrie and Chase's (2000) model in which uplift of the Colorado Plateau and Southern Rocky Mountains is directly related to the preceding Cretaceous Sevier orogeny. Figure 8.20A shows the overthickened crust of the Sevier orogenic belt at the end of the Cretaceous Period. This overthickening resulted in a sufficiently high temperature in the middle crust to produce a viscous state. Figure 8.20B shows the viscous middle crust being squeezed by high pressure, causing it to inject laterally into the foreland. This thickened the crust to the east, pushing the Colorado Plateau and Southern Rocky Mountains to a higher elevation.

to 300-foot increase in elevation. This renders Laramide crustal deformation on the Plateau an insignificant factor in its uplift, although it cannot be completely ignored.

The negligible uplift from crustal thickening led Spencer (1996) to evaluate possible changes in the mantle lithosphere during the low-angle subduction that drove the Laramide orogeny. According to Spencer, the low-angle trajectory of the subducting Farallon plate sliced horizontally into the mantle lithosphere, cleaving a large part of it from the overlying North American plate (see Fig. 8.4). As the decoupled lower lithosphere slowly collapsed into the viscous asthenosphere, it was replaced by hot asthenosphere that flowed upward to the base of the remaining lithosphere. The replacement of cool mantle lithosphere with relatively hot mantle heated the crust and remaining mantle lithosphere from beneath, driving thermal uplift and raising the overlying Colorado Plateau and Southern Rocky Mountains. According to Spencer's model, this process could account for the bulk of Plateau uplift. An important component of this scenario is the preservation of the uppermost mantle lithosphere immediately beneath the crust. This remaining mantle lithosphere explains the continued isotopic chemical signature of old mantle lithosphere in post-Laramide igneous rocks, even though it was presumably removed during low-angle subduction.

Another model developed by McQuarrie and Chase (2000) suggests that Laramide uplift of the Plateau was directly related to the preceding Sevier orogeny. They propose that during the Late Cretaceous, crust in the Sevier orogenic belt immediately to the west reached a maximum thickness of 70 km. The heat generated in this overthickened crust would have produced a 15 km thick viscous layer in the middle crust of the Sevier belt (Fig. 8.20). They calculated that in crust this thick the interplay of heat and pressure at mid-crustal levels would render the crust viscous and enable it to flow slowly. Above this layer the temperature was too low for such partial melting; below this layer, in the lower crust, where the temperature was even greater, increased pressure elevated the melting temperature, and it remained solid. According to this model, the great pressure exerted on the viscous middle crust by the overlying crust squeezed it laterally eastward, where it was injected into the middle of the 30 km thick crust of the Plateau and eventually farther east to generate the Southern Rocky Mountains (Fig. 8.20b). This mid-crustal flow would have thickened the Plateau crust to its present-day 40 to 45 km and increased its elevation by 2 km. Its movement would also have transmitted enough stress into the overlying brittle crust to cause faulting on preexisting weaknesses. This would have produced the monoclines and fault-bounded mountains that characterize the Laramide orogeny. In this way, the Colorado Plateau could have experienced large-scale uplift without the presence of large bounding faults to accommodate such a rise. This model does not preclude low-angle subduction at deeper levels. It would, however, minimize its effect on the upper crust, as the subducting oceanic plate would have been decoupled from the upper crust by the viscous middle crust.

Problems persist with Early Tertiary uplift and some of the proposed explanations. Although the development of a viscous middle layer in overthickened crust is plausible, the lateral intrusion of the layer far into relatively cold crust is problematic (Morgan 2003). If this injection had indeed occurred, the hot viscous intruding layer, upon entering the relatively cool crust of the Plateau, would have transferred its heat into the surrounding host crust. This would have lowered the temperature of the intruding layer and decreased its viscosity, reducing its ability to flow. After entering the middle crust of the Plateau, the viscous layer simply would not have maintained the temperature required to remain viscous. Thus the eastward transfer of middle crust would have stalled early in the process.

A longer-standing problem stems from studies on the modern drainage network of the Colorado Plateau that are interpreted by some workers to indicate Late Tertiary rather than Early Tertiary uplift. All the available evidence gathered so far for the evolution of an external drainage and attendant canyon cutting, including the development of the Grand Canyon, indicates that it began less than 6 million years ago (Longwell 1946, Lucchitta 1972, Spencer and others 2001, Ranney 2005). The corollary to this is that many believe that external drainage development was stimulated by renewed uplift of the Plateau. While this may not necessarily be true, it is an assumption that drives many geologists to continue to seek evidence and explanations for Late Tertiary uplift of the region, even in the absence of structural evidence or obvious tectonic activity. These arguments and the evidence behind them are explored in the following section.

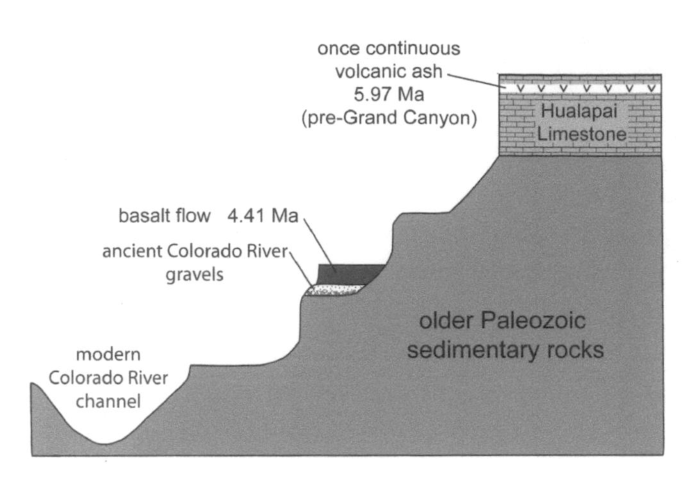

FIG. 8.21. Schematic cross section showing the lower Grand Canyon area and significant radiometric dates that bracket the evolution of the modern Colorado River drainage. The Hualapai Limestone at the top was deposited in a closed lake basin before the Colorado River drainage had developed. In contrast, the ancient Colorado River gravel and sand deposit on the high terrace is the first unequivocal record of the modern Colorado River drainage. Overlying basalt with an age of 4.41 Ma provides an important maximum age constraint for this drainage. Age data and relations are from Spencer and others 2001 and Faulds and others 2001.

Late Tertiary Evolution of the Colorado Plateau

The key debate concerning Late Tertiary evolution of the Colorado Plateau hinges on whether the Plateau rose, the surrounding Basin and Range province dropped, or some possible combination of the two. The most telling evidence is preserved along the southwest margin, where the Colorado River exits the Plateau and cuts southward across the Basin and Range to empty into the Gulf of California. A brief review of this evidence is warranted to frame the arguments for and against later uplift of the region.

First and foremost, the best evidence obtained so far indicates that the Colorado River drainage did not exist before 6.0 Ma, but was undoubtedly established by 4.4 Ma, placing its inception near the Miocene-Pliocene boundary (5.3 Ma). These data come from the western limit of the Grand Canyon in the Lake Mead region of northwest Arizona and southern Nevada. The ages come from $^{40}Ar/^{39}Ar$ radiometric dates from volcanic rocks interbedded with sediments whose origin are critical to the evolution of the Grand Canyon and the vast Colorado Plateau drainage. The 6.0 Ma age (5.97 ± 0.07 Ma) was obtained from a volcanic ash layer within the lacustrine Hualapai Limestone (Spencer and others 2001). The Hualapai was deposited in a closed, fault-bounded lake basin in the transition zone between the Colorado Plateau to the east and the Basin and Range to the west (Fig. 8.21). These deposits are the youngest feature identified so far to unquestionably predate the Colorado River, which now bisects the basin. The 4.4 Ma date (4.41 ± 0.03 Ma) comes from a basalt flow in the same area that flowed over early Colorado River gravels (Faulds and others 2001) (Fig. 8.21). Collectively, the rocks and their relations in this small area demonstrate that the Colorado River and related external drainage across the Plateau evolved within the time period of 6.0 and 4.4 Ma. If the cutting of the Grand Canyon, as well as all the canyons of the Plateau, occurred such a short time ago (relatively speaking!), what was going on prior to external drainage?

Taking a step back in time, the most telling record discovered so far for the pre-6.0 Ma setting on the Colorado Plateau that has a bearing on drainage evolution lies in Early to Middle Tertiary river gravels dispersed in channels cut across the tops of plateaus in northwest Arizona. These *rim gravels*, as they are known, occur on both the south and north sides of the Grand Canyon. Furthermore, they show with certainty that before ~25 Ma a major river system flowed north and northeast from Laramide-generated highlands located in central Arizona, traversing the western Grand Canyon region of today (Young and McKee 1978, Young 2001). Although subsequent erosion has removed most of these deposits, their scattered remnants demonstrate that the rivers were directed northward, possibly feeding into the southern margin of the lacustrine Claron basin in southwest Utah. The 25 Ma minimum age for this river system is provided by a basalt flow associated with the gravels that yielded a K-Ar radiometric age of 24.7 ± 3.5 Ma (Young and McKee 1978). The significance of this setting is the realization that the regional paleoslope was to the north, with headwaters in highlands to the south. The position of these highlands today is the much lower elevation setting of the Basin and Range province.

Putting the known history of the Colorado River drainage into a chronological context, from Paleocene to Late Oligocene time (~25 Ma), rivers drained north from Laramide highlands in central Arizona and *across* the present trace of the Grand Canyon. Sometime after 25 Ma this drainage pattern was disrupted by the onset of Basin and Range extension, in which the overthickened crust to the west and south began to pull apart and collapse. Extension led to the development of isolated fault-bounded basins, including that in which the Hualapai Limestone was deposited. Not only did extension disrupt the earlier drainage pattern, but the highland source area in central Arizona was down-faulted well below the level of the Colorado Plateau. It was not until sometime between 6.0 and 4.4 Ma that the complex internal drainage that had endured on the Plateau since the Early Tertiary was unified into a single system that was able to breach its boundaries. This new path cut westward through the transition zone and, upon reaching the Basin and Range, ran southward to an outlet in the Gulf of California. There are numerous hypotheses for the origin of Colorado Plateau drainage integration—almost as many as the geologists that have considered the problem. Because there are so many ideas, but still no consensus, we return instead back to the timing and mechanisms for Plateau uplift. The reader interested in the specifics of Grand Canyon origin is directed to Wayne Ranney's excellent book *Carving Grand Canyon* (2004) and the collection of more technical papers in *Colorado River Origin and Evolution* (Young and Spamer 2001).

Geophysicist Paul Morgan (2003) proposes an interesting model for Late Tertiary Plateau uplift based on phase changes in the mafic metamorphic rocks that comprise the crust/mantle boundary in the lithosphere. Studies of sedimentary rocks preserved across the Colorado Plateau show that it remained at or near sea level for most of the Paleozoic and Mesozoic eras—an astounding period of 500 million years! In order to maintain this elevation, subsidence must have been continuous. This is notable because eventually the combined growing sedimentary succession and underlying basement rock (collectively the crust) would become so thick that, based on isostatic principles, it would have to rise above sea level. Yet it didn't. According to this model, the continuous sedimentation and attendant subsidence depressed the underlying mantle lithosphere into a deepening realm of higher pressure. At some elevated pressure, the original mantle rock, **garnet granulite** (density of 2.9 g/cm^3), transformed by a phase change into **eclogite,** with a greater density of 3.6 g/cm^3. This density increase further stimulated subsidence, prolonging it. While this may explain the unexpected 500 million years of subsidence, what does it have to do with the much later uplift of the Colorado Plateau?

According to Morgan (2003), Late Tertiary uplift of the Colorado Plateau ultimately was the result of a regional heating event. This thermal event, which occurred at about 25 Ma, generated the laccolith clusters (La Sal, Abajo, and Henry mountains) on the Plateau, as well as the San Juan and Marysvale volcanic fields on its margins; however, simple regional heating and an accompanying decrease in crustal density was not enough to account for the total post-Laramide uplift that has been suggested for the Plateau. Instead, Morgan suggests that the mafic rock at the crust/mantle boundary, which was particularly sensitive to this temperature change, experienced a reversal of the earlier phase change with a conversion from dense eclogite to less dense garnet granulite. Modeling shows that this metamorphosis could produce 5000 feet (1500 m) of isostatically driven uplift across the Colorado Plateau.

Evidence for this Late Tertiary phase change lies in inclusions found in volcanic rocks erupted on the Plateau. These **xenoliths**, as the inclusions are known, are pieces of solid rock that become incorporated into magma either at its point of origin or as it ascends through the lithosphere. Xenoliths are important because they offer a glimpse of the actual rocks that make up the lower crust and upper mantle. Xenoliths from Colorado Plateau volcanics older than 25 Ma consist mostly of eclogite, although some garnet granulite also occurs. Conversely, no eclogite xenoliths are known from volcanic rocks younger than 25 Ma (Morgan 2003). Although the absence of something does not provide the most satisfying evidence, the abundance of eclogite before 25 Ma and its absence afterward fit the proposed phase change at the crust/mantle boundary as a consequence of the 25 Ma thermal event.

The final hypothesis presented here attributes rapid downcutting on the Colorado Plateau to Late Tertiary extension and the resulting topographic drop of the adjacent Basin and Range province (Lucchitta 1972). Thus, widespread canyon cutting across the Plateau may be due to downdropping in the Basin and Range rather than Plateau uplift, although relatively speaking, the Plateau ends up high. This implies that the Colorado Plateau did not necessarily rise in the Late Tertiary, but instead likely was elevated with the surrounding region sometime earlier, probably during the Early Tertiary Laramide orogeny.

As discussed previously, during Early and Middle Tertiary time the future Basin and Range region to the south and west stood high and shed sediment northward across the Plateau. This setting endured to at least 25 Ma, after which these Laramide highlands were downfaulted during the extension that produced the modern Basin and Range province. This fundamental shift from the compressional tectonics of the Sevier and Laramide orogenies to an extensional regime was caused by a relaxation of compressional forces to the west and the collapse of hot, overthickened crust in the earlier-formed mountain belts. It was about this time that the lower Colorado River drainage began to develop as the north-south-trending valleys of the Basin and Range focused runoff southward to the opening Gulf of California. The Colorado Plateau at this time remained internally drained. Eventually, however, the incipient drainage that would become the Colorado River cut eastward onto the Plateau. The low elevation of the Basin and Range, and the high elevation of the Plateau energized this expansion and hastened the process of headward erosion. A major obstacle to this drainage expansion was the Laramide-age Kaibab upwarp in northern Arizona. Upon breaching this broad highland, the growing drainage system captured relict channels of the previous internal drainage network. Slowly, over a few million years, the new drainage evolved into the vast system of today, which funnels high-mountain snowmelt from the west side of the Southern Rocky Mountains into the Green and Colorado rivers, which meet to become the Colorado at the Confluence, in the heart of Canyonlands. From there the water roils through Cataract Canyon before dumping its sediment load into Lake Powell, which covers Glen Canyon. Water emitting from Glen Canyon dam enters the Grand Canyon where its southern course shifts westward and drops off the Colorado Plateau onto the Basin and Range, where it again courses south along the Arizona-California border. Soon after crossing the international border into Mexico, the river (or what is left of it today) empties into the Gulf of California.

In this brief review it becomes apparent that the timing and mechanism(s) for Colorado Plateau uplift remain controversial, but are they unknowable? The only certainty is that during the Late Cretaceous the shoreline of the sea spread across the region, and today these sea-level deposits sit at an elevation of ~7,000 feet, leading to the inescapable conclusion that sometime in the last 65 million years these deposits were pushed upward. This has been known for many years. The other, more recent certainty is that before 6 Ma the Colorado River drainage as we know it did not exist, yet the Grand Canyon was well established by 4.4 Ma, an incredibly short period of time by geologic standards (Spencer and others 2001, Faulds and others 2001). While further constraints for this uplift are being refined by the army of geologists working directly and indirectly on the problem, the shortage of evidence caused by erosion makes this difficult. If the puzzle is solvable, however, the caliber and sheer numbers of geologists working in the region today will undoubtedly resolve it. It is exactly these types of problems that motivate geologists.

The Quaternary Period

Canyon Cutting, Mammoths, and the Appearance of Humans

The final period in Earth history, the Quaternary, spans from 1.6 Ma to the present. It is subdivided into two epochs. The older of the two, the Pleistocene, covers most of the Quaternary and ranges from 1.6 Ma to 10,000 years before present (or 10 ka). The subsequent Holocene Epoch, which we are now in, stretches from 10,000 years to the present. We now enter into geologic time that typically is measured in thousands rather than millions of years. The abbreviation *ka*, which stands for kilo-annums, or thousands of years, is thus used for many discussions of absolute age.

Although the features seen across the Colorado Plateau are largely the product of Tertiary processes, Quaternary activity put the finishing touches on the modern landscape. It is, however, an ongoing process.

Abundant Quaternary deposits are recognized across the Colorado Plateau, but they occur in discontinuous patches. Just as in late Tertiary history, this is because erosion rather than deposition continues to dominate the region. Most of these deposits have not been cemented and consist of loose sand, gravel, and mud. Techniques for dating these young sediments are currently evolving, but while new techniques have been developed in the past decade, many require further testing and refinement. Those techniques that have been successful have yet to see widespread application to Colorado Plateau sediments. As a result, detailed knowledge of the Quaternary history of the Plateau is sparse, and well-studied areas are widely separated.

The ensuing discussion of the Quaternary history of southeast Utah is limited to some of the better-studied events and processes. This includes the Pleistocene glacial history of the La Sal Mountains. Quaternary salt deformation is another important process. The best-known features are the Fisher Valley salt anticline, a continuation of earlier salt-related activity, and the development of the horst and graben system that makes up the Needles District of Canyonlands. The diverse Quaternary megafauna of the Colorado Plateau, including mammoths, sloths, and camels, will be reviewed, as will the gradual change in flora, which is related to Pleistocene and Holocene climate change. Finally, the complex interplay between Quaternary erosion and sedimentation will be reviewed. Where available, this will include recent rates of downcutting in various southeast Utah drainage systems. This final chapter closes with a speculative look to the future.

The Quaternary History of Southern Utah

The Glacial History of the La Sal Mountains

The Quaternary glacial history of the La Sal Mountains has been studied by USGS geologist Gerald M. Richmond (1962) and geologists John Shroder and Robert Sewell (1985). Richmond first ventured into the high La Sal Mountains in the 1950s in an effort to map Quaternary deposits and landforms. This work, combined with numerous analyses of the soil zones that mantled the Quaternary deposits, led him to identify five discrete glacial episodes, each separated by an interglacial interval (during which glaciers melted completely or shrank to insignificant sizes). Subsequent, more detailed work by Shroder and Sewell (1985) built on Richmond's work with the goal of differentiating glacial moraine deposits from nonglacial mass movement deposits (e.g., landslides, earthflows, and debris flows). Detailed field mapping was used to identify other glacial features such as glacially polished bedrock and cirques. Upon close inspection Shroder and Sewell concluded that many of Richmond's glacial deposits were, in fact, the product of mass movements. Moreover, many of the glacial events interpreted by Richmond were unrecognizable. The following overview comes dominantly from the more recent work of Shroder and Sewell.

Following their study, Shroder and Sewell (1985) recognized that the Quaternary record in the La Sal Mountains was dominated by mass movements, and that glacial activity had played a relatively minor role in the sculpting process. The dominant triggering factor in these mass movements is the type of sedimentary rock that hosted the laccolith intrusions. The most common sedimentary units in these mountains, the Jurassic Morrison Formation and the Cretaceous Mancos Shale, are composed of unstable shale that, when saturated with water, collapses on a large scale. Most movement occurred during the Pleistocene, when the region was wetter. When saturated hillsides flow, the product may resemble glacial deposits in their lobe-shaped morphology and unsorted mixture of clay- to boulder-sized fragments. In their reexamination, Shroder and Sewell recognized only nine small glaciated areas situated in the highest valleys of the La Sal Moun-

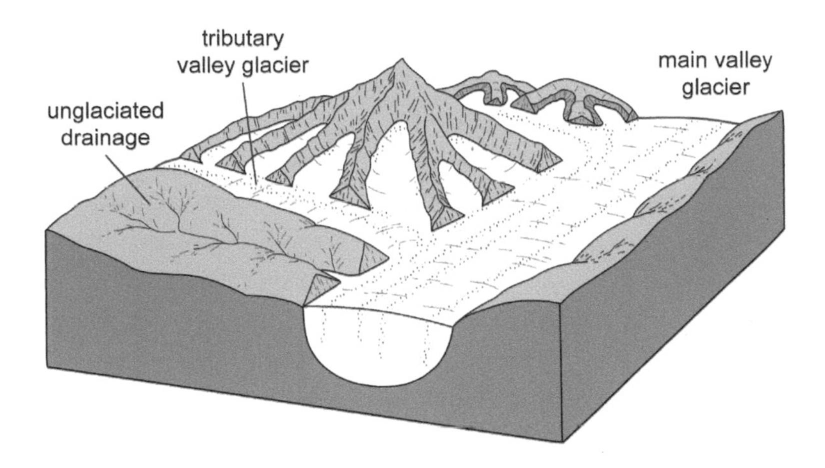

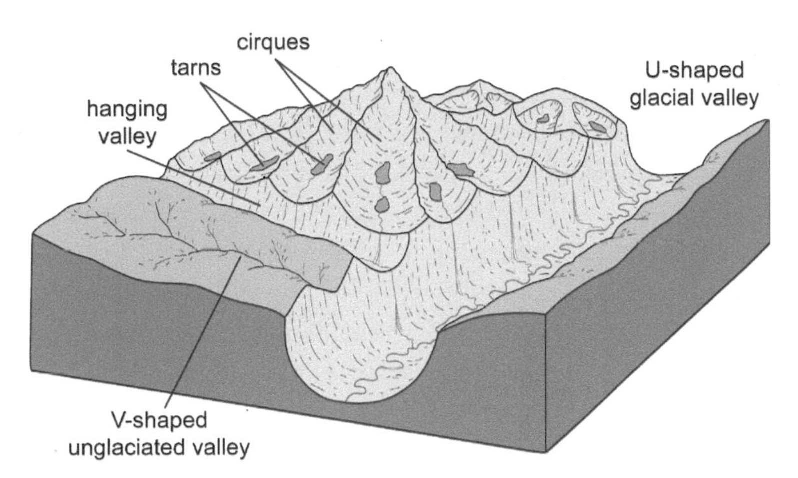

FIG. 9.1. Diagrams showing the various erosional features of glaciation. Figure 9.1A shows the carving action of glaciers and the resulting U-shaped valley. Not shown is the precursor to glaciation in which the river-carved valleys were V-shaped, like the unglaciated drainage on the lower left. Figure 9.1B shows postglaciation erosional features, including the tarns, or glacial lakes, that form when glaciers scoop deeply into the bedrock.

tains, whereas Richmond (1962) had outlined a region about twenty times larger. He should not, however, be faulted: the pioneering efforts of such geologists in large regions such as the La Sal range are by necessity of a reconnaissance nature. It is only after these pioneers lay the groundwork that others can build on that foundation and focus on smaller areas in more detail.

In their study of the three discrete mountain groups that comprise the La Sal Mountains, Shroder and Sewell (1985) found unequivocal signs of glaciation in the North and Middle mountain groups. The South Mountain Group apparently holds no glacial features. Evidence for glaciation includes the presence of **cirques**: broad, bowl-shaped valley heads excavated into bedrock by the insistent grinding action of glaciers (Fig. 9.1). These typically are located immediately below high peaks, on the north and east sides, where the cirque is shaded from the sun. Glacier-sculpted valleys below cirques are broad, with a characteristic U-shaped cross-section. This is a contrast to the typical V-shaped, river-carved valley, where the erosive agent is confined to a narrow ribbon of the valley floor (Fig. 9.1). Within glacial cirques and U-shaped valleys is ice-polished bedrock with striations that record the downvalley passage of the ice mass. Striations are linear gouges scraped into bedrock by pieces of rock embedded into the ice at the base of a glacier. As these rocks are dragged across the bedrock with the downhill flow of the glacial ice, the linear striations record the trajectory of movement. In addition to eroding, glaciers leave distinct deposits. These are commonly in the form of **moraines:** large, sinuous, linear ridges composed of unsorted and unstratified sediment ranging from clay to boulders, called **glacial till**. **End,** or **loop, moraines** are arc-shaped ridges in map view that extend from wall to wall across glacial valleys (Fig. 9.1). These moraines arc downvalley and mark the terminus of the glacier, where much of the sediment incorporated into the glacial ice from above and below ultimately accumulates. Glacial valleys may contain a succession of several end moraines, each left behind as a glacier shrinks and recedes upvalley, pausing for geologic moments to allow the sediment to pile up. **Lateral moraines** are another telltale sign of glaciation. These linear ridges of till are deposited upvalley of the end moraines adjacent and parallel to the steep valley walls. They may be preserved on both sides of the valley. The sediment comes from detritus that has been pried off the bedrock walls above the glacier through frost-wedging processes to land on the glacier margin. As the glaciers shrink and recede, this debris is left behind as linear mounds of loose sediment. All of these features in various combinations were recorded by Shroder and Sewell (1985) to document Pleistocene glacial activity in the La Sals. They also noted the presence of nonglacial features to show that other parts of the range had been left unscathed by glaciers.

The paucity of datable material in Quaternary glacial deposits has driven scientists mostly to relative dating techniques in piecing together glacial histories. The relative ages of glacial deposits can be interpreted by comparing the degree of weathering of boulders in glacial till, or the thickness of soil zones developed on moraines. More deeply weathered boulders or thicker, better developed soils are a function of the time that the deposit has been exposed. Another useful feature is the preservation of moraines and the degree to which their original morphology has been degraded by erosion. Recently, some innovative isotopic dating methods have been applied to Quaternary deposits, although they have not yet been used in the La Sal Mountains. One complication specific to glacial deposits is that each succeeding glacial event may destroy or render unrecognizable the preceding glacial deposits, especially if later glaciers are more extensive and override earlier moraines. As a result, most alpine glacial chronologies go back only to the last few glacial periods.

The oldest preserved glacial deposits in the La Sal Mountains are the product of Bull Lake Glaciation, named for the Bull Lake area in the Wind River Mountains, Wyoming. All available dates on Bull Lake deposits suggest this glacial period ranged from ~200 to 130 ka. This was separated from the ensuing Pine-

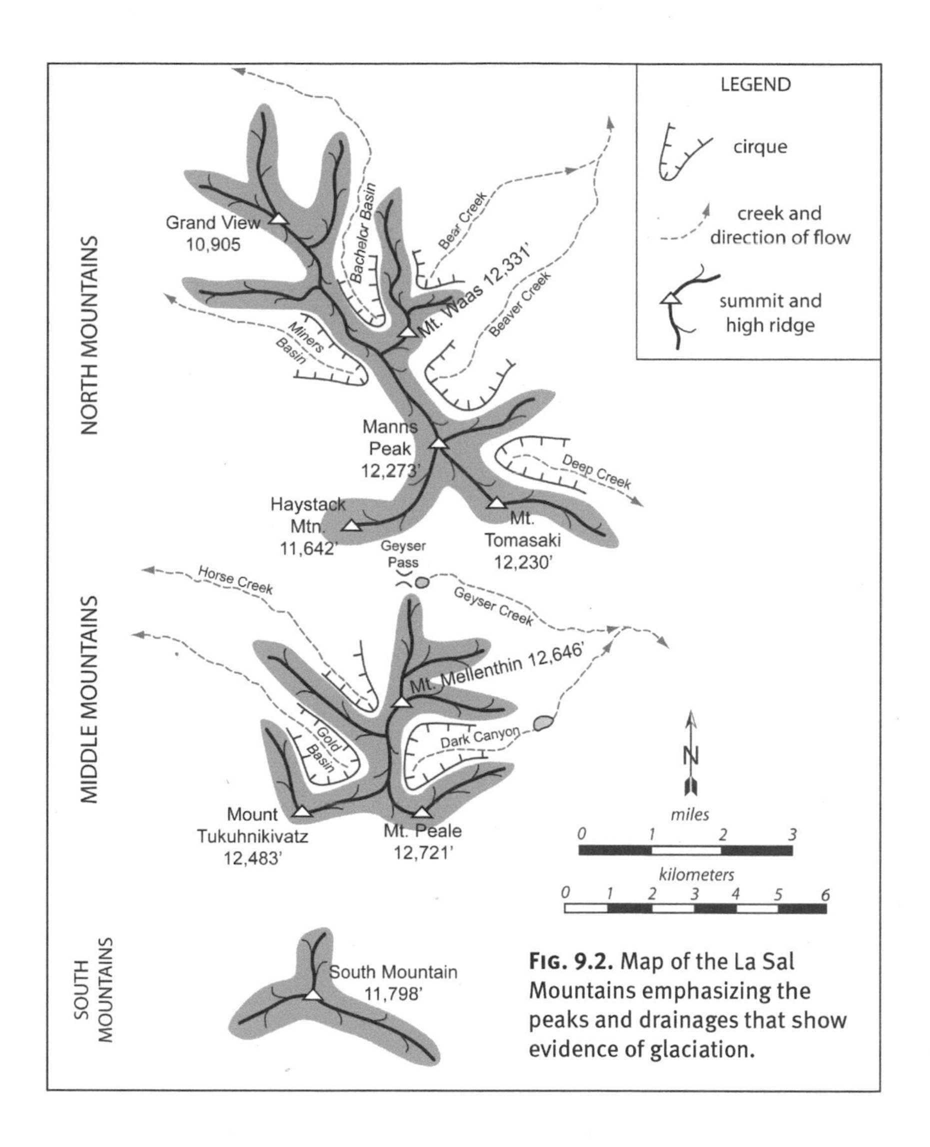

Fig. 9.2. Map of the La Sal Mountains emphasizing the peaks and drainages that show evidence of glaciation.

dale Glaciation by a warmer interglacial period during which the glaciers receded or disappeared completely. The Pinedale Glaciation began ~70 ka and terminated ~16 to 14 ka. This most recent glacial period is the best preserved. While some smaller-scale glaciers may have held out locally in the La Sal Mountains, Bull Lake and Pinedale deposits form the bulk of the glacial record.

Glaciation was most extensive in the North Mountain Group of the La Sal Mountains, with six valleys holding strong evidence for glacial activity. One of the most extensive, Beaver Creek, is a long, northeast-facing valley that drains a wide cirque (Fig. 9.2). This cirque is shadowed by the high, hump-backed peaks of Mt. Waas (12,331 feet) and Manns Peak (12,273 feet). Three discrete glacial episodes have been identified in Beaver Creek valley, providing one of the most complete records in the La Sals (Shroder and Sewell 1985). The oldest evidence consists of two lateral moraines preserved downvalley that are of likely Bull Lake age. Younger end moraines of probable Pinedale age are present in the high cirque area. The youngest glacial deposit is a post-Pinedale end moraine situated high in the northern part of the cirque.

Deep Creek, which lies immediately south of Beaver Creek, is a smaller drainage, with its upper reaches in the shadows of Mt. Waas and Mt. Tomasaki (12,230 feet) (Fig. 9.2). Alpine glaciers at the foot of these peaks excavated two deep cirques, which together constitute a steep-walled amphitheater at the upper end of the valley. Downvalley three lateral moraines of unknown age line the north side. Dense forest makes the recognition of other moraines, if present, difficult.

Geyser Creek occupies a broad valley that heads at Geyser Pass, a low spot between the North Mountain and Middle Mountain groups (Fig. 9.2). Geyser Creek drops eastward from the pass. In the upper reaches a poorly developed cirque, grooved bedrock, and a series of three end moraines record the stepwise, upvalley recession of a glacier. The moraines are interpreted as Pinedale age based on soil development and morphology (Shroder and Sewell 1985).

Bear Creek is a small, northeast-facing valley just north of Beaver Creek, headed by Mt. Waas (Fig. 9.2). The valley morphology is indistinct because it is mostly mantled by the rubble of more recent avalanches. A possible moraine is present in the upper basin, and a loop moraine has been identified downvalley in the forest (Shroder and Sewell 1985).

Bachelor Basin is located at the north tip of North Mountain Group and is drained by Castle Creek, which feeds through Castle Valley (Fig. 9.2). The basin hosts two cirques at its upper end, and each has a small end moraine at its outer margin.

Miners Basin is on the west side of the North Group in a drainage that also faces Castle Valley (Fig. 9.2). This drainage originally was interpreted by Richmond (1962) to have been widely glaciated. A reexamination by Shroder and Sewell, however, found that Richmond's lengthy lateral moraines are benches of more-resistant Paleozoic and Mesozoic sedimentary rock mantled with a thin rubble layer resembling glacial till. Convincing evidence for glaciation is present in the high cirque, where a well-preserved closed loop moraine was identified.

Shroder and Sewell also suggested that the Dry Fork and West Fork of Mill Creek had been glaciated, but they were unable to find unequivocal evidence in the heavily forested drainages.

The Middle Mountain Group is smaller in area than North Mountain, but not in stature, and includes the highest peaks of the La Sals: Mt. Mellenthin (12,645 feet), Mt. Tukuhnikivatz (12,483 feet), and the highest in the range, Mt. Peale (12,721 feet) (Fig. 9.2). Three valleys associated with these peaks hold significant evidence for glaciation; all are headed by well-developed, north-facing cirques.

Dark Canyon cirque originates on the north face of Mt. Peale before the drainage arcs eastward (Fig. 9.2). The cirque contains a **tarn**, a glacially carved basin filled by a small lake. Polished boulders are scattered across this high bowl, and recessional end moraines are evident a short distance downvalley.

Gold Basin occupies the northern flank of Mt. Tukuhnikivatz (Fig. 9.2). Brumley Creek, which drains it, is fed by a trio of large cirques. A series of both lateral and end moraines occur high in the basin, and there presently are two active rock glaciers. **Rock glaciers** are remnants of glacial ice covered and insulated by a veneer of loose rock shed from surrounding cirque walls. A loop moraine immediately below the cirques is interpreted to be of Pinedale age. Several lateral moraines downcanyon are believed to represent Bull Lake Glaciation.

Finally, Horse Creek drains from the west flank of Mt. Mellenthin, originating in a wide, northwest-facing cirque (Fig. 9.2). A lateral and a recessional loop moraine are preserved about 1 km down the valley. Three recessional moraines are recognized approximately 3 km from the cirque, in a flat area where the Geyser Pass road crosses Horse Creek (Shroder and Sewell 1985).

Glacial deposits in the La Sals with no age designation are likely related to the more recent Pinedale Glaciation, and possibly are younger, although more detailed work is needed. In the absence of at least two generations of deposits, relative dating techniques are not valid. There simply is no "relative to. . . ."

Glaciation in this eastern part of the Colorado Plateau apparently was confined to the La Sal Mountains. The only other candidate for glaciation is the Abajo Mountains, and there is no record of glaciation there. It should be noted that the adjacent Southern Rocky Mountains in southwest Colorado—including the Elk Range, San Juan Mountains, and La Plata Mountains—hold abundant evidence for widespread glaciation.

The Quaternary History of the Fisher Valley Salt Anticline

The upper reaches of the Fisher Valley salt anticline hosts a Quaternary sedimentary basin with ~480 feet of Late Tertiary to Quaternary sediments (Colman and others 1988). This succession contains one of the best records of Quaternary conditions in this part of the Colorado Plateau, including two far-traveled and well-dated volcanic ash deposits, providing crucial time constraints. Besides being one of the most complete Quaternary sequences on the Plateau, these sediments document the recent rise of a salt diapir into the valley center. Quaternary sediments were deformed by the ascent of this large blob of Paradox Formation, which is squeezing onto the valley floor from the depths. The Onion Creek diapir, as it is known, also obstructed the earlier simple, downvalley drainage pattern, forcing a dramatic reorganization.

Data for the Quaternary history of Fisher Valley are held in the strata, which have been studied and dated by a variety of methods (Colman 1983, Colman and others 1986, Colman and others 1988). Three informal Late Cenozoic units have been delineated in this unique basin.

Pliocene(?) gravel

The basal unit is 80 feet of Pliocene gravel. Where exposed along the north end of the basin, where Onion Creek has cut into the succession, the gravel lies on chaotically folded gypsum, shale, and limestone of the Paradox Formation. The Paradox is brought up in the Onion Creek diapir. The contact is an angular unconformity, and the gravels locally have been infolded into the diapir. Gravel contains a sand matrix and exhibits planar bedding and cross-stratification. Clasts consist of igneous rock from the nearby La Sal Mountains and Mesozoic sedimentary rock from the surrounding, valley-bounding escarpment.

The age of these basal gravels is interpreted from paleomagnetic analysis of the sequence (Colman and others 1986). Periodically through geologic time, Earth's magnetic field has switched poles; that is, the North Pole becomes the south and the South Pole becomes the north. These reversals, as they are known, are typically recorded in rocks with high iron content and a magnetic signature. The timing of the reversals is well-dated from igneous rocks around the world, and so the signature in other rock types can be correlated with the dated magnetic reversals, providing otherwise elusive age constraints. Today's magnetic field is considered normal. The normal polarity we are presently in ranges back to 730 ka and is known as the **Brunhes** normal polarity chron of the paleomagnetic time scale. The chronozone that preceded the Brunhes is the **Matuyama** reversed polarity chron, which spanned 0.73 Ma (730 ka) to 2.48 Ma. The **Gauss** normal polarity chron precedes the Matuyama and ranges from 2.48 to 3.40 Ma.

The sedimentary succession in Fisher Valley records all three chrons. These are calibrated here by radiometric dates from interbedded volcanic ash and, for younger sediments, ^{14}C dates on organic material. Paleomagnetic analysis indicates that basal gravels in Fisher Valley were deposited during the Gauss chron, establishing their age at between 3.4 and 2.48 Ma.

Basin fill deposits

The basin fill deposits overlie Pliocene gravel and are subdivided into lower and upper units separated by an angular unconformity. The base of the upper unit is marked by the Lava Creek ash marker bed, dated at 0.61 Ma.

The lower basin fill unit reaches up to 330 feet thick and consists of several cycles of cross-bedded fluvial sand and gravel overlain by structureless eolian sand. Cycles are separated by buried soil zones, signifying breaks in sedimentation.

The age of the lower basin fill is constrained from the analysis of a variety of datable materials (Colman and others 1986). Within the upper part of this lower unit is the well-dated Bishop ash, yielding an age of 0.73 Ma (730 ka). This extensive blanket of airborne ash was blasted from a colossal eruption that resulted in the formation of the Long Valley caldera, a large, collapsed volcanic feature in eastern California. Bishop ash has been identified as far east as Nebraska, providing a significant time line for Quaternary sediments across western North America. Paleomagnetic analyses show that the lower basin fill spans all the previously discussed polarity chrons: the Gauss, Matuyama, and Brunhes. The base of the lower basin fill falls in the Gauss normal polarity chron, indicating an age of more than 2.48 Ma. The top of the lower basin fill falls in the Brunhes normal chron. Thus, the lower basin fill ranges more than 2.48 Ma and is older than 0.61 Ma based on the age of the overlying Lava Creek ash.

The upper basin fill reaches up to 165 feet thick and, like the underlying lower unit, consists of fluvial sand with minor gravel lenses. Along the basin margins coarse gravel with clasts up to 5 feet in diameter is common. The Lava Creek ash defines the base of the upper unit and locally reaches 3 feet thick. The contact between the ash and the lower unit is an angular unconformity. At least one other angular unconformity has been recognized within the upper basin fill along the basin margins. Like the Bishop ash, the Lava Creek ash is regionally extensive and has been dated by the K-Ar method at 0.61 Ma (610 ka) (Izett 1981). This ash was blown across a large part of the West during a particularly explosive eruption in the Yellowstone area.

Six buried soils are recognized in this succession, and the well-developed $CaCO_3$-bearing soil zones are the basis for establishing a minimum age for the upper basin fill. By analyzing the amount of $CaCO_3$ in the various buried soils, and calibrating this with the well-dated volcanic ash units and polarity chron boundaries through the basin fill, the accumulation rate for $CaCO_3$ can be estimated. Through this complicated process, the age of the buried soil at the top of the basin fill is estimated at 0.25 Ma (250 ka) (Colman and others 1986). Collectively, these dates indicate that the upper basin fill was deposited over the period 610 to 250 ka. The age of the entire basin fill succession ranges from more than 2,480 ka to 250 ka.

Holocene deposits

The Holocene deposits that cap the Fisher Valley succession consist of up to 15 feet of massive, red-brown sand and silt of mostly eolian origin. Locally this loose sediment is piled into dunes. Throughout the basin this deposit blankets the 250 ka soil zone at the top of the upper basin fill. In one small area this unit includes a 30-foot-thick lens of yellow silt to coarse sand that contains thin layers of carbonaceous plant material yielding a ^{14}C date of 9330 ± 155 years (Colman and others 1986). Gastropods and rare bivalves, coupled with the thin, carbon-rich peat horizons, suggest a marsh environment, probably fed by localized freshwater springs (Colman and others 1988).

The comprehensive analysis of Cenozoic strata preserved in the upper reaches of the Fisher Valley salt anticline by Steve Colman and colleagues provides an exceptionally detailed reconstruction of activity that is unavailable from other

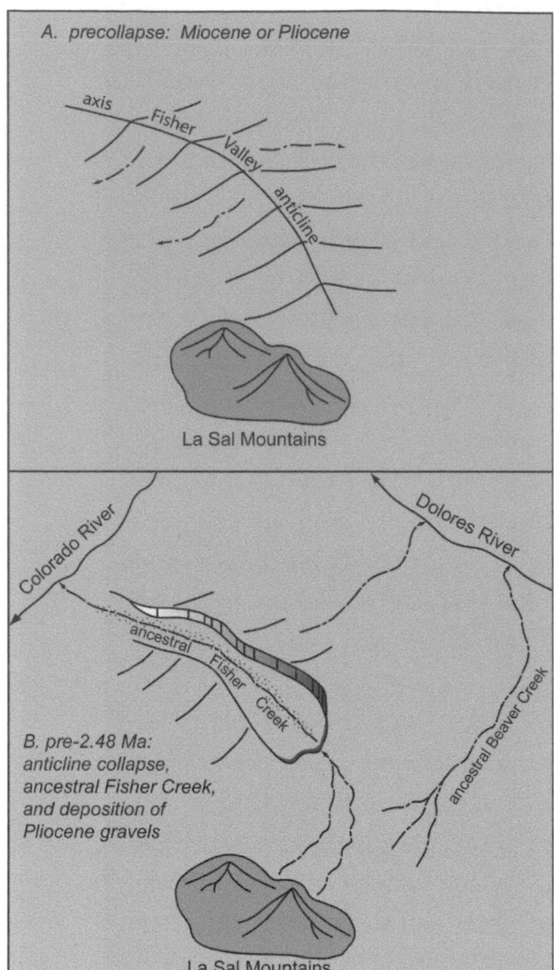

FIGS. 9.3A AND 9.3B. Diagrams illustrating the evolution of the Fisher Valley salt anticline from an uncollapsed anticline to its current configuration. Figure 9.3A shows a hypothetical precollapse state during the Miocene or Pliocene with drainages only coming off the anticline flanks. Figure 9.3B shows a postcollapse salt anticline and development of the ancestral Fisher Creek before 2.48 Ma. The drainage originates in the La Sal Mountains laccoliths. This phase is represented by the Pliocene gravels containing igneous clasts from the La Sal Mountains.

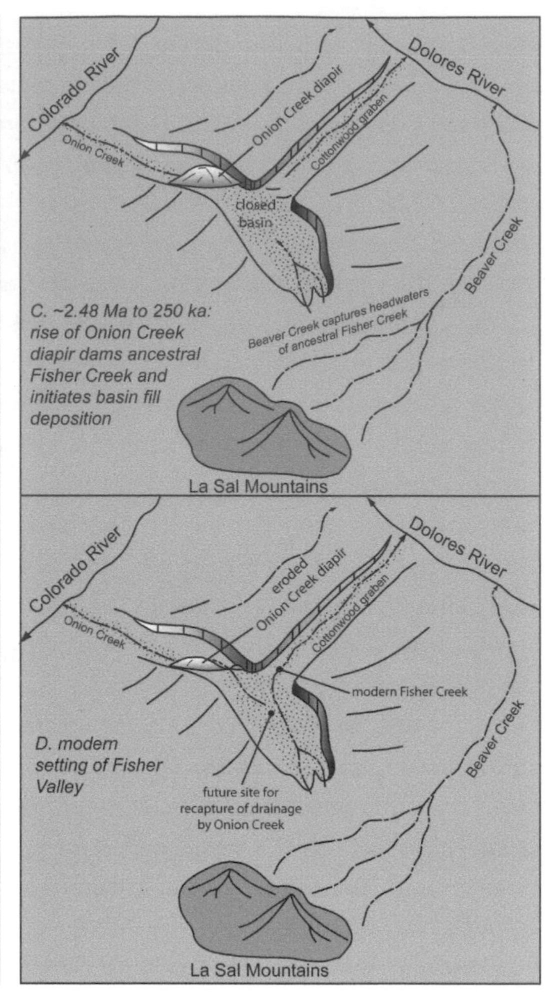

FIGS. 9.3C AND 9.3D. Figure 9.3C shows the setting for the period ~2.48 Ma to 250 ka. During this time Beaver Creek captured the headwaters of Fisher Creek, cutting off the supply of La Sal igneous clasts. The rise of the Onion Creek diapir at this time isolated upper Fisher Valley from the lower part and formed a closed basin. Onion Creek drained into the Colorado River from its headwaters on the north flank of the diapir. Cottonwood graben developed on the northeast flank of Fisher Valley, but headward erosion had not yet tapped the closed basin. Figure 9.3D shows the modern setting. The Cottonwood graben drainage has captured the closed basin drainage to form modern Fisher Creek, which drains to the Dolores River. Onion Creek has also cut headward through the diapir and eventually will capture Fisher Creek, restoring the original drainage.

collapsed salt anticlines in the region. This history begins prior to the Late Miocene or Early Pliocene, before the Fisher Valley anticline had collapsed to form a valley and was, in fact, a highland of moderate relief. At this time Mesozoic strata arced upward to the anticline axis, and small streams ran off either side of the axis, to the northeast and southwest (Fig. 9.3a). Some of these small drainages are preserved on the remnant limbs of the anticline. These have been beheaded, however, by the subsequent collapse of the axial region of the anticline as the salt beneath was dissolved in the shallow subsurface. Collapse of the axial reaches of the anticline created Fisher Valley and led to the evolution of the ancestral Fisher Creek drainage (Fig. 9.3b). This juvenile stream had headwaters in the La Sal Mountains and flowed down Fisher Valley, cutting a canyon through the uncollapsed lower part of the anticline to drain into the Colorado River in Professor Valley. This drainage pattern is told by Pliocene(?) gravels that contain telltale clasts of plutonic igneous rock from the La Sal Mountains.

Sometime before ~2.5 Ma the Onion Creek diapir pushed up through the floor of Fisher Valley, forcing a substantial reorganization of Fisher Creek drainage. Mobilization of salt likely was triggered by unloading of the Paradox Formation after anticline collapse. The rise of the diapir created a closed basin in the upper reaches of the valley as it blocked the earlier path of Fisher Creek. At about this same time, Beaver Creek, situated high above the south end of Fisher Valley, captured the La Sal Mountains headwaters source of Fisher Creek (Fig. 9.3c). This effectively isolated Fisher Creek from the La Sals and cut off the source for igneous clasts. It also reduced considerably the volume of water and sediment entering the basin. It was in this small basin that the basin fill sediments were deposited. The maximum of age of 2.48 Ma for these sediments provides timing constraints for the rise of the diapir, which split the earlier drainage and prompted basin fill deposition. The lack of igneous clasts in basin fill deposits also indicates the capture of the headwaters by Beaver Creek. While upper Fisher Valley was dammed by the rising salt mass, Onion Creek evolved on the north side of the diapir as a shortened continuation of the ancestral Fisher Creek drainage, flowing northward at a much reduced flow to the Colorado River.

Basin fill deposition was influenced by a combination of salt movement and climate change. The multiple angular unconformities concentrated along the basin margin were caused by the interplay between salt withdrawal from beneath the immediate basin area to produce subsidence, and the rise of salt into the adjacent diapir, which caused uplift along the northern basin margin (Colman and others 1988). Eventually this transfer of salt slowed, and sedimentation caught up. The basin was filled and stabilized by ~250 ka, the minimum age of basin fill sediments. At this time a tributary drainage to the Dolores River had formed in Cottonwood graben, a northeast-trending, downfaulted valley on the northeast limb of Fisher Valley anticline (Fig 9.3c). As this drainage cut headward into the basin of upper Fisher Valley, it captured Fisher Creek and diverted it to the Dolores River via Cottonwood graben. Today Fisher Creek heads in the amphitheater that forms the south end of Fisher Valley and flows northward through the basin fill, only to curve northeast down through Cottonwood graben and the Dolores River (Fig. 9.3d).

The most recent phase in the evolution of the Fisher Valley salt anticline is a work in progress. Currently Onion Creek, a memento of ancestral Fisher Creek, runs north to the Colorado River; Fisher Creek, which was blocked earlier by the intruding diapir, escapes the upper valley by exiting through Cottonwood graben. Onion Creek, however, has cut headward through the salt plug and extends another 3 miles upvalley into basin fill sediments. Onion Creek continues its headward erosion into the easily eroded sediments, and today is less than 1.5 miles from the channel of Fisher Creek. Eventually Onion Creek will tap into the upper reaches of Fisher Creek, capturing it and restoring much of the original drainage of ancestral Fisher Creek (Fig. 9.3d).

Quaternary History of the Salt Valley Salt Anticline in Arches National Park

Evidence for Quaternary salt-related deformation in other areas of the Paradox basin has been reported by Dane (1935) and Cater (1970) and is documented in some detail by Charles Oviatt (1988) for the Salt Valley anticline, which includes Arches National Park. Oviatt's findings are augmented by the presence of Bishop ash (730 ka) and Lava Creek B ash (620 ka) in the deformed Quaternary sequence. An overview of this study is provided below.

Although Oviatt documents several instances of Quaternary deformation in the Salt Valley anticline, the most detailed and instructive account comes from a 1 x 2 km area in Arches National Park. This small area is centered on the intersection between the main paved Arches road and the road to Delicate Arch trailhead. Here a thick (up to 100 feet) sequence of Quaternary sediments has been subdivided into five distinct units. The Bishop and Lava Creek B volcanic ash beds occur within these sediments as important marker beds and vital time lines. Sediments consist of gravel, sand, and marl (carbonate mud) deposited in fluvial settings and isolated bodies of standing water. Deposits range in age from more than one million years to 128 ka.

Basal gravels

Two distinct generations of basal gravels overlie folded Mesozoic rock in this part of the collapsed Salt Valley anticline. In the east part of this area the basal gravel unit contains Bishop ash, whereas in the west part of the study area the Bishop ash lies in sandy deposits situated 60 feet *above* the basal gravel unit. This shows that the gravel to the west is older (Oviatt 1988), a relationship that would be unrecognizable in the absence of the Bishop ash. Different ages in identical gravels also suggest dramatic changes in depositional setting over short distances. Because the well-dated Bishop ash is 730 ka, Oviatt concluded that the basal gravel to the east that contains the ash is less than 1 Ma, and the gravel unit to the west likely is more than 1 Ma. Clasts in both gravels are dominated by Mesozoic sedimentary rocks that are exposed in the tilted flanks of the collapsed salt anticline, although sparse pebbles of chert, igneous, and metamorphic rock are also present. Mesozoic clasts derive from the walls of the collapsed salt anticline valley, while the other clasts are interpreted as the more-resistant detritus of eroded Cretaceous and Tertiary strata that were part of an earlier incarnation of the Book Cliffs a short distance north. As this great palisade of Cretaceous and Tertiary rock was whittled back northward by erosion, the resistant pebbles of the conglomeratic units were spread to the south, only to be recycled and funneled into subsequent drainages. The basal gravel units are interpreted to record an earlier version of Salt Valley Wash, which presently drains the southern part of Salt Valley (Oviatt 1988).

Lower basin-fill sediments

Lower basin-fill sediments overlie basal gravels and consist of unconsolidated sand with thin marl beds (Oviatt 1988). These sediments contain ostracodes (small bivalves), diatoms, and traces of plant roots, suggesting deposition in standing water. Like the gravels on which they sit, lower basin-fill sediments represent two generations of deposition. The older of the two, to the west, contains the 730 ka Bishop ash and, in the upper 10 feet, the Lava Creek B ash dated at 620 ka. To the east, the younger strata overlie the gravel that contains the Bishop ash and thus are younger than 730 ka. Rapid lateral changes in coeval sediments indicate deposition under fluctuating conditions. Moreover, these basin-fill sediments, as well as underlying gravels, are folded and faulted. Thickening of some beds in synclines and thinning of those same beds in adjacent anticlines shows that deformation was concurrent with deposition. Abrupt facies changes and deformation were associated with activity in the Paradox salts, due either to the rise of salt diapirs or the dissolution of salt in the shallow subsurface (Oviatt 1988). Either scenario would generate localized highs and lows, folding and faulting of the sediments, and disrupt drainage. Sedimentation and deformation were succeeded by an interval of erosion in which deformed sediments were eroded flat, preparing the stage for the next episode of deposition.

Upper basin-fill sediments

Upper basin-fill sediments form the top of the Quaternary succession and are separated from underlying Quaternary and Mesozoic strata by a pronounced angular unconformity (Oviatt 1988). These younger sediments consist dominantly of sand but grade toward the valley margins to sandy gravel. Clasts in gravel are of the same Mesozoic sandstones that form the escarpment of the collapsed valley. Upper basin-fill sediments are undisturbed by the folding and faulting that so obviously affect underlying strata. These sediments unconformably overlie the 620 ka Lava Creek B ash, providing a maximum age of less than 620 ka. The minimum age is estimated by the degree of calcium carbonate development in soil at the top of the deposit. Oviatt (1988) assigned a Middle Quaternary age to the unit, which translates to an age range of less than 620 ka to a minimum of 128 ka. Upper basin-fill sediments accumulated in a small closed basin that formed slightly north of earlier depositional centers.

Overview

Basal gravels in the Quaternary succession of the lower Salt Valley were deposited in a south-flowing drainage, essentially a precursor to the modern Salt Valley Wash. These ~1 Ma deposits indicate that the Salt Valley anticline had experienced large-scale collapse prior to this time due to dissolution of the underlying Paradox salts in the subsurface. Sand and marl of the lower basin-fill overlie the gravels and are in part coeval with them. These finer-grained sediments were deposited in small ponds that formed during the deformation due either to salt dissolution in the shallow subsurface, the diapiric rise of salt, or both. Lower basin-fill sediments range in age from more than 730 ka to ~620 ka and record salt deformation both during and after deposition. Sedimentation was followed by an interval of erosion in which the earlier fold- and fault-generated topography was planed flat. Upper basin-fill sands were deposited on this erosion surface. These finer-grained deposits range from ~620 to 128 ka in age. They are mostly undeformed, suggesting a cessation of salt-related deformation by this time.

Incision History and Canyon Evolution in Southeast Utah and Adjacent Areas

The rate of **downcutting,** or the **incision rate,** of a river to form a canyon is an estimate of the amount of erosion that has occurred per thousand (ka) or million (Ma) years. This rate can be controlled by various combinations of climate, river discharge, tectonics, and sediment supply. Simply stated, this is an estimate of the *average* rate at which a river cuts downward over a given duration of time, and provides a number that can be compared to other parts of a river system and other drainages in the region. In reality, however, the rate likely fluctuated considerably over certain short-term periods. Many methods for dating river terrace deposits and other features are currently being applied to Colorado Plateau drainages and the Colorado River across the West in a concentrated effort to obtain incision rates for the Late Tertiary and Quaternary periods.

Problems with determining incision rates include the scarcity of young datable materials and the difficulty of dating such young rocks. Other problems lie in the nature of the average rates that are calculated. As a specific example, ancient terrace deposits situated 200 m above the modern Colorado River at Bullfrog basin in Glen Canyon have yielded a ^{10}Be isotopic age of 479 ± 12 ka. An average incision rate of 0.42 m/ka is obtained by dividing the 200 m of incision by 479 ka (Davis and others 2004). Over this 479 ka period, however, river discharge certainly fluctuated greatly, with processes alternating between erosion and deposition. Time constraints simply are not available for shorter periods in order to pinpoint these short-term fluctuations. Moreover, while focusing on incision or erosion rates over a particular period, it is the intervals of deposition that provide the valuable age constraints and datums in the form of perched gravel deposits left high and dry as the channel continued to cut downward. In fact, the alternating glacial/interglacial intervals that define the Quaternary Period drove intense fluctuation in discharge and associated processes. The sparse data that are available for these processes on the Colorado Plateau are constantly being tested and refined. As more scientists focus on particular problems such as incision rates across canyons of the West, our knowledge of the rates of their development will become better constrained and more precise.

Quaternary incision rates have been deciphered for three segments of the Colorado River drainage network in the southeast Utah region. These include multiple data sets from the Glen Canyon area, one of which is from the previously mentioned Bullfrog basin, and rates from the Lees Ferry area in northern Arizona, which forms the boundary between Glen Canyon upstream and Grand Canyon downstream. A third locality has been calculated from data along the San Juan River at Bluff, Utah. Rates from these three areas provide a framework for discussion of incision rates throughout the region, and the basis for comparison to rates farther south in Grand Canyon.

An investigation into the incision history of Glen Canyon by Thomas Hanks and others (2004) used sediments shed from Navajo Mountain and deposited near the confluence of the Colorado and San Juan rivers to estimate ~250 m of downcutting since deposition. This number was obtained from the sediments preserved on benches elevated above the present-day canyons that radiate from the flanks of Navajo Mountain. By extrapolating from the slopes of Navajo Mountain, across the scattered, gradually sloping, gravel-mantled benches to Glen Canyon, a reasonable estimate of 250 m was reconstructed (Fig. 9.4). This assumes that this remnant grade was a smooth, slightly concave-upwards slope that extended to the Colorado River at the time of deposition. Estimates of the age of these deposits were made on several lines of evidence, including the volume of $CaCO_3$ accumulated in the soil formed in these sediments, paleomagnetic analysis, and exposure ages for the sediments using abundances of the isotopes ^{10}Be and ^{26}Al. After evaluating the various age data, the time of deposition was estimated at 500 ka. Taken together, 250 m of incision over the past 500 ka produces an average rate of 0.5 m/ka for this part of the Colorado River (Hanks and others 2004). These results

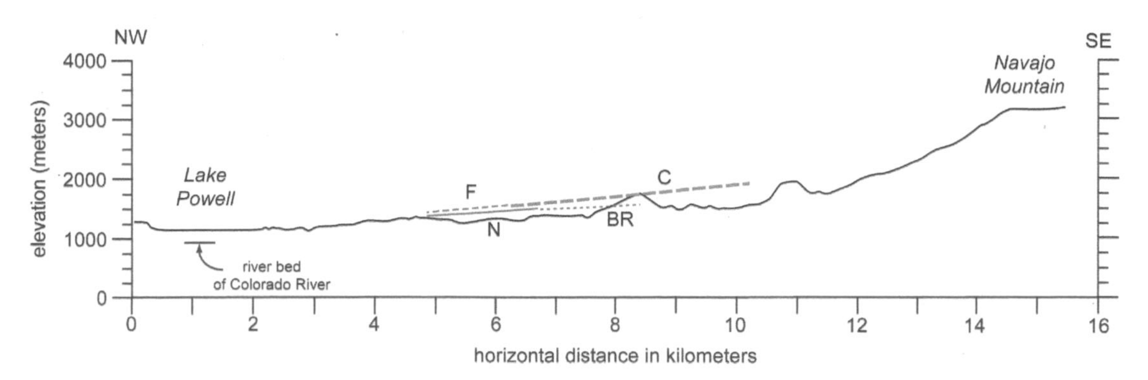

FIG. 9.4. Topographic profile from Navajo Mountain to Lake Powell (formerly Glen Canyon). The gray, dashed lines in the center show various gravel-bearing erosion surfaces that grade from Navajo Mountain to the rim of Glen Canyon. F is the Forbidding surface; C is the Cha surface; N is the Nasja surface; and BR is the Bald Rock surface. Each of these individual surfaces extends radially away from Navajo Mountain within 110 degrees of each other, yet collectively they are remarkably consistent. The rim of Glen Canyon is at ~1240 m, whereas the bed of the Colorado River beneath Lake Powell is at ~990 m, indicating 250 m of incision since these various surfaces were active. From Hanks and others 2001.

are consistent with the previously discussed incision rate of 0.42 m/ka reported by Davis and others (2004) 60 miles (96 km) upstream at Bullfrog basin. These rates for downcutting in Glen Canyon are up to four times higher than those for the rest of the river system.

Another investigation into incision within the local Colorado River drainage network was conducted on San Juan River gravels ~150 miles (240 km) upstream (east) of the Colorado/San Juan confluence (Wolkowinsky and Granger 2004). Coarse river gravels on a terrace 150 m above the present San Juan River bed at the town of Bluff yield a ^{10}Be and ^{26}Al isotopic age of 1.36 Ma. These numbers provide an average incision rate of 0.11 m/ka over the past 1.36 million years. This rate is considerably lower than the 0.42 m/ka calculated for Glen Canyon downstream, although this may be in part due to the much longer time interval being used in this study.

Lees Ferry, Arizona, ~65 miles (104 km) downstream from the mouth of the San Juan River in Glen Canyon, marks the beginning of Grand Canyon. As in Glen Canyon, terraces at Lees Ferry have yielded a high incision rate of 0.4 m/ka (Davis and others 2004). These rates are estimated from carbonate-bearing soils in several terraces in the area.

Incision rates farther downstream in Grand Canyon provide a stark contrast to the high rates found at Lees Ferry and Glen Canyon. At Granite Park 208 miles downstream from Lees Ferry, incision rates ranging from 0.12 to 0.16 m/ka have been calculated for the past 500 ka (Lucchitta and others 2000, Davis and others 2004).

In an effort to evaluate the role of faulting in the incision process in Grand Canyon, Pederson and others (2002) used high-resolution isotopic dating of a variety of rock types associated with terraces and their deposits at three localities in Grand Canyon. They found incision rates of .072 to .092 m/ka at Granite Park, the farthest downstream locality. Granite Park is situated immediately downstream of the Toroweap fault, on the downthrown side of this active, north-trending normal fault. Upstream, on the upthrown side of this fault near Toroweap overlook, an incision rate of .136 m/ka was calculated, almost two times the rate on the west side of the fault. Much farther upstream, between Toroweap and Lees Ferry, incision occurred at a rate of .144 m/ka. Except for the Granite Park rate, these data support incision rates obtained by other workers for the Grand Canyon region and confirm a relatively low rate compared with those of Glen Canyon upstream. The reduced rate of incision in Granite Park is attributed to Quaternary movement on the Toroweap fault (Pederson and others 2002). As the west side of the fault dropped downward, incision in this stretch stalled, while incision upstream, on the upthrown block, accelerated in an attempt to achieve a smooth gradient. This study documents the local effect of faulting and tectonism on downcutting by rivers.

The relatively high incision rates for the Glen Canyon/Lees Ferry region (~0.4 m/ka) must be explained in the context of the low rates downstream in the Grand Canyon (~0.10 m/ka) and upstream on the San Juan River near Bluff (0.11 m/ka). What this distills to is the concept of river equilibrium—what it means and how rivers work to achieve it.

Equilibrium in a river system refers to an idealized river with a smooth, concave-up longitudinal profile (cross section parallel to flow). This profile begins with a steep gradient in the mountainous headwaters region; the gradient decreases gradually downstream until it becomes flat at its mouth, where it empties into the sea or a lake (Fig. 9.5a). A river in equilibrium has no abrupt drops where the gradient suddenly increases, nor alternations of flat and steep stretches, only a gradual downstream decrease in gradient. The idealized equilibrium profile is divided into three distinct segments based on gradient and the dominant activity of the river in that stretch (Fig. 9.5a). The headwaters in the steep upper reaches of the river have excess energy, so erosion dominates the activity of this segment. Downstream, the middle segment has a lower, moderate gradient; in this part neither deposition nor erosion occurs. Instead, sediment produced by erosion in the headwaters bypasses the middle segment and is transported downstream. The lower segment has a very low gradient and includes the mouth of the river. Here the gradient is so low that the river changes to a deficient-energy river, meaning that it does not have enough energy to move the sediment supplied to it by the upper reaches. Deposition dominates in this segment. Keep in mind that a river in equilibrium is an idealized river that rarely, if ever, occurs in the real world. Because of tectonic activity in the form of uplift and downdropping, changes in rock type and related changes in resistance to erosion along a river bed, and variations in sediment supply due to a number of factors,

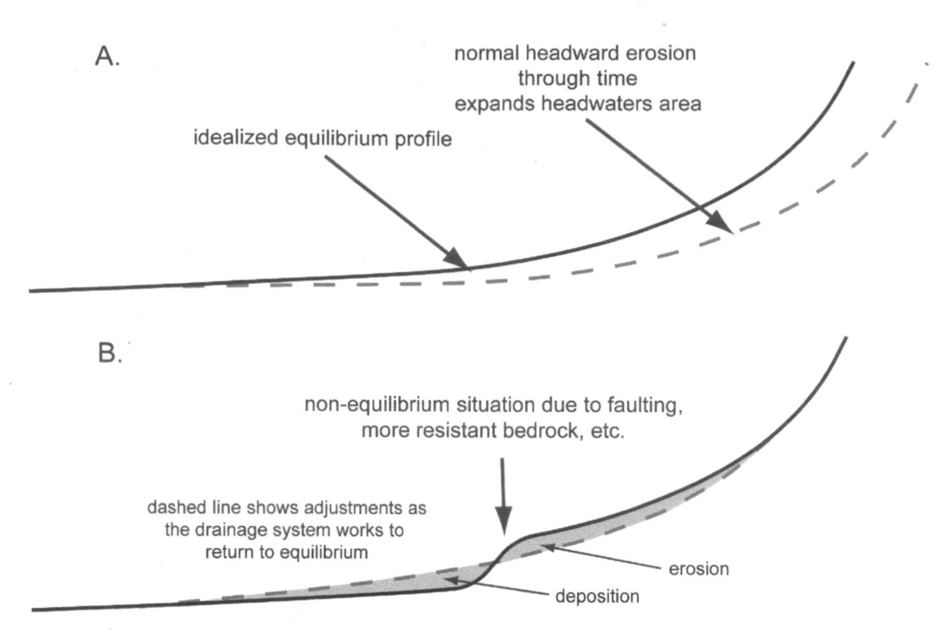

Fig. 9.5. Figure 9.5A shows an idealized (and rarely realized) profile for a river in equilibrium. In such a river the steeper headwaters would be experiencing downcutting and erosion. The middle reaches would be stable, with neither erosion nor deposition taking place. Instead, the sediment produced in the headwaters would simply bypass this segment. In the river's lower reaches, where the gradient is much reduced, the river would deposit sediment as it loses energy and is thus unable to transport its sediment load. The most common activity of an equilibrium river is the gradual headward erosion into the headwaters area where there is sufficient energy to actively erode. This ultimately expands the system's drainage area. Figure 9.5B shows how a river that is not in equilibrium, for whatever reason, works to reach such a state by the processes of erosion and/or deposition in an effort to smooth its gradient.

rivers throughout their extent more commonly alternate between flat and steep stretches. Taken on a large scale, rivers mimic an equilibrium profile, but in detail they do not. Still, they work constantly toward the unachievable: a smooth profile and equilibrium.

Because the headwaters region, even in non-equilibrium rivers, possesses a steep gradient and excess energy, erosion continually carves headward, deeper into the mountainous source terrane (Fig. 9.5a). Similarly, abrupt increases in gradient (e.g., waterfalls) may occur in any part of a real river system (Fig. 9.5b). This local increase in gradient, known as a **knickpoint**, creates a localized zone of excess energy and prompts the river to cut through that steep stretch in an effort to create a smooth profile. Knickpoints may form where rivers encounter relatively hard, resistant layers, or where a river crosses an active fault. A well-documented example of adjustment across a fault is the previously discussed study of the Colorado River in Grand Canyon above and below the Toroweap fault (Pederson and others 2002).

Scientists seeking to explain the high incision rates at Lees Ferry and Glen Canyon have suggested a combination of movement on the Toroweap and Hurricane faults in western Grand Canyon and upstream knickpoint migration as the river cut headward (Davis and others 2002, Lucchitta and others 2002). It is generally agreed, however, that increased incision due to uplift on the east (or upstream) side of these faults is responsible for only a fraction of the incision upstream (Pederson and others 2002).

Most of the increased incision at Lees Ferry and Glen Canyon in the past 500 ka has been attributed to headward erosion as it pushed a prominent knickpoint upstream (Davis and others 2002, Lucchitta and others 2002). Upon initiation of the integrated Colorado River drainage ~5 million years ago, the river began to carve headward from western Grand Canyon. Although the drainage system was well established by this time, the young river likely had a bumpy channel with numerous knickpoints of various scales and origins. As the river ground headward to establish a smooth, uninterrupted grade, knickpoints were slowly pushed upstream. Meanwhile, the downstream reaches approached a relatively smooth profile (Fig. 9.6). By ~500 ka the river had carved a "smooth" path upstream through Grand Canyon to near Lees Ferry. In the past 500 ka, however, the river has labored intensely to whittle a knickpoint down to a smooth grade. It is this effort that has generated the higher incision rate for this stretch. Earlier, the river apparently had reached a stable grade in most of Grand Canyon, resulting in reduced incision rates there for the past 500,000 years. The exception to this is the minor increase immediately upstream from the Toroweap and Hurricane faults.

Quaternary Life

Pleistocene life was similar to modern life, with the main difference being the presence of a Pleistocene megafauna of large and unique mammals. Many of these died out ~11,000 years ago, contributing to the definition of the inexact Pleistocene/Holocene boundary. Pleistocene plants mostly were the same species as those seen today on the Colorado Plateau, but their elevation ranges were lower today due to a cooler Pleistocene climate. Humans appeared in North America near the end of the Pleistocene but had little effect on the region until the past 60 years. The most profound change at this time was the end of the Ice Age, a warming of the Earth that gradually melted the massive ice sheets and extensive mountain glaciers.

Most fossils of the Pleistocene megafauna and flora on the Colorado Plateau are found in the deep, dry caves and alcoves gouged into canyon walls. The bones of large mammals that characterize Pleistocene life are found in the thick layers of sand and clay that partially fill these deep recesses. At least as important as the actual bones is the more abundant dung of these animals, which in addition to offering evidence for their presences provides detailed information on their diet

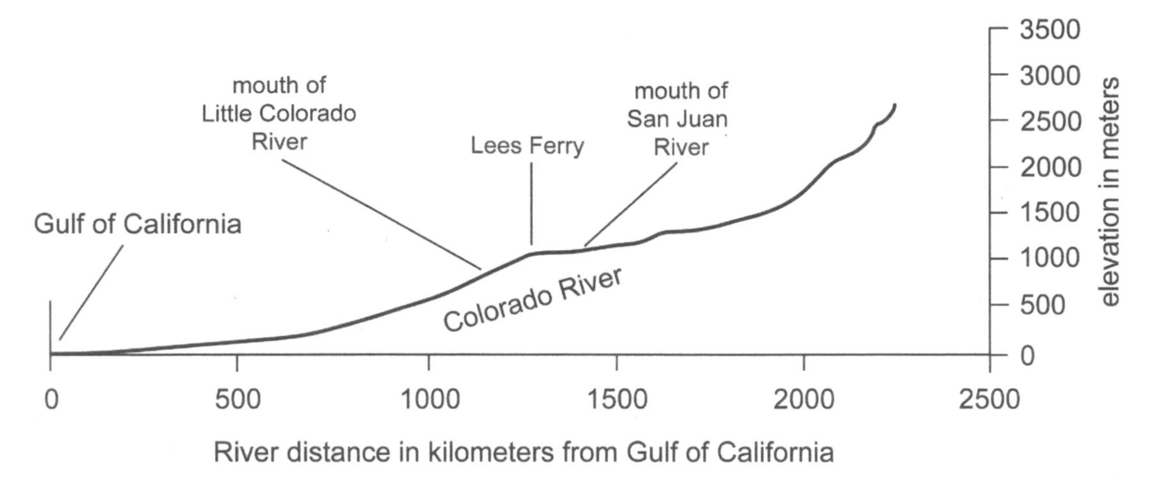

FIG. 9.6. A longitudinal profile of the Colorado River channel showing its gradient in the various areas discussed in the text. Modified from Wolkowinsky and Granger 2004.

and thus the flora that surrounded that particular area 15,000 to 10,000 years ago. The most significant plant record, which includes leaves, seeds, pollen, and twigs, is well preserved in packrat middens, essentially stored in stacks of plant material and viscous packrat urine. Bones, dung, and vegetation are datable by ^{14}C methods, providing a high-quality chronology for Pleistocene life on the Colorado Plateau.

Pleistocene Megafauna

Close relatives of some large Pleistocene mammals endure in modern mountainous regions and northern latitudes, but many members of the megafauna became extinct ~11,000 years ago, not to be seen again except in the fossil record. Fossil bones and dung of extinct mammals from the sediments of dry caves on the Colorado Plateau include those of the mammoth (*Mammuthus*), Shasta ground sloth (*Nothrotheriops*), shrub-ox (*Euceratherium collinum*), and the extinct Harrington's mountain goat (*Oreamnos harringtoni*). Fossil evidence for relatives of modern camels, horses, and bison has also been recovered. Like many of their Pleistocene contemporaries, the horse became extinct in North America ~11,000 years ago but was reintroduced much later to the continent by Spanish explorers.

The most spectacular of the megafauna were the mammoths, great elephant-like animals with large, curving tusks. Mammoth remains, including tusks and bones, have been documented at forty-one locations on the Colorado Plateau (Agenbroad and Mead 1989). Mammoth dung is also a common and important component of the record and is well preserved in many dry alcoves of the region (Mead and others 1986). ^{14}C ages on mammoth remains on the Plateau reach a maximum of 28,290 ± 2100 years BP from Grobot Grotto in the Escalante River drainage. The youngest age is 10,350 ± 110 years BP from Professor Valley, along the Colorado River between the Fisher and Castle salt anticline valleys. A mammoth tusk recovered from Bartlett Wash, immediately west of Arches National Park, yielded a ^{14}C age of 12,880 ± 370 years BP (Woodward-Clyde Consultants 1984).

In an effort to determine as accurately as possible the latest existence of mammoths on the Colorado Plateau, Agenbroad and Mead (1989) compiled thirty-eight radiocarbon dates on mammoth remains and dung obtained so far. From these, they used the four youngest ^{14}C dates to calculate a weighted average minimum date of 11,270 ± 65 years BP. The youngest is the previously mentioned 10,350 year date from Professor Valley. The other three youngest ages come from a layer of dung discovered in Bechan Cave in the Glen Canyon–Colorado River area. The cave's name comes from the Navajo word for "big feces," a reference to the abundant mammoth dung (Mead and others 1986). Out of twenty ^{14}C dates from this dung blanket, the three youngest range from 11,670 to 11,870 years BP.

The weighted average minimum age for mammoth extinction on the Colorado Plateau of 11,270 years fits well with the minimum age of several other extinct mammals in the region (Agenbroad and Mead 1989). Remains of Harrington's mountain goat have been recovered from caves in the Grand Canyon and Natural Bridges National Monument. A weighted average minimum age for the Grand Canyon remains provides a date of 11,160 ± 125 years BP (Mead and others 1986). Shasta ground sloth remains from a Grand Canyon cave yield a date of 11,018 ± 50 years BP (Martin and others 1985). Finally, shrub-ox dung from the dung blanket in Bechan Cave provides a date of 11,630 ± 150 years BP (Kropf and others 2007). Collectively, the minimum ages of these various members of the Pleistocene megafauna on the Colorado Plateau, and thus the time of their extinction, currently appears to have occurred between 11,000 and 12,000 years BP. Additional discoveries and dates in the future may further constrain the age of this significant event.

Mammoth dung from Bechan Cave yield ^{14}C dates ranging from 13,505 to 11,670 years BP and was analyzed for plant remains in an effort to determine their diet and the makeup of plants that surrounded the cave (Mead and others 1986, Agenbroad and Mead 1987). Identification of an assortment of grasses, rushes, and sedges indicate that riparian vegetation dominated the mammoth diet. Plant macrofossils recovered from the dung include birch, rose, saltbush, sagebrush, blue spruce, wolfberry, and dogwood (Davis and others 1985). This plant assemblage is considerably different from the plants that occupy the canyon today.

The plant content in mammoth dung from Bechan Cave suggests that ~12,000 years ago the area was a steppe upland dominated by grasses, sagebrush, and oak, with a riparian area along the canyon floor (Agenbroad and Mead 1987). Today, however, this particular plant assemblage is found at elevations 4000 feet (1200 m) higher than Bechan Cave, in the nearby Henry Mountains. This shift in vegetation zones is attributed to warming and seasonal precipitation changes in the region over the past 12,000 years and is recognized across the Colorado Plateau based on an abundance of other evidence. The following section looks briefly at vegetation changes on the Colorado Plateau since the Pleistocene.

Pleistocene/Holocene Plant Assemblages and Climatic Implications

In high-relief regions such as the Colorado Plateau different plant assemblages occur at different elevations, producing distinct plant zones. Plant zonation is caused by variations in temperature and precipitation at different altitudes. The modern Colorado Plateau hosts several plant zones that are used to investigate and compare with zones of the past, and thereby interpret climates from thousands of years ago. The best preserved of these ancient vegetation records come from packrat middens found in the deep dry caves and alcoves in canyons of the Colorado Plateau.

Packrat middens are the dwelling structures of these small mammals (*Neotoma* sp.) constructed of mounds of plant material "cemented" by the packrat's crystallized urine and feces. These middens contain passageways and chambers where the animal lives, but when humidity rises, it may collapse onto itself, solidifying the pile. The living area is subsequently rebuilt on this mass, producing a growing, stratified pile with the oldest material on bottom and youngest on top. If a packrat dies, its midden may be taken over by another, providing a somewhat continuous occupation and plant record. Packrats forage within 100 m of their middens, collecting leaves, twigs, and seeds from the local plant population. In addition, wind-transported pollen adheres to the sticky midden material. All provide a snapshot of vegetation in that area, at that time, which can be compared with the modern assemblage. Although the middens are stratified, providing a relative age, radiocarbon dates from plant material provide a high-resolution sequence of ^{14}C dates.

In southeast Utah two well-documented packrat midden sequences that range from latest Pleistocene (> 12,000 years BP) to recent have been studied by Julio Betancourt (1984, 1990). Both are located on the east side of Comb Ridge monocline in deep alcoves eroded into the hogback of Jurassic Navajo Sandstone. The lower of the two is Fishmouth Cave, at an elevation of 5200 feet (1585 m) (Fig. 9.7). The second is Allen Canyon Cave, situated at an elevation of 7200 feet (2195 m) on the forested southwest flanks of the Abajo Mountains, where the range meets Comb Ridge. Betancourt's work provides the opportunity to evaluate two similar settings at different elevations. These two localities track the profound change in vegetation across the Pleistocene/Holocene boundary, which can be used to interpret the paleoclimate. Combined, these data track the evolution of the modern plant communities that occupy this part of the Colorado Plateau today.

Fishmouth Cave lies in the lower elevations between the two highways that slice through Comb Ridge, Utah Highway 95 to the north and U.S. Highway 163 to the south (Fig. 9.7). The data of Betancourt (1984) for Fishmouth Cave are shown

Fig. 9.7. Aerial photograph of Fishmouth Cave, on the east-facing slope of the Comb Ridge monocline. The deep alcove has formed in the Navajo Sandstone that forms the dip slope of the ridge. Comb Wash can be seen in the background.

in Figure 9.8 and are summarized below. The oldest midden, dated at 12,770 years BP, contains abundant fragments of limber pine (*Pinus flexilis*) and Douglas fir (*Pseudotsuga menziesii*), as well as Rocky Mountain and common juniper (*Juniperis scopulorum* and *J. communis*). Today this community is limited to elevations of 8000 feet or higher. In a later midden dated at 9700 years BP, limber pine and both juniper species are absent, and Douglas fir has decreased. In contrast, Utah juniper (*Juniperus osteosperma*) first appears in this midden to dominate the area around the cave; it continues to dominate today. Douglas fir had a continued but reduced presence through 6100 years BP but apparently disappeared soon after. A notable appearance of piñon pine (*Pinus edulis*) occurs briefly in a midden dated at 3740 years BP, but by 3550 years BP it was gone. Piñon presumably was migrating through the area on its way to the slightly cooler and wetter environment it presently occupies. Fishmouth Cave today lies in a Utah juniper-grassland setting, just below the higher-elevation piñon-juniper woodland that dominates nearby Cedar Mesa (see Fig. 9.7). The pre-Holocene community of limber pine–Douglas fir indicates a vertical shift of ~2800 feet considering the higher elevations at which this community is found today.

Allen Cave, at a loftier elevation of 7200 feet, lies below a mesa occupied by a piñon-juniper woodland. The cave mouth area is now surrounded by Douglas fir

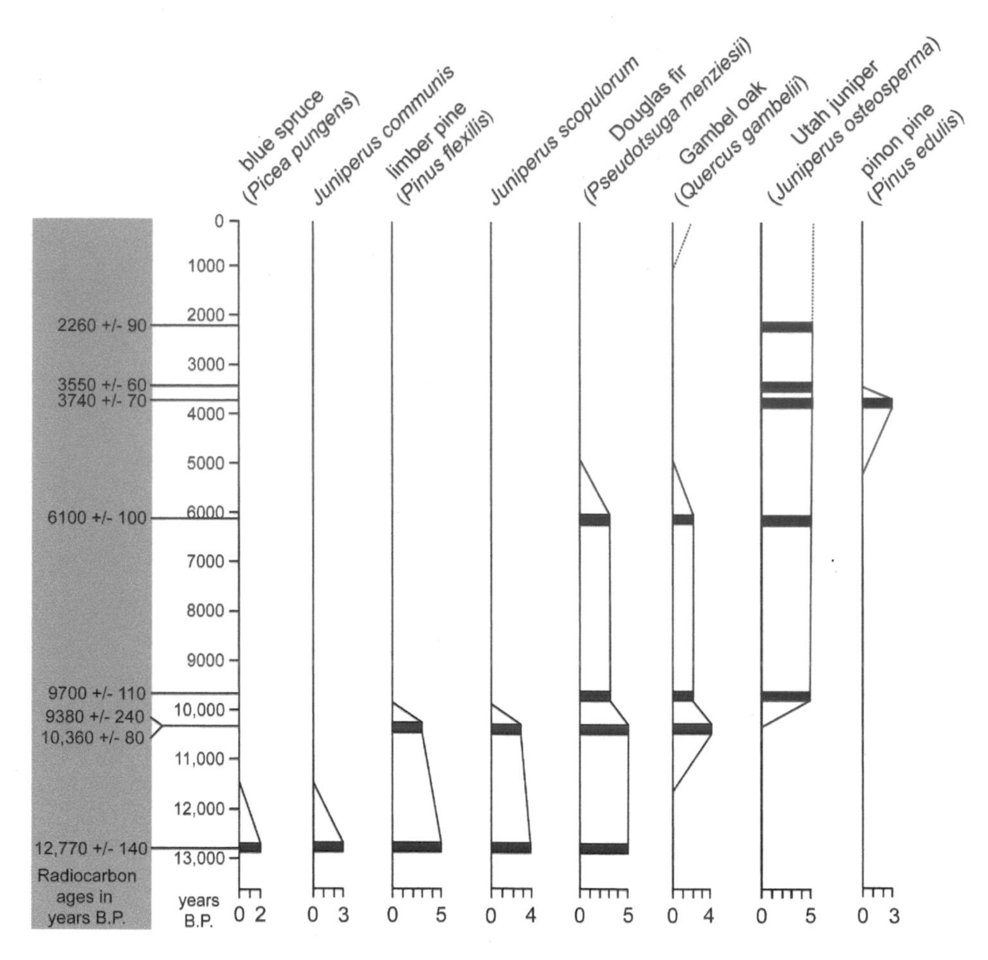

FIG. 9.8. Chronosequence of various significant plant macrofossils and their relative abundances in fossil packrat middens from Fishmouth Cave, San Juan County, Utah, on the east side of Comb Ridge. A scale of 1 to 5 is used to show relative abundances (1 = rare, 2 = uncommon, 3 = common, 4 = very common, 5 = abundant). After Betancourt 1984.

and ponderosa pine, with a grove of aspen (*Populis tremuloides*) a short distance downslope (Betancourt 1984). Nine middens analyzed by Betancourt (1984, 1990) have yielded ten radiocarbon ages ranging from 11.3 to 1.8 ka, providing an age range comparable to that of Fishmouth Cave, but 2000 feet higher in elevation; thus Allen Cave covers the same time span, but with a record of somewhat different vegetation zones.

Betancourt's data (1984) from the Allen Cave middens are shown in Figure 9.9. Here the period 11,300 to 9660 years BP was dominated by Engelmann spruce (*Picea engalmanii*), limber pine, subalpine fir (*Abies lasioscarpa*), and common juniper, with blue spruce (*Picea pungens*) appearing as a minor component in the middle of the interval. Ponderosa pine arrived later in this period as the earlier species waned and then disappeared before 7200 years BP. At 7200 years BP ponderosa pine dominated, and Utah juniper appeared; both endure to the present.

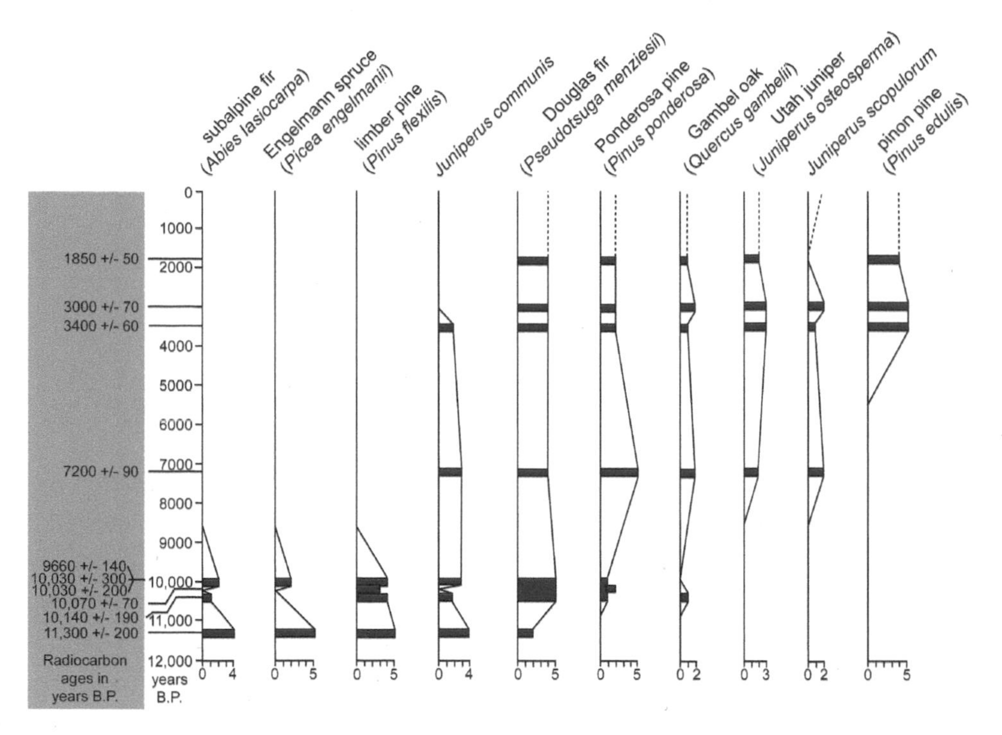

FIG. 9.9. Chronosequence of significant plant macrofossils and their relative abundances in fossil packrat middens from Allen Cave, San Juan County, Utah, on the flanks of the Abajo Mountains. Allen Cave is situated at an elevation of 7200 feet (2195 m). A scale of 1 to 5 is used to show the relative abundances of plant material (1 = rare, 2 = uncommon, 3 = common, 4 = very common, 5 = abundant). After Betancourt 1984.

Piñon pine, which today dominates with Utah juniper, did not show up until the next midden, which is dated at 3400 years BP. It is the most abundant component in middens dated at 3400 and 3000 years BP. Moreover, these two middens represent the beginning of the modern flora. In summary, the Pleistocene/Holocene boundary at Allen Cave is marked by a transition from a spruce-fir forest to the modern piñon-juniper woodland with localized ponderosa pine and Douglas fir.

In addition to data from Fishmouth and Allen caves in southeast Utah, Betancourt (1990) used data from across the region to reconstruct Late Pleistocene/Holocene climate conditions for the Colorado Plateau. Additional midden localities used in this reconstruction include Glen Canyon and the San Juan River canyon in southern Utah, the Little Colorado River and Grand Canyon (two localities) in Arizona, and Chaco Canyon in New Mexico. Late Pleistocene vegetation zones are compared with modern zones in Figure 9.10, which shows a pronounced upward shift in plant communities across the Colorado Plateau, presumably in response to a gradual warming trend. This change began at ~10,000 years BP with the notable arrival of ponderosa pine, Gambel oak (*Quercus gambelii*), and Utah juniper in the region (Betancourt 1984). Piñon pine is not evident until Middle Holocene time, after which it proceeded to spread across the Plateau. Although most of these major components of modern flora appeared near the Pleistocene/Holocene boundary, they did not become established at their present elevations until ~2000 years BP.

Pleistocene and Holocene floras revealed by packrat middens record the initiation of profound climatic changes at ~14 ka as warming occurred and Rocky Mountain glaciers began to melt (Betancourt 1990). The Holocene increase in the elevations of plant zones across the Colorado Plateau allowed Betancourt to calculate Pleistocene summer temperatures that were at least 6.3°C (11.3°F) cooler than today. Some specific vegetation changes allow for more detailed speculation on the Pleistocene/Holocene climate change. The absence of ponderosa pine and summer-flowering herbs and grasses in Pleistocene middens suggests that summers were dryer than those today. Modern summer rainfall supplies 20 to 30 percent of the total annual precipitation on the Colorado Plateau. The absence of ponderosa pine with flowering herbs and grasses suggests that less than 10 percent of the annual precipitation during the Pleistocene fell during the summer. Instead, more precipitation fell during the winter. The present monsoon precipitation pattern that accounts for much of the summer rainfall on the Colorado Plateau developed between 10 and 8 ka.

The Arrival of Humans

The first evidence of humans on the Colorado Plateau—or in North America, for that matter—goes back ~12,000 years to two cultures of early peoples. Their recognition comes not from their remains, as none have been recovered from the Colorado Plateau, but by the distinctive styles of their rock projectile points (Fig. 9.11). The earliest, the Clovis people, hunted giant mammoths and other Late Pleistocene megafauna of the Ice Age. This has been documented in several sites in the Southwest and Great Plains where Clovis points and butchering tools have been found with mammoth skeletons (Meltzer 2003). Besides the unique projectile points, several petroglyphs depicting mammoths have been discovered in canyons in southeast Utah. These almost certainly were pecked into the sandstone by Clovis age people.

At ~11,000 years BP, as the Clovis people expanded across North America, the Folsom culture descended from them. Although they overlapped slightly in time with the Clovis, the Folsom apparently split off to adapt to the Rocky Mountain and Great Plains regions. Their projectile points differ from those of Clovis in that they were both smaller and more finely crafted (Fig. 9.11). In addition, all available evidence suggests that the Folsom people hunted bison, as the mammoths and other Pleistocene megafauna that the Clovis had hunted became scarce or extinct. It is believed that at this time the northward recession of the continental ice sheet that heralded the end of the Ice Age led to the evolution of huge expanses of grasslands; with this came the abundant bison (Meltzer 2003).

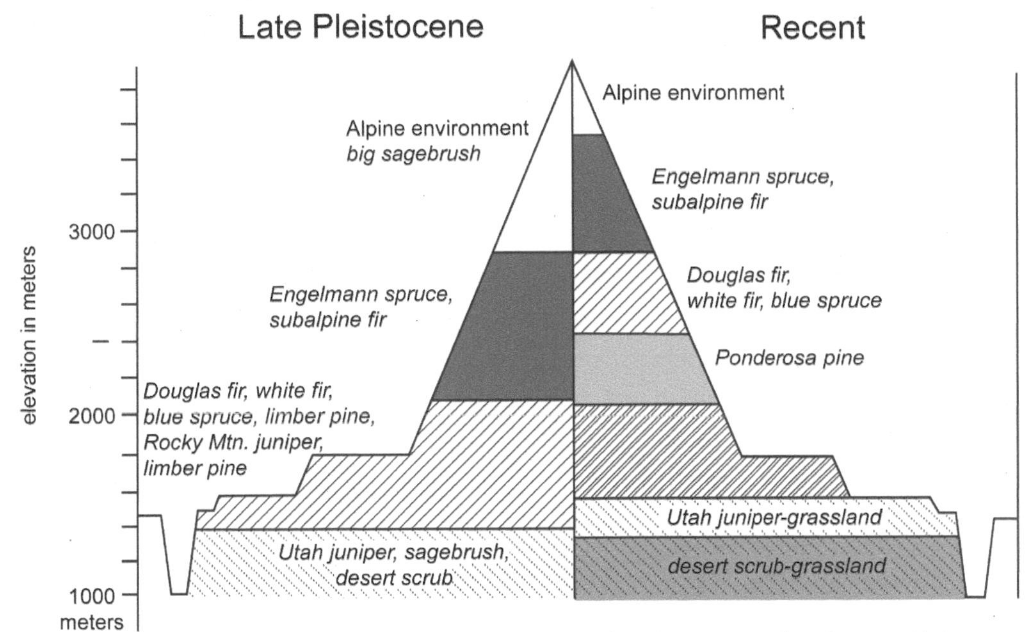

FIG. 9.10. Generalized vegetation and their elevations comparing the Late Pleistocene zones on the left to modern zones on the right. Modified after Betancourt 1990.

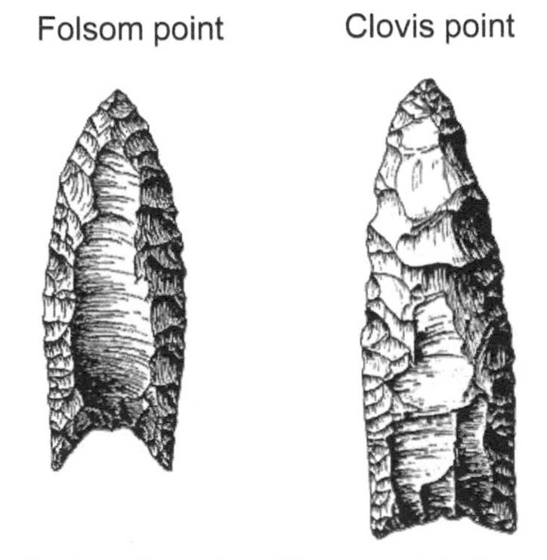

FIG. 9.11. Examples of Folsom and Clovis points from North America.

The oldest ^{14}C date from any Clovis site in the Great Plains and Southwest is 11,570 years BP (Meltzer 2003), providing an approximation for the arrival of the first humans in North America. It must have been intimidating: an entire, seemingly endless continent, barren of humans but wild and diverse. The Clovis people, however, apparently rose to this exceptional challenge, spreading to all corners of the continental United States within a few hundred years.

All evidence suggests that the Clovis people migrated initially from eastern Siberia into Alaska and spread southward from there. Although Siberia and west-central Alaska today are separated by a sea, the immense volume of water frozen in Pleistocene glaciers lowered the sea level almost 400 feet (120 m) relative to today. This exposed the Bering land bridge, a narrow swath of land that connected Siberia with west-central Alaska from ~25,000 to about 10,000 years BP, creating an avenue for migration not only for humans but also animals. Upon reaching western Alaska, however, these hunter-gatherers would have been blocked from further passage by the vast North American ice sheet that prohibited southward movement. It is speculated that as the glaciers melted and contracted northward, routes opened allowing people to migrate southward into the North American interior. The first route opened along the Pacific coast, which, by ~13,000 years BP was mostly free of ice, although formidable obstacles would have endured along the rugged coastline. Even today valley glaciers collapse into the sea along this coast. The second corridor opened along the east side of the Rocky Mountains at ~11,500 years BP. This route similarly held many dangers immediately after the recession of the ice sheet. A complete lack of vegetation, including trees for fuel, coupled with a landscape of boggy ponds and violent rivers filled with glacial meltwater portray a difficult setting. There is no evidence along either possible corridor for a Clovis presence, and at this point the exact route remains an unsolved problem (Meltzer 2003).

Plate 1. Oblique view aerial photo of the Needles District of Canyonlands looking north toward the Colorado River canyon. The sandy wash on the right is Lake Canyon graben, which turns west at a right angle to drain into the Colorado via Lower Red Lake Canyon.

Plate 2. Aerial view of Upheaval dome in the Island in the Sky District of Canyonlands. The orange, cliff-forming sandstone that bounds the central crater is the Jurassic Wingate Sandstone. The outer ring of resistant sandstone is the Jurassic Navajo Sandstone. The light-colored and red strata in the central crater are the Triassic Moenkopi (red) and Chinle (light-colored) formations. Upheaval Canyon drains the central crater on the upper left side of the photo. Note the road to the trailhead in the lower left corner.

Plate 3. Aerial view of Upheaval dome showing the radial folds in the Kayenta Formation and Wingate Sandstone.

Plate 4. Convolutions in the Dewey Bridge Member of the Carmel Formation in Arches National Park and surrounding areas may be related to shock waves that rippled through the region during the meteor impact that formed Upheaval dome in the Island in the Sky District of Canyonlands National Park.

Plate 5. Lower part of the Chinle Formation showing the obvious angular unconformity along the Colorado River between Castle Valley and Moab, Utah.

Plate 6. Outcrop of Navajo Sandstone showing its typical large-scale cross-stratification and dome-shaped weathering.

Plate 7. The cliffs in the shaded foreground are Wingate Sandstone, and the fins in the middle ground, in the Behind the Rocks area, are Navajo Sandstone. The La Sal Mountains loom in the background.

Plate 8. Aerial view to the north of the north end of Comb Ridge monocline. The light-colored strata on the left (west) side are the Permian Cedar Mesa Sandstone. Note how the strata flatten to horizontal on the west side to form Cedar Mesa, the type locality for this sandstone unit. The western edge of the Abajo Mountains can be seen in the background to the east.

PLATE 9. Aerial view to the north of Comb Ridge. The main spine of Comb Ridge is Navajo Sandstone, overlain by the Carmel Formation and Entrada Sandstone. The beds in this photo are dipping toward the lower right (east). Note also the parallel fractures in the upper surface of the Entrada Sandstone.

PLATE 10. View to the south across the Castle Valley salt anticline of Round Mountain, an igneous plug in the middle of the valley, and the north cluster of the La Sal Mountains laccoliths.

PLATE 11. Aerial view to the west of Castleton Tower, the slender pinnacle to the left, and the Castle Valley area. To the right is the Rectory, with the Priest forming the right side. These towers are cut from the Wingate Sandstone, which also forms the prominent cliff band in the middle ground on the west side of Castle Valley. The Wingate is underlain by the Triassic Chinle and Moenkopi formations. The prominent white sandstone band in the cliffs across the valley is an outlier of the Permian White Rim Sandstone. Note that it pinches out to the left (south).

PLATE 12. Aerial view of the contact between the diapir of Paradox evaporite rocks, in this case white gypsum, and the red-orange Cutler Formation in upper Fisher Valley. Note the road to the left, in the bottom of the valley, for scale.

PLATE 13. The strike valley of Comb Wash, which has been eroded into the less-resistant rocks of the Triassic Moenkopi Formation. Resistant rocks of Comb Ridge on the right (east) consist of the Jurassic Glen Canyon Group (Wingate, Kayenta, and Navajo). Cedar Mesa, on the left (west), is dominated by the resistant Permian Cedar Mesa Sandstone.

PLATE 14. Aerial view of lower Grand Gulch on Cedar Mesa, the type locality for the orange and white Permian Cedar Mesa Sandstone that makes up the deeply incised canyon.

PLATE 15. Jacobs Chair (on the right) is composed of resistant orange Wingate Sandstone. The snow-clad Henry Mountains loom in the distance, to the west.

Plate 16. The depths of White Canyon, cut into the Cedar Mesa Sandstone immediately upstream from the Black Hole.

Plate 17. Fisher Towers, located east of Moab, Utah. The small upper (darker) part is formed from the Moenkopi Formation. The larger lower (lighter) part is formed from the Cutler Formation.

Plate 18. The erosion of Entrada Sandstone, a member of the San Rafael Group, has created many arches, fins, and walls visible in Arches National Park. This view shows the Three Gossips (left), Sheep Rock (center), and the Organ (right).

Plate 19. The White Rim Sandstone, a formation of the Cutler Group, outlines the Green River gorge in Canyonlands National Park.

Plate 20. Aerial view of the confluence of the Green and Colorado rivers in the heart of Canyonlands. The view is from the west and shows the Green River in the lower left with the Colorado River coming in at the upper left. Rocks at river level are mostly the lower Cutler beds with Cedar Mesa Sandstone at the top.

Plate 21. Mexican Hat, a rock formation in southeastern Utah from which the town of Mexican Hat gets its name.

Arches National Park

Although this road log through Arches National Park begins at the north end of the Moab salt anticline, where the entrance station and visitor center are situated, most of the park's scenic attractions are centered on the Salt Valley anticline to the east. The road log begins as one exits from the visitor center parking lot. The following is an introduction to the park entrance area to get the reader oriented to the immediate surroundings.

While waiting to pay the entrance fee, note the view ahead to the north. The strata directly ahead on the east side of the highway appear to collapse downward, toward the road, by dipping to the west. Here the highway cuts across the main trace of the north-trending Moab fault, a large normal fault. The younger rocks on the east side have been dropped hundreds of feet relative to the older rocks on the west. The entire park entrance area is, in reality, a **fault zone**, a cluster of north-northwest-trending parallel normal faults, all associated with salt movement in the subsurface. The main fault follows the trace of the highway here, but many smaller faults parallel it to the west and east. As the Moab salt anticline dies out here at its northern terminus, the two main faults that to the south bound either side of the wide valley converge into this half-mile-wide fault zone. The apparent collapse of the strata ahead is the result of displacement along these faults. The strata on either side of the highway, although all of similar color, are of very different ages and origins. In fact, the rocks at road level on the east are 100 million years younger than those on the west. Stratigraphically, the rocks on the east (directly ahead) would normally lie above the highest cliff of red sandstone on the skyline to the west.

A review of the strata in this entrance area is helpful at this point (Fig. A.1). On the east side of the Moab fault, within the boundary of Arches National Park, the Lower Jurassic Navajo Sandstone is the white, intensely fractured and jointed sandstone at road level east of the entrance road and visitor center. The Navajo is overlain by the Dewey Bridge Member of the Carmel Formation. The

Fig. A.1. View of Arches National Park entrance showing strata found in the immediate area of the visitor center and throughout most of the park.

dark red Dewey Bridge is easily recognized throughout the park by its interbedded sandstone and shale characterized by large-scale, contorted, and wavy stratification. These deformed strata are succeeded by the steep orange-red cliffs and rounded benches of the Slick Rock Member of the Entrada Formation. The resistant sandstone of the Slick Rock is the prominent cliff-former and host of most of the arches in the park. It is overlain by white sandstone of the Moab Member of the Middle Jurassic Curtis Formation. Although the Moab Member is an eolian sandstone similar to the underlying Slick Rock, they are easily discerned by a shift from the orange Slick Rock to the white Moab Member.

The main trace of the Moab fault, which is very well exposed in the wash below the highway (Fig. A.2), separates the dominantly Middle Jurassic package from the older rocks to the west. To the west is the jumbled and intensely fractured Pennsylvanian age Honaker Trail Formation. Exposed in the wash are fossiliferous limestone beds that include abundant crinoid columnals, brachiopods, and bryozoans (Fig. A.3). This is overlain by the orange-red Permian Cutler Formation, a thick, slope-forming sequence of fluvial sandstone and conglomerate. The remaining steep, ledge-lined slopes near the top, but beneath the towering cliffs, are the Triassic Moenkopi and Chinle formations, both slope-forming red beds similar to

FIG. A.2. Surface of the Moab fault in the wash directly west of the Arches road. Note the numerous grooves on the fault surface and the fractured nature of the brittle Honaker Trail Formation. Because the Cutler contains shale that bends rather than breaks, it has folded and twisted in a plastic manner.

FIG. A.3. Marine invertebrate fossils from the Pennsylvanian Honaker Trail Formation along the Moab fault.

the Cutler. The angular, blocky red cliffs that cap the slopes are the Lower Jurassic Wingate Sandstone (main cliff) and the overlying Kayenta Formation, which weathers into a more broken, ledge-forming cliff.

For a sense of how much displacement the Moab fault has accommodated, the Navajo Sandstone at road level on this east side overlies the Kayenta Formation that lines the cliff top to the west!

Road Log

0.0 Pull out of the visitor center complex and drive into the park. The first rocks encountered are riddled with closely spaced fractures caused by proximity to the Moab fault. This is common near faults. In the new highway roadcut to the west, numerous small displacement fault blocks of the Honaker Trail Formation can be seen (Fig. A.4).

0.1 Here is the contact between the red siltstone of the Dewey Bridge (lower) and the overlying sandstone of the Slick Rock Member. Both are Middle Jurassic in age and part of the San Rafael Group.

0.5 The white Navajo Sandstone is at road level and is overlain by convoluted siltstone of the Dewey Bridge Member. Large-scale crossbedding and well-sorted, rounded sand in the Navajo is a testament to its eolian origin.

The rapid changes in rock units here is due to a combination of their westward dip and the rise of the road through the switchbacks. As the road flattens at the top of the switchbacks, it mostly follows the resistant top of the Navajo Sandstone.

0.7 As another switchback is approached, the folded siltstone and sandstone of the Dewey Bridge is directly ahead. The penguin-like pinnacle rising above this fin is in the Slick Rock Member.

1.1 Moab fault and overlook. The view of the highway below offers an excellent overview of the main trace of the Moab fault, which passes through the narrow part of the highway where the convergence of resistant rock types constricts its path (Fig. A.5). The east side of this normal fault has moved down relative to the upthrown west side. Fault movement, like the formation of Moab valley to the south, is driven by the slow flow of thick salt deposits in the subsurface.

Also seen from this vantage is another, smaller fault. Uphill to the east are the lower white slopes of the Navajo Sandstone that rise to a towering, orange-red wall of the Slick Rock Member. The Dewey Bridge Member, which normally separates these two units, has been eroded from the top of the Navajo here. The contact between the Navajo and Slick Rock at the base of the wall is another normal fault in which the Slick Rock on the east side has dropped down against the Navajo on the west. Like the main Moab fault, this is a northwest-trending, down-to-the-east normal fault, but with much less offset. As the wall is traced north, the Dewey Bridge Member can be seen juxtaposed against the Slick Rock Member.

1.4 The alcove to the north hosts two small faults that can be seen from the road (by passengers!). The trace of one is covered by large boulders, but offset in the Slick Rock Member is still apparent (Fig. A.6).

Fig. A.4. A roadcut through the Moab fault zone on the west side of Highway 191 as seen from the Arches visitor center parking lot. The multiple faults shown are in the Pennsylvanian Honaker Trail Formation. The relative motion between the larger faults is shown with arrows, but many other, smaller-scale faults are visible as well.

Fig. A.5. View to the north of Highway 191 and the entrance road to Arches National Park showing the Moab fault in the area. Note that the fault juxtaposes the Pennsylvanian Honaker Trail Formation on the upthrown (west) side against the Jurassic Dewey Bridge Member of the Carmel and the Slick Rock Member of the Entrada on the downthrown east side.

1.9 A clear view of the contact between the Dewey Bridge and Slick Rock members.

2.1 Park Avenue viewpoint and trailhead. This is a good place to stretch your legs, take a short hike, and examine the rocks more closely. The trail goes 1 mile down to the northeast, to another trailhead at its terminus, where it intersects the main road at the Courthouse Towers viewpoint (Fig. A.7).

The strata in this area show a pronounced dip to the northeast, toward the axis of the Courthouse syncline, another product of salt movement in the subsurface. As the salt flowed deliberately toward the Moab salt anticline to the southwest and the Salt Valley anticline to the northeast,

FIG. A.6. View to the north of two parallel normal faults in an alcove along the road. The lines above the outcrop show the angle and trace of the faults below. The arrows at the bottom point to marker beds that have been offset by the fault.

it flowed *away* from the region between the two anticlines to form the downfolded syncline.

The Park Avenue trail follows the bumps and hollows of the Navajo Sandstone and Dewey Bridge Member of the Carmel Formation. The overlying Slick Rock Member is the great cliff-former through this narrow, pinnacle-lined corridor. Locally capping the Slick Rock is the white eolian sandstone of the Moab Member of the Curtis Formation. Numerous vertical fractures through these rocks lend a pronounced angularity to the landscape. Like all the deformation features in Arches, these fractures formed from the stresses associated with anticline/syncline development during salt movement.

2.4 La Sal Mountains viewpoint. The view south from this short loop, across the top of the Navajo Sandstone, provides a spectacular panorama with the La Sal Mountains as the centerpiece. This range, unlike all the other

Fig. A.7. View from the trailhead to the north down Wall Street. The floor of the canyon is in the Navajo Sandstone. The walls consist of the Dewey Bridge and Slick Rock members. The slope of the canyon floor and the strata that line it are to the northeast, toward the axis of the Courthouse syncline.

rocks in this view, is composed of igneous plutonic rock. These mountains formed ~30–28 million years ago when magma rose from great depths in the crust to invade the sedimentary rocks. The succession of sedimentary rocks was much thicker at that time, and all activity occurred in the subsurface. As the molten rock pushed up through the strata, overlying sedimentary rocks were domed upward and tilted around the flanks of the mushroom-shaped intrusions. Although water and the ice of glaciers have stripped away the less-resistant sedimentary rocks from the summits, tilted strata still surround their flanks.

To the north and east of the viewpoint are buttes and fins carved from the Dewey Bridge (lower) and Slick Rock members (Fig. A.8).

3.0 As the road descends the southwest limb of the Courthouse syncline, the Three Gossips, carved from the fractured Slick Rock Member, are prominent ahead (Fig. A.9). The butte on the right is the Organ, which has a foundation of wavy-bedded Dewey Bridge Member but is mostly hewn from the Slick Rock Member.

3.7 Courthouse Towers viewpoint. This is the other end of the Park Avenue trail, where hikers dropped off at the Park Avenue trailhead can be picked up. The Organ looms above this viewpoint. To the west are the Three Gossips, and immediately north is the appropriately named Sheep Rock.

Fig. A.8. An unnamed butte in the foreground and the Organ in the background. The lower, wavy-bedded sandstone and siltstone on this butte comprise the Dewey Bridge Member of the Carmel Formation. The smooth sandstone of the upper part is the Slick Rock Member of the Entrada Sandstone. Both are of Jurassic age.

4.4 The road crosses the axis of Courthouse syncline. To the north the upper part of Courthouse Wash faithfully follows the axis of the syncline. It is common for drainages to follow geologic features such as folds, faults, and changes in rock type. From here the barely discernible dip of the strata changes from northeast (previous) to southwest.

4.6 Bridge across Courthouse Wash. To the south a wonderful hike can be taken through the cottonwood-lined glen of lower Courthouse Wash. It is about 5 miles south to the mouth of Courthouse Wash, where a trailhead parking area is located on Highway 191. Most of the shaded canyon slices through the resistant Navajo Sandstone.

To the north the wash changes to a wide, sandy canyon, the result of being situated in the more easily eroded Dewey Bridge Member. There are some fantastic petroglyph panels along this section, but the hiking is not as intimate as in the lower reaches.

4.7 Parking area on the north for access to lower and upper Courthouse Wash.

4.8 The wavy bedding in the Dewey Bridge Member is especially conspicuous in the lower part of the cliff ahead. The fractured Slick Rock Member overlies it. When viewing the Slick Rock parallel to the trend of the fractures, it is blocky and riddled with vertical cracks. The view perpendicular to

FIG. A.9. The Three Gossips, at the north end of the Park Avenue trail, are formed from the Slick Rock Member of the Entrada Sandstone. Vertical fractures that form during salt deformation erode to the blocky pedestals that make up this feature.

the fracture trend, however, produces spectacular, unbroken walls embellished only by great curving arcs of conchoidal surfaces and black desert varnish. The conchoidal patterns that accent these surfaces are typical of fracture or joint surfaces in homogeneous material such as this sandstone. They are the same as those formed in chert, obsidian, or even glass, when fractured, but on a colossal scale.

5.2 The ragged pinnacles next to the road are in the Dewey Bridge Member.

5.8 On the north (left) the Slick Rock Member lives up to its name, forming characteristic slickrock benches and slopes, and rounded cliffs.

6.1 Petrified Dunes overlook. The view south is across the top of the Navajo Sandstone, which has weathered into rounded slickrock domes. Although the Navajo was deposited as sand dunes, these domes are not the forms of ancient dunes. Their shape instead is due to a combination of fractures and the homogeneous nature of the windblown sand that comprises these deposits. These landforms are erosional in origin rather than depositional.

7.3 The field of bizarre pinnacles to the north (left) is the result of fractures and weathering. Fractures become more abundant and better developed

FIG. A.10. Pinnacles in the Dewey Bridge Member of the Carmel Formation. As the Salt Valley salt anticline is approached to the east, the rocks become increasingly jointed, which leads to erosion into bizarre fields of towers and pinnacles such as these. Note also the contorted bedding in the Dewey Bridge, a characteristic of the unit throughout Arches National Park.

as the collapsed central part of the Salt Valley anticline is approached (Fig. A.10).

8.9 Balanced Rock viewpoint. Balanced Rock and surrounding pinnacles have a base of Dewey Bridge Member of the Carmel Formation topped with a bulbous caprock of the Slick Rock Member of the Entrada Formation (Fig. A.11). This area is floored by Navajo Sandstone. Note that the Park Service sign that points out Navajo and Entrada sandstones is outdated and incorrect. Prior to 2000 the Dewey Bridge was a member of the Entrada Sandstone; however, Hellmut Doelling (2000) properly redefined it as a member of the Carmel Formation. The overlying Slick Rock Member remains the sole member of the Entrada in this region.

9.1 Turnoff to the Window section of the park and a loop road with short hikes to several arches. See short auxiliary road log.

10.1 Panorama Point turnoff. Take this short loop road for an overview of the Salt Valley salt anticline.

10.4 Panorama Point. To the north the concept of a collapsed salt anticline is clearly etched into the landscape. The trend of this one, the Salt Valley anticline, as in all the other salt anticlines in the region, is northwest-southeast. Today only the heavily fractured southwest and northeast

FIG. A.11. Balanced Rock has a foundation of the Dewey Bridge Member of the Carmel Formation. This unlikely rock formation has a bulbous top made of the Slick Rock Member of the Entrada Sandstone.

flanks of the anticline remain. The central or axial part of the large fold has collapsed to form the broad, undulating valley in the center of this view. At an earlier time in the history of this fold the flanks, or limbs, would have risen to the center to converge at the high point of the fold axis. It should be noted that much erosion has taken place regionally, removing thousands of feet of overlying strata to exhume the Jurassic rocks that now dominate the landscape.

The precollapse salt anticline developed incrementally through time. Almost as soon as Permian strata were laid down across the Pennsylvanian salt of the Paradox Formation, the salt began to flow. As it sought relief from the pressure of the overlying sediments, it flowed laterally and then upward to form the salt anticlines. As later sediment piled on top, the flow of salt continued, and folding progressed. During folding, the overlying layers became lithified and then fractured. It was through these abundant fractures that water percolated down, eventually reaching the top of the salt, which it readily dissolved. The resulting void beneath the domed layers eventually allowed the axial part of the anticline to collapse. Continued salt dissolution has produced the landscape we see today.

The flanks of the anticline to the southwest (left) and northeast (right) are separated by the collapse valley in the center, which is partly mantled by a thin blanket of sediment. In the foreground of the valley are both the oldest and the youngest rocks exposed in the region, juxta-

Fig. A.12. View northwest up the center of the collapsed Salt Valley anticline. The light-colored mounds in the middle of the collapse valley are diapirs of the Pennsylvanian Paradox Formation, which pushed development of the anticline. Upon nearing the surface, the more-soluble halite or salt dissolved, and the crest of the anticline collapsed. The diapirs shown here consist of thin, highly contorted beds of dark shale and dolomite and lighter gypsum, all the less-soluble components of the Paradox. The Klondike Bluffs in the back are on the southwest limb of the anticline and are formed of sandstones of the Middle Jurassic Carmel, Entrada, and Curtis formations.

posed by faults. On the older side, the salt-bearing Pennsylvanian Paradox Formation lies just beneath the valley floor and is exposed in the blue-gray mounds in the valley center (Fig. A.12). The mounds of Paradox, which have pierced the valley floor, are composed of gypsum, sandstone, shale, and limestone--all insoluble components of the formation. Although halite (or salt) is the dominant rock type in the Paradox, and its viscous flow is the driving force behind this fold, it is completely dissolved in these exposures, leaving this chaotic assemblage of less-soluble lithologies to represent the formation at the surface.

In the northeast part of the valley, faulted against the flanks of the anticline, are brown and green mounds of shale, and blocky brown buttes of very different rocks. These are the much younger rocks of the uppermost Jurassic Morrison Formation, the Cretaceous Cedar Mountain Formation, Dakota Sandstone, and Mancos Shale. These great blocks are bounded on all sides by normal faults that accommodated their collapse

FIG. A.13. View to the north across Salt Valley of the east flank of the Salt Valley collapsed salt anticline. Note the northwest-trending fins of the Fiery Furnace. Formed in the Slick Rock Member, these fins developed during the erosion of parallel northwest-trending fractures. As erosion progresses, the resistant sandstone between fractures endures as parallel fins, which is where most of the park's arches have formed.

into the valley center, as they essentially sank into the top of the underlying Paradox.

On the northeast limb of the anticline (to the right) the heavily fractured orange Slick Rock and white Moab sandstones are well exposed (Fig. A.13). The fractures have an obvious trend that is parallel to the fold axis. The collaboration of these fractures with water and time has transformed uniform layers of resistant sandstone into a unique field of closely spaced parallel fins. It is in these fins that the world-famous arches of the park develop. The prominent zone of fins seen from here forms the Fiery Furnace, a wonderful area of deeply shadowed recesses hemmed in by rounded towers and fins. This intensely jointed sandstone trends northward into the Devil's Garden, where a great concentration of arches has been hewn into these narrow fins.

The same resistant strata define the southwest limb: the resistant Slick Rock and Moab members. In the distance, along this limb, are the prominent cliffs of the Klondike Bluffs, a maze of fins formed in the Dewey Bridge, Slick Rock, and Moab members. A rough dirt road leads to the base of this isolated area.

To the south are the green and gray (and often white) alpine slopes of the La Sal Mountains, which culminate at 12,271 feet with Mt. Peale. Most of the higher peaks in this intrusive center reach more than 12,000 feet.

In the shadow in front of and east (left) of the range can be seen the slender pinnacle of Castle Rock (also known as Castleton Tower). This tower and adjacent buttes mark the northeast limb of the Castle Valley salt anticline. To the west (right) of this cluster is the gently southwest-sloping Porcupine Rim, which defines the southwest limb of the Castle Valley anticline. The void between the view of these limbs is the collapsed axial part of the anticline. Out of sight in this flat-bottomed valley is the community of homes and small ranches that make up Castle Valley.

10.9 The brown mound ahead is a collapsed block that has dropped down along faults into the valley center. The lower part is the Jurassic Morrison Formation with a caprock of Cretaceous Dakota Sandstone, a unit that in a normal stratigraphic context would be located high above all the rocks seen so far.

11.4 As the road drops steeply off the southwest limb of the Salt Valley anticline, it cuts through the red Jurassic Wingate Sandstone, which lies below the Kayenta Formation and Navajo Sandstone.

11.7 Cache Valley overlook. Looking east across downfaulted mounds of gray Mancos Shale at the jointed platform of Slick Rock and Moab sandstones, two prominent but small-scale faults can be seen. These east-west-trending faults dropping down to the south are expressed as large steps. The view to the southeast is of the south end of Salt Valley as it swings from a trend of northwest to east before curving northwest again into Cache Valley. Cache Valley is simply the southern extension of the Salt Valley anticline, but it has not collapsed as completely. The "valley" in this case is a chaotic jumble of downdropped, folded, and faulted bocks.

As the road curves north from Cache Valley overlook, several small-scale normal faults can be seen along the road. These faults define the northeast flank of the Salt Valley anticline and accommodate the collapse of the numerous fault blocks on this side of the valley. The rock units exposed in these blocks include the Morrison and Cedar Mountain formations, Dakota Sandstone, and Mancos Shale, as well as the Middle Jurassic strata that dominate much of the park landscape.

11.9 Delicate Arch turnoff. The road log continues ahead on the main road; however, because Delicate Arch is the icon of both the park and the state of Utah, a brief overview of its geology is warranted.

The road to Delicate Arch initially passes through sandy, gray Mancos Shale, which is part of one of the fault-bounded blocks that occupy this part of the valley. The parking area for the trailhead is at Salt Wash, where a homestead site was established by John Wesley Wolfe in 1898. The trail to Delicate Arch first traverses the colorful shales of the Morrison Formation, but after crossing one of the region's many faults, it

FIG. A.14. Delicate Arch, the icon of Arches National Park, seen from the south at Delicate Arch overlook. Note the people under the arch and to the left on the skyline for a sense of scale.

climbs the sandstone surface of the Moab Member. The Slick Rock Member is encountered as the trail nears the arch, and it makes up the great slickrock bowl that holds the unique landmark (Fig. A.14). Delicate Arch formed along the contact between the Slick Rock and Moab members. It is the remnant of a fin generated by the regional joint system that formed in response to anticline development.

12.5 Drab brown and gray rocks on the east side (right) are part of the Upper Cretaceous Mancos Shale that was dropped into the axial valley as it formed. One of the numerous small-scale faults that assisted in the collapse can be seen directly behind the closest shale mounds as a thin, northwest-trending fin of brown rock. This is the fault surface; the rock on either side has been eroded. Fault zones create narrow bands of pulverized rock as the two blocks grind in opposite directions. This porous and permeable fault zone may then become a conduit through which chemical-rich fluids can move freely. In some cases the precipitation of minerals from the passing fluids creates fault-zone rocks that are more resistant (harder) than the surrounding rocks. Here the recemented fault-zone rocks are much more resistant than the surrounding Mancos Shale.

13.6 The fault that crosses the road is expressed in the tombstone-like fins on either side of the road.

FIG. A.15. Sand Dune Arch, reached by a short hike, is hidden in a shaded alcove between fins of Slick Rock Sandstone.

13.7 The road rises up through the Morrison Formation and into the white sandstone of the Moab Member. The fracture pattern of the Moab member into discrete blocks is due to its proximity to several faults.

14.2 As the road turns east, the view is of the orange walls of the Slick Rock Member perpendicular to the joint orientation.

14.6 As the road curves northward, the parallel orientation of the fins and joints becomes obvious.

14.9 Their proximity to the road make these fins an appealing area to explore with a short hike into the towering walls, halls, and alcoves found within this maze-like area.

16.2 Trailhead for Sand Dune and Broken arches. These arches have formed in the narrow fins of the Slick Rock Member that are seen paralleling the road for the previous 2 miles (Fig. A.15).

ORIGIN OF ARCHES IN ARCHES NATIONAL PARK

Several accounts of the origin of the numerous arches in Arches National Park have been published in response to the intense interest in these unique landforms (e.g., Lohman 1975; Doelling 1985, 2000). The arches are hosted exclusively by the now-familiar Middle Jurassic succession of the Dewey Bridge, Slick Rock, and Moab members. The great concentration of arches in the park is a direct consequence of the closely spaced joints and narrow fins in these resistant sandstone units (Fig. A.16). The ubiquitous narrow sandstone fins are the dominant controlling factor (Fig. A.17).

A.

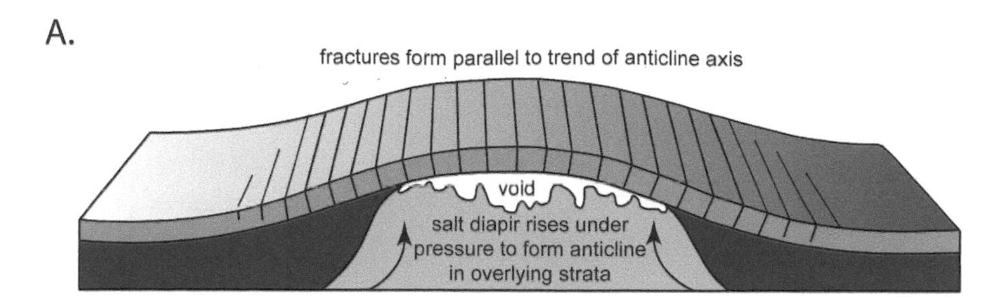

B.

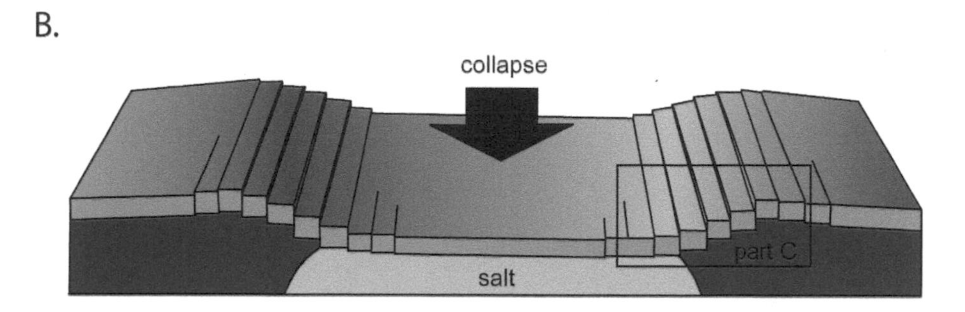

Figs. A.16A and A.16B. Evolution of collapsed salt anticline valleys and arches in Arches National Park. Figure A.16A shows the diapir of Paradox salt rising under the pressure exerted by overlying sediments and forming an anticlinal fold in overlying strata. These brittle rocks fracture, allowing the percolation of meteoric water downward to Paradox salt, dissolving the salt to create a void. Figure A.16B depicts the dissolution of underlying salt, causing blocks of overlying strata to collapse into the void and subside into the salt. This produced the salt anticline valleys of the Arches National Park area. Box shows area of detail in Figure A.16C.

The second major factor in arch formation is water--not the scouring action of running water, but the chemical action of the fluid, namely its ability to dissolve the cement in sandstone. Arches formed solely in the Slick Rock Member typically develop in the narrowest of fins, those ranging from 10 to 20 feet wide, although they sometimes reach up to 50 feet (Lohman 1975). As the joints that separate these sandstone fins are widened by erosion, the narrow floor between the fins fills with loose sand from the breakdown of the sandstone. This loose sand holds water in the form of moisture. While the wet sand sits against the base of the fin, the H_2O slowly and methodically eats into the sandstone. This was (and is being) accomplished by the dissolution of the $CaCO_3$ that cements the sandstone, reducing it to loose sand. Eventually the dissolution wears a hole into the base of the fin. Erosion and a further deepening of the adjacent joint slowly bring the hole to light, and it grows into a noticeable arch. In the long-term, fracturing and a continued reaction with water leads to a slowly evolving arch that eventually collapses. Older arches tend to have smoother, rounded outlines, whereas younger ones (relatively speaking) are more angular, with sharper edges. Arches formed

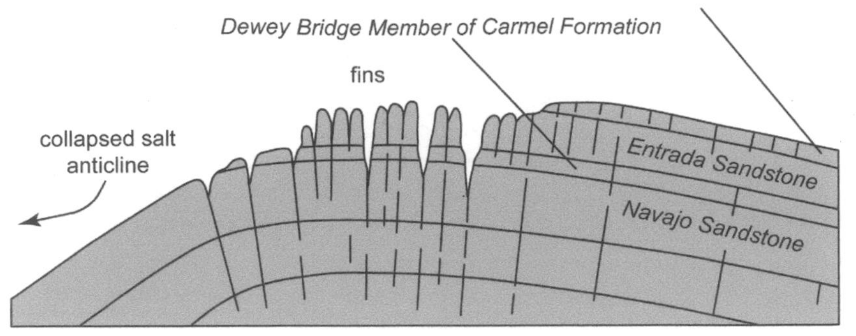

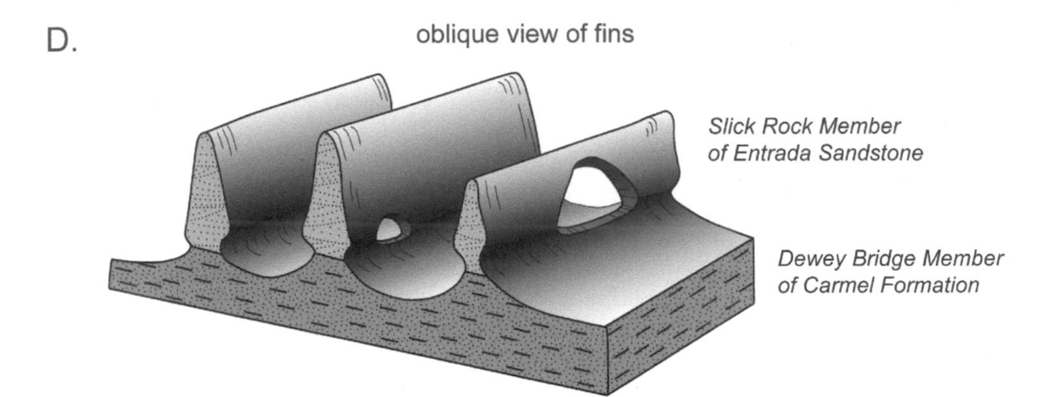

Figs. A.16C and A.16D. After collapse of the axial region of the salt anticline (Fig. A.16C), the limbs begin to collapse into the void, producing a subsidiary anticline and accentuating the fractures that parallel the trend of the anticlinal fold. Preferential erosion along the weaknesses produced by the fractures produces parallel fins. An oblique view (Fig. A.16D) shows the evolution of arches in the fracture-induced fins. A combination of spring-sapping at the contact between two units (dissolution of the cement that holds the sand grains together), wet sand at the base of the fins (also dissolving cement), and exfoliation thins the base of the fins. These thinned areas eventually evolve into arches. Figure A.16C is after Doelling 1985.

exclusively in the orange sandstone of the Slick Rock Member include Broken Arch at the north end of the Fiery Furnace, and Tunnel Arch and most of the others located in Devil's Garden.

The contact between the Slick Rock and the underlying Dewey Bridge Member is also a common locus for the development of arches. Horizontal breaks such as shale or siltstone within sandstone-dominated successions create weaknesses in several ways. Preferential erosion along these weaker beds initiates recesses that deepen with time. In areas where joints are not common, the slow, downward percolation of water through the overlying porous sandstone eventually reaches the less-permeable, fine-grained strata, forcing it to move laterally until emitting from the rock as a spring. Where water issues from the rock, the $CaCO_3$ cement in the rock is slowly dissolved, and an alcove, or cave-like recess, will develop. This process is known as spring-sapping: the spring water saps the cement from the

FIG. A.17. Fins of sandstone such as these in the Slick Rock Member host most of the arches in the park.

sedimentary rock. In the narrow fins of jointed regions the combination of preferential erosion and spring-sapping centered on these weak and impermeable units eventually leads to development of an arch. Double Arch and most of the neighboring arches in the Windows section have formed along the contact between the Dewey Bridge and overlying Slick Rock members (Fig. A.18). The relatively weak, impermeable silty beds at the top of the Dewey Bridge were the catalyst for the evolution of these amazing rock spans.

16.6 On the west (left) is a dirt road to Klondike Bluffs, a mirror image of the Slick Rock fins here, but forming on the southwest limb of Salt Valley anticline. Because this road is not paved, the Klondike Bluffs remain a

Fig. A.18. Double Arch, in the Windows section of Arches. Both arches shown here formed along the weakness at the contact between the lower Dewey Bridge Member and the overlying massive sandstone of the Slick Rock Member.

Fig. A.19. Skyline Arch formed in a fin of Slick Rock Member sandstone. Note that the base of the arch corresponds with a bedding plane in the sandstone that is an inherent weakness in the fin.

beautiful, mostly isolated, and uncrowded area. The road ends abruptly at a parking area for a trail that winds through the fins to Tower Arch.

16.9 Viewpoint and trailhead for Skyline Arch, formed in a narrow fin of the Slick Rock Member (Fig. A.19).

17.4 Entrance to the loop drive to Devil's Garden and the campground.

17.6 The turnoff to the campground, situated in a delightful setting within the great fins and within a short distance of a few scattered arches. The campground, unfortunately, is often crowded and noisy, as most national park campgrounds are these days.

17.8 Devil's Garden trailhead and the end of the road. The trail winds northwestward through fins of the Slick Rock Member, occasionally cutting down into the underlying Dewey Bridge. The trail takes hikers on a 2-mile stroll past eight arches, all carved from the Slick Rock Member of the Entrada Sandstone.

From the Junction of Utah Highway 191 and Route 313 to Island in the Sky, Canyonlands National Park

This road log begins at the turnoff from Utah Highway 191 and follows the less-hurried Route 313 to its terminus at Grand View Point in Canyonlands National Park. Several important spurs off this road are addressed in subsequent road logs, including the road to Dead Horse Point State Park and the spur to Upheaval dome in Canyonlands.

Road Log

0.0 Junction of Utah State Highway 191 and Route 313. Drive west on 313. The sandstone slabs at ground level are part of the Upper Jurassic Salt Wash Member of the Morrison Formation.

0.2 Cross the railroad tracks. This crossing approximately coincides with the trace of the Moab fault. Here the large normal fault is a subtle feature that places the Upper Jurassic Morrison Formation on the downdropped east side against the Permian Cutler Formation on the upthrown west side (Fig. B.1). The offset along the fault here is considerable, effectively displacing the entire Triassic and most of the thick Jurassic sequence by the juxtaposition of the Morrison and Cutler formations. In fact, looking northwest at the great rampart of rock, the Morrison would be located well above all the rocks exposed there.

The base of the escarpment to the northwest is made of the Permian Cutler Formation, characterized by thick orange and red beds that have been sculpted into knobby cliffs. Although most of the Cutler is of fluvial origin, the arid Permian climate is indicated by the thick, cross-stratified eolian dune deposit at the top of the formation. Overlying the Cutler

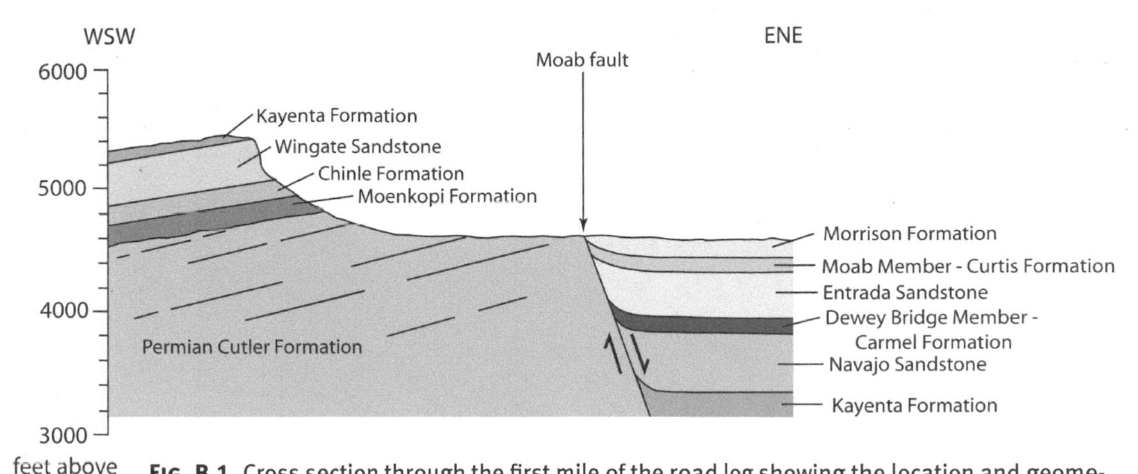

FIG. B.1. Cross section through the first mile of the road log showing the location and geometry of the Moab fault and the various strata both exposed and buried in the subsurface. Note the location of the Jurassic Kayenta Formation on either side of the fault, which provides a reference point with respect to the magnitude of offset.

is a succession of thin-bedded, dark red siltstone, mudstone, and sandstone of the Lower Triassic Moenkopi Formation. The green slopes and thin black and brown cliffs above are the Upper Triassic Chinle Formation. Traces of crumbling roads that traverse these slopes are the paths of 1950s era prospectors seeking uranium in the lower sandstone and conglomerate lenses of the formation. Massive palisades of the Lower Jurassic Wingate Sandstone crown the sequence.

Turning south towards Moab, one can see the great rampart of the same succession sloping westward. This pronounced dip to the west is due to the rise of salt in this northern extent of the Moab–Spanish Valley salt anticline and the Moab fault, which is the local expression of this uplift.

1.0 The Cutler Formation dives below the surface on the north side of the canyon. The overlying Moenkopi Formation is thin here at ~100 feet, possibly the consequence of salt uplift during its deposition. The white sandstone bed midway up the lower slope designates the base of the Chinle. Along these Chinle slopes can be seen the decomposing remnants of prospector jeep trails from the uranium boom days of the 1950s.

1.2 The Moenkopi disappears into the subsurface.

2.0 The sharp Chinle-Wingate contact drops to near road level on the north. Upcanyon the thin ledges of Kayenta Sandstone cap the Wingate cliffs. Large-scale cross-stratification of the Wingate along these walls are cross sections through the giant dunes of Jurassic windblown sand, a testament to the arid climate of that time.

2.9 The monolithic Wingate is replaced by the ledges and cliffs of the fluvial Kayenta Formation. These ledges mark the passage of northwest-flowing sandy rivers that temporarily displaced the vast Wingate dune field.

FIG. B.2. View to the north of Merrimac Butte. All of the strata exposed here are of Jurassic age.

3.3 Details of the sandy Kayenta river channels are etched into the steep roadcuts on the left (south) side. The channels are predominantly sandstone, with thin, wispy mudstone horizons that were laid down at slack water times or on the sheltered lee side of sand bars that migrated across the ancient channel floor. The multiple, stacked large lenses are floored by large, concave-up scours, each marking the high-energy erosional event that was required to initiate each successive channel. White sandstone knobs to the northwest are part of the overlying Navajo Sandstone and indicate the regained toehold of the eolian sand blanket that spread across the region.

3.4 The roadcuts through the switchback expose murals of Kayenta on both sides of the road.

4.5 Rest area on the right (north) side of the road. The white and orange slickrock knobs are in the Navajo Sandstone. Note the large-scale eolian cross-stratification throughout the Navajo.

6.2 Looking to the north across the top of the ancient dune deposits of Navajo Sandstone and modern sand piles, one can see a cliff band that exposes the entire Middle Jurassic San Rafael Group. The base of the cliff is the Dewey Bridge Member of the Carmel, easily recognized by its large-scale wavy and irregularly folded nature (Fig. B.2). This is overlain by a massive orange cliff of Entrada Sandstone, yet another Jurassic eolian

FIG. B.3. The Behind the Rocks Wilderness Study Area, in the middle ground, consists of fins of Jurassic Navajo Sandstone formed from the weathering of fractures produced by the stresses of salt anticline uplift. The Moab–Spanish Valley salt anticline is out of sight, immediately behind these fins. In the foreground, the lower slopes are of the Chinle Formation, and the cliffs above are of the Wingate Sandstone. In the shadowed background are the high peaks of the La Sal Mountains.

deposit. The thinner white cap rock is the Moab Member of the Curtis Formation, the remnant of a final dune field in this area that was more limited in extent than its precursors. Rubble-covered slopes above these cliff-formers are the Summerville Formation, which is the top of the San Rafael Group and part of the Upper Jurassic Morrison Formation.

8.3 Turnoff to the left (south) to scenic viewpoint. This vantage point provides a commanding view to the southeast. In the distance are the La Sal Mountains, a great cluster of igneous intrusive rocks in the form of laccoliths. These crystallized after the injection of magma into the thick succession of Mesozoic strata over the period of 27.9 to 25.1 Ma (Nelson and others 1992). The fins of red rock in the middle ground, in front of the La Sals, make up the Behind the Rocks area that forms the western flank of the Moab-Spanish Valley salt anticline (Fig. B.3). The fins have been exhumed from the erosion of parallel fractures in the Navajo Sandstone. As water (and thus erosion) was concentrated along these fractures, they were slowly widened, leaving the massive rounded fins.

FIG. B.4. Alcoves in various stages of development formed by spring-sapping along the impermeable shale bed that caps the Dewey Bridge Member. As water percolates down through the permeable Slick Rock Member, it encounters the impermeable shale at the top of the Dewey Bridge and is forced to flow laterally to emit as a spring. Through time the water dissolves the calcium carbonate cement that holds these sedimentary rocks together, and alcoves slowly develop. Black wet spots in the back of these alcoves mark the springs.

To the northeast is a broad mesa rooted by the contorted Dewey Bridge strata, which are overlain by steep walls of Entrada and Moab Member sandstones. The yawning alcoves concentrated along the Dewey Bridge–Entrada contact near the east end of the mesa have been carved by a process known as spring-sapping (Fig. B.4).

Spring-sapping is a type of ground water erosion that involves the dissolution of the $CaCO_3$ cement that holds the sedimentary rocks together. The process begins with the percolation of rainwater or snowmelt downward through the porous eolian sandstones. Upon reaching the impermeable siltstone that marks the top of the Dewey Bridge, the downward seep of the water halts, and it flows horizontally across the top of the siltstone barrier. Because the water will preferentially flow toward an unconfined area such as a canyon wall, the water eventually emerges from the rock at the surface as a spring along the contact between the two different rock types. This continuous outward flow of water slowly dissolves the $CaCO_3$ cement in the sedimentary rock, effectively reducing the outer rock of the canyon wall to unconsolidated sand and mud. Eventually this dissolution process opens up an alcove, or deep recess, in the canyon wall. This is the origin of most of the deep alcoves throughout the Colorado Plateau. Blackened water streaks are a common feature in the shadowed recesses of these alcoves, as is vegetation in the form of columbines and other alpine wildflowers that take advantage of the shade and reliable water source in this otherwise parched landscape.

9.8 The uplift of the San Rafael Swell can be seen in the distance to the northwest.

11.6 To the left (east) are petroleum storage tanks. Scattered across this high plateau can be seen pumps and, more recently, the towers of drill rigs, all part of the exploration for and extraction of petroleum from deeply underlying strata. Producers in this area include the small Big Flat, Bartlett Flat, and Long Canyon oil fields, which produce from the Mississippian and Pennsylvanian strata. While these fields today mostly are abandoned or in decline, there has been a recent resurgence in activity in the form of exploratory drilling and seismic surveys in the area.

12.3 Roadcuts are in the Kayenta Formation.

13.2 From this high point the laccoliths of the La Sal Mountains can be seen to the southeast and those of the Henry Mountains to the southwest. Intrusive rocks in the Henry Mountains have yielded radiometric dates ranging from 31.2 to 23.3 Ma, comparable to the range of ages obtained from the La Sal Mountains intrusive cluster.

14.5 The road to the left (east) is the continuation of State Route 313 that ends at Dead Horse Point State Park and connects to the Pucker Pass/Long Canyon road. The white knob is made of Navajo Sandstone. This road log continues straight (south) to the Island in the Sky District of Canyonlands National Park.

16.6 To the right (west) is a dome of Navajo Sandstone. Behind the dome, in the distance, loom the Henry Mountains. Also seen from this stretch of road are the La Sal Mountains to the southeast and the Abajo Mountains to the south. The Abajos are the third of three major clusters of intrusive laccoliths to pierce the thick sedimentary succession in this southeast part of the Utah Colorado Plateau. The Abajo Mountain intrusions have been dated from 28.6 to 22.6 Ma, within the range documented for both the La Sal (27.9 to 25.1 Ma) and the Henry mountains (31.2 to 23.3 Ma) intrusive centers (Nelson and others 1992). The coincidence in timing, style of intrusion, and the composition of these three plutonic centers strongly suggests a common origin for the magma that generated these great, blisterlike laccolithic mountains.

18.8 Enter Canyonlands National Park.

21.9 Cross "the Neck," a slender edge of Navajo Sandstone that plunges precipitously to both the east and west (Fig. B.5). The Neck is being undermined on both sides by the heads of large canyons. On the west is the head of Taylor Canyon, which drains toward the Green River. The drainage to the

Fig. B.5. View to the east from "the Neck" of Island in the Sky, looking down to the Shafer trail. The Shafer trail intersects with the White Rim Road at the far end, as seen in this photo.

east is Shafer Canyon, which leads a short distance to the Colorado River. The incredibly situated Shafer Trail road hugs the walls of this canyon in a series of dizzying switchbacks.

22.1 Shafer Trail viewpoint is on the east side of the road with a small parking area. From this vantage point the numerous switchbacks of Shafer Trail road can be seen below, winding through the escarpment (Fig. B.6). The viewpoint area and surrounding rocks at this upper level are in the Navajo Sandstone. Immediately below the Navajo the road cuts through the underlying Kayenta Formation. Beneath the Kayenta are the vertical red cliffs of the Wingate, where the dropoffs on the outer edge of the road become breathtaking. The upper part of the road, however, is surprisingly good so long as your car remains on it. This section is followed by the final descent through the slopes of the Chinle and Moenkopi formations. The prominent white bench onto which the road levels out is the top of the Permian White Rim Sandstone. Near the bottom, slopes to the left of the road contain more scars of prospector's trails. These old jeep trails traverse the Chinle wherever it is exposed in this region. They were bulldozed into the shale slopes in the quest for valuable uranium ore, which sometimes was concentrated in the ancient, sandstone-filled river channels of the Chinle.

FIG. B.6. The upper switchbacks of the Shafer trail as it drops from Island in the Sky. Here the road cuts through the uppermost cliffs of Navajo Sandstone, the steep, ledgy Kayenta Formation, and the underlying cliffs of Wingate Sandstone. In the lower left corner is the slope-forming Chinle Formation, which underlies the Wingate.

23.1 After crossing the Neck, you have entered the Island in the Sky. The traverse across its undulating surface winds through rounded knobs and domes of Navajo Sandstone. They are slowly disintegrating into the sandy soil that threatens to engulf them. The La Sal Mountains make sporadic appearances, far to the southeast, while the steep rims that bound this high mesa provide brief glimpses of the Kayenta and Wingate formations.

27.5 On the right (west) is the turnoff to Upheaval dome and Green River overlook. This road log continues straight (south) and ends at Grand View Point. A subsequent road log covers the various destinations down this spur road, including Upheaval dome.

28.4 The Green River shimmers far below, where it is hemmed in by the lower canyon walls of Permian strata. The White Rim Sandstone forms the prominent white bed, exposed over an increasingly broad bench as the overlying red Moenkopi is peeled back by the forces of erosion. The red Organ Rock Formation lies beneath the resistant white cap rock and extends down to river level.

33.1 Orange Cliffs overlook. From here the Green River can be seen winding its way through the tranquil meanders of Stillwater Canyon toward its eventual rendezvous with the Colorado River, ~10 miles to the south (Fig. B.7). The dominantly eolian White Rim Sandstone forms the white

Fig. B.7. View of the Green River from Orange Cliffs overlook. The Permian White Rim Sandstone forms the prominent mesa top of its namesake, the White Rim. The Orange Cliffs are a continuous escarpment of Wingate Sandstone. Ekker Butte is an isolated erosional remnant of Wingate Sandstone.

bench above the river. Across the river loom the imposing Orange Cliffs, a lengthy barrier of Wingate Sandstone that rims the western edge of Canyonlands. The Orange Cliffs can be traced southward to form one of several escarpments that drop toward the Maze, the district of Canyonlands National Park that occupies the vast region west of the Green and Colorado rivers. The isolated tower of Wingate on the far side of the river, but in front of the distant Henry Mountains, is Ekker Butte.

33.3 Grand View Point overlook, elevation 6080 feet. The bluish, forested Abajo Mountains lie to the south. In front of them, and slightly west, is the Needles District of Canyonlands, a great jumble of fins and towers carved into the intensely jointed Permian Cedar Mesa Sandstone. Immediately below the overlook is Monument basin, a red, crenulated landscape of slender, undulating towers sheltered by the overlying White Rim Sandstone (Fig. B.8). These pinnacles are hewn from the Permian Organ Rock Formation, and it is only the umbrella-like protection of the resistant White Rim that keeps them from reverting to mud and sand. The obvious road that wraps around the edge of Monument basin is the White Rim Road, an old jeep trail that has evolved into a popular multiday mountain bike ride. The White Rim Sandstone marks the top of the Paleozoic succession and is overlain by the benches and slopes of the red Moenkopi Formation, followed by the multicolored slopes of the Chinle,

FIG. B.8. View to the south from Grand View Point of the Needles District of Canyonlands and the laccoliths of the Abajo Mountains on the skyline. Monument basin, the pinnacle-lined bowl in the foreground, is rimmed by the Permian White Rim Sandstone, a white eolian deposit. The pinnacles of Monument basin are in the Permian Organ Rock Formation, composed mostly of low-energy fluvial deposits.

both Triassic in age. The massive red, fractured cliffs of Wingate Sandstone form the great palisade that rises to this overlook. Here at Grand View Point, however, we stand on a thin carapace of the succeeding Kayenta Formation, of fluvial origin.

Somewhere deep within the jumble of rocks that lies between the Needles District and Monument basin is the confluence of the two major rivers of the Colorado Plateau and western North America: the Colorado and Green. From here, however, this great junction is concealed in the red folds of some of the most rugged canyon country on Earth.

In the distance, to the east, are the massive humps of the La Sal Mountains. Also to the east, but much closer, runs the Colorado River, carrying the water and sediment of the western half of Colorado southward toward the Pacific Ocean. Here, for the most part, the Colorado is hidden from view by the rims of the steep canyon that it has cut.

Utah Route 279 along the Colorado River to Potash

This route begins in the center of the Moab–Spanish Valley salt valley anticline. As it passes through the Portal into the Colorado River canyon, it cuts the southwest limb of this large fold. The road winds along the north bank of the Colorado River and faithfully follows this spectacular canyon. The road log ends at the surprising industrial complex of Potash, where potassium salts are being extracted from the subsurface of the Cane Creek anticline. The road, however, continues from here first as a gravel road and eventually as a rough, four-wheel-drive road called the White Rim Road, a popular multiday bike route through Canyonlands National Park.

Road Log

0.0 As you turn west onto Route 279, a great escarpment looms ahead. This is the faulted southwest flank of the Moab–Spanish Valley salt valley anticline. This route begins at the north end of the valley, where the two faulted valley sides converge, constricting to a close at the entrance to Arches National Park. Downvalley to the south, the great collapse feature widens, and the valley opens up. The strata on this southwest side can be seen dipping to the southwest, away from the valley.

Focusing back on the escarpment, the lower part directly ahead is a jumble of faulted slivers of rock as old as the Pennsylvanian Honaker Trail Formation. The prominent upper, continuous part of the scarp is held up by the massive red cliffs of Jurassic Wingate Sandstone capped by a thin veneer of Kayenta Formation.

To the left (south) is the open sand pile of the old Atlas Uranium mill site. In the 1950s uranium ore extracted from across the Colorado Plateau was trucked here for processing. Today the old buildings have been

FIG. C.1. View to the west of the Colorado River canyon immediately west of the Portal. The westward dip of the strata here is the result of the Moab–Spanish Valley salt anticline immediately to the east. These strata are also part of the east limb of the Kings Bottom syncline, whose west limb can be seen in the distance dipping back toward the east (labeled "east-dipping Navajo Sandstone").

leveled and cleared. The toxic radioactive tailings are buried out of sight, beneath the great sand pile, but they remain dangerously perched on the bank of the Colorado River, the water source for millions of people downstream in Arizona, Nevada, and California. The process of removal of these tailings began in 2002 and is ongoing. Presently the plan is to have them completely removed by 2019, but that will depend on continued federal funding.

0.2 The Triassic Moenkopi Formation is exposed at road level. Here it consists of thinly bedded dark red shale and orange sandstone of tidal origin.

0.9 To the left (east) are the extensive wetlands of the Colorado River. This broad wetlands area is the product of ongoing salt dissolution in the shallow subsurface and subsidence of fluvial deposits. Recent drill hole data indicate that more than 400 feet of Quaternary fluvial deposits overlie the top of the Paradox Formation (Smith and Goodknight 2005). To the south, along the center of the collapsed salt anticline, the beds on this side dip southwest, while those on the other side, against the backdrop of the La Sals, dip northeast. At some time prior to collapse, these beds met high above the current valley floor to form the axis of the anticline.

Fig. C.2. West-northwest-trending fractures in the Navajo Sandstone along the Wall Street climbing area cause it to weather into giant, rounded fins.

1.8 The Colorado River is off to the left (east). To the right, high above, are the Wingate, Kayenta, and Navajo formations, better seen in the walls of the Portal, directly ahead. There the ledge-riddled slopes of the Chinle make up the base of the steeply tilted succession. Red Wingate and Kayenta cliffs and bald domes of Navajo Sandstone succeed the Chinle.

2.6 Enter the Portal (Fig. C.1). The remainder of the route described here is in the Colorado River canyon. Above the road are huge blocks of sandstone fallen from the high cliffs, victims of gravity. Beneath this mantle of detritus and exposed in the roadcut are tilted beds of the Chinle Formation.

2.8 Massive, dark red Wingate Sandstone is exposed at road level.

3.0 Mile marker 12.

3.2 The Kayenta Formation appears at road level as the underlying Wingate slickrock dives into the subsurface. The Kayenta is the same color as the Wingate, but the interbedded fluvial sandstone and shale weather into a choppy ledge and recess pattern.

3.6 Massive, rounded cliffs of eolian Navajo Sandstone appear above the broken Kayenta as its southwestward dip sends it below the surface. Unconsolidated mounds of rounded pebbles and cobbles ~20 feet above road level spill down to the road edge. These ancient Colorado River gravels are many thousands of years old and mark the bottom of the river channel at that earlier time.

^ **FIG.** C.3. Three-toed dinosaur tracks along the road on a tilted slab of Jurassic Kayenta Formation.

› **FIG.** C.4. View of Behind the Rocks Wilderness Study Area from a terrace along the Colorado River. The terrace here is about 150 feet above the modern river level.

4.0 Mile marker 11. Navajo Sandstone cliffs jut from the roadside. Proceed with caution because the Wall Street rock climbing area extends along this regularly jointed headwall, and many of the routes begin just a few feet from the road (Fig. C.2). Many of the climbers and spectators forget that cars also use this narrow road.

4.6 Here the canyon cuts the axis, or "trough," of the Kings Bottom syncline. The southwest limb of the Moab anticline that begins at the Portal is also the northeast limb of this adjacent syncline. This axis is the low point of the broad downwarp. The beds have been dipping southwest up to this point, in approximately the direction of travel. For this reason younger and younger rocks have been exposed at road level. After crossing the axis of the syncline, the southwest limb will be traversed, and the strata will gradually rise, with older rocks emerging from the subsurface in a mirror image of the previous limb.

5.0 Mile marker 10. Directly ahead the Kayenta and Navajo can be seen dipping back in this direction.

5.4 The Kayenta is exposed at road level on the right (north).

5.6 On the right is Poison Spider Mesa trailhead and a viewpoint for dinosaur tracks. A mounted metal tube at the pullout provides a view of a tilted limestone block with two obvious three-toed dinosaur tracks on its bedding surface (Fig. C.3). This block fell from the Navajo Sandstone cliff behind it. The limestone was deposited in a small pond that formed within the largest dune field ever recognized on Earth. This pond and others like it likely provided essential watering holes in the arid Jurassic erg that drew animals, large and small, to drink.

Approximately 150 feet above the road is a terrace veneered with rounded river cobbles that were deposited on the floor of an older Colorado River, before it cut down to its present level (Fig. C.4). Similar terraces upstream near the Portal have an estimated age of 30 ka (thousand years) (Smith and Goodknight 2005).

6.2 Exposed at road level is the massive, large-scale, cross-stratified Wingate Sandstone. Across the river Wingate cliffs seemingly rise out of the water. The blockier Kayenta overlies it.

7.3 Strata again dive downward, but this is not another fold. Instead, the road, which follows the winding trace of the river, again intersects the axis of the Kings Bottom syncline. The syncline axis trends consistently northwest-southeast, whereas the river canyon meanders sinuously, cutting in and out of the axial region.

7.7 The Kayenta returns to road level.

8.6 Strata dip into the subsurface, so the Navajo descends to road level, where it looms as steep cliffs. Ahead can be seen bedding dipping *towards* the road. The syncline axis is immediately ahead.

9.2 Thick red shale and orange sandstone beds of the Kayenta are well exposed in the roadcuts.

9.5 The Pinto and Corona arches trailhead is on the right. The 1 mile trail winds across slickrock to the arches. Rock along the road was cut for the railroad that originates at the Potash plant a few miles to the south. The railway here leaves the main canyon to enter Bootlegger Canyon. At the head of that canyon the railroad enters a tunnel cut through the rock to emerge above the Atlas Minerals tailings pile, where this road log begins. From there the railroad extends northward, parallel to Highway 191, to a rail junction along I-70.

9.7 Gold Bar BLM recreation site.

10.2 The Wingate Sandstone is exposed at road level. The railroad is immediately above.

10.3 The railroad cut above the road provides a rare clear view of the sandstone-filled river channels and thin shale beds that comprise the Kayenta Formation.

10.6 Strata continue to climb as the road follows the southwest limb of the Kings Bottom syncline. This is also the northwest limb of the approaching Cane Creek anticline.

FIG. C.5. Unlike most arches in the region, Jug Handle Arch is a detached exfoliation flake of Wingate Sandstone.

11.5 Rubble-strewn slopes at road level are underlain by the Triassic Chinle Formation, its first appearance since entering the Portal.

12.5 The sharp Chinle/Wingate contact plunges temporarily back below the road level, and the road is shadowed by angular columns of vertical Wingate Sandstone.

12.9 Jug Handle Arch has been formed from a curved slab of Wingate partially detached from the main wall (Fig. C.5). The Chinle is exposed at road level. From here the strata arc up toward the axis of the Cane Creek anticline.

13.3 As the less-resistant, shaley strata of the Chinle and underlying Moenkopi formations emerge from the subsurface, the canyon widens considerably.

13.5 The Potash industrial facility comes into view (Fig. C.6). The large green metal buildings and belching smoke offer an incongruent picture after one has followed the placid Colorado River along its seemingly remote canyon. Potassium salts (potash) are extracted from the Paradox Formation in the subsurface by the method of solution mining. In this process water from the Colorado River is pumped into the Paradox in the subsurface via drill holes. After circulating in the subsurface and picking up dissolved salts, the brine is pumped to the surface and placed in large,

FIG. C.6. The Potash industrial facility lies on the northeast flank of the Kane Creek anticline. The final processing is done in the large buildings on the right, next to the Colorado River.

open evaporation ponds. These ponds are located immediately south of this complex, just out of sight. As the water evaporates, the salts become increasingly concentrated and eventually precipitate as a white crystalline solid. The final processing and loading onto railcars takes place here. This entire process is economically feasible because of the rise of the Paradox salts in the subsurface and resultant formation of the Cane Creek anticline. This, coupled with the erosion by the adjacent Colorado River, has made the salts more accessible and relatively cheap to extract.

13.9 The base of the Upper Triassic Chinle Formation is clearly displayed on the promontory across the river as a thin but prominent white sandstone bed approximately a third of the way up the slope and cliff couplet. The Lower Triassic Moenkopi Formation forms the red slope beneath the Chinle. Just above river level the top of the Permian Cutler Formation emerges from the subsurface.

14.1 The road enters the broad amphitheater occupied by the Potash industrial complex. The amphitheater has been excavated by the Colorado River from interbedded sandstone and siltstone of the Cutler Formation. These rocks, the oldest seen so far, were brought to the surface first by uplift on the Cane Creek anticline and were later exhumed by the Colorado River.

14.8 The Intrepid Potash facility. The railroad seen earlier originates at this complex. The piles of white granular material scattered about consist of potassium salts. This is the final processing plant before the salts are shipped out on railcars. The paved public road continues on the left for a couple miles before turning to dirt/gravel.

Fig. C.7. View to the north from Anticline overlook of the Cane Creek anticline. The photo is centered on its approximate axis. The strata can be seen dipping to the southwest (or left) on the left, and to the northeast (or right) on the right. The Colorado River cuts the anticline in the middle of the photo, exposing the older Pennsylvanian and Permian strata. The cliffs on the skyline are the Jurassic Wingate Sandstone.

14.9 Directly ahead, the red beds of the upper Cutler Formation rise toward the axis of the Cane Creek anticline (Fig. C.7). Below the Cutler and lining the river just downstream is the gray limestone of the Pennsylvanian Honaker Trail Formation. This fossiliferous unit was deposited in a warm, shallow marine environment about 300 million years ago.

15.5 Rounded cobbles and pebbles on the left are terrace gravels from an older Colorado River. These are disturbed as they have been bulldozed and used for building roads in the Potash facility.

16.2 On the left (south) is the put-in point for river trips down Cataract Canyon of the Colorado River. This marks the end of the paved road and this road log. The graded dirt road that continues ahead climbs to the axis of the Cane Creek anticline and continues for several more miles across the large complex of evaporation ponds and associated infrastructure, part of the expansive Potash facility (Fig. C.8). After crossing this area, the road enters National Park Service land and the beginning of the lengthy White Rim Road, which mostly follows the top of the Permian White Rim Sandstone (Fig. C.9).

FIG. C.8. Giant evaporation ponds in the middle of the Kane Creek anticline are used to concentrate potassium salts.

FIG. C.9. Resistant beds of crossbedded Permian White Rim Sandstone create an overhanging caprock that protects the more easily eroded Organ Rock Formation along the White Rim Road. This four-wheel-drive and mountain bike route follows the top of the White Rim Sandstone over most of its extent.

Utah Highway 128 from Moab to Cisco via Dewey Bridge

This road log follows Utah Highway 128 east from its intersection with U.S. Highway 163 north of Moab. The road runs along the south bank of the Colorado River for ~30 miles before crossing it at Dewey Bridge. From there the road diverges from the river and heads northeast to Cisco, where the road log ends.

Along this segment through the deeply incised Colorado River canyon are four different salt anticlines, and the road passes through strata ranging from Permian to Cretaceous in age. Most of the canyon cuts transverse to the axes of the great anticlines and the associated intervening synclines, providing fantastic, true-scale cross sections through these salt-induced folds.

Road Log

0.0 At its intersection with U.S. Highway 63, turn east onto Utah Highway 128. Ahead on the right is an excavation that exposes an angular unconformity within the Upper Triassic Chinle Formation. The dark red fluvial sandstone and shale of the Chinle is overlain by broken cliffs of orange Lower Jurassic Wingate Sandstone. Both the angular unconformity and the intense fracturing in the Wingate are due to the proximity to the collapse zone of the Moab–Spanish Valley salt anticline, which forms the wide valley directly behind you and hosts the town of Moab. The angular unconformity developed as the rise of Pennsylvanian salt into the evolving anticline tilted the lower Chinle strata eastward. This was closely followed by deposition of the uppermost Chinle on top of the tilted strata. Salt-driven uplift continued afterward as well, as shown by the gentle northeastward tilt of all the surrounding strata. The canyon enters the east limb of the Spanish Valley salt anticline along the zone of collapse, accounting for the localized fracturing in the Wingate. Across the

Fig. D.1. Water flows from a pipe at the side of the road at Matrimony Spring along the contact between the shale of the Chinle Formation and the porous sandstone of the overlying Wingate Sandstone. The recent discovery of coliform bacteria has resulted in closure of this source.

river the Lower Jurassic Glen Canyon Group--composed of the Wingate, Kayenta, and Navajo formations (in ascending order)--is tilted 14° to the northeast. If the anticline were restored to a precollapse geometry, these strata would continue to climb higher to the west, culminating at the fold axis somewhere high above the center of the valley. Continuing west from this axis the beds would slope instead to the southwest, forming the west limb of the great fold. The remnants of this west limb are spectacularly exposed across the valley at the Portal, where the Colorado River again enters a narrow canyon.

0.1 Matrimony Spring is to the right, beneath the overhang, where water pours from the rock along the contact between the Chinle and the Wingate Sandstone (Fig. D.1). The spring forms as water percolates down through the porous Wingate until stalling at the impermeable shale at the top of the Chinle. Upon reaching this barrier, the water is forced laterally toward an area of reduced pressure, which in this case is the base of the canyon wall. Similar springs emit from canyon walls throughout the region, but rarely do they flow as voluminously and regularly as Matrimony Spring.

FIG. D.2. The fluvial Kayenta Formation, just above river level, displays stacked lenses of sandstone that represent individual river channels. The Wingate lies below, and the Navajo overlies the Kayenta. These three Lower Jurassic units make up the Glen Canyon Group.

0.4 Across the river are excellent exposures of the Kayenta Formation, which overlies part of the Wingate. Although both formations are dark red, the underlying Wingate forms smooth, continuous cliffs, whereas the Kayenta that caps the cliff forms irregular, broken cliff bands with numerous recesses. The fluvial Kayenta here displays wide lenses of sandstone and conglomerate separated by thin strands of siltstone (Fig. D.2). These lenses are cross sections of Kayenta river channels that drained mountains in Colorado (the Uncompahgre highlands), transporting their detritus west to a shallow seaway in Nevada.

The great, smooth cliffs of Wingate, which line most of this canyon, are composed of homogeneous, fine-grained eolian sandstone. The lack of any other rock types in the Wingate promotes the development of unbroken, vertical cliffs.

1.3 Across the river, the ledgy-weathering Kayenta plunges eastward into the subsurface to be replaced by the massive, concave walls of eolian Navajo Sandstone. Small, seeplike springs form horizontal black lines along these walls, their drips marked by black, mossy streaks and, in some places, patches of vegetation that cling to minute ledges. These seeps form as water percolates down through the porous Navajo. The lateral migration is likely driven by a thin silt or clay layer impeding the downward flow at the horizon that the water seeps from the wall (Fig. D.3).

Fig. D.3. A deep alcove formed in a dome of Navajo Sandstone by spring-sapping, an ongoing process in which water trickling out of seeps emitting from the sandstone dissolves the calcium carbonate cement, eventually crumbling the sandstone and developing recesses and alcoves where springs are located. These alcoves are common in Navajo Sandstone.

1.5 On the right (south), immediately above the road, are terraces of Colorado River gravel preserved on Kayenta ledges. These gravel deposits attest to an earlier time when the river bottom was at that level.

2.1 The roadcut on the right (south) side at eye level is the Kayenta-Navajo contact. This contact represents a change from the fluvial conditions of the Kayenta to the hyperarid, eolian conditions of the Navajo Sandstone.

2.4 As the road passes through the axis of the Courthouse syncline, the Kayenta rises back from the subsurface. Ahead it is seen dipping west, the opposite direction of when it was last seen. The Courthouse syncline is a major fold that, like the salt anticlines, has been intermittently active since the Permian. The west limb of this large downwarp is shared with the Spanish Valley anticline to the west, which you just passed through. The approaching east limb similarly is part of the Salt Valley anticline, which is coming up to the east and makes up most of Arches National Park. Evidence for the development of the syncline at the same time as the salt anticlines is in the large variations in unit thickness across this relatively small area. For instance, the Chinle is 334 feet thick on the southwest flank of the Salt Valley anticline and thickens to ~800 feet only 5 miles to the west in the axis of the syncline. Downwarping concurrent with adjacent uplift is explained by the lateral and vertical flow of the thick underlying Paradox salt. Salt flowed laterally to the southwest and northeast from beneath the Courthouse syncline, causing overlying strata to be slowly folded downward. That same salt rose to form the early-stage upwarps of the Spanish Valley and Salt Valley anticlines. Specific evidence for such activity during the Late Triassic is provided by the dramatic thickness changes in the Chinle. Similar variations are seen in Permian and Jurassic strata.

3.0 The trailhead and parking area for Negro Bill Canyon is on the right (south) side of the road. The trail begins in ledges of the Kayenta Formation. The steep walls and domes towering over this canyon are part of the overlying Navajo Sandstone.

3.9 Across the river looms a steep wall of Wingate capped by Kayenta that exhibits well-defined lenses of ancient river channels.

4.6 Due to the westward dip of the strata, increasingly older rocks are visible here, with patches of dark red Lower Triassic Chinle sandstone and siltstone appearing on the lower slopes between mounds of talus from overlying strata.

5.0 The Chinle Formation makes up the lower red slopes on both sides of the canyon. More-resistant fluvial sandstone and conglomerate form ledges; siltstone and shale, the remnants of floodplains, weather into slopes. The westward dip of these strata reflect gentle tilting on the west flank of the approaching Castle Valley salt anticline.

6.1 Across the river the lower half of the slope is composed of the Lower Triassic Moenkopi Formation, which from a distance looks identical to the overlying Chinle. These two formations are easily divisible only where a thick white sandstone unit marks the base of the Chinle. Although this sandstone can be seen halfway up the slope, it is not as obvious elsewhere in the area.

7.0 Hal Canyon BLM recreation area. The massive promontory across the river is cut by a small fault at its tip but provides a clear review of the Mesozoic stratigraphy seen so far. The dark red Moenkopi Formation—the product of rivers, tidal flats, and sabkhas—makes up the lower part of the steep slope. The base of the overlying Chinle, also a slope-former, is evident by the white sandstone several hundred feet above the river. The Chinle gives way abruptly to the broken cliffs of Wingate Sandstone, which here is shattered due to the adjacent fault. Fluvial deposits of the Kayenta top the Wingate cliffs.

7.8 There is a popular bouldering area on the right side of the road in a jumble of blocks of Wingate Sandstone. The source for these blocks can be

Fig. D.4. This pronounced angular unconformity within the Upper Triassic Chinle Formation indicates salt movement and associated deformation during Chinle deposition.

seen in the large recess in the cliffs directly above this boulder pile. Such **mass wasting** processes, as the large-scale downhill movement of earth under the influence of gravity is known, is a vital process in canyon formation. As rivers slice downward, deepening the canyon, the weakened and unsupported canyon walls recede through large-scale mass movements such as this rockfall.

9.1 A small fault bisects the canyon wall directly ahead and across the river. The fault is marked by a small recess in the cliff. On this minor, northwest-trending fault the northeast side has dropped slightly relative to the southwest side.

9.6 The small side canyon across the river marks a second northwest-trending fault, this one on the southwest side. Together these two related faults form a **graben**, a central, downdropped block bounded on each side by a fault and an upthrown block. Like the pervasive tilting of strata throughout the area, faulting likely is related to the movement of salt in the subsurface.

10.2 Takeout beach. The Chinle Formation is well exposed at road level. The formation has thickened in the axial part of the syncline with the preservation of earlier basal strata that are not present elsewhere in the area. This fluvial sandstone and conglomerate form thick, resistant ledges at the road and river level.

10.8 Across the river at road level is a spectacular angular unconformity separating the lower and the upper parts of the Chinle Formation (Fig. D.4), attesting to the forces of salt deformation. Here the lower Chinle was deposited and then locally forced upward by salt movement, likely soon after deposition. Subsequent erosion beveled the deformed surface flat. This was followed closely by deposition of the upper Chinle over the top. Such angular relations *within* a formation are uncommon elsewhere and indicate a rapid rate of salt deformation during the Late Triassic. Although these types of relations are evident throughout the salt anticline region in strata ranging from Permian to recent age, nowhere are they so clearly displayed.

11.2 The side canyon tributary of Salt Wash enters across the river to the north. Salt Wash drains a large part of Arches National Park. On the south side of the road are exposures of the basal mottled strata of the Chinle Formation. These rocks consist of varicolored fluvial conglomerate and sandstone, and associated floodplain siltstone and shale. The intense mottling in various shades of red, purple, gray, and white was caused by seasonal fluctuations in the Late Triassic water table during alternating wet and dry periods. Locally these rocks contain great concentrations of vertical, tubelike burrows 3 to 4 cm in diameter. The bleached nature of these burrows contributes to the mottled appearance. These trace fossils are thought to have been the dwellings of lungfish or crayfish that burrowed deeply into the unconsolidated sediments in an effort to remain below the fluctuating water table (see Chapter 5 for a detailed discussion).

13.0 The Chinle is exposed directly ahead in the steep slopes on the north side of the river. The obvious westward dip of the Chinle is the result of salt uplift in the approaching Cache Valley and Castle Valley anticlines. The contact between the Chinle and the overlying Wingate Sandstone is yet another angular unconformity.

13.3 Just ahead, the slightly folded mass of thinly bedded red rocks is composed of the Moenkopi Formation, although a small part of the peak is a remnant of the Chinle. These rocks form the north end of the Castle Valley salt anticline and were deformed by the rise and subsequent collapse of that structure. The broad, flat-bottomed Castle Valley is just south of these Moenkopi hills. Looming to the east, behind the hills, is the gently north-sloping rampart of Parriott Mesa. Strata exposed on this mesa are, from bottom to top, slopes of Moenkopi and Chinle formations capped by vertical cliffs of Wingate Sandstone and Kayenta Formation.

14.0 Bridge crosses Castle Creek, which emerges from the north end of Castle Valley salt anticline. The creek cuts through the Moenkopi.

14.4 The road hugs the dark red Moenkopi Formation, here composed of thinly bedded sandstone and shale of probable tidal flat origin.

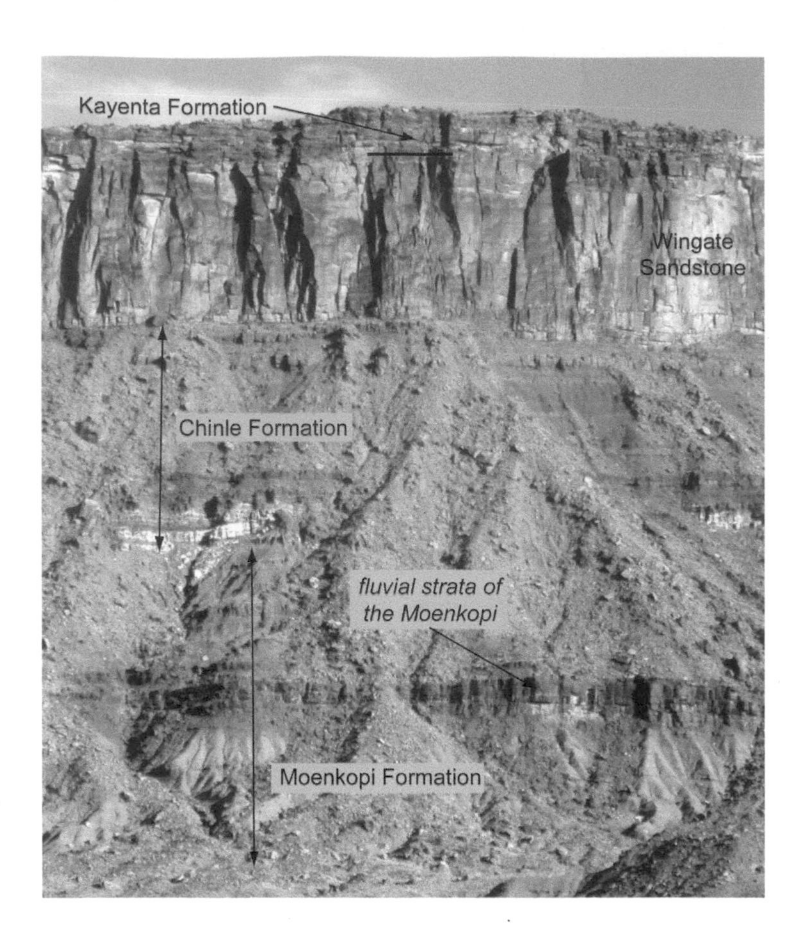

FIG. D.5. View of Dry Mesa, across the Colorado River to the north, exhibiting the Mesozoic strata for this part of the road log.

15.4 The turnoff to the south leads to Castle Valley and the La Sal Mountain loop road, which takes a circuitous route back to Moab. The loop road traverses the north and west flanks of the La Sal Mountains and provides access to trailheads and four-wheel-drive roads into this unique alpine environment. Castle Valley is a large collapsed salt anticline valley that is located ~2 miles south, just over the red Moenkopi hills. Looking back to the southwest, at the striated red hills that you just passed beneath, the buckled Moenkopi can now be seen. These strata mark the northern end of the Castle Valley salt structure. They were folded and faulted as the underlying Paradox salt was confronted by groundwater and partially dissolved. This northern terminus of partially collapsed Moenkopi differs from the mature, flat-bottomed Castle Valley. These steep, deformed hills are still in the early stages of salt valley anticline evolution.

Left of the Castle Valley road, to the southeast, is the towering bulk of Parriott Mesa. To the north, across the Colorado River, is its mirror image, Dry Mesa. Both clearly display the Mesozoic succession familiar from the earlier, narrow parts of the canyon. Looking north at the bold ramparts of Dry Mesa, the lower half is occupied by the Lower Triassic Moenkopi Formation (Fig. D.5). The basal rounded, pale orange slopes

Fig. D.6. Profile of Castleton Tower (right) and the Rectory (left) in an evening thunderstorm. Both features are hewn from the Lower Jurassic Wingate sandstone of eolian origin.

consist of gypsum-rich, thinly bedded sandstone and mudstone of tidal flat origin. The prominent black-stained, red, narrow cliff band above is fluvial sandstone that splits the Moenkopi. Recent work by Western State College geology students indicates that these rivers ran northwest, parallel to and along the flanks of the Castle Valley salt anticline, suggesting that it was a positive area at this time. Colonnades of thin-bedded Moenkopi resume above this pronounced break. The basal sandstone of the overlying Chinle Formation is the obvious white unit that stands out in the mass of orange-red, ledgy slopes. The mostly red Chinle ledges and slopes rise to meet the base of the vertical red Wingate cliffs. Capping the Wingate is a relatively thin veneer of the more broken, cliff-forming Kayenta Formation.

16.0 Mounds of Moenkopi on the south side of the road are topped by well–rounded cobbles from the Colorado River of many thousands of years ago, before it had cut down to its present level. On the north side is a gravel pit in these deposits. Such pits are common along the Colorado River because they provide clean, readily accessible material for road construction.

FIG. D.7. View to the north of the south end of the Cache Valley salt anticline. In this particular segment the collapse is incomplete, providing a snapshot of the process. The dashed line at the base of the Jurassic Wingate Sandstone provides a stratigraphic reference point.

16.5 The Rocky Rapid turnoff to the left (north) leads to the banks of the Colorado River. To the south, between Parriott Mesa to the west and the unnamed butte on the east, the long, discontinuous ridge of Wingate Sandstone forms the rock formations of the Priest and the Nuns, the Rectory, and the slender finger of Castle Rock (a.k.a. Castleton Tower), all popular rock climbing destinations (Fig. D.6 and Plate 11).

To the north, across the Colorado River, the Moenkopi, Chinle, Wingate, and Kayenta have experienced large-scale faulting and folding. These mountain-sized blocks of twisted strata are in the slow process of collapsing into the dissolving underlying salt of the Cache Valley salt anticline (Fig. D.7). As the salt pushed its way up to the near surface, the heavily fractured overlying rocks allowed water to trickle down and slowly dissolve the uppermost salt. The east-west trend of Cache Valley differs in orientation from the other, northwest-trending salt anticlines (Fig. D.8). Cache Valley provides a continuous link between the Salt Valley anticline to the north and the Fisher Valley/Sinbad anticline to the southeast. As at the termination of Castle Valley, these rocks are a younger, partially collapsed part of a large-scale salt anticline. Cache Valley has not yet evolved into a mature, flat-bottomed valley. Instead, it furnishes a snapshot of the

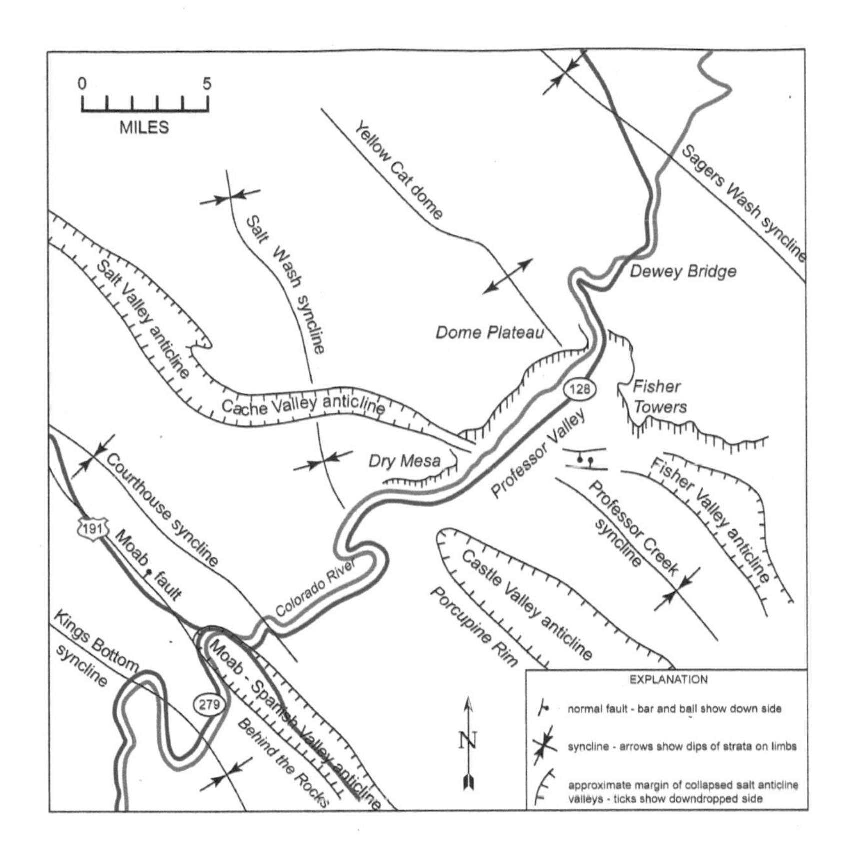

FIG. D.8. Map showing the road log route and important salt tectonic elements, including the collapsed salt anticlines and associated faults and salt withdrawal synclines. Geographic features are labeled in italics. Modified after Doelling and Ross 1998.

collapse process, one that took place long ago in the more mature Salt Valley and Castle Valley anticlines.

17.6 The large hills directly north are folded and faulted Moenkopi Formation in a continuation of the Cache Valley collapse zone.

17.9 Thinly bedded red sandstone, shale, and mudstone of the Moenkopi are well exposed in cross section on the north side canyon wall. Gracefully sinuous ripples mark the bedding surfaces of these rocks, attesting to deposition in shallow water with currents that traveled back and forth, probably a tidal flat margin of the shallow sea that lay immediately to the west (Fig. D.9).

18.4 The ranch road on the right (south) provides a convenient place to pull over and get oriented to the new surroundings. After emerging from the disruption of the Cache Valley salt anticline, the Colorado River can be seen below the walls to the northeast. This river corridor is known as Professor Valley, named for an early settler, Dr. Sylvester Richardson, who was known locally as "Professor." Prior to the construction of Dewey Bridge or any reliable road, supplies for the community of Castle Valley were brought by railroad to Cisco. From there supplies were hauled to the Colorado River and then floated on rafts to the mouth of Professor Creek, where Richardson built a cabin in 1879. After Richardson moved his home farther up Professor Creek, the old house became a store, and a hotel was

Fig. D.9. Symmetric ripples on a bedding plane in the Triassic Moenkopi Formation in a wash below Parriott Mesa.

added. Many of the features seen from this vista were named for Dr. Sylvester Richardson.

Along the river is the Permian Cutler Formation, discerned from overlying red beds by its light orange color. The Cutler is the oldest unit seen so far and consists predominantly of fluvial sandstone and conglomerate, and lesser eolian sandstone. During Cutler deposition in the Permian Period, this view would have been of the mountainous Uncompahgre highlands looming several miles to the east. In the foreground would have been vigorous whitewater streams tumbling from the heights, shuffling the detritus of erosion westward, ever closer to the shifting shoreline of the sea somewhere to the west. Dispersed between these foamy rivers were small, patchy dune fields, a testimony to the arid Permian climate.

The contact here between the Cutler and the overlying Moenkopi is an angular unconformity, the result of tilting of the Cutler by the upward push of salt prior to Moenkopi deposition. The Cutler beds exposed along the lower slopes of Professor Valley are tilted slightly to the west and are truncated by the darker Moenkopi strata. The overlying rocks are the Chinle, Wingate, Kayenta succession that is the motif for this area. White slickrock knobs of Navajo Sandstone can be seen capping the Wingate-Kayenta cliffs in some places along this unbroken rampart.

The broad, cliff-rimmed bowl that surrounds the road is known as the Richardson Amphitheater, also named for "the Professor." To the east and southeast is the upwarp of Fisher Valley anticline in an uncollapsed segment. Its crest is sliced by the spectacular canyon of Onion Creek. The bristling fins and pinnacles along the crest and flanks of the fold are carved from the sandstone and conglomerate of the Cutler. Known as the

FIG. D.10. Stratigraphy of the Fisher Towers area. Most of the actual towers are composed of the Permian Cutler Formation. The taller towers are capped by remnants of the darker Moenkopi Formation.

Fisher Towers, these massive pinnacles seem to appear and vanish with the changing light (Fig. D.10).

Behind the salt-cored fold, farther southeast, the upper reaches of Fisher Valley widen into a mature, flat-bottomed salt collapse valley. This particular segment is unique because the Paradox evaporates have actually punched up through the valley floor to be exposed as intricately folded gray-white mounds (Fig. D.11 and Plate 12). Although it was salt that drove this ascent, no salt is present in these mounds. As the Paradox rose to the surface, the salt was easily dissolved, leaving the less-soluble interbedded limestone, gypsum, and shale exposed in the mounds. The name "Onion Creek" comes from the bad taste of the water caused by the dissolved solutes.

Immediately west (right) of the slot of Onion Creek are two large tilted knobs capped by buff-colored Wingate Sandstone. These blocks were separated and downdropped from the adjacent Fisher Mesa by normal faults. These downfaulted blocks delineate the southwest margin of the Fisher Valley salt anticline.

South and west of Fisher Mesa rise the forested slopes and rocky alpine summits of the La Sal Mountains. The range culminates at Mt.

Fig. D.11. Aerial view of the contact between the diapir of Paradox evaporite rocks, in this case white gypsum, and the red-orange Cutler Formation in upper Fisher Valley. Note the road to the left, in the bottom of the valley, for scale.

Peale with an elevation of 12,721 feet, although from this vantage point it is hidden by the northern cluster of La Sal peaks. Mt. Waas (12,331 feet) is the prominent high point seen from here.

The La Sal Mountains are composed of a variety of igneous intrusive or plutonic rock types in the form of laccoliths. These mushroom-shaped plutonic bodies of magma were injected from below into shallow levels of the sedimentary succession. According to radiometric dates obtained from the La Sal Mountains by Nelson and others (1992), these rocks were emplaced between 27.9 and 25.1 Ma. Since intrusion and crystallization, erosion has eaten away most of the overlying sedimentary rocks to expose the resistant granitic core of the laccoliths.

The laccoliths of the La Sal Mountains are the same age and composition as those that make up the nearby Abajo and Henry mountains and share a similar origin. During the same interval that these plutonic bodies were being injected into the shallow subsurface (~29 to 25 Ma), large volcanoes of the San Juan volcanic field in southwest Colorado were spewing immense volumes of hot ash across the region. The origin of all this magma is a topic of much debate, but it is likely that these seemingly discrete events have a shared genesis.

18.8 Crossing of Professor Creek. The headwaters of the creek, tucked between Fisher Mesa (east) and Adobe Mesa (west) to the south, are in Mary Jane Canyon. The floor of the surrounding Richardson Amphitheater is almost completely within the Cutler Formation.

20.1 To the south is Onion Creek Road, a graded dirt road that cuts southeast through the Fisher Valley anticline axis and into the wide collapsed valley behind it. The road is often rerouted as parts are washed out by the occasional flashflood that gushes through the narrow corridor of Onion Creek. Its foul-tasting water picks up dissolved ions as it flows through the Paradox evaporates. The water, which contains a high sulfur and arsenic content, rises up as groundwater through the core of the collapsed part of the anticline. It makes its way to the surface at a point called Stinking Spring, another clue that the water should not be consumed.

The maze of fins and pinnacles in the distance, to the southeast, very clearly shows the anticline geometry. The Cutler on the west limb (right side) dips to the southwest, while that on the east (left) side dives to the northeast. The bristling array of towers that extend from the wall along the northeast limb of the fold are known collectively as Fisher Towers. Numerous deep canyons descend from this area to intersect with Onion Creek and are readily accessed along this road. Most of the intricately fluted towers are hewn from the Cutler Formation, although some of the higher ones are capped by the Moenkopi. The contact between the two formations can be seen in the upper part of the tallest tower, known as the Titan. The pale red of the Cutler is overlain by the darker, brown-red beds of the Moenkopi Formation.

The search for petroleum associated with these salt anticlines led to the drilling of several wells a few miles down this road. Although no petroleum was discovered, these wells have provided a vital glimpse into the subsurface of this part of the Paradox basin. One of the wells was drilled through an astounding 12,360 feet of coarse arkosic sandstone (Chenoweth and Daub 1983)—more than 2 vertical miles of the same type of sediments currently exposed in the Cutler Formation at the surface! Below this was a 40 foot zone of disrupted sediments interpreted to be a horizon that was formerly occupied by salt. At some point it is thought that salt was squeezed laterally by the pressure of the overlying sediments and rose in the adjacent Fisher Valley anticline. Some thin tongues of limestone were identified within this arkose, suggesting that a shallow sea crept in from the west temporarily.

Another well in this area drilled through more than 5000 feet of arkosic limestone below the Cutler arkose. This interval is interpreted as the easternmost facies of the Pennsylvanian Honaker Trail Formation. It is underlain by another 1700 feet of Paradox salt. The extraordinary thickness of the Honaker Trail in this area is interpreted by some to be due to the lateral flow of salt during its deposition, locally increasing subsidence and therefore providing vertical space for sediment accumulation during Honaker Trail deposition (Chenoweth and Daub 1983).

20.6 Crossing of Onion Creek, one of the rare perennial streams to flow in this region. The white crust along the red sand banks consists of salts precipitated from the mineral-laden streamwater.

21.0 Turnoff to the south. This road leads to the BLM campground and the Fisher Towers Trail. This spectacular 3-mile trail (one way) winds in and out of the fluted fins and towers, offering a close-up look at the Permian Cutler Formation. The view to the west of the trail, across the maze of Onion Creek tributaries, is a panorama of Castleton Tower and the Nuns and Priest, all hewn into the narrow ridge of fractured Wingate Sandstone lining the northeast flank of the Castle Valley salt anticline.

22.0 As the road dips eastward, the roadcuts and surrounding erosional remnants expose sandstone and conglomerate of the Cutler Formation.

23.2 Hittle Bottom boat launch and BLM campground. In the early 1900s Tom Kitsen and his family lived here. Kitsen used a team of horses to carry the mail from the post office in Cisco to Castleton. This place was the half-way stop where he changed his team of horses. The grave site at the entrance to the campground was that of his mother, who fell in the river, caught pneumonia, and died. It later became the site of Hittle ranch, for which the campground is named.

Straight ahead is one of the last continuous exposures of the Cutler through Kayenta succession before the road exits Richardson Amphitheater. The pale orange-red Cutler at the base is overlain by the red, horizontally striped, ledgy cliffs of the Triassic Moenkopi Formation. Recent work on this Moenkopi section by Western State College geology students has shown that the rivers that deposited the middle part of the formation (Ali Baba Member) flowed north-northwest, parallel to the trend of the salt anticlines. This is contrary to the southwest flow that would be expected from rivers that spilled off the last vestige of the Uncompahgre highlands. This deviation in flow direction is attributed to the topographic presence of the Fisher Valley salt anticline during the Early Triassic, deflecting the rivers to the northwest around the structure. Ongoing studies are looking at the effect of salt uplift on the Moenkopi elsewhere in the region. The contact between the Moenkopi and overlying Chinle Formation is obscure due to the absence here of the Chinle's white basal sandstone marker; however, the fluvial Chinle Formation begins in the middle of the talus-covered slope above the Moenkopi cliffs. The

Chinle slopes are overshadowed by the red and black desert-varnished walls of eolian Jurassic Wingate Sandstone. Perched atop these vertical walls is the ledge-forming fluvial Kayenta Formation. Ahead the river bends north, as does the road, entering a narrow canyon hemmed in by the towering Wingate. At this point the salt anticlines are left behind.

24.0 As the road enters the canyon, the lighter Cutler gradually disappears beneath a cover of detritus that has fallen from the cliffs above. The thin-bedded nature and numerous ripple marks in the lower part of the overlying Moenkopi attest to its tidal influence. Before continuing, pause to look back down the road for one of the more spectacular views in the region, with the Colorado River, Fisher Towers, and the La Sal Mountains dominating the scene. This panorama was used in a tobacco ad several years ago, complete with a cigarette-smoking cowboy strategically placed next to a glowing campfire on the sand bar. Unfortunately, the scale of the superimposed cowboy required that he be ~200 feet tall to fit the true scale of the river and the surrounding cliffs.

24.6 Across the river is the last exposure of the Cutler before it plunges into the subsurface. The bedding in the surrounding walls and ahead in the canyon shows an obvious northeast dip. This is the southwest limb of the northwest-trending Sagers Wash syncline, whose axis lies farther to the east, east of Dewey Bridge. It is difficult to tell if the syncline is related to salt withdrawal or the Tertiary uplift on the Uncompahgre highlands, which lay a short distance east. This is, however, the zone that during the Pennsylvanian Period would have been flooded by an influx of coarse clastic sediment off the rising highlands. This would have prevented significant evaporite deposition along this eastern margin of the Paradox basin (see Fig. D.12). It is more likely that the Sagers Wash syncline was generated by Tertiary rejuvenation of the Uncompahgre highlands.

26.5 The Chinle Formation is at road level here, but none of the members that comprise it elsewhere can be recognized in this area. Instead, it is a succession of red sandstone and siltstone derived from the low hills of the Uncompahgre highlands. Across the river the desert-varnished walls of Wingate Sandstone display the large-scale cross-stratification that characterizes these eolian dune deposits.

26.9 Across the river the vertical Wingate walls are interrupted by the shadowy slot of Yellowjacket Canyon.

29.0 On both sides of the river at road level is the red Kayenta Formation. Its smaller-scale cross-stratification and uneven cliff and ledge style of erosion highlights the stacked lenses of river channels that were responsible for its deposition. These lenses represent sand-choked, braided rivers that swept westward off the nearby remnants of the Uncompahgre highlands.

FIG. D.12. The Jurassic succession is well exposed at Dewey Bridge, the type locality for the Dewey Bridge Member of the Carmel Formation. The view is to the west, across the Colorado River and Dewey Bridge, which is no longer in use for vehicles. The white wooden bridge recently burned when a fire was started in the vegetation below it.

Buff-colored domes of eolian Navajo Sandstone overlie the Kayenta, indicating a return to the regional eolian system that dominated earlier during Wingate time. The Navajo thins here as it begins its eastward pinchout against the highlands to the east.

Directly ahead, across the river, the surviving formations of the Middle Jurassic San Rafael Group rest on the Navajo (Fig. D.12). Their obvious northeast dip is toward the axis of the Sagers Wash syncline. Pinnacles extending from the butte are the Dewey Bridge Member of the Carmel Formation. This is the type locality, where the member was defined and for which it was named. The base of this orange and white unit exhibits the convolutions that are characteristic of these rocks throughout this region; the upper part displays weathered vertical joints. The Dewey Bridge is succeeded by the eolian-derived yellow-white Entrada Sandstone, which weathers to a rounded slickrock bench. The thin white band of the Moab Member of the Curtis Formation crowns the butte. Above this should lie shale, sandstone, and siltstone of first the Summerville Formation and then the Tidwell Member of the Late Jurassic Morrison Formation. Here, however, these are covered by detritus from the overlying blocky brown ledges of the Salt Wash Member of the Morrison. Fluvial deposits of the Morrison are world-renowned for dinosaur fossils and commonly host uranium mineralization.

29.3 Looming over the road here is the Kayenta Formation. A close look reveals cross-stratification of various scales, types, and geometries.

29.7 The Dewey Bridge BLM campground is on the north side of the road.

29.8 Colorado River crossing. The old Dewey Bridge, built in 1916, is the white suspension bridge immediately to the east. Prior to its construction, this was one of the main river crossings between Cisco and Moab, and it was serviced by a ferry. Even after the bridge was built for a price of $25,000, its use was restricted by its 8 foot width and a weight limit. In 1985 $2.7 million was provided by the state to build the new concrete, multilane bridge in use today (Firmage 1996).

30.0 To the left is Kokopelli's Trail, a popular mountain bike route that winds east to the Fruita/Grand Junction area. This area immediately north of the river was a failed real estate venture called "Rio Colorado." Several years ago the land was purchased by a developer who proceeded to build a model home next to the river and a water treatment plant, and carved a loop road through the subdivision. After several years of losses, the enterprise went bankrupt. In a remarkable collaborative effort between concerned local residents, Grand Canyon Trust, the Nature Conservancy, and Utah Open Lands the subdivision was eventually purchased for preservation. This group has since worked hard to erase all traces of the subdivision and restore the land to its earlier, less-disturbed state.

30.3 On the left is a terrace of coarse gravel that butts against sandstone of the Dewey Bridge Member. These rounded cobbles evoke a time when the terrace was the riverbed of the Colorado River and the wall of Dewey Bridge would have formed the steep north bank.

To the south, across the river, the buttress-forming sandstone walls continue to show the pronounced northeast dip associated with the Sagers Wash syncline. The Dewey Bridge Member, the Entrada Sandstone, and the Moab Member make up the steep sandstone cliff. Once again the Summerville and Tidwell are concealed beneath blocks of the Salt Wash Member of the Morrison.

31.0 The Dewey Bridge-Entrada contact at road level is marked by a recess between the two slickrock-forming sandstones. The recess is formed by the easily eroded silty sandstone beds at the top of the Dewey Bridge Member.

31.2 The resistant sandstone that makes up the Moab Member of the Curtis Formation dips below the rubble at road level. It is replaced by the rubble-covered slopes of the Summerville Formation, which marks the top of the San Rafael Group, and the lithologically similar Tidwell Member of the Morrison Formation. This combined succession is a heterogeneous assemblage of ledge-forming sandstone, limestone, and siltstone, coupled with beds of slope-forming red, green, and purple shale. Sitting above these slopes are discontinuous brown cliffs of conglomeratic sandstone that make up the fluvial Salt Wash Member of the Morrison Formation. The regulated trickle of the Dolores River enters the Colorado on the south side.

31.6 The northeast dip of the Salt Wash brings it to road level.

32.1 The valley widens as the bright green shale of the Brushy Basin Member of the Morrison is exposed at road level. The shale beds represent floodplain deposits of low-energy rivers, and the associated sandstone beds mark the position of river channels.

32.5 The basal sandstone of the Lower Cretaceous Burro Canyon/Cedar Mountain Formation drops to road level. Conglomerate in these blocky exposures contains pebbles of green mudstone, presumably from erosion of the underlying Brushy Basin Member.

33.2 The valley again opens as the road approaches the Upper Cretaceous strata, including the Dakota Sandstone, Mancos Shale, and Ferron Sandstone.

33.7 On the left (west) side of the road are blocks of tan, cross-stratified Dakota Sandstone. This basal part of the Dakota was deposited in east-flowing streams that originated in the Sevier orogenic belt, along what is today the Utah-Nevada border. By this time the Uncompahgre highlands, which earlier had such an influence on sedimentation, were buried. To the east the Cretaceous strata can be seen dipping towards the road, to the southwest. This is the northeast limb of the Sagers Wash syncline. The axis of this large structural trough lies between this point and the southwest-dipping strata, and follows the upcoming drainage of Sagers Wash.

34.1 Mounds of carbonaceous siltstone and shale on the left (west) side of the road are in the middle part of the Dakota and were laid down in a coastal swamp environment.

34.4 Sagers Wash and the axis of the syncline. The wash below the bridge is floored by white and yellow Dakota sandstone beds with wispy, black carbonaceous laminations. When broken, the sandstone contains clots of black macerated plant material that have been reduced to carbon. The abundant plant material suggests a sandy delta-swamp environment for this part of the Dakota.

La Sal Mountain Loop Road from Castle Valley (Utah Highway 128) to the Junction with U.S. Highway 191

This paved road heads south from the Colorado River through Castle Valley and eventually climbs onto the north flanks of the La Sal Mountains. It continues along the west flank of the extensive sky island, providing numerous entry routes into these mountains before descending westward to U.S. Highway 191, where this road log ends. From this junction, Highway 191 can be taken north to Moab or south to the Needles District of Canyonlands or the towns of Monticello, Blanding, and Bluff.

Geologically this road log begins at the north end of the collapsed salt anticline of Castle Valley and traverses its length southward. From Castle Valley it rises through deformed and metamorphosed Mesozoic strata on the flanks of the La Sal Mountains intrusive center. The road traverses the bench that forms the western flank of these laccoliths before dropping into the Moab–Spanish Valley salt anticline valley.

Road Log

0.0 Turn south off of Utah Highway 128 onto the La Sal Mountain loop road. To the right (southwest) is the folded Triassic Moenkopi Formation, which forms the north end of the Castle Valley salt anticline. This deformation is related to the rise and subsequent collapse of subsurface Paradox evaporates. Ahead on the left (southeast) is the truncated prow of Parriott Mesa. The strata that comprise this massive façade from base to top include the Moenkopi Formation at road level and part way up the basal slopes, overlain by upper slopes of the Triassic Chinle Formation. The headwall of vertically fractured red rock that makes up the main

FIG. E.1. Stratigraphy of the northwest part of the western flank of the Castle Valley salt anticline showing the prominent eolian facies at the top of the Cutler Formation. This localized eolian unit pinches out to the south over a distance of less than 1 mile and is not present on the east side of the valley. For scale, houses at the base of the Cutler cliff are denoted by an H.

escarpment is the eolian Wingate Sandstone of Jurassic age. The mesa top is veneered by terraced benches of fluvial sandstone that make up the Jurassic Kayenta Formation.

1.7 The road drops into Castle Valley, a large northwest-trending salt anticline valley. The town of Castle Valley spreads across the valley floor and dots the lower parts of the western rampart. Unlike most of the canyons throughout the region, this wide valley was not carved by the incessant flow of water. Instead the central axial part of the fold collapsed as the shallowly buried core of rising salt was dissolved by groundwater. Thus, what began as a regional-scale linear highland evolved through time into a valley by large-scale collapse. The crest of the imposing western wall is Porcupine Rim; the dissected slickrock tableland behind it hosts numerous mountain bike trails. Below the undulating crestline the wall provides a panoramic mural that reveals some interesting and unique stratigraphic relations. The lowermost red and white slickrock cliffs are thick eolian deposits at the top of the Permian Cutler Formation (Fig. E-1). Similar eolian deposits are missing only 1 mile across the valley, on the east flanks, and they are absent to the west in sporadic exposures between Castle Valley and the Island in the Sky District of Canyonlands. These

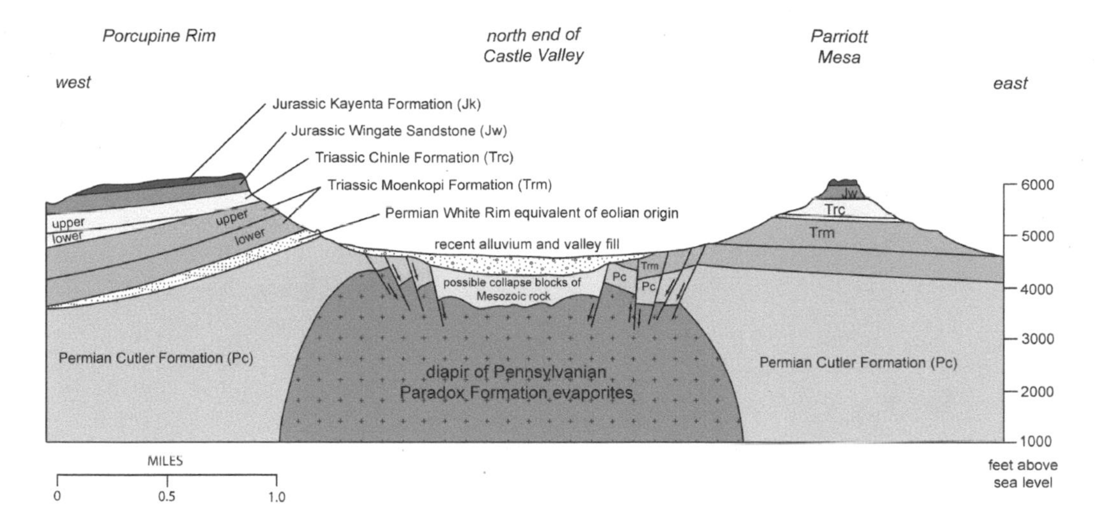

FIG. E.2. Geologic cross section through the north end of the Castle Valley collapsed salt anticline. Note that the strata presently bounding the collapsed valley previously extended as an anticline upward and across the valley; however, because the Paradox evaporites were dissolved in the subsurface as water infiltrated downward through the fractured Mesozoic strata, the axial part of the anticline gradually collapsed into the underlying void. Modified after Doelling and Ross 1998.

deposits are in the same position as the eolian White Rim Sandstone, which forms the top of the Permian succession 35 miles to the southwest, in Island in the Sky. They are likely an isolated outlier related to the larger White Rim erg and represent the eastern margin of a separate but related dune field that formed as sand blew eastward on the prevailing winds. These deposits were preserved in the subsiding syncline that developed between the rising Castle Valley anticline to the east and the Moab–Spanish Valley anticline to the west. As the sand bounced across the crest of the Moab–Spanish Valley upland, it became trapped in the sinking trough between the folds. This handily explains the absence of eolian deposits along the east limb of the Castle Valley fold. The sand simply was unable to make it over the crest of the Castle Valley anticline. If these eolian beds are traced southward along the west flank wall, they can be seen to pinch out beneath the Moenkopi Formation in an angular relationship, another common product of salt uplift and erosion. The Permian strata were tilted by the continued push of salt from beneath and subsequently beveled off by erosion before being blanketed by the Lower Triassic Moenkopi Formation. The Moenkopi is overlain by the similar red, slope-forming Chinle Formation, the continuous crimson escarpment of Wingate Sandstone, and a thin veneer of discontinuous Kayenta Formation (Fig. E.2).

1.9 Directly ahead looms the North Mountain group of the La Sal Mountains—a large, isolated cluster of laccoliths composed mostly of the igneous rock diorite. The magma from which this rock crystallized was

Fig. E.3. Round Mountain, a satellite intrusion of resistant igneous rock in the middle of the Castle Valley salt anticline. The low mounds forming a rim around the intrusion are composed of Pennsylvanian and Permian sedimentary rocks.

injected into the sedimentary rocks of the Castle Valley/Paradox Valley salt anticline ~25 to 29 million years ago. This lengthy salt anticline and the others throughout the region were controlled by faulted weaknesses in the deep basement rocks. In this case, these same weaknesses were exploited by the ascent of molten rock through the thick, otherwise stable crust of the Colorado Plateau. In front of the La Sals, protruding from the middle of flat-bottomed Castle Valley, is the gray, tree-covered knob of Round Mountain (Fig. E.3). This lone hump of igneous rock is a small intrusive body associated with the main La Sal Mountains mass. The more-resistant igneous rock rises from the easily weathered sedimentary rocks of the salt anticline valley.

2.7 The north end of Parriott Mesa rises to the skyline on the left (east). Directly ahead is the slender monolith of Castleton Tower, its vertical cracks often speckled with rock climbers. Left (east) of this free-standing pillar is the serrated butte that includes, from left to right, the Priest, Nuns, and the Rectory (Fig. E.4). All the monuments and spires along this ridgeline are hewn from Wingate Sandstone, with their greatest heights capped by a thin catwalk of Kayenta.

3.4 To the left (east) the lowermost orange slopes are in the Cutler Formation. The Cutler shows a pronounced tilt to the northeast and is overlain by the darker red slopes of the Moenkopi. Here, as across the valley, the contact is an angular unconformity, with the base of the Moenkopi made

Fig. E.4. From left to right, the Priest, Nuns, the Rectory, and the slender pinnacle of Castleton Tower on the east limb of the Castle Valley salt anticline. These towers are composed of the cliff-forming Wingate Sandstone with a thin cap of Kayenta Formation. The base of the Moenkopi Formation is marked by a thin, white bed of gypsum. Similarly, the base of the Chinle Formation is shown by a thin, white sandstone bed.

clear by a thin marker bed of white gypsum on the Cutler. The angular relationship is distinct as Cutler beds rise to the south, causing increasingly older strata to be truncated by the gypsum in that direction.

4.0 Lower craggy outcrops on the left (east) are fluvial sandstone and conglomerate of the Cutler. Above them is an excellent view of Castleton Tower.

5.5 Crossing of Castle Creek, the dominant drainage in the valley. Most of the creek's water is diverted upstream into an irrigation ditch that provides water throughout the valley. Castle Creek joins with Placer Creek about 3 miles downstream. Placer Creek skirts the west side of the valley below Porcupine Rim.

5.6 To the right (west) is Round Mountain, elevation 6184 feet. The top bristles with fractured diorite and is fringed by steep talus slopes dotted with piñon and juniper trees. Low, undulatory mounds on the south side are rocks of the Moenkopi, Cutler, and Pennsylvanian evaporates of the Paradox Formation. These rocks were dragged up with the rising magma that formed Round Mountain. On the left (east) side of the road is Adobe Mesa, built on strata that range from Permian Cutler Formation on the valley floor to the Jurassic Kayenta Formation along its tree-covered top.

6.3 Surficial gravels on both sides of the road are in the Placer Creek Formation and are derived from the La Sal highlands that tower ahead. The

Placer Creek Formation makes up the floor of this part of the valley. Boulders and cobbles that comprise it consist of diorite from the igneous core of the La Sals and an approximately equal component of Mesozoic sandstone clasts. These sandstones were hardened by contact metamorphism during the injection of magma that upon crystallization formed the laccoliths. The heat of the magma essentially baked the surrounding host sedimentary rocks, hardening them. The clasts in the Placer Creek Formation are only slightly rounded due to their hardness and short travel distance.

7.6 Mounds of gravel on the right (west) are in the Quaternary age Harpole Mesa Formation. Ahead and to the left, the south end of Adobe Mesa rolls downward so the Wingate dips southward, toward the La Sal Mountains.

7.7 On the right (west) side of the road is a gray-green slice of gypsum and shale from the Paradox Formation that has squeezed up, probably along a fault related to the collapse of Castle Valley.

7.9 The imposing peak that dominates the south end of Castle Valley is Grand View Peak, elevation 10,905 feet. The bulk of this laccolith is composed of diorite.

8.6 Town of Castleton. Though today it is more of a community than a town, in its early incarnation in the late 1800s, it was a boomtown. Initially settled in the late 1870s by prospector Doby Brown, by 1882 Castleton had an established post office. It was growing steadily but boomed in 1888 when gold was discovered in Miners Basin, a high valley in the La Sal Mountains above the town. For a few decades following this discovery, Castleton was a point for supplies and recreation for the miners, complete with a hotel and saloon. Despite this development and growth, data from the 1930 census documented a population of only six (Firmage 1996). Today Castleton is primarily a ranching community.

9.4 Conglomerate in the roadcut on the left (east) side of the road is composed of angular, pebble-sized fragments of igneous rock. This lithified deposit also exhibits a well-defined horizontal stratification.

9.7 To the northeast, upper Castle Creek hosts a large, downwarped fold or syncline. As is typical of such folds, the fold axis floors the valley. The northeast limb of the fold (on the left) is displayed by the downdropped and fractured south end of Adobe Mesa. The thick, resistant sandstone that drops into the valley is the Wingate-Kayenta couplet. The overlying Navajo and Entrada sandstones can be seen above and behind this plunging cliff band as tree-clad slickrock benches. Across the valley, to the south (on the right), the southwest limb is exposed as flatirons of Wingate that dive northward off the flanks of the Grand View Mountain laccolith.

10.5 To the northeast is upper Castle Valley and the south terminus of Adobe Mesa. High on the flanks of the mesa is orange Navajo Sandstone overlain by a thick strip of lighter orange Entrada Sandstone. The gray mounds high atop the mesa are remnants of the Quaternary Harpole Mesa Formation, which was deposited as an apron around this part of the La Sals and is associated with glaciation higher in the massif. These deposits precede much of the canyon cutting around the edges of the range and are the earliest Quaternary deposits recognized in the area (Richmond 1962).

10.6 Turn right (south) onto the La Sal Mountain loop road. The road straight ahead winds eventually to Gateway, Colorado, along the Dolores River.

10.8 To the left (east) is the hogback of resistant Wingate Sandstone that defines the southwest limb of the syncline. From this view the Wingate can be seen to be underlain by red slopes of the silty Chinle Formation. Directly ahead are the talus and tree-covered slopes of Grand View Mountain.

11.4 The road cuts through talus from Grand View Mountain. These blocks offer the chance to examine the diorite up close. White, rectangular crystals in this rock are the mineral plagioclase, and the black, elongate crystals are hornblende. Also within these talus blocks are dark, pebble-sized inclusions of basement rock. These xenoliths, as they are known, fell into the magma as it rose through the basement rock on its way to crystallization at shallow levels in the crust.

11.7 A sweeping view opens to the north down Castle Valley (Fig. E.5). From left to right (west to east) the view encompasses the unbroken crest of Porcupine Rim with its west-tilting strata forming the western rampart of the valley, the flat valley floor, and the east margin of the wide salt valley in which all the strata dip eastward. This eastern palisade is defined by Parriott Mesa in the distance to the north, the prominent landmarks of Castleton Tower and the Rectory in the middle, and Adobe Mesa at its south end. The Round Mountain intrusion juts from the center of the valley and is surrounded by several hundred feet of sand and gravel that also veneers the entire valley floor. This unconsolidated valley fill hosts an aquifer that is a vital water source for the ranches and communities in Castle Valley. The aquifer is recharged by snowmelt from the La Sal Mountains that feeds down Castle and Placer creeks. Deeper beneath the valley fill sediments are brecciated remnants of the same Mesozoic rocks that make up the surrounding walls. Prior to collapse of the anticline, these strata would have arced high over the modern valley, connecting the west side strata with those of the eastern margin (see Fig. E.2). As the underlying salt was slowly dissolved, the crest of the fold settled into the evolving void to form the valley. Brecciated Mesozoic rocks are underlain by an

Fig. E.5. View to the north of the east flank of the Castle Valley salt anticline from the north flanks of the La Sal Mountains. From left to right are Parriott Mesa, Castleton Tower, the Rectory, and the long cliff band of Wingate Sandstone that bounds Adobe Mesa. Note the tilted panel of rock at the south end of Adobe Mesa.

estimated 9000-foot-thick column of Paradox salt, most of which flowed laterally from adjacent areas into the anticline during its episodic rise (Doelling and Ross 1998). Thus, while the wide valley floor was not carved by the incessant overland flow of rivers, it ultimately was forged by water. This water, however, percolated slowly downward through the fractures, methodically dissolving salt and undermining the great arc of overlying rocks. While the streamlets of Castle and Placer creeks and their more turbulent, glacier-fed precursors may have distributed the gravel that now lines the valley, they had no major role in its genesis.

12.4 Entrance to the Manti–La Sal National Forest.

12.6 On the right (northwest) is Harpole Mesa, capped by a remnant fringe of glacial outwash of the Harpole Mesa Formation. This is the type locality and namesake for the formation, which is the oldest of the Quaternary deposits with a glacial affinity.

13.5 Roadcuts in the saddle expose outcrops of diorite. This ridge, which emanates from the laccolith cluster, is composed of resistant igneous intrusive rock.

14.8 To the right (northwest) is Pinhook Valley, which drains into Castle Valley as Placer Creek. The dirt road continues 2 miles down the valley to the Pinhook battle site, where one of the region's bloodiest settler-Indian conflicts culminated. On June 15, 1881, after a long chase from the vicinity of modern-day Dolores, Colorado, a posse of cowboys and miners caught up with a band of Ute Indians on Boren Mesa, a few miles south of here. There they killed seven Indian men. The surviving Utes fled north to Pinhook Valley, where they ambushed the posse, killing ten men in return, including eight from the posse and two cattlemen that had ridden up from Castle Valley to see what had prompted all the gunfire. Unreliable reports of Indian casualties at Pinhook range from two to eighteen. After two days of fighting, the remaining Indians escaped and were never captured (Firmage 1996).

15.0 The broken red sandstone in the roadcuts is part of the Jurassic Summerville or Morrison Formation. The uncertainty stems from poor exposure and the fact that it is bounded by faults. Parts of the two formations are very similar.

15.2 The dirt road on the left (east) climbs steeply for ~3 miles to Miners Basin. Gold was discovered there in the late 1880s, and silver and copper were also produced, although none in great quantities. In its heyday the town had a hotel, store, and several restaurants and saloons. The boom lasted until 1907, when a financial panic gripped the nation, affecting even this isolated settlement (Firmage 1996).

18.6 The gravel road on the right (west) leads to Sand Flats and the Slickrock Trail, popular with mountain bikers.

19.1 The brush-covered hills on the left (east) are held up by the Cretaceous Dakota Sandstone and are underlain by the shale-dominated Brushy Basin Member of the Jurassic Morrison Formation. The road and the brushy, undulating landscape to the right (west) are on slumps and large landslide complexes that are a common feature of the Brushy Basin Member.

19.9 Directly ahead (to the southeast) is a clear view of the middle cluster of the La Sal Mountains, which include the highest peaks in the range. The prominent ridge-topped peak on the right side of the cluster is Mt. Tukuhnikivatz, elevation 12,483 feet. According to Van Cott (1990), the word *Tukuhnikivatz* translates to "where the sun sets last" in Paiute. Mt. Peale, the highest peak in the La Sal Mountains, is hidden from view by the mass of Mt. Tukuhnikivatz. It is directly behind it, however, and rises to an elevation of 12,721 feet. The high peak to the left (northeast) of Tukuhnikivatz is Mt. Mellenthin, elevation 12,646 feet. The deep, bowl-like cirque that separates these two peaks is Gold Basin. The southern cluster of the La Sal Mountains looms a short distance to the south.

20.3 The red shale and sandstone visible in roadcuts are in the Brushy Basin Member of the Morrison.

21.9 As the road drops into Mill Creek Canyon, the resistant cliff band is sandstone and conglomerate of the Salt Wash Member of the Morrison. The thick sandstone cliffs are the channel deposits of east-flowing rivers; these are separated by red, sandy shale beds, the floodplain deposits of the rivers. The colorful sandstone and shale at road level are the basal Tidwell Member of the Morrison and the underlying Summerville Formation. The two are lumped together here because their similarities and poor exposure make them difficult to differentiate. Both are of shallow marine origin.

22.4 Crossing of Mill Creek. The gravel road on the left (east) leads up the canyon to Oowah Lake. Below the bridge is the Moab Member of the Curtis Formation, which underlies the Summerville Formation. Here the resistant eolian sandstone of the Moab Member has been cleaved and polished into a deep slot canyon by the energetic waters of Mill Creek.

23.1 Crossing of Horse Creek. For the next 7 miles the road traverses various members of the Morrison Formation, despite the fact that the road drops considerably in elevation over this distance. Continuity of the Morrison is accommodated by several down-to-the-west faults that drop the Morrison in a stepwise fashion as the road drops into the Moab–Spanish Valley collapsed salt anticline. These faults will be pointed out as they are encountered, although with the exception of intense fracturing and noticeable tilting of the strata, they are difficult to pinpoint from a moving car.

23.7 To the left is a graded dirt road to Gold Basin and Geyser Pass. The road crosses the La Sal Mountains and reaches a high point at Geyser Pass, elevation 10,600 feet. This road also provides the most convenient access for hikers wanting to climb the cluster of highest peaks in the range: Mounts Peale, Mellenthin, and Tukuhnikivatz.

25.8 Several northwest-trending faults cut the Morrison Formation through this area.

27.8 On the right (north) is a bowl of colorful, multihued shale of the Brushy Basin Member, providing the first clear view of this member of the Morrison. This slope-former is normally covered by vegetation at this elevation.

28.8 The Spanish Valley overlook, at the point of the switchback. Immediately below this overlook is another northwest-trending fault that hosts some of the collapse of the Moab–Spanish Valley salt anticline. Below and to the west is the broad, flat floor of this collapsed salt anticline valley. Far to the west on the skyline are the three humps of the Henry Mountains, laccolithic intrusions of the same age and approximate composition as the La Sal laccoliths.

30.0 West-tilted Morrison Formation is exposed in the roadcut on the right (north).

31.1 On the right (north) is west-tilted resistant sandstone and conglomerate of the Cretaceous Burro Canyon Formation, which overlies the Morrison.

32.8 The road traverses the flat-bottomed floor of the Moab–Spanish Valley collapsed salt anticline. Prior to this collapse, the red sandstone walls bounding the valley on the east and west rose high into an anticline. Hills to the left (west) are composed of the Burro Canyon Formation.

34.3 On the left side of the valley is a palisade of orange Lower Jurassic Glen Canyon Group sandstones. This southwest margin of the valley is riddled with faults that parallel the wall. The lower unit is the eolian Wingate Sandstone, overlain by the fluvial Kayenta Formation. The cap of lighter sandstone is the dome- and fin-forming Navajo Sandstone.

34.9 Turn left (west) on Old Airport Road.

35.5 U.S. Highway 191. Turn right (north) to Moab.

Utah Route 211 to the Needles District, Canyonlands National Park

This road log begins at the intersection of U.S. Highway 191 and Utah Route 211, and continues west on 211 to end in the spectacular Needles District of Canyonlands National Park. Initially the route parallels the north flanks of the Abajo Mountain laccoliths and passes by Newspaper Rock State Park, the world-famous Indian Creek rock climbing area, and North and South Six-Shooter peaks, all hewn from the resistant, cliff-forming Jurassic Wingate Sandstone. The road log terminates at a dead end at the Big Spring Canyon trailhead, situated at the contact between the Permian lower Cutler beds and the overlying Cedar Mesa Sandstone. The strata encountered along the route are shown in Figure F.1.

Road Log

0.0 The conspicuous landmark of Church Rock rises from the sage plain on the east side of U.S. 191 at the junction with Utah Route 211, the road to the Needles District (Fig. F.2). The red base of this beehive-shaped rock is in the Dewey Bridge Member of the Jurassic Carmel Formation. The steep, middle, buff-colored sandstone is the Slick Rock Member of the Entrada Sandstone, which is capped by a knob of the Moab Member sandstone of the Curtis Formation. The smaller knob at the top is a remnant of Jurassic Wanakah Formation, composed of interbedded sandstone and thin shale beds.

To the west, the mound of buff-colored sandstone on the left (south) side of the road is mostly eolian Slick Rock Member and Moab Member capped by a thin remnant of the Wanakah Formation.

1.0 The slickrock knolls ahead on the right (north) side of the road consist of, in the lower part, the Slickrock, Moab, and Wanakah units. Here they are capped by the red siltstone and sandstone of the basal Tidwell Member,

AGE		FORMATION	GENERAL ENVIRONMENT
Jurassic	Upper	Morrison Formation: Brushy Basin Member	fluvial and lacustrine
Jurassic	Upper	Morrison Formation: Salt Wash Member	fluvial
Jurassic	Upper	Morrison Formation: Tidwell Member	shallow marine
Jurassic	Middle	Wanakah Formation	shallow marine
Jurassic	Middle	Slick Rock Member Entrada Sandstone	eolian
Jurassic	Middle	Dewey Bridge Member Carmel Formation	shallow marine
Jurassic	Lower	Navajo Sandstone	eolian
Jurassic	Lower	Kayenta Formation	fluvial
Jurassic	Lower	Wingate Sandstone	eolian
Triassic		Chinle Formation	fluvial and lacustrine
Triassic		Moenkopi Formation	shallow marine and tidal flat
Permian		Organ Rock Formation	low energy fluvial
Permian		Cedar Mesa Sandstone / Cutler Formation	eolian / fluvial
Permian		lower Cutler beds	fluvial and marine

FIG. F.1. Stratigraphic column of units seen along Utah Highway 211 and in the Needles District of Canyonlands National Park and surrounding areas.

which makes up the base of the Jurassic Morrison Formation. The tan, ledge-forming caprock above that is the fluvial Salt Wash Member of the Morrison, the youngest unit seen along this road.

2.9 The small cluster of buildings on the right (north) side of this low pass is a restored remnant of a long-defunct religious colony called the Home of Truth, established in 1933 by Mrs. Marie Ogden from New Jersey (Fig. F.3) (Bjørnstad 1988). Ramshackle shacks and building foundations are also present on the left (south) side of the road, back to the east on the flats. Marie Ogden purportedly received messages from the spirit world and used these to direct the approximately 100 members of the colony. In an increasingly bizarre turn, in 1935 a woman member of the colony died of cancer, and Marie claimed to feel vibrations from the dead woman's body, leading her to announce that the woman would soon return to life. The woman's body was transferred to a nearby deep, cool alcove, where it was prayed over and washed daily in a pickling brine. Eventually word of the strange corpse and activity leaked out, and after several months the sheriff came to investigate. By that time the body was mummified, and it was decided that it posed no health risk to the colonists. The body remained in the cave, and the colonists waited, but the woman never returned to life. Gradually the colony members drifted away, and by 1942 the popu-

Fig. F.2. Church Rock, at the intersection of U.S. Highway 191 and Utah Route 211. This erosional remnant illustrates clearly the stratigraphy of the Middle Jurassic San Rafael Group.

Fig. F.3. Site of the Home of Truth religious colony, established in 1933 by Mrs. Marie Ogden of New Jersey. The buildings have been recently refurbished, and a sign has been put up.

FIG. F.4. Newspaper Rock. The petroglyphs blanketing this sheltered wall of varnished Wingate Sandstone were etched into the rock by Native Americans over several hundred years.

lation had dwindled to seven. The buildings at the top of the pass have recently been refurbished and appear to be sporadically occupied.

3.2 White sandstone outcrops on the right (north) side are the eolian Jurassic Navajo Sandstone, which lies stratigraphically below the Dewey Bridge Member of the Carmel.

5.1 The Abajo Mountains rise directly ahead. These Oligocene age igneous intrusions invaded the Jurassic and Cretaceous sedimentary rocks ~30 million years ago. Subsequent erosion has exhumed the igneous cores of these laccoliths.

The tree-covered ridge immediately south (left) is, from bottom to top, the Slick Rock and Moab members, the Wanakah Formation, and the slope-forming red siltstone of the Tidwell Member.

6.7 Mile marker 12. The base of the tree-covered ridge to the south is the fault-bounded south edge of the Shay graben. A **graben** is a down-dropped block bounded on both sides by parallel normal faults.

9.2 The road to Hart's Draw forks to the left (south). This road rises to the forested slopes of the Abajo Mountains. The northern border fault of Shay graben lies immediately north of the road and is expressed as a resistant fin of white Navajo Sandstone.

9.7 The road winds down through the white Navajo Sandstone into Indian Creek.

10.9 Where the Indian Creek bed is encountered, the red fluvial deposits of the Kayenta Formation are exposed beneath the white Navajo Sandstone.

Fig. F.5. The north end of Bridger Jack Mesa consists of pinnacles of Wingate Sandstone. The upper, ledgy-weathering caprock consists of the Kayenta Formation. The Triassic Chinle Formation makes up the basal slopes and sandstone mounds in the foreground.

11.3 A normal fault in the gully to the right (east) is the northern bounding fault on the Shay graben. The fault raises the red Kayenta Formation on the left (north) against the white Navajo Sandstone on the right (south). At the old corral the Wingate Sandstone has risen to road level. Here the more broken and discontinuous Kayenta can be seen above the smooth walls of Wingate. The Navajo is above but out of sight from the road. Collectively these three Lower Jurassic formations constitute the Glen Canyon Group.

11.7 Mile marker 7. The orange Wingate hangs over the road.

11.9 Newspaper Rock is on the right (east) side of the road. Multiple generations of petroglyphs that go back to the indigenous Anasazi have been pecked into the Wingate Sandstone (Fig. F.4).

12.3 The Wingate along the road weathers to bulging slickrock. The abundant large-scale cross-stratification attests to its eolian origin as large sand dunes.

15.2 The lower red slopes are sandstone and siltstone of the Triassic Chinle Formation capped by the red cliff line of Wingate. The overlying ledgy red slopes are Kayenta.

Proceed with caution: the next few miles of vertically fractured Wingate Sandstone form the rock climbing mecca known as Indian Creek.

FIG. F.6. Mesozoic stratigraphy along Indian Creek canyon on the east side of the road. The red shale and sandstone of the Triassic Moenkopi Formation are exposed at road level. The top of the formation is marked by the basal Moss Back Member of the Chinle Formation, which forms a blocky cliff between the two shale-dominated formations. The Wingate is a massive cliff-former, attracting rock climbers from around the world. It is overlain by the ledgier sandstone of the Kayenta Formation. The eolian Navajo Sandstone is the white slickrock dome-former that caps the entire succession.

Pullouts are usually crammed with cars and people, and it is common to round a corner only to come upon a car stopped in the road with its occupants peering and pointing at fractures in the cliffs. The numerous exceptional cracks that split these steep walls are likely the result of fractures produced during the mild warping of the strata by the intrusions of the nearby Abajo Mountains.

18.2 As the valley widens due to the presence of the easily eroded Chinle shales at creek level, a fantastic panorama emerges. Ahead on the left (northwest) is Bridger Jack Mesa, its northern end bristling with pinnacles of Wingate Sandstone, some capped by horizontal ledges of Kayenta (Fig. F.5). Farther north, as the valley continues to widen, are the isolated fingers of South and North Six-Shooter peaks. The fingers are Wingate and rest on slopes of Chinle (upper) and the underlying red Triassic Moenkopi Formation (lower slopes). The Island in the Sky District of Canyonlands can be seen atop the distant red cliff band on the northern horizon. The closer view, directly to the right (east), from bottom to top, is of the multicolored shale slopes of the Chinle, red Wingate cliffs riddled with vertical fractures, and the red, horizontally lined strata of the Kay-

Fig. F.7. Stratigraphy of the north side of North Six-Shooter Peak.

enta Formation on top. Domes of white Navajo Sandstone tower above the entire succession (Fig. F.6).

19.6 The graded dirt road on the left (west) eventually leads to Beef Basin and Elk Ridge, two beautiful and remote parts of the higher canyon country.

21.0 The ledge-forming, brown-gray sandstone at road level on the right (east) is the Moss Back Member of the Chinle Formation. This resistant unit forms the easily recognized base of the formation. The Moss Back represents channel deposits of northwest-flowing rivers.

22.8 On the right (east) the Moss Back ledges rise, and the red-brown slopes and thin ledges of the Triassic Moenkopi Formation emerge from beneath it.

25.2 The free-standing spire of North Six-Shooter Peak is especially prominent from this vantage point (Fig. F.7).

25.4 Crossing of Indian Creek. At the top of the hill, just past the bridge, is a rough dirt road on the left (west) to Davis and Lavender canyons.

25.8 The route is now in the Permian Cutler and Cedar Mesa Sandstone. The alternating dark red and lighter orange strata represent a zone of intertonguing between the darker fluvial beds of the Cutler and the lighter orange eolian deposits of the Cedar Mesa Sandstone. The Cutler was

Fig. F.8. Detail of Cedar Mesa Sandstone stratigraphy on Squaw Butte. The eolian sandstone shows large-scale cross-stratification and forms steep cliffs and slickrock benches, while the waterlain deposits consist of thin deposits of laminated, sandy shale that are more easily eroded to form recesses. The bioturbated sandstone forms when the top of the eolian sandstone becomes wet and colonized by plants. Most of the bioturbation is caused by plant roots extending down into the wet sand.

deposited in west-flowing rivers that originated in the ancestral Uncompahgre highlands to the east. These rivers dispersed westward into the wide dune field of eolian sand before reaching the sea that lay to the west, in central Utah. This is the zone where the coarse-grained rivers met the dune field.

28.1 The ribbed mesa escarpment on the left (west) clearly shows the alternations between the dark red fluvial Cutler deposits and the lighter eolian sandstone as the Cedar Mesa erg is approached. Here the dark fluvial deposits can be seen cutting down into the eolian sandstone. The Cedar Mesa erg was extensive, and these eolian deposits extend across the entire Needles area and westward across most of the Maze District.

30.4 Over this short distance eolian beds have become more abundant, although the darker fluvial beds can still be seen. Ahead to the west, on the skyline, are the pinnacles and fins of jointed Cedar Mesa Sandstone for which the Needles District is named.

30.8 Canyonlands National Park boundary.

31.9 The Cedar Mesa Sandstone dominates the landscape here and all along the road through the park.

32.2 A short distance down the road on the right is the Needles Outpost, the only store in the area. They carry food, drinks, books, and other supplies.

33.0 National park entrance station. Pay your dues.

33.1 The visitor center is on the right. This is an excellent place to learn more about the park and obtain permits for backpacking and backcountry camping. Maps and books are also available for purchase.

33.7 Roadside ruin overlook. The surrounding rocks are Permian Cedar Mesa Sandstone. The abundant large-scale cross-stratification is a characteristic of eolian dune deposits.

35.0 The sprawling, terraced formation on the left (south) is Squaw Butte. The clear, continuous exposure of Cedar Mesa Sandstone in its walls offers the opportunity for a detailed discussion of its deposition (Fig. F.8). The thick orange and white cliffs etched with cross-stratification are eolian dune deposits, which dominate the succession. Thin recesses that generate the obvious horizontal lines are waterlain deposits of less-resistant purple mudstone and silty sandstone. The extensive Cedar Mesa erg was bordered to the west by the sea and on the east by west-flowing rivers that dissipated into the giant sand sea. Mudstone and siltstone beds were deposited when the ground water table rose, due either to a rise in sea level to the west or an influx of water from fluvial systems to the east. Regardless of the water's origin, this rise temporarily halted eolian dune deposition. Ponded water between the dunes trapped finer sediment. The upper, dry parts of the dunes continued to move, but the sand supply eventually diminished; the once high dunes were then reworked into a thin layer over a large area, resulting in the widespread exposure of the damp water table at the surface. The cohesive nature of wet sediments would have prohibited further erosion, although the flat region was likely dotted with small ponds of ground water. Slowly a regionally extensive layer of mud and silty sand blanketed this surface. When the water table subsequently dropped, the damp surface disappeared, and dry sand again was available to pile into dunes. If the lowered water table was the result of a drop in sea level to the west, it would have had the dual effect of exposing large tracts of shoreline sand to erosion, providing an abundant supply of sediment for the renewed erg.

35.4 Woodenshoe Butte is on the left (southwest) (Fig. F.9). The main part of the formation is massive eolian sandstone underlain by a thin unit of darker mudstone and siltstone. Some of this less-resistant rock has eroded away, leaving a small window between the toe and heel of the "shoe." The formation of this window was also promoted by the narrow fin geometry of this part of the butte.

Fig. F.9. Woodenshoe Butte, formed from a narrow fin in the Cedar Mesa Sandstone. The butte's window developed from the erosion of the thin, less-resistant siltstone bed that can be seen as a narrow, dark band beneath the "shoe." All the rock in this view is Cedar Mesa Sandstone.

35.8 The turnoff to Salt Creek, a delightful hike now that this delicate desert riparian area has been closed to motorized vehicles.

36.1 On the left (west) is the turnoff to Squaw Flat campground, Elephant Hill jeep trail, and a trailhead that accesses most of the hiking trails that penetrate the fascinating backcountry of the Needles. Squaw Flat is one of the more appealing national park campgrounds, with campsites spread around the sprawling butte immediately to the west in isolated alcoves surrounded by a pygmy forest of piñon and juniper trees. The flanks of the butte also provide ready access to vantage points in all directions, as well as a close-up look at the Cedar Mesa Sandstone.

36.4 Ahead to the north, in the distance, is the continuous red cliff line of Wingate Sandstone that holds up Island in the Sky. Immediately left

FIG. F.10. Cedar Mesa Sandstone in the Needles District of Canyonlands. In the middle ground, the eolian caprock (white sandstone) is more resistant than the underlying silty sandstone, producing the mushroom-like weathering pattern seen here. The pinnacles in the back are fins of Cedar Mesa Sandstone formed by the enhanced weathering of parallel fractures. The Needles District is named for the spires created by this unique weathering pattern.

(west) of this sheer-walled mesa is the isolated palisade of Junction Butte.

37.0 The mushroom shape of the buttes of Cedar Mesa Sandstone along the next few miles of road is attributed to alternations of resistant and less-resistant rock. The resistant caprock of tan, cross-stratified eolian sandstone hangs over a less-resistant pedestal of red silty sandstone (Fig. F.10).

38.1 Pothole Point trailhead. This half-mile trail (1 mile roundtrip) traverses the flat upper surface of the tan eolian sandstone. The potholes along this trail fill with water after rain or snowmelt. These ephemeral cisterns have evolved from low-relief depressions and undulations on the surface of the sandstone (Fig. F.11). As water rolled into these low-relief hollows over thousands of years, the natural acidity of the water methodically

FIG. F.11. Potholes filled with rainwater are common on the tops of buttes and mesas capped by Cedar Mesa Sandstone. They form as water fills low-lying areas and slowly dissolves the calcium carbonate cement in the sandstone, deepening the potholes over time through this ongoing process.

dissolved the calcium carbonate ($CaCO_3$) that cements the sand grains together. As sand grains are loosened and washed out or blown out by the wind, the depressions enlarge and deepen. When filled with water, long-dormant organisms spring from their burial site in the mud at the bottom of the pothole, creating an ecosystem of microscopic life.

39.5 Big Spring Canyon trailhead and the end of the road. One of the most interesting and varied trails in the park begins here and ends at the Confluence Overlook, a sublime vantage point ~1000 feet above the junction of the Green and Colorado rivers, where they merge to create the giant artery of the desert West, the Colorado River. It is a spectacular 5 mile (10 mile roundtrip) hike to the Confluence, but even a short stroll leads to incredible panoramas. The bottom of Big Spring Canyon here is floored by gray, fossiliferous limestone and varicolored sandstone and mudstone of the lower Cutler beds, an informally named Pennsylvanian-Permian unit that underlies the Cedar Mesa Sandstone. The current informal designation stems from a disagreement among geologists on the formal name and definition of this interval of rock. The now-familiar Cedar Mesa forms the landscape at road level.

The varied lithologies of the lower Cutler beds were deposited near the fluctuating shoreline of a sea that deepened to the west. As sea level rose and fell, offshore marine limestone, nearshore mudstone, fluvial conglomeratic sandstone, and eolian sandstone competed for space and shifted east or west with the fluctuating sea level. Finally, as the shoreline experienced a major shift to the west, exposing a vast coastal plain, southeast-directed winds pushing off the sea prompted the development of the Cedar Mesa erg.

West on Utah Highway 95 across Comb Ridge, Cedar Mesa, and White Canyon to Lake Powell

This route traverses west along one of the most scenic roads in southeast Utah. The log begins at the junction of U.S. Highway 191 with Utah Highway 95, heading west. The road initially crosses White Mesa, eventually encountering the Comb Ridge monocline. From this fantastic landform, the road climbs onto the Monument upwarp, which has been a mild highland sporadically throughout geologic time. The modern version of this uplift forms Cedar Mesa, and Highway 95 provides numerous opportunities to access the deep canyons that incise this highland. The spectacular canyons of Cedar Mesa include Grand Gulch, the Owl and Fish canyons system, Road Canyon, Lime Canyon, and numerous others. Cedar Mesa is rich in both canyons and the ruins of Anasazi cliff dwellings. Also along this route is a short spur road to Natural Bridges National Monument, whose spectacular geologic features are addressed in a separate road log. The monument's natural bridges are carved in the Cedar Mesa Sandstone in White Canyon. The latter half of the road log parallels White Canyon and eventually crosses it before descending to Reservoir Powell (Colorado River). The log continues several miles past the reservoir and terminates at an overlook of the impoundment.

Road Log

0.0 Junction of U.S. Highway 191 and Utah Highway 95. Go west on Highway 95 past the convenience store and gas station. The road travels across the top of White Mesa, which is mantled with a thick, red sandy soil derived from the weathering of surrounding Mesozoic sedimentary rocks.

0.9 The road drops into Westwater Canyon through a cliff band of tan, conglomeratic sandstone that makes up the base of the Lower Cretaceous Burro Canyon Formation. The Burro Canyon was deposited by northeast-flowing rivers with a source to the southwest in western Arizona and/or

Fig. G.1. View to the north of the Bears Ears on Elk Ridge. These buttes can be seen from many vantage points in southern Utah and provide a valuable landmark. They are capped by erosional remnants of the Jurassic Wingate Sandstone forming resistant cliff bands. The lower slopes are in the Triassic Chinle Formation.

Nevada. Conglomerate consists mostly of rounded pebbles of chert and quartzite.

1.0 Here the road descends through the green shale of the Brushy Basin Member of the Upper Jurassic Morrison Formation. This colorful shale represents the floodplain deposits of low-energy, northeast-flowing rivers in a setting geographically similar to that of the overlying Burro Canyon Formation.

1.4 As the road climbs out of the canyon, roadcuts reveal green and purple shale of the Brushy Basin Member.

1.5 Cross-bedded pebble conglomerate of the basal Burro Canyon is well exposed in roadcuts.

3.1 To the west-northwest two prominent buttes known as the Bears Ears rise above Elk Ridge on the skyline (Fig. G.1). The Bears Ears provide a good landmark because they are visible from tens of miles distant. These high buttes are floored by slopes of Upper Triassic Chinle Formation. The irregular cliffs of caprock are cut from the Lower Jurassic Wingate Sandstone.

3.7 Roadcuts at the top of Cottonwood Canyon are in loose, rounded pebbles and cobbles deposited in relatively recent times (within the last 2 million years) as part of an apron of sediment shed from the Abajo Mountains immediately to the north. This great blanket of detritus was shed in a radial pattern from the Abajo Mountains and was deposited here on the Burro Canyon Formation before the deep dissection of the surrounding canyons that is seen today. These clasts are mostly composed of quartz diorite, the hard igneous intrusive rock that forms the laccoliths of the Abajos. Hardened quartz sandstone clasts, probably metamorphosed by the heat of the intrusive bodies ~28 million years ago, make up a subordinate population of these waterlain deposits. All the clasts are encrusted with a thin, white rind of calcium carbonate, and the entire roadcut of young sediments is mildly cemented by this mineral. This **caliche**, as it is known, is a common feature of shallow soil zones in arid to semiarid climates. It forms in the loose sediment by the downward percolation of meteoric waters (rain or snow melt) into the soil horizon. As the water trickles down through the voids in the sediment, it dissolves and accumulates calcium ions. Because of the dry climate the water only reaches shallow levels in the soil or sediments before it begins to evaporate. As the water evaporates, $CaCO_3$ precipitates from the moisture, coating the surfaces of the sediment particles. Through time this precipitation may build up to a considerable volume, eventually filling all the pore spaces and cementing the sediments.

3.9 The pillars and cliffs on the right (north) side of the road are carved in conglomeratic sandstone of the Burro Canyon Formation.

4.2 The road descends through green and purple shale of the Brushy Basin Member. The canyon below is lined with sandstone and conglomerate of the Salt Wash and Westwater Canyon members of the Morrison Formation.

5.8 The Salt Wash Member is exposed in roadcuts near the bottom of Cottonwood Creek. Brushy Basin Wash, the type locality for the overlying Brushy Basin Member, converges with Cottonwood Wash a short distance upstream to the north. Both drain from the flanks of the Abajo Mountains.

6.3 As the road climbs from the drainage, the Salt Wash is again well exposed in the roadcuts. The various types of cross-bedding and cross-stratification attest to its origin in the channels of high-energy rivers.

7.1 Roadcuts on the right (north) expose brown-red and olive-green shale with thin sandstone beds, all part of the marginal marine Tidwell Member of the Morrison. Overlying this is a tongue of the eolian Bluff Sandstone Member of the Morrison. This is the northern limit of the Bluff dune field, and the sandstone beds interfinger with the marine strata of the Tidwell. Just 35 miles south, near its namesake town of Bluff, the eolian-dominated member thickens to ~300 feet. As the road continues west, it begins to cut across east-dipping strata of the Comb Ridge monocline. As the road traverses this regional structure, increasingly older strata are encountered.

8.6 The roadcut exposes the Tidwell Member (lower shale and thin sandstone beds) overlain by the eolian Bluff and fluvial Salt Wash members. The fine-grained, large-scale, cross-stratified sandstone of the Bluff interfingers here with conglomeratic sandstone of the Salt Wash. Directly ahead, to the southwest, are the great east-dipping slabs of the Lower Jurassic Navajo Sandstone outlining the toothlike crest of the Comb Ridge monocline.

8.9 On the left (south) side of the valley the resistant red and white Middle Jurassic Entrada Sandstone weathers to prominent slickrock benches.

9.8 To the left (south) the thinner-bedded, silty sandstone of the Dewey Bridge Member of the Middle Jurassic Carmel Formation peeks out from beneath an overhang of Entrada Sandstone at wash level.

10.4 The turnoff on the right (north) goes a short distance to some of the more accessible Butler Wash Anasazi ruins.

11.4 To the east, across the alluvium of Butler Wash, is a good view of the Dewey Bridge with a cap of white sandstone and underlying slopes of thin-bedded, red siltstone. The top of the white Navajo Sandstone is barely visible below, at wash level.

11.8 To the west the white, east-sloping slickrock of the Navajo Sandstone rises to form the serrated crest of Comb Ridge monocline. The obvious parallel joints on the Navajo slopes parallel the dip of the monocline, and were formed under the same stresses as those that formed the monocline (Fig. G.2). These joints have been deepened and their appearance enhanced by the flow of water off the crest. They are also accentuated by the growth of trees and shrubs in these fractures, which take advantage of the concentration of moisture in this desiccated land.

12.2 The road curves westward into the Navajo Sandstone.

12.7 The roadcut enters a deep corridor that has been blasted through Comb Ridge. The contact between the eolian Navajo and the red rocks of the underlying Kayenta Formation occurs at the entrance to this manmade slot.

12.8 The road enters the Lower Jurassic Wingate Sandstone, a massive, red cliff-former of eolian origin. The Wingate forms the base of the Jurassic succession and is the base of the Glen Canyon Group, which includes, in ascending order, the Wingate, Kayenta Formation, and Navajo Sandstone.

12.9 At the exit of the colossal roadcut is the contact between the Upper Triassic Chinle Formation and the overlying Wingate. The red mudstone and sandstone of the Chinle degrades to steep slopes that offer the opportunity for a straight road. Across this straight segment the road descends

FIG. G.2. Dip slope bedding surface of the Jurassic Navajo Sandstone on the back (east) side of Comb Ridge showing well-developed parallel fractures. As water runs down the dip slope, it is channeled into the fractures, which are then widened by erosion. The concentration of water in these cracks promotes the growth of vegetation, as seen here accentuating the fractures.

through the Chinle to the lush cottonwood groves of Comb Wash; the ominous slabs of Wingate slickrock loom above. To the west, across the wash, is the red- and white-striped Permian age Cedar Mesa Sandstone, which appears to rise from the depths of Comb Wash.

14.0 The bumpy dirt road of Comb Wash on the left heads south from here, eventually connecting with Highway 163 between Bluff (to the east) and Mexican Hat (to the west). This road and Comb Wash follow the strike valley that has been cut into the nonresistant Triassic Moenkopi and Permian Organ Rock formations. A **strike valley** follows the strike or trend of stratification in tilted sedimentary rocks. Technically the **strike** of tilted sedimentary rocks is a horizontal line drawn across the bedding surface that accurately describes the trend of large-scale structures such as Comb Ridge monocline. This part of the monocline *strikes* north-south. This is clearly shown in both the linear trend of the resistant spine of sandstone (Navajo and Wingate) that forms Comb Ridge and the incised strike valley of Comb Wash, which is cut into the soft, shale-dominated units (Moenkopi and Organ Rock formations).

Fig. G.3. The strike valley of Comb Wash, which has been eroded into the less-resistant rocks of the Triassic Moenkopi Formation. Resistant rocks of Comb Ridge on the right (east) consist of the Jurassic Glen Canyon Group (Wingate, Kayenta, and Navajo). Cedar Mesa, on the left (west), is dominated by the resistant Permian Cedar Mesa Sandstone.

Comb Ridge monocline

Comb Ridge monocline is a regional-scale fold that forms the east boundary of the Monument upwarp and can be traced for ~80 miles (Fig. G.3 and Plate 13). From here the monocline extends a short distance northward, where it fades into the western flank of the Abajo Mountains. To the south, however, Comb Ridge continues across the San Juan River and into Arizona, where it arcs southwest to merge with the Organ Rock monocline near the town of Kayenta.

Like most of the Colorado Plateau monoclines, Comb Ridge is interpreted to be the surface expression of a reverse fault in the deep subsurface. This fault originates in the older Precambrian basement complex of igneous and metamorphic rocks that form the foundation for the thick sedimentary succession. Although this fault cannot be viewed directly, its existence is inferred based on the geometry of the exposed monocline and its similarities with monoclines elsewhere—namely the Grand Canyon. There the East Kaibab monocline exposed in the Paleozoic sedimentary succession is rooted by a deep-level reverse fault in Precambrian rocks that has been exhumed and put on display by the Colorado River. Based on this and other observations, the basement fault that formed Comb

FIG. G.4. Aerial view of lower Grand Gulch on Cedar Mesa, the type locality for the orange and white Permian Cedar Mesa Sandstone that makes up the deeply incised canyon.

Ridge is interpreted as a reverse fault formed by compression that pushed the west side up and over the east side.

The present incarnation of the Comb Ridge monocline and Monument upwarp, which it bounds, was pushed up during the Tertiary Laramide orogeny, a widespread pulse of compression that affected much of western North America from ~65 to 50 Ma. This mountain-building event also formed the modern Southern Rocky Mountains in Colorado and northern New Mexico. This was not the only pulse of uplift on Monument upwarp, just the most recent. In fact, regional stratigraphic relations in Pennsylvanian through Jurassic rocks point to sporadic recurrent activity on Monument upwarp and the faults that bound it. Basement faults that underlie and control many of the monoclines on the Colorado Plateau, including Comb Ridge, have been intermittently active since their inception during the Late Precambrian. Since that time, any tectonic stress that rippled through the region, whether extensional or compressional, generated movement on some of these structures. It is much easier to reactivate some preexisting weakness, such as these early faults, than to initiate a completely new fault in unbroken rock. Because of this, many of the monoclines across the Plateau have long but sporadic movement histories.

14.1 Good campsites beneath canopies of cottonwood trees can be found a short distance up the dirt road running north along Comb Wash.

14.5 The road enters the Permian Cedar Mesa Sandstone, composed of white sandstone with thin beds of red, sandy siltstone (Figure G.4 and Plate 14). Thick white sandstone units exhibit large-scale cross-stratification of eolian origin. In contrast, the thin beds of sandy siltstone are horizontally stratified, suggesting deposition in a shallow water setting. Shallow water in the midst of an arid dune field may have appeared as groundwater levels rose to form ponds in low-lying areas between large dunes. An alternative version of this process involves the periodic rise and fall of sea level. The sea bordered the dune field to both the south and the west; a sea level rise would have allowed seawater to infiltrate the porous dune sand, temporarily flooding the extensive dune field with shallow water. The wet surface would have trapped silt that otherwise would have been blown away.

17.0 The road crests on Cedar Mesa, the geographic feature for which the Permian sandstone takes its name, and follows the top of the Cedar Mesa Sandstone for the next 30 to 40 miles.

19.1 Directly ahead are the two Permian units exposed across Cedar Mesa: the lower red and white Cedar Mesa Sandstone and the overlying fluvial Organ Rock Formation, which weathers to steep red slopes.

20.4 On the right (north) side of the road is Cedar Mesa Sandstone clearly displaying its characteristic large-scale eolian cross-stratification.

24.2 Salvation Knoll. This vantage point is in the thin-bedded, dark red sandstone and mudstone of the Organ Rock Formation. On the hillside immediately to the north is a dark red, knobby weathering cliff. This is the basal Hoskinnini Member of the Lower Triassic Moenkopi Formation. The Organ Rock and Moenkopi are similar in appearance and from a distance are difficult to tell apart; however, the distinctive knobby, cliff-forming nature of the overlying Hoskinnini is maintained throughout this region and provides a convenient marker.

From this pullout many distant landmarks of the Four Corners region can be seen. The nearest feature to the southeast is the serrated backbone of Comb Ridge. Behind that, in the distance, is the broad silhouette of the Carrizo Mountains on the Navajo reservation in Arizona. To the east is Sleeping Ute Mountain in Colorado, with the prominent knob at its north end. North of Sleeping Ute is the snow-clad cluster of Colorado's La Plata Mountains. Farther back and slightly to the north hover the ghostlike San Juan Mountains.

Salvation Knoll was named on Christmas Day, 1879, by four members of an advance scout party for the Mormon Hole-in-the-Rock expedition. Composed of 250 men, women, and children with livestock and wagons, the expedition was charged with finding an acceptable route from Escalante, Utah, to the southeast corner of the state, which they were

Fig. G.5. Stratigraphy on the north side of the road illustrating the characteristic ribbed cliff band of the Hoskinnini Member of the Moenkopi Formation and its surrounding units.

instructed to settle. On Christmas Day the members of the advance party, dejected and unsure of their location, climbed this high point and saw the Abajo Mountains a short distance to the north. Although they were out of food, the sight cheered them considerably (Miller 1959).

The new route pioneered by this expedition crossed 200 miles of some of the most daunting canyon country in existence. More than a year after leaving the town of Escalante, and literally cutting through rock, crossing rivers, and dealing with abundant, axle-deep sand, the pioneers established the settlement of Bluff. The brutal Hole-in-the-Rock route was followed by several smaller parties in the following year but was then abandoned.

24.6 On the right (north) side is a clear view of the contact between the Organ Rock Formation (lower) and the Hoskinnini Member (upper). The mesas ahead to the west are held up by the cliff-forming Wingate and Kayenta sandstones. In the distance, to the southwest, is the shadowed hump of Navajo Mountain at an elevation of 7155 feet. It forms a towering landmark along the Utah-Arizona border from most vantage points in the region.

Navajo Mountain is a laccolith that was intruded at approximately the Cretaceous-Tertiary boundary (~70 to 60 Ma). This age falls within the range of the Carrizo Mountain intrusions (~74–60 Ma), the La Plata Mountain intrusives (~67 Ma), and Sleeping Ute Mountain (~72–64 Ma) (Semken and McIntosh 1997). These ages put this Four Corners intrusive cluster into a distinctly different age range than the Abajo, La Sal, and Henry mountains laccolithic clusters.

25.0 To the left (south) is the juniper- and piñon-dotted Cedar Mesa. Although it appears to be an endless flat, numerous deep canyons incise this great mesa, including Fish, Owl, Road, and Lime canyons. These drain into Comb Wash and eventually empty into the San Juan River. Farther west is the immense canyon of Grand Gulch and its sinuous tributaries, which drain directly into the San Juan River.

25.8 To the right (north) are the lower red slopes of the Organ Rock Formation overlain by the steep horizontal rib of the Hoskinnini (Fig. G.5). This basal unit of the Moenkopi is succeeded by more red slopes of the upper, undivided part of the formation. The thin, brown ledge at the top of the mesa is sandstone and conglomerate of the basal Shinarump Member of the Upper Triassic Chinle Formation. The Shinarump, which is of fluvial origin, fills large, swale-like channels that were cut into the upper surface of the underlying Moenkopi Formation.

28.2 To the left (south) is Utah Route 261. This mostly paved road heads south past various trailheads to the rugged canyons of Cedar Mesa and the incredible overlook at Muley Point. After 35 miles the road turns to graded dirt for a few miles as it winds down the hair-raising switchbacks of the Mokee Dugway, which was chiseled from the steep south rim of Cedar Mesa. The road cuts down through the multitiered Cedar Mesa Sandstone and the upper Cutler Formation and terminates at Highway 163 near the town of Mexican Hat.

29.0 The roadcuts are in the Organ Rock Formation.

30.0 On the right (north) is the turnoff to Natural Bridges National Monument and access to Elk Ridge and Dark Canyon Plateau, the high but deeply dissected plateau on which the Bears Ears sit (see Road Log H).

35.8 White Canyon, notched into the Cedar Mesa Sandstone, rims the road on the right (south). The road parallels this rim for the next 30 miles. This is the canyon and that hosts Natural Bridges.

37.5 To the left (south) is Utah Route 276. This paved road dead-ends at Lake Powell, at Halls Crossing. A regular ferry transports cars across the reservoir, providing access to the nearby Waterpocket Fold monocline of Capitol Reef and the Henry Mountains. On the way, however, is a spur to Clay

Fig. G.6. Cheese Box Butte, an erosional remnant of the Triassic Moenkopi Formation resting on a pedestal of Permian Organ Rock Formation.

Hills Crossing, where river trips through the deep lower gorge of the San Juan River take out before reaching the dead water of Lake Powell.

Directly ahead are slopes of Organ Rock topped by a cliff of the Hoskinnini Member. For some distance from here the road closely follows the white sandstone bench formed by the top of the Cedar Mesa Sandstone, although it occasionally rises into the lower red beds of the Organ Rock.

40.6 Straight ahead on the left (south) side of the highway is an old dirt road carved up through the Organ Rock and Moenkopi formations to a uranium mine in the lower part of the Chinle Formation.

43.1 Just ahead is Cheese Box Butte (Fig. G.6). The lower slopes of this circular erosional remnant are in the Organ Rock Formation; the steep sides are the Hoskinnini Member, and the upper slopes are eroded from the upper Moenkopi. The summit block is a knob of resistant Moenkopi sandstone.

45.0 High on the south rim are loose gray slopes spilling over the mesa top. These are tailings from a uranium mine in the basal part of the Chinle Formation. The Chinle, which caps this rim, was explored extensively during the 1950s uranium boom, and several rich mines were developed. Most of the uranium in this area is associated with copper and is concentrated in sandstone and conglomerate of the basal Shinarump Member.

47.4 The Henry Mountains, a cluster of granitic laccoliths, loom ahead. The Colorado River canyon, however, lies between here and these high mountains.

Fig. G.7. Jacobs Chair is composed of resistant orange Wingate Sandstone.

49.1 Crossing of Fry Canyon.

49.5 Fry Canyon Lodge, with a motel, cafe, and gas station, is on the left. This is the only such outpost between Blanding (to the east) and Lake Powell (closed in 2009).

50.8 The large butte to the northwest is Jacobs Chair, carved from the resistant, red eolian Wingate Sandstone (Fig. G.7 and Plate 15). The massive chairlike monolith is named for a local cattleman, Jacob Adams, who died while trying to cross White Canyon before floodwaters receded.

55.7 The knobby red pinnacles on the left (south) side of the road are in the Hoskinnini Member of the Moenkopi. The rest of the Moenkopi overlies it. The red slopes below, at road level, are interbedded sandstone and siltstone of the Organ Rock Formation. The road continues to follow the contact between the Organ Rock Formation and Cedar Mesa Sandstone, into which White Canyon cuts. As it continues west, the steep cliffs recede from the road, and the lavender and red slopes of the Chinle Formation can be seen above the Moenkopi. These are overlain by a continuous cliff of Wingate Sandstone, which to the west is capped by sandstone ledges of the Kayenta. The numerous dirt roads that split southward from this segment of the road climb to uranium mines in the Chinle. This area lies within the White Canyon uranium district, and these mines are some of the richer ones in the area.

56.8 The red- and white-striped cliffs above the road to the left (south) are in the Hoskinnini, which has expanded with the addition of thick, white

FIG. G.8. Close-up view of the conglomerate in the Hoskinnini Member of the Triassic Moenkopi Formation. The coarse, angular fragments are pieces of chert derived from the erosion of the Permian Kaibab Limestone. The matrix is dominantly quartz sand from erosion of the Permian White Rim Sandstone. The lens cap, for scale, is ~10 cm diameter.

FIG. G.9. The depths of White Canyon, cut into the Cedar Mesa Sandstone immediately upstream from the Black Hole.

conglomerate beds. The conglomerate, which can be easily examined in the blocks strewn across the slopes below, is composed of angular white chert pebbles in a fine quartz sand matrix (Fig. G.8). These sediments came from the west in the Circle Cliffs upwarp (modern Capitol Reef area). During the earliest Triassic the Permian White Rim Sandstone and Kaibab Limestone were exposed to erosion as part of this upwarp. Quartz sand came from the White Rim, and chert derives from the Kaibab. Neither is present in this immediate area, although the White Rim appears above the Organ Rock a short distance to the west near the Colorado River.

64.7 The dark mountain island of the Henrys looms prominently in the distance. In front of this great cluster of laccoliths lie orange, humpbacked mounds of Jurassic Navajo Sandstone. Concealed between here and these mounds are the Colorado River and Glen Canyon, both submerged beneath the still water of Lake Powell.

66.1 The butte ahead on the right (north) side of the road is hewn from the Organ Rock Formation. The light-colored unit midway up is an eolian

Fig. G.10. An exposure near Hite, Utah. All of the strata below the White Rim Sandstone are part of the Permian Organ Rock Formation, which is dominantly fluvial but contains thin eolian beds. The Permian White Rim Sandstone succeeds the Organ Rock. Here it is a very thin eolian unit that pinches out a short distance to the east. The blocky cliff band above the White Rim is the Hoskinnini Member of the Moenkopi Formation, a prominent basal conglomerate limited to this area.

sandstone, indicating an arid, windswept fluvial plain environment for this upper part of the formation.

67.2 The bridge spans the narrow slot of lower White Canyon. This deep, shadowed incision is part of the Black Hole, a demanding stretch of the canyon that can be traversed only by swimming through deep pools of very cold water (Fig. G.9 and Plate 16).

69.1 Directly ahead, the finger of eolian sandstone that was encountered several miles back forms a faded orange slickrock ledge that is easily distinguished from the dominantly dark red fluvial facies of the Organ Rock. This bed thickens noticeably westward. To the right (east) the bed exhibits its large-scale eolian cross-stratification.

70.1 The lower part of the butte ahead on the left (west) side of the road clearly shows an interfingering relationship between the white eolian sandstone and the red fluvial deposits (Fig. G.10). In this exposure it looks like a short-lived, shallow Permian river sliced across the eolian sand pile, only to be buried beneath a blanket of fine eolian sand.

71.4 An incredible view can be seen to the northwest as the road emerges from the roadcut in the Organ Rock. The white Cedar Mesa Sandstone forms the foundation of this vista, succeeded by the red Organ Rock with its conspicuous sliver of orange eolian sandstone, continuing to thicken. This is the first glimpse of the overlying Permian White Rim Sandstone, an eolian unit that is absent to the east. Westward the easily recognized White Rim separates the underlying Organ Rock from the similar overlying Moenkopi Formation. A thinner version of the White Rim can be seen to the left (south) in the cliffs high above this pullout. Its characteristic large-scale cross-stratification can also be seen in this high exposure.

73.4 Crossing of the Colorado River, the main artery of the Colorado Plateau canyon country (Fig. G.11). It is this great river that has driven the erosion that ultimately is responsible for the canyons, buttes, plateaus, and other unique landforms that characterize this vast and diverse region. Several years of drought had caused the water level of Lake Powell to drop dramatically by 2004, returning this stretch to a turbulent, free-flowing river with water the color of chocolate milk. The resurrected river cut a channel into the soft, muddy, canyon-rimmed delta that has been building since the early 1960s. In the decades since then, the sediment-laden Colorado River met the slack water of the reservoir here. The abrupt decrease in water power forced the deposition of huge quantities of sand and mud. Today these sediments, once the floor of the reservoir, form a wide, vegetated floodplain that borders the river. What will it be like as you look from this bridge--a free-flowing river or slack-water reservoir? Politicians and bureaucrats would have the public believe that the water will soon return in abundance, and that the recent drought cannot continue. Those who understand the vagaries of the natural world would argue, "Why not?" The denial of serious drought is not just foolish, it is dangerous. When those in charge of water supplies refuse to accept the reality of drought, even as they should be planning for that eventuality, society will suffer. Geologic history tells us that western North America has experienced some of the most long-lived arid conditions in Earth's history. The widespread eolian sandstones that form the immediate surroundings here attest to this. From this vantage point, five units of eolian sandstone that collectively represent tens of millions of years can be seen. The notion that because we are here this will not happen again is ridiculous.

74.5 The exposures here offer the chance to review the different types of rock that characterize this region. The road cuts through the cream-colored

FIG. G.11. View to the west of the free-flowing Colorado River from the bridge on Utah Highway 95 near Hite. This photo was taken in spring 2006, after a prolonged drought. Normally Lake Powell backs upstream past this area. The grassy areas on either side of the river are in mud and sand that were deposited in the reservoir. As the water level dropped, the river reestablished its channel by cutting into these deltaic sediments. The Henry Mountains can be seen in the background.

Permian Cedar Mesa Sandstone with its fine sand and obvious large-scale cross-stratification. The overlying Organ Rock Formation is marked by thin, dark red beds and here is split by the thick, light orange eolian sandstone noted earlier. The overlying yellow-white White Rim Sandstone indicates a wholesale return to an arid eolian setting and marks the end of the Permian. The basal Hoskinnini Member of the Triassic Moenkopi Formation forms a blocky caprock to this Permian succession.

75.6 Crossing of the Dirty Devil River. This trickle of water, which drains a large but mostly dry area, was named by John Wesley Powell during his first expedition down the Colorado River. The discovery of the creek and the origin of its name is best explained in the following excerpt from Powell's classic book *The Exploration of the Colorado River and Its Canyons* (1865):

> As we go down to this point we discover the mouth of a stream which enters from the right. Into this our little boat is turned. The water is exceedingly muddy and has an unpleasant odor. One of the men in the boat following, seeing what we have done, shouts to Dunn and asks whether it is a trout stream. Dunn replies, much disgusted, that it is a "dirty devil," and by this name the river is to be known hereafter.

Later in the expedition Powell named a clear, clean stream in the Grand Canyon Bright Angel Creek in contrast to the Dirty Devil River.

75.9 The Permian strata just reviewed can be seen on both sides of the road. Looking west, however, overlying Triassic and Jurassic rocks can be clearly seen. Above the White Rim Sandstone are the thinly bedded, chocolate brown and dark red Triassic rocks of the Moenkopi Formation. Its horizontal striations are succeeded by debris-cloaked slopes through which purple mudstone of the Chinle Formation can be seen. Standing tall above the Chinle slopes are the orange and black colonnades of Jurassic Wingate Sandstone. Capping these resistant cliffs are thinner-bedded fluvial sandstones of the Kayenta Formation. Set back and out of sight from this vantage point are the domes of Navajo Sandstone. Collectively the Wingate, Kayenta, and Navajo make up the Glen Canyon Group, named for this area.

78.9 As the highway climbs out of the canyon, the eolian bed in the Organ Rock is exposed at road level.

79.2 This deep roadcut reveals the contact between the fluvial Organ Rock and eolian White Rim Sandstone.

79.3 Exposure of an unconformable erosion cut into the top of the White Rim by the Hoskinnini Member of the Moenkopi (Fig. G.12).

79.4 The road enters the red, thinly bedded strata of the Moenkopi Formation. The abundance of symmetrical ripples and mud cracks affirms its tidal flat origin.

79.7 Turn left (east) to the overlook of Lake Powell and Hite Marina, where this route ends. The short spur to the overlook cuts through the Moenkopi, whose well-exposed bedding surfaces display abundant ripples.

80.2 A short walk leads to a precipitous drop straight down to the Colorado River. To the northeast is the junction between the shallow Dirty Devil River and the much larger Colorado. In 2004, after a period of drought, both were recognizable and flowing in reestablished channels. Normally, however, this view is of Lake Powell. The cluster of buildings to the southeast is Hite Marina, which was rendered unusable by the low water level. The boat docks have been moved downstream, where there is still a lake. Hite Marina was named for the settlement of Hite, which was located

Fig. G.12. An exposure of the basal Hoskinnini Member of the Moenkopi Formation on the west side of the Colorado River along Utah Highway 95. The Hoskinnini consists of angular chert pebbles in a matrix of well-rounded quartz sand. The chert is derived from erosion of the Permian Kaibab Limestone, and the quartz sand comes from erosion of the Permian White Rim Sandstone underlying the Hoskinnini Member. Note camera lens cap for scale.

a few miles downstream on the west bank of the river, across from the mouth of White Canyon.

The town of Hite has a much richer and more interesting history than the marina.The now submerged settlement was founded in 1883 by a gold prospector named Cass Hite. Prior to coming to this rugged canyon country, Hite was a member of William Quantrill's Civil War guerillas, a murderous group of unlawful Confederate misfits, and thus he was officially branded an outlaw (Van Cott 1990). Hite had been prospecting in the area since the end of the Civil War and had befriended the local Indians. In fact, the Navajo Chief Hoskinnini told him that there was gold along the Colorado River. Hite's subsequent discovery of small amounts of placer gold in the gravel bars of Glen Canyon led him to build a cabin at the future Hite town site. Eventually he opened a store to supply local miners and others who passed through. A post office was established in 1889, with mail arriving from Green River, a desolate horseback trip over a one-way distance of ~100 punishing miles.

During the Great Depression another prospector, Arthur Chaffin, bought the settlement, and in 1946 he set up a ferry across what had already become known as Dandy Crossing. It was powered by an automobile engine and was kept from being swept downstream by a cable spanning the river. The town became especially busy during the uranium boom of the 1950s as it was the only crossing on the Colorado that was served by a ferry for some distance. This boom stimulated the Atomic Energy Commission to build a uranium-processing mill at the mouth of White Canyon, across the river from Hite. This short-lived mill was active from 1949 to 1954 and resulted in the founding of the town of White Canyon, which had a post office and a thirty-seat schoolhouse. After the mill closed, however, the town site was abandoned and fell into ruin. Finally, in 1964, after the gates of Glen Canyon Dam were closed and the water began to back up, the settlements of Hite and White Canyon, as well as tons of toxic uranium mill tailings, were inundated beneath the water of Lake Powell.

Utah Route 275 to Natural Bridges National Monument

This short road log begins at the intersection of Utah Routes 95 and 275, and describes the geology along 275 into Natural Bridges National Monument. The visitor center and monument headquarters, 10 miles from the intersection of Routes 95 and 275, mark the beginning of a loop road to various trailheads and overlooks. The road closely follows the contact between the Permian Cedar Mesa Sandstone, which lies below, and the overlying Permian Organ Rock Formation. The loop road follows the top of the Cedar Mesa Sandstone but dips down into the formation where it crosses shallow drainages.

Road Log

0.0 Turn north onto Utah Route 275 and enjoy the wonders of upper White Canyon. This 10-mile-long road ends at Natural Bridges National Monument and a loop of overlooks.

0.6 To the right (north) is San Juan County Road 228, a graded dirt road that climbs to the mesa top and Elk Ridge. The road ascends, in succession, the Permian Organ Rock, the Triassic Moenkopi, and the Chinle formations. At the top the road passes between the prominent landmarks of the Bears Ears, which are held up by the cliff-forming Jurassic Wingate Sandstone.

1.0 Another graded dirt road, County Road 254, on the right (north) leads to Deer Flat and numerous uranium prospects.

1.7 The ribbed bluff above the road on right (north) is floored by lower slopes of Organ Rock overlain by a cliff of the Hoskinnini Member of the Lower Triassic Moenkopi Formation.

2.0 The high mesa in the distance to the south is Moss Back Butte, capped by the characteristic red cliff band of the Wingate and Kayenta formations.

3.9 Natural Bridges National Monument boundary.

4.4 Monument headquarters and visitor center. Facilities include restrooms, water, an interpretive center, and a bookstore that also sells maps. The loop road begins at the exit to the visitor center parking lot. The entire loop is in the Cedar Mesa Sandstone.

6.4 Sipapu Bridge overlook. A short stroll provides a spectacular view of both White Canyon and the Cedar Mesa Sandstone that makes up its walls. Large-scale eolian cross-stratification is prominent in the etched sandstone cliffs. The recesses that separate the bulging cliffbands are in the more easily eroded siltstone beds that separate the thick dune deposits. The head of White Canyon, which drains a huge area, is in the saddle that separates the Bears Ears. When there is water in this ephemeral stream, it flows west into the Colorado River or Lake Powell.

Sipapu Bridge, like the other stone "bridges" in this canyon, resembles an arch but has a very different genesis. The evolution of a bridge begins with a stream that meanders across a wide valley cut into soft, easily eroded rock. As the sinuous river channel is cut downward into a harder, more-resistant layer, its snakelike path becomes inscribed into the hard rock in the form of a meandering canyon. At this time a long, necklike ridge of rock extends into the arc of the meander. If the neck becomes eroded enough at its base, as floodwaters crash into it on the outside of a sharp river bend, the neck will thin at its base, and the river will eventually erode through. At this point the river will push through the hole to establish a shorter path, and a true bridge is formed. The old meander is abandoned. This is the scenario envisioned for the evolution of all the bridges at Natural Bridges.

To the northwest, the top of the broad mesa across the canyon is Deer Flat. The steep slopes that rise to the vast tabletop are made of the Organ Rock and Moenkopi formations. The resistant caprock is the basal sandstone in the Chinle Formation.

7.0 Sipapu Bridge trailhead. A 1.2 mile (round-trip) trail winds down through the canyon walls to the bridge. The trail includes several ladders to aid passage through cliff bands to gain the canyon floor. The short hike is well worth it.

7.3 Horsecollar Ruin overlook. A short hike along the rim lead to a spot where the remnants of a kiva and two granaries can be seen below in the canyon.

9.3 Kachina Bridge trailhead and overlook. This bridge is another cut-off meander within the sinuous folds of upper White Canyon. The tributary that enters from the south is from Armstrong Canyon. A steep drop has formed where this drainage enters White Canyon. Called a **pourover,** it is usually a dry waterfall. Because Armstrong Canyon is a much smaller drainage and has less water than White Canyon, it has not been cut as deeply. Erosion in there has not kept up with that of White Canyon, creating the steep drop at their confluence.

11.2 Owachomo Bridge trailhead and overlook. The origin of Owachomo Bridge differs slightly from the other bridges. It spans a hole cut into a narrow fin of rock just downstream of the confluence of Armstrong and Tuwa canyons. The hole is perched above the modern floor of Armstrong Canyon and appears to have been hewn by the flow of water through the canyon. Upon carving the hole, however, the stream did not flow through it. Instead it continued on its original path, where it cut down to its present level, leaving the area beneath the bridge, literally, high and dry.

12.5 View of the Bears Ears to the northeast. These two buttes sit Elk Ridge, a plateau that is held up by the resistant basal sandstone of the Chinle Formation. The lower slopes of the Bears Ears are in the Chinle shale, and they are capped by freestanding pedestals of resistant Wingate Sandstone. The Elk Ridge road passes through the saddle that separates these isolated buttresses.

13.7 End of loop road.

Bluff to Mexican Hat via Utah Highways 191 and 163

The town of Bluff, Utah, was established April 6, 1880, by the worn and exhausted members of the Hole-in-the-Rock expedition after eight grueling months of traversing the rugged wilderness between Escalante, Utah, and this new town site. These families—men, women, children, and babies—had been called on by the Mormon Church to settle the San Juan River country in this empty part of southeast Utah. Their new route dropped to the Colorado River in Glen Canyon through the Hole in the Rock, a steep and narrow defile that was blasted, carved, and constructed through the precipitous cliffs on the west side of the canyon. Even after completing this crossing, these dauntless pioneers still had a long, arduous journey ahead, searching out routes to bypass the rugged canyons of Cedar Mesa and the final obstacle, the Comb Ridge monocline. This was the roughest and most difficult route possible between Escalante and Bluff, and it was used only a few times the following year before being abandoned.

Bluff, initially known as Bluff City, was constructed on the San Juan River floodplain as a closely spaced cluster of one-room log cabins arranged to create a fortlike geometry as protection from the local Utes and Navajos, who had long roamed the region. The broad, flat floodplain, coupled with an abundance of reliable water for irrigation, made the area good for farming, but sporadic flooding of the San Juan required periodic rebuilding. Today this idyllic, cliff-lined town is protected, to a large extent, by upstream dams that control the degree of flooding.

Road Log

0.0 The road log begins at the east edge of town at the intersection of Utah Highway 191 (from Blanding) and Highway 163 (from Montezuma and Aneth). Regardless of the route into town, the road is hemmed in by the slickrock walls of the Jurassic Bluff Sandstone, named for this locality.

Fig. I.1. An outcrop of the Jurassic Wanakah Formation and Bluff Sandstone at the intersection of Highways 191 and 163, where the road log begins.

Here the eolian Bluff Sandstone forms the resistant caprock of the local mesas. Directly beneath it is the Middle Jurassic Wanakah Formation, composed of brown-red, thinly bedded shale, siltstone, and sandstone (Fig. I.1). At ground level, the orange, silty Jurassic Entrada Sandstone emerges beneath the Summerville.

On the southeast corner of this intersection is the stone building of Cow Canyon Trading Post, which sells local Native American rugs, pottery, and other art. It also houses a bookstore and an excellent restaurant. As is evident when one passes through town, today Bluff is inhabited mostly by river and backcountry guides, artists, and those who cater to the tourist industry.

0.9 Cottonwood Creek comes in from the north. This creek drains a huge area and extends into the flanks of the Abajo Mountains, but it is often dry. To the south is a profile of cliffs lining the south side of the San Juan River. Near the base is the orange Entrada Sandstone, overlain by the brown-red slopes of the Wanakah Formation. The smooth cliff face of Bluff Sandstone caps the succession. A finger of thin, less-resistant siltstone can be seen in the middle of the Bluff cliff. This disappears a short distance west as it thins to a complete pinchout.

2.5 The road rises to a bench in the Dewey Bridge that forms a terrace veneered by rounded pebbles and cobbles of an earlier San Juan River.

3.8 To the left (south) is a short road to Sand Island campground and the boat launch for San Juan River trips. The campground area at river level is lined by cliffs of Jurassic Navajo Sandstone with abundant petroglyphs. To the right (north), in the middle of the Bluff Sandstone cliff, is the siltstone interval that can be seen to pinch out quickly to the west. Slopes of Wanakah are beneath.

4.6 Another, higher river terrace with rounded gravel of the old San Juan River. The road that goes left (south) is Highway 191; it crosses a bridge to the Navajo reservation on the other side of the river. The mining operation provides sand and river gravels for building roads and other construction projects.

5.4 The undulating ridge of white sandstone on the skyline is the Lower Jurassic Navajo Sandstone of eolian origin. The Navajo comprises the top of the Lower Jurassic Glen Canyon Group, which also includes the Kayenta Formation and Wingate Sandstone. Here the resistant Navajo marks the crest of the Comb Ridge monocline, which snakes southward to Kayenta, Arizona, and north to the Abajo Mountains. The Navajo Sandstone plunges eastward into the subsurface, as do all the strata affected by this regional tectonic element. Comb Ridge marks the eastern boundary of the Monument upwarp, the most recent incarnation of a broad, uplifted region that exhumes older Pennsylvanian and Permian rocks immediately to the west. This significant upwarp has risen intermittently throughout geologic history, affecting sedimentation in a subtle but recognizable way.

5.9 The contact between the red and gray Carmel Formation and underlying Navajo Sandstone is exposed in the roadcut. As the road drops into Butler Wash, the surrounding rock is Navajo Sandstone, characterized by large-scale cross-stratification.

7.0 A deep roadcut blasted through Comb Ridge in Navajo Sandstone.

7.1 Contact between the Navajo Sandstone and the underlying, dark red Kayenta Formation.

7.2 Contact between the Kayenta Formation and Wingate Sandstone, a massive, red cliff-former.

7.4 Red shale marks the appearance of the Triassic Chinle Formation, which underlies the Wingate.

7.7 Comb Wash is cut into the soft Triassic Chinle and Moenkopi shales and parallels the sinuous barrier wall of Comb Ridge, which can be traced from here to the north and south. This spectacular palisade is held up by

Fig. I.2. The view north from the top of Comb Ridge, showing its sinuous trace. The cliffs that form the upper part of the ridge are the Glen Canyon Group, which are, from bottom to top, the Wingate Sandstone, Kayenta Formation, and Navajo Sandstone (not labeled).

the resistant Glen Canyon Group sandstones seen in previous roadcuts (Fig. I.2).

8.0 Thin, brown-red beds of the Triassic Moenkopi here dip up to 50° eastward. Thin siltstone, shale, and sandstone beds contain abundant ripples, attesting to a tidal-influenced shallow marine environment.

8.3 Contact between the Moenkopi and various Permian formations that change lithology and/or thin in this area. Most of the lighter-colored Permian strata here are in the Cedar Mesa Sandstone, which is a prominent, sandstone-dominated cliff-former just a few miles to the northwest. Here, however, it has become a nonresistant assemblage of gypsum, limestone, and crumbly orange and white sandstone and siltstone. The Cedar Mesa Sandstone is overlain by the Permian Organ Rock Formation, which consists of dark red shale, sandstone, and siltstone that closely resembles the underlying Moenkopi.

8.7 Red beds throughout this section are part of the Permian age lower Cutler beds. Because geologists cannot reach an agreement as to what these strata fit with, or should be called, they currently go by the informal designation of "lower Cutler beds." As the road travels up the east-plunging dip slope formed on resistant beds, the level top of Monument upwarp is reached.

9.3 The resistant, dark gray rock that forms the surface is marine limestone in the lower Cutler beds. The few limestones here, interbedded with the dominant fluvial red beds, mark the transition from a long period of

Fig. I.3. Panoramic view of Valley of the Gods. The sandstone, siltstone, and shale of the lower Cutler beds make up the steep lower slopes, and the eolian Cedar Mesa Sandstone makes up the upper, cliff-forming part.

marine deposition in the Pennsylvanian to mostly continental deposition in the Permian. As the sea regressed westward, the nonmarine fluvial and eolian red beds expanded across the area.

11.4 The buttes, pinnacles, and mesas of Monument Valley can be seen directly ahead on the distant skyline. To the northwest is the southern edge of Cedar Mesa, a vast area of deep canyons and wooded mesa tops held up by the eolian Cedar Mesa Sandstone.

15.0 To the right (north) are the pinnacles and towers of the Valley of the Gods, hewn from lower Cutler beds with remnant caps of Cedar Mesa Sandstone. The buttes directly ahead are carved from the same rocks: the lower, dark red strata are low-energy fluvial deposits of lower Cutler beds with a lighter-colored caprock of Cedar Mesa Sandstone.

16.7 To the right (north) is a graded dirt road that winds through the towers and pinnacles of Valley of the Gods (Fig. I.3). Near the west end of this dirt road is the isolated Valley of the Gods Bed and Breakfast at the base of Cedar Mesa. The 16-mile-long road comes out on Highway 261 at the base of Mokee Dugway, a series of spectacular switchbacks that climb to the top of Cedar Mesa. Highway 261 can be taken southward to the Goosenecks of San Juan State Park (no fee) or back to this road, Highway 163.

17.1 Lower Cutler beds are exposed to the south, capped by a resistant limestone, and rise onto the steep flank of the Raplee anticline, a large asymmetric fold.

Fig. I.4. View of the south edge of Cedar Mesa from Mokee Dugway showing slopes of fluvial Permian lower Cutler beds at the base and the overlying, cliff-forming Permian Cedar Mesa Sandstone of eolian origin.

18.6 To the north is Cedar Mesa. The lower Cutler beds make up the basal slopes, while the upper cliffs are composed of Cedar Mesa Sandstone (Fig. I.4).

19.5 A spectacular view of the Raplee anticline opens to the south. This striped incline of tilted red and gray strata is cut deeply by steep furrows that plunge from its flat top to the San Juan River. Exposed in the outer layers are the red and gray strata of the lower Cutler beds. Deeper in the chute-like canyons is gray limestone of the underlying Pennsylvanian Honaker Trail Formation, shoved up from the depths by the anticlinal uplift.

20.7 On the right (north) is the turnoff to State Highway 261, which leads first to a spur to Goosenecks Overlook. Here the cliff-rimmed San Juan River has incised three consecutive, deep meander loops into the somber gray limestone of the Pennsylvanian Honaker Trail and Paradox formations. Farther up the highway is the previously discussed intersection with the Valley of the Gods road and Mokee Dugway. At the top of Mokee Dugway is the juniper- and piñon-covered expanse of Cedar Mesa. As Route 261 cuts across the mesa top, deep canyons incised into the Cedar Mesa Sandstone are viewed momentarily through the trees. On the east are Lime, Road, Owl, and Fish canyons, and on the west are Slickhorn Gulch and Grand Gulch. Eventually this road intersects Utah Highway 95 near Natural Bridges National Monument.

21.6 To the left (south) is a view of the San Juan River as it cuts into the base of the Raplee anticline.

Fig. I.5. Mexican Hat rock, an unusual erosional remnant in the lower Cutler beds, capped by the base of the Cedar Mesa Sandstone.

22.2 To the left (south) is an excellent view of Mexican Hat rock with a backdrop of the Raplee anticline (Fig. I.5). The basal part of Mexican Hat rock is in the lower Cutler beds. Differential weathering of the interbedded, resistant sandstone and nonresistant shale has resulted in the wide and narrow ribs. The "cap" rock is the base of the resistant Cedar Mesa Sandstone.

23.6 Enter the "town" of Mexican Hat.

24.3 The San Juan River and bridge come into view ahead.

24.5 The San Juan Trading Post, restaurant, and motel. The restaurant provides great views of the river and serves the best Navajo tacos in the region. (This is my opinion rather than scientific fact.) Across the river is the Navajo reservation, and the road continues to Monument Valley, but the road log ends here.

References

Chapter 2

Alvarez, W., E. Staley, D. O'Connor, and M. A. Chan. 1998. A synsedimentary deformation in the Jurassic of southeastern Utah—A case of impact shaking? *Geology*, v. 26, pp. 579–582.

Atchley, S. C., and D. B. Loope. 1993. Low-stand aeolian influence on stratigraphic completeness. Upper member of the Hermosa Formation (latest Carboniferous), southeast Utah, USA. In K. Pye and N. Lancaster, eds., *Aeolian Sediments Ancient and Modern*. International Association of Sedimentologists, Special Publication 16, pp. 127–149.

Baars, D. L. 1976. The Colorado Plateau aulacogen: Key to the continental-scale basement rifting. *Proceedings of the Second International Conference on Basement Tectonics*, pp. 157–164.

Baars, D. L., and C. M. Molenaar. 1971. *Geology of Canyonlands and Cataract Canyon*. Four Corners Geological Society, 6th Field Conference Guidebook, 99 pp.

Baars, D. L., and G. M. Stevenson. 1981. Tectonic evolution of the Paradox basin, Utah and Colorado. In D. L. Wiegand, ed., *Geology of the Paradox Basin*. Rocky Mountain Association of Geologists, 1981 Field Conference, pp. 23–31.

———. 1984. The San Luis uplift, Colorado and New Mexico—An enigma of the Ancestral Rockies. *Mountain Geologist*, v. 21, pp. 57–67.

Baker, A. A. 1933. Geology and oil possibilities of the Moab District, Grand and San Juan Counties, Utah. *U.S. Geological Survey Bulletin 841*, 95 pp.

Biggar, N. E., D. R. Harden, and M. L. Gillam. 1981. Quaternary deposits in the Paradox Basin. In D. L. Wiegand, ed., *Geology of the Paradox Basin*. Rocky Mountain Association of Geologists, 1981 Field Conference, pp. 129–145.

Buchner, E., and T. Kenkmann. 2008. Upheaval Dome, Utah, USA. Impact origin confirmed. *Geology*, v. 36, pp. 227–230.

Chidsey, T. C., Jr., D. E. Eby, and D. M. Lorenz. 1996. Geological and reservoir characterization of small shallow-shelf carbonate fields, southern Paradox basin, Utah. In A. C. Huffman, Jr., W. R. Lund, and L. H. Godwin, eds., *Geology and Resources of the Paradox Basin*. Utah Geological Association Guidebook 25, pp. 39–56.

Choquette, P. W., and J. D. Traut. 1963. Pennsylvanian carbonate reservoirs, Ismay field, Utah and Colorado. In R. O. Bass, ed., *Shelf Carbonates of the Paradox Basin*. Four Corners Geological Society, 4th Field Conference Guidebook, pp. 157–184.

Condon, S. M. 1995. Geology of pre-Pennsylvanian rocks in the Paradox Basin and adjacent areas, southeastern Utah and southwestern Colorado. *U.S. Geological Survey Bulletin* 2000-G, 53 pp.

———. 1997. Geology of the Pennsylvanian and Permian Cutler Group and Permian Kaibab Limestone in the Paradox Basin, southeastern Utah and southwestern Colorado. *U.S. Geological Survey Bulletin* 2000-P, 46 pp.

Doelling, H. H. 1985. Geologic map of Arches National Park and vicinity, Grand County, Utah. Utah Geological Survey, Map 74, scale: 1:50,000.

———. 1988. Geology of Salt Valley anticline and Arches National Park. In *Salt Deformation in the Paradox Basin*. Utah Geological and Mineral Survey Bulletin 122, pp. 1–58.

———. 2000. Geology of Arches National Park, Grand County, Utah. In D. A. Sprinkel, T. C. Chidsey, Jr., and P. B. Anderson, eds., *Geology of Utah's Parks and Monuments*. Utah Geological Association Publication 28, pp. 11–36.

———. 2001. Geologic map of the Moab and eastern part of the San Rafael Desert 30' x 60' quadrangles, Grand and Emery Counties, Utah, and Mesa County, Colorado. Utah Geological Survey, Map 180, scale: 1:100,000.

Elias, G. K. 1963. Habitat of Pennsylvanian algal bioherms, Four Corners area. In R. O. Bass, ed., *Shelf Carbonates of the Paradox Basin*. Four Corners Geological Society, 4th Field Conference Guidebook, pp. 185–203.

Elston, D. P., E. M. Shoemaker, and E. R. Landis. 1962. Uncompahgre front and salt anticline region of Paradox Basin, Colorado and Utah. *American Association of Petroleum Geologists Bulletin*, v. 46, pp. 1857–1878.

Evans, J. E., and J. M. Reed. 1999. Reinterpretation of the Pennsylvanian Molas Formation (San Juan basin) as a loessite, not as a terra rosa paleosol. *Geological Society of America Abstracts with Programs*, v. 31, no. 7, p. 160.

Frahme, C. W., and E. B. Vaughn. 1983. Paleozoic geology and seismic stratigraphy of the northern Uncompahgre front, Grand County, Utah. In J. D. Lowell, ed., *Rocky Mountain Foreland Basins and Uplifts*. Rocky Mountain Association of Geologists Guidebook, pp. 201–211.

Franczyk, K. J., G. Clark, D. C. Brew, and J. K. Pitman. 1995. Chart showing lithology, mineralogy, and paleontology of the Pennsylvanian Hermosa Group at Hermosa mountain, La Plata County, Colorado. U.S. Geological Survey Miscellaneous Investigations Series, Map I-2555. 18 pp.

Gianniny, G. L., and J. A. T. Simo. 1996. Implications of unfilled accommodation space on mixed carbonate-siliciclastic platforms: An example from the Lower Desmoinesian (Middle Pennsylvanian), southwestern Paradox basin, Utah. In M. W. Longman, and M. D. Sonnenfeld, eds., *Paleozoic Systems of the Rocky Mountain Region*. Rocky Mountain Section, SEPM (Society for Sedimentary Geology), pp. 213–234.

Ginsburg, R., R. Rezak, and J. L. Wray. 1971. Geology of calcareous algae—notes for a short course. Sedimenta I, University of Miami, Division of Marine Geology and Geophysics, 71 pp.

Goldhammer, R. K., E. J. Oswald, and P. A. Dunn. 1991. Hierarchy of stratigraphic forcing: Example from Middle Pennsylvanian shelf carbonates of the Paradox basin. In E. K.

Franseen, W. L. Watney, C. G. St. C. Kendall, and W. Ross, Sedimentary modeling: Computer simulations and methods for improved parameter definition. Kansas Geological Survey, Bulletin 233, pp. 361–413.

———. 1994. High frequency glacio-eustatic cyclicity in the Middle Pennsylvanian of the Paradox basin; an evaluation of Milankovitch forcing. In P. L. deBoer and D. G. Smith, eds., *Orbital Forcing and Cyclic Sequences*. International Association of Sedimentologists Special Publication 19, pp. 243–283.

Grammer, M. G., G. P. Eberli, F. S. P. Van Buchem, G. M. Stevenson, and P. Homewood. 1996. Application of high-resolution sequence stratigraphy to evaluate lateral variability in outcrop and subsurface—Desert Creek and Ismay intervals, Paradox basin. In M. W. Longman and M. D. Sonnenfeld, eds., *Paleozoic Systems of the Rocky Mountain Region*. Rocky Mountain Section, SEPM (Society for Sedimentary Geology), pp. 235–266.

Harrison, T. S. 1927. Colorado-Utah salt domes. *American Association of Petroleum Geologists Bulletin*, v. 11, pp. 111–133.

Heyman, O. G. 1983. Distribution and structural geometry of faults and folds along the northwestern Uncompahgre uplift, western Colorado and eastern Utah. In W. R. Averett, ed., *Northern Paradox Basin—Uncompahgre Uplift*. Grand Junction Geological Society, 1983 Field Trip, pp. 45–57.

Hite, R. J. 1975. An unusual northeast-trending fracture zone and its relation to basement wrench faulting in northern Paradox basin, Utah and Colorado. In J. E. Fassett, ed., *Canyonlands Country*. Four Corners Geological Society, 8th Field Conference, pp. 217–223.

Hite, R. J., D. E. Anders, and T. G. Ging. 1984. Organic-rich source rocks of Pennsylvanian age in the Paradox Basin of Utah and Colorado. In J. Woodward, F. F. Meissner, and J. L. Clayton, eds., *Hydrocarbon Source Rocks of the Greater Rocky Mountain Region*. Rocky Mountain Association of Geologists, pp. 255–274.

Hite, R. J., and D. H. Buckner. 1981. Stratigraphic correlations, facies concepts, and cyclicity in Pennsylvanian rocks of the Paradox Basin. In D. L. Wiegand, ed., *Geology of the Paradox Basin*. Rocky Mountain Association of Geologists, 1981 Field Conference, pp. 147–159.

Huffman, A. C., Jr., and Condon, S. M. 1993. Stratigraphy, structure, and paleogeography of Pennsylvanian and Permian rocks, San Juan Basin and adjacent areas, Utah, Colorado, Arizona, and New Mexico. *U.S. Geological Survey Bulletin* 1808-O, 44 pp.

Huntoon, P. W. 1979. The occurrence of ground water in the Canyonlands area of Utah, with emphasis on water in the Permian section. In D. L. Baars, ed., *Permianland*. Four Corners Geological Society, 9th Field Conference Guidebook, pp. 39–46.

———. 1982. The Meander anticline, Canyonlands, Utah: An unloading structure resulting from horizontal gliding on salt. *Geological Society of America Bulletin*, v. 93, pp. 941–950.

———. 1988. Late Cenozoic gravity tectonic deformation related to the Paradox salts in the Canyonlands area of Utah. In Salt Deformation in the Paradox Basin. *Utah Geological and Mineral Survey Bulletin* 122, pp. 82–93.

———. 2000. Upheaval Dome, Canyonlands, Utah: Strain indicators that reveal an impact origin. In D. A. Sprinkel, T. C. Chidsey, Jr., and P. B. Anderson, eds. *Geology of Utah's Parks and Monuments*. Utah Geological Association Publication 28, pp. 619–628.

Huntoon, P. W., G. H. Billingsley, and W. J. Breed. 1982. Geologic map of Canyonlands National Park and vicinity. Canyonlands Natural History Association, scale: 1:62,500.

Huntoon, P. W., and H. R. Richter. 1979. Breccia pipes in the vicinity of Lockhart Basin, Canyonlands area, Utah. In D. L. Baars, ed., *Permianland*. Four Corners Geological Society, 9th Field Conference, pp. 47–53.

Huntoon, P. W., and E. M. Shoemaker. 1995. Roberts rift, Canyonlands, Utah, a natural hydraulic fracture caused by comet or asteroid impact. *Ground Water*, v. 33, pp. 561–569.

Jackson, M. P. A., D. D. Schultz-Ela, M. R. Hudec, I. A. Watson, and M. L. Porter. 1998. Structure and evolution of Upheaval Dome: A pinched-off salt diapir. *Geological Society of America Bulletin*, v. 110, pp. 1547–1573.

Johnson, S. Y., M. A. Chan, and E. A. Konopka. 1992. Pennsylvanian and Early Permian paleogeography of the Uinta-Piceance Basin region, northwestern Colorado and northeastern Utah. *U.S. Geological Survey Bulletin* 1787-CC, 35 pp.

Kenkmann, T. 2003. Dike formation, cataclastic flow, and rock fluidization during impact cratering: An example from the Upheaval Dome structure, Utah. *Earth and Planetary Science Letters*, v. 214, pp. 43–58.

Kenkmann, T., A. Jahn, D. Scherler, and B. A. Ivanov. 2005. Structure and formation of a central uplift: A case study at the Upheaval Dome impact crater, Utah. In T. Kenkmann, F. Horz, and A. Deutsch, eds., Large meteorite impacts III. *Geological Society of America*, Special Paper 384, pp. 85–115.

Kluth, C. F., and P. J. Coney. 1981. Plate tectonics of the Ancestral Rocky Mountains. *Geology*, v. 9, pp. 10–15.

Kosanke, R. M. 1995. Palynology of part of the Paradox and Honaker Trail Formations, Paradox Basin, Utah. *U.S. Geological Survey Bulletin* 2000-L, 7 pp.

Kriens, B. J., E. M. Shoemaker, and K. E. Herkenhoff. 1997. Structure and kinematics of a complex impact crater, Upheaval Dome, southeast Utah. In P. K. Link, and B. J. Kowallis, Mesozoic to recent geology of Utah. *Brigham Young University Geology Studies*, v. 42, part 2, pp. 19–31.

———. 1999. Geology of the Upheaval Dome impact structure, southeast Utah. *Journal of Geophysical Research*, v. 104, pp. 18867–18887.

Leeder, M. 1999. *Sedimentology and Sedimentary Basins: From Turbulence to Tectonics*. Oxford: Blackwell, 592 pp.

Lohman, S. W. 1974. The geologic story of Canyonlands National Park. *U.S. Geological Survey Bulletin* 1327, 126 pp.

Loope, D. B. 1984. Eolian origin of Upper Paleozoic sandstones, southeastern Utah. *Journal of Sedimentary Petrology*, v. 54, pp. 563–580.

———. 1988. Rhizoliths in ancient eolianites. *Sedimentary Geology*, v. 56, pp. 301–314.

Loope, D. B., and Z. E. Haverland. 1988. Giant dessication fissures filled with calcareous eolian sand, Hermosa Formation (Pennsylvanian), southeastern Utah. *Sedimentary Geology*, v. 56, pp. 403–413.

Louie, J. N., S. Chavez-Perez, and G. Plank. 1995. Impact deformation at Upheaval Dome, Canyonlands National Park, Utah revealed by seismic profiles. *EOS*, v. 76, p. 337.

Mallory, W. W. 1960. Outline of Pennsylvanian stratigraphy of Colorado. In R. J. Weimer and J. D. Haun, eds., *Guide to the Geology of Colorado*. Rocky Mountain Association of Geologists, pp. 23–33.

Marshak, S., K. Karlstrom, and J. M. Timmons. 2000. Inversion of Proterozoic extensional faults: An explanation for the pattern of Laramide and Ancestral Rockies intracratonic deformation, United States. *Geology*, v. 28, pp. 735–738.

Mattox, R. B. 1968. Upheaval Dome, a possible salt dome in the Paradox basin, Utah. In R. B. Mattox, ed., *Saline Deposits*. Geological Society of America Special Paper 88, pp. 331–347.
McGill, G. E., and A. W. Stromquist. 1975. Origin of graben in the Needles District, Canyonlands National Park. In J. E. Fassett, ed., *Canyonlands Country*. Four Corners Geological Society, 8th Field Conference, pp. 235–243.
McKee, J. W., N. W. Jones, and T. H. Anderson. 1988. Las Delicias basin: A record of late Paleozoic arc volcanism in northeastern Mexico. *Geology*, v. 16, pp. 37–40.
———. 1999. Late Paleozoic and early Mesozoic history of the Las Delicias terrane, Coahuila, Mexico. In C. Bartolini, J. L. Wilson, and T. F. Lawton, eds., *Mesozoic Sedimentary and Tectonic History of North-Central Mexico*. Geological Society of America Special Paper 340, pp. 161–189.
McKnight, E. T. 1940. Geology of area between Green and Colorado Rivers, Grand and San Juan Counties, Utah. *U.S. Geological Survey Bulletin* 908, 147 pp.
Melton, R. A. 1972. Paleoecology and paleoenvironment of the upper Honaker Trail Formation near Moab, Utah. *Brigham Young University Geology Studies*, v. 19, part 2, pp. 45–88.
Merrill, W. M., and R. M. Winar. 1958. Molas and associated formations in the San Juan Basin–Needle Mountains area, southwestern Colorado. *American Association of Petroleum Geologists Bulletin*, v. 42, pp. 2107–2132.
Miser, H. D. 1924a. Geologic structure of San Juan Canyon and adjacent country, Utah. *U.S. Geological Survey Bulletin* 751-D, 80 pp.
———. 1924b. *The San Juan Canyon, Southeastern Utah*. U.S. Geological Survey Water Supply Paper 538, 80 pp.
Nuccio, V. F., and S. M. Condon. 1996. Burial and thermal history of the Paradox basin, Utah and Colorado, and petroleum potential of the Middle Pennsylvanian Paradox Formation. *U.S. Geological Survey Bulletin* 2000-O, 41 pp.
Okubo, C. H., and R. A. Schultz. 2007. Compactional deformation bands in Wingate Sandstone: Additional evidence of an impact origin for Upheaval Dome, Utah. *Earth and Planetary Science Letters*, v. 256, pp. 169–181.
Peterson, J. A., and R. J. Hite. 1969. Pennsylvanian evaporite-carbonate cycles and their relation to petroleum occurrence, southern Rocky Mountains. *American Association of Petroleum Geologists Bulletin*, v. 53, pp. 884–908.
Phillips, M. 1975. Cane Creek Mine solution mining project, Moab potash operations, Texasgulf Inc. In J. E. Fassett, ed., *Canyonlands Country*. Four Corners Geological Society, 8th Field Conference Guidebook, 261 pp.
Pray, L. C., and J. L. Wray. 1963. Porous algal facies (Pennsylvanian) Honaker Trail, Jan Juan Canyon, Utah. In R. O. Bass, ed., *Shelf Carbonates of the Paradox Basin*. Four Corners Geological Society, 4th Field Conference Guidebook, pp. 204–234.
Prommel, H. W. C., and H. E. Crum. 1927. Salt domes of Permian and Pennsylvanian age in southeastern Utah and their influence on oil accumulation. *American Association of Petroleum Geologists Bulletin*, v. 11, pp. 373–393.
Raup, O. B., and R. J. Hite. 1992. Lithology of evaporite cycles and cycle boundaries in the upper part of the Paradox Formation of the Hermosa Group of Pennsylvanian age in the Paradox Basin, Utah and Colorado. *U.S. Geological Survey Bulletin* 2000-B, 37 pp.

———. 1996. Bromine geochemistry of chloride rocks in the Middle Pennsylvanian Paradox Formation of the Hermosa Group, Paradox Basin, Utah and Colorado. *U.S. Geological Survey Bulletin* 2000-M, 117 pp.

Rueger, B. F. 1996. Palynology and its relationship to climatically induced depositional cycles in the Middle Pennsylvanian (Desmoinesian) Paradox Formation of southeastern Utah. *U.S. Geological Survey Bulletin* 2000-K, 22 pp.

Scherler, D., T. Kenkmann, and A. Jahn. 2006. Structural record of an oblique impact. *Earth and Planetary Science Letters*, v. 248, pp. 43–53.

Scotese, C. R., and J. Golonka. 1992. PALEOMAP Paleogeographic Atlas, PALEOMAP Progress Report #20. Department of Geology, University of Texas at Arlington.

Stevenson, G. M., and D. L. Baars. 1986. The Paradox: A pull-apart basin of Pennsylvanian age. In J. A. Peterson, ed., *Paleotectonics and Sedimentation in the Rocky Mountain Region*. American Association of Petroleum Geologists, Memoir 41, pp. 513–539.

Stokes, W. L. 1948. *Geology of the Utah-Colorado Salt Dome Region with Emphasis on Gypsum Valley, Colorado*. Utah Geological Society Guidebook no. 3, Geology of Utah, 50 pp.

Stone, D. S. 1977. Tectonic history of the Uncompahgre uplift. In H. K. Veal, ed., *Exploration Frontiers of the Central and Southern Rockies*. Rocky Mountain Association of Geologists, pp. 23–30.

Stromquist, A. W. 1976. *Geometry and Growth of Grabens, Lower Red Lake Canyon Area, Canyonlands National Park, Utah*. Department of Geology and Geography, University of Massachusetts, Contribution no. 28, 118 pp.

Timmons, J. M., K. E. Karlstrom, C. M. Dehler, J. W. Geissman, and M. T. Heizler. 2001. Proterozoic multistage (ca. 1.1 and 0.8 Ga) extension recorded in Grand Canyon Supergroup and establishment of northwest- and north-trending tectonic grains in the southwestern United States. *Geological Society of America Bulletin*, v. 113, pp. 163–181.

Trudgill, B. D., and J. A. Cartwright. 1994. Relay ramp forms and normal fault linkages—Canyonlands National Park, Utah. *Geological Society of America Bulletin*, v. 106, pp. 1143–1157.

Tuttle, M. L., T. R. Klett, M. Richardson, and G. N. Breit. 1996. Geochemistry of two interbeds in the Pennsylvanian Paradox Formation, Utah and Colorado: A record of deposition and diagenesis of repetitive cycles in a marine basin. *U.S. Geological Survey Bulletin* 2000-N, 86 pp.

Wengerd, S. A., and M. L. Matheny. 1958. Pennsylvanian system of the Four Corners region. *American Association of Petroleum Geologists Bulletin*, v. 42, pp. 2048–2106.

White, M. A., and M. I. Jacobson. 1983. Structures associated with the southwest margin of the Ancestral Uncompahgre uplift. In W. R. Averett, ed., *Northern Paradox Basin—Uncompahgre Uplift*. Grand Junction Geological Society, 1983 Field Trip, pp. 33–39.

Williams, M. R. 1996. Stratigraphy of Upper Pennsylvanian cyclical carbonate and siliciclastic rocks, western Paradox basin, Utah. In M. W. Longman and M. D. Sonnenfeld, eds., *Paleozoic Systems of the Rocky Mountain Region*. Rocky Mountain Section, SEPM (Society for Sedimentary Geology), pp. 283–304.

Williams, R. L. 1964. Geology, structure, and uranium deposits of the Moab quadrangle, Colorado and Utah. U.S. Geological Survey Miscellaneous Investigation Series, Map I-360, scale: 1:250,000.

Williams-Stroud, S. C. 1994. The evolution of an inland sea of marine origin to a non-marine lake: The Pennsylvanian Paradox salt. In R. W. Renaut and W. M. Last, eds.,

Sedimentology and Geochemistry of Modern and Ancient Saline Lakes. Society for Sedimentary Geology (SEPM) Special Publication 50, pp. 293–306.

Ye, H., L. Royden, C. Burchfiel, and M. Schuepbach. 1996. Late Paleozoic deformation of interior North America: The greater Ancestral Rocky Mountains. *American Association of Petroleum Geologists Bulletin*, v. 80, pp. 1397–1432.

Chapter 3

Baars, D. L. 1962. Permian system of Colorado Plateau. *Bulletin of the American Association of Petroleum Geologists*, v. 46, pp. 149–218.

———. 1987. The Elephant Canyon Formation revisited. In J. A. Campbell, ed., *Geology of Cataract Canyon and Vicinity*. Four Corners Geological Society, 10th Field Conference Guidebook, p. 81–90.

———. 1991. The Elephant Canyon Formation—for the last time. *Mountain Geologist*, v. 28, pp. 1–2.

Baars, D. L., and W. R. Seager. 1970. Stratigraphic control of petroleum in White Rim Sandstone (Permian) in and near Canyonlands National Park, Utah. *American Association of Petroleum Geologists Bulletin*, v. 54, pp. 709–718.

Baker, A. A., and J. B. Reeside, Jr. 1929. Correlation of the Permian of southern Utah, northern Arizona, northwestern New Mexico, and southwestern Colorado. *Bulletin of the American Association of Petroleum Geologists*, v. 13, pp. 1413–1448.

Blakey, R. C. 1980. Pennsylvanian and Early Permian paleogeography, southern Colorado Plateau and vicinity. In T. D. Fouch and E. R. Magathan, eds., *Paleozoic Paleogeography of the west-central United States*. Rocky Mountain Section, Society of Economic Paleontologists and Mineralogists, pp. 239–258.

———. 1996. Permian eolian deposits, sequences, and sequence boundaries, Colorado Plateau. In M. W. Longman and M. D. Sonnenfield, eds., *Paleozoic Systems of the Rocky Mountain Region*. Rocky Mountain Section, SEPM (Society for Sedimentary Geology), pp. 405–426.

Bush, A. L., C. S. Bromfield, and C. T. Pierson. 1959. Areal geology of the Placerville quadrangle, San Miguel County, Colorado. *U.S. Geological Survey Bulletin* 1072-E, pp. 299–384.

Campbell, J. A. 1979. Lower Permian depositional system, northern Uncompahgre basin. In D. L. Baars, ed., *Permianland*. Four Corners Geological Society Guidebook, 9th Field Conference, pp. 13–21.

———. 1980. Lower Permian depositional systems and Wolfcampian paleogeography, Uncompahgre basin, eastern Utah and southwestern Colorado. In T. D. Fouch and E. R. Magathan, eds., *Paleozoic Paleogeography of West-central United States*. Rocky Mountain Paleogeography Symposium 1. Rocky Mountain Section, Society of Economic Paleontologists and Mineralogists, pp. 327–340.

———. 1981. Uranium mineralization and depositional facies in the Permian rocks of the northern Paradox basin, Utah and Colorado. In D. L. Wiegand, ed., *Geology of the Paradox Basin*. Rocky Mountain Association of Geologists, 1981 Field Conference, pp. 187–194.

Campbell, J. A., and H. R. Ritzma. 1979. *Geology and Petroleum Resources of the Major Oil-impregnated Sandstone Deposits of Utah*. Utah Geological and Mineralogical Survey Special Studies 50, 24 pp.

Cater, F. W. 1970. *Geology of the Salt Anticline Region in Southwestern Colorado*. U.S. Geological Survey, Professional Paper 637, 80 pp.

Chan, M. A. 1989. Erg margin of the Permian White Rim Sandstone, SE Utah. *Sedimentology*, v. 36, pp. 235–251.

Condon, S. M. 1997. Geology of the Pennsylvanian and Permian Cutler Group and Permian Kaibab Limestone in the Paradox basin, southeastern Utah and southwestern Colorado. *U.S. Geological Survey Bulletin* 2000-P, 46 pp.

Cook, D. A. 1991. Sedimentology and shale petrology of the Upper Proterozoic Walcott Member, Kwagunt Formation, Chuar Group, Grand Canyon, Arizona. Master's thesis, Northern Arizona University, Flagstaff, Arizona, 158 pp.

Cross, W., and E. Howe. 1905. Red beds of southwestern Colorado and their correlations. *Geological Society of America*, v. 16, 476 pp.

Cross, W., E. Howe, and L. Ransome. 1905. Silverton folio. U.S. Geological Survey Atlas, Folio 120, 40 pp.

Cross, C. W., and A. C. Spencer. 1900. *Geology of the Rico Mountains, Colorado*. U.S. Geological Survey Annual Report 21, pp. 37–88.

Dane, C. H. 1935. Geology of the salt anticline and adjacent areas, Grand County, Utah. *U.S. Geological Survey Bulletin* 863, 179 pp.

Dubiel, R. F., J. E. Huntoon, S. M. Condon, and J. D. Stanesco. 1996. Permian deposystems, paleogeography, and paleoclimate of the Paradox basin and vicinity. In M. W. Longman and M. D. Sonnenfield, eds., *Paleozoic Systems of the Rocky Mountain Region*. Rocky Mountain Section, SEPM (Society for Sedimentary Geology), pp. 427–444.

Gregory, H. E. 1938. *The San Juan Country*. U.S. Geological Survey Professional Paper 188, 123 pp.

Huntoon, J. E., and M. A. Chan. 1987. Marine origin of paleotopographic relief on eolian White Rim Sandstone (Permian), Elaterite basin, Utah. *American Association of Petroleum Geologists*, v. 71, pp. 1035–1045.

Huntoon, P. W., G. H. Billingsley, Jr., and W. J. Breed. 1982. Geologic map of Canyonlands National Park and vicinity, Utah. Moab, Utah, Canyonlands Natural History Association, scale 1:62:500.

Jackson, J. A., ed. 1997. *Glossary of Geology*. 4th ed. American Geological Institute, 769 pp.

Kamola, D. L., and M. A. Chan. 1988. Coastal dune facies, Permian Cutler Formation (White Rim Sandstone), Capitol Reef National Park area, southern Utah. *Sedimentary Geology*, v. 56, pp. 341–356.

Kunkel, R. P. 1958. Permian stratigraphy of the Paradox Basin. In A. F. Sanborn, ed., *Guidebook to the Geology of the Paradox Basin*. Intermountain Association of Petroleum Geologists, 9th Annual Field Conference, pp. 163–168.

Langford, R., and M. A. Chan. 1988. Flood surfaces and deflation surfaces within the Cutler Formation and Cedar Mesa Sandstone (Permian), southeastern Utah. *Geological Society of America Bulletin*, v. 100, pp. 1541–1549.

Lewis, G. E., and P. P. Vaughn. 1965. *Early Permian Vertebrates from the Cutler Formation of the Placerville Area, Colorado*. U.S. Geological Survey Professional Paper 503-C, pp. C1-C50.

Loope, D. B. 1984. Eolian origin of upper Paleozoic sandstones, southeastern Utah. *Journal of Sedimentary Petrology*, v. 54, pp. 563–580.

———. 1985. Episodic deposition and preservation of eolian sands: A late Paleozoic example from southeastern Utah. *Geology*, v. 13, pp. 73–76.

———. 1988. Rhizoliths in ancient eolianites. *Sedimentary Geology*, v. 56, pp. 301–314.
Loope, D. B., G. A. Sanderson, and G. J. Verville. 1990. Abandonment of the name "Elephant Canyon Formation" in southeastern Utah: Physical and temporal implications. *Mountain Geologist*, v. 27, pp. 119–130.
Mack, G. H., and K. A. Rasmussen. 1984. Alluvial-fan sedimentation of the Cutler Formation (Permo-Pennsylvanian) near Gateway, Colorado. *Geological Society of America Bulletin*, v. 95, pp. 109–116.
McKnight, E. T. 1940. Geology of the area between Green and Colorado Rivers, Grand and San Juan counties, Utah. *U.S. Geological Survey Bulletin* 908, 147 pp.
Sanderson, G. A., and G. J. Verville. 1990. Fusulinid zonation of the General Petroleum No. 45-5-G core, Emery County, Utah. *Mountain Geologist*, v. 27, pp. 131–136.
Sanford, R. F. 1995. Ground-water flow and migration of hydrocarbons to the Lower Permian White Rim Sandstone, Tar Sand Triangle, southeastern Utah. *U.S. Geological Survey Bulletin* 200-J, 24 pp.
Schultz, A. W. 1984. Subaerial debris-flow deposition in the upper Paleozoic Cutler Formation, western Colorado. *Journal of Sedimentary Petrology*, v. 54, pp. 759–772.
Stanesco, J. D., and J. A. Campbell. 1989. Eolian and noneolian facies of the Lower Permian Cedar Mesa Sandstone Member of the Cutler Formation, southeastern Utah. *U.S. Geological Survey Bulletin* 1808-F, 13 pp.
Stanesco, J. D., R. F. Dubiel, and J. E. Huntoon. 2000. Depositional environments and paleotectonics of the Organ Rock Formation of the Permian Cutler Group, southeastern Utah. In D. A. Sprinkel, T. C. Chidsey, Jr., and P. B. Anderson, eds., *Geology of Utah's Parks and Monuments*. Utah Geological Association, Publication 28, pp. 591–605.
Steele, B. A. 1987. Depositional environments of the White Rim Sandstone Member of the Permian Cutler Formation, Canyonlands National Park, Utah. U.S. *Geological Survey Bulletin* 1592. 20 pp.
Terrell, F. M. 1972. Lateral facies and paleoecology of Permian Elephant Canyon Formation, Grand County, Utah. *Brigham Young University Geology Studies*, v. 19, pp. 3–44.
Tidwell, W. D. 1988. A new Upper Pennsylvanian or Lower Permian flora from southeastern Utah. *Brigham Young University Geology Studies*, v. 35, pp. 33–56.
Uphoff, T. L. 1997. Precambrian Chuar source rock play: An exploration case history in southern Utah. *American Association of Petroleum Geologists Bulletin*, v. 81, pp. 1–15.

Chapter 5

Baker, S. P., and J. E. Huntoon. 1996. Depositional analysis of the Black Dragon Member of the Triassic Moenkopi Formation, southeastern Utah. In A. C. Huffman, W. R. Lund, and L. H. Goodwin, eds., *Geology and Resources of the Paradox Basin*. Utah Geological Association Guidebook 25, pp. 173–190.
Blakey, R. C. 1973. Stratigraphy and origin of the Moenkopi Formation (Triassic) of southeastern Utah. *Mountain Geologist*, v. 10, pp. 1–17.
———. 1974. Stratigraphic and depositional analysis of the Moenkopi Formation, southeastern Utah. *Utah Geological and Mineralogical Survey Bulletin* 104, 81 pp.
———. 1977. Petroliferous lithosomes in the Moenkopi Formation, southern Utah. *Utah Geology*, v. 4, pp. 67–84.

Blakey, R. C., E. L. Basham, and M. J. Cook. 1993. Early and Middle Triassic paleogeography of the Colorado Plateau and vicinity. In M. Morales, ed., Aspects of Mesozoic geology and paleontology of the Colorado Plateau. *Museum of Northern Arizona Bulletin* 59, pp. 13–26.

Blakey, R. C., and R. Gubitosa. 1983. Late Triassic paleogeography and depositional history of the Chinle Formation, southern Utah and northern Arizona. In M. W. Reynolds and E. E. Dolly, eds., *Mesozoic Paleogeography of West Central United States*. Rocky Mountain Section, Society of Economic Paleontologists and Mineralogists, Rocky Mountain Paleogeography Symposium 2, pp. 57–76.

Blodgett, R. H. 1984. Nonmarine depositional environments and paleosol development in the Upper Triassic Dolores Formation, southwestern Colorado. In D. C. Brew, ed., *Field Trip Guidebook*. Geological Society of America, Rocky Mountain Section, 37th Annual Meeting, Durango, Colorado, Fort Lewis College, pp. 46–92.

———. 1988. Calcareous paleosols in the Triassic Dolores Formation, southwestern Colorado. In J. Reinhardt and W. R. Sigleo, eds., *Paleosols and Weathering through Geologic Time: Principles and Applications*. Geological Society of America Special Paper 216, pp. 103–121.

Cater, F. W., Jr. 1970. *Geology of the Salt Anticline Region in Southwestern Colorado*. U.S. Geological Survey Professional Paper 637, 80 pp.

Cross, W., and C. W. Purrington. 1899. Telluride Folio, Colorado. U.S. Geological Survey Geologic Atlas of the United States, Folio 57, 18 pp.

Deacon, M. W. 1990. Depositional analysis of the Sonsela Sandstone bed, Chinle Formation, northeast Arizona and northwest New Mexico. M.S. thesis, Northern Arizona University, Flagstaff, 127 pp.

Dubiel, R. S. 1994. Triassic deposystems, paleogeography, and paleoclimate of the Western Interior. In M. V. Caputo, J. A. Peterson, and K. J. Franczyk, eds., *Mesozoic Systems of the Rocky Mountain Region, USA*. Rocky Mountain Section, SEPM (Society for Sedimentary Geology), pp. 133–168.

Dubiel, R. F., R. H. Blodgett, and T. M. Bown. 1987. Lungfish burrows in the Upper Triassic Chinle and Dolores formations, Colorado Plateau. *Journal of Sedimentary Petrology*, v. 57, pp. 512–521.

Dubiel, R. F., J. T. Parrish, J. M. Parrish, and S. C. Good. 1991. The Pangaean megamonsoon—Evidence from the Upper Triassic Chinle Formation, Colorado Plateau. *Palaios*, v. 6, pp. 347–370.

Fillmore, R. P. 2006. A salt anticline-controlled fluvial system: A preliminary study of the Ali Baba Member of the Triassic Moenkopi Formation, eastern Utah and western Colorado. *Geological Society of America Abstracts with Programs*, v. 38, no. 6, p. 7.

Gilluly, J., and J. B. Reeside. 1928. Sedimentary rocks of the San Rafael Swell and some adjacent areas in eastern Utah. *U.S. Geological Survey Professional Paper* 150-D, pp. 61–110.

Gregory, H. E. 1917. Geology of the Navajo country—a reconnaissance of parts of Arizona, New Mexico, and Utah. *U.S. Geological Survey Professional Paper* 93, 161 pp.

Hasiotis, S. T., and C. E. Mitchell. 1989. Lungfish burrows in the Upper Triassic Chinle and Dolores formations, Colorado Plateau—Discussion: New evidence suggests origin by a burrowing decapod crustacean. *Journal of Sedimentary Petrology*, v. 59, pp. 871–875

Hazel, J. E., Jr. 1994. Sedimentary response to intrabasinal salt tectonism in the Upper Triassic Chinle Formation, Paradox Basin, Utah. *U.S. Geological Survey Bulletin* 2000-F, 34 pp.

Huntoon, J. E., R. F. Dubiel, and J. D. Stanesco. 1994. Tectonic influence on development of the Permian-Triassic unconformity and basal Triassic strata, Paradox basin, southeastern Utah. In M. V. Caputo, J. A. Peterson, and K. J. Franczyk, eds., *Mesozoic Systems of the Rocky Mountain Region, USA*. Rocky Mountain Section, SEPM (Society for Sedimentary Geology), pp. 109–131.

Irwin, C. D. 1971. Stratigraphic analysis of Upper Permian and Lower Triassic strata in southern Utah. *American Association of Petroleum Geologists Bulletin*, v. 55, pp. 1976–2007.

Lucas, S. G., A. B. Heckert, J. W. Estep, and O. J. Anderson. 1997. Stratigraphy of the Upper Triassic Chinle Group, Four Corners region. In O. J. Anderson, B. S. Kues, and S. G. Lucas, eds., *Mesozoic Geology and Paleontology of the Four Corners Region*. New Mexico Geological Society Guidebook, 48th Field Conference, pp. 81–107.

McKee, E. D. 1954. Stratigraphy and history of the Moenkopi Formation of Triassic age. *Geological Society of America Memoir* 61, 133 pp.

McKnight, E. T. 1940. Geology of the area between Green and Colorado rivers, Grand and San Juan Counties, Utah. *U.S. Geological Survey Bulletin* 908, 147 pp.

Ochs, S., and M. A. Chan. 1990. Petrology, sedimentology and stratigraphic implications of Black Dragon Member of the Triassic Moenkopi Formation, San Rafael Swell, Utah. *Mountain Geologist*, v. 27, pp. 1–18.

Pipiringos, G. N., and R. B. O'Sullivan. 1978. Principal unconformities in Triassic and Jurassic rocks, western interior United States—a preliminary report. *U.S. Geological Survey Professional Paper* 1035-A, 29 pp.

Powell, J. W. 1876. Report on the geology of the eastern portion of the Uinta Mountains and a region of country adjacent thereto. *U.S. Geological and Geographical Survey of the Territories*, 218 pp.

Riggs, N. R., T. M. Lehman, G. E. Gehrels, and W. R. Dickinson. 1996. Detrital zircon link between the headwaters and terminus of the Upper Triassic Chinle-Dockum paleoriver system. *Science*, v. 273, pp. 97–100.

Robeck, R. C. 1956. Temple Mountain Member—New member of Chinle Formation in San Rafael Swell, Utah. *American Association of Petroleum Geologists Bulletin*, v. 40, pp. 2499–2506.

Shoemaker, E. M., and W. L. Newman. 1959. Moenkopi Formation (Triassic? and Triassic) in salt anticline region, Colorado and Utah. *Bulletin of the American Association of Petroleum Geologists*, v. 43, pp. 1835–1851.

Stewart, J. H. 1957. Proposed nomenclature of Upper Triassic strata in southeastern Utah. *American Association of Petroleum Geologists Bulletin*, v. 53, pp. 1866–1879.

———. 1959. Stratigraphic relations of Hoskinnini Member (Triassic?) of Moenkopi Formation on Colorado Plateau. *Bulletin of the American Association of Petroleum Geologists*, v. 43, pp. 1852–1868.

Stewart, J. H., F. G. Poole, and R. F. Wilson. 1972a. Stratigraphy and origin of the Triassic Moenkopi Formation and related strata in the Colorado Plateau region. U.S. *Geological Survey Professional Paper* 691. 195 pp.

———. 1972b. Stratigraphy and origin of the Chinle Formation and related Upper Triassic strata in the Colorado Plateau region. *U.S. Geological Survey Professional Paper* 690, 336 pp.

Stewart, J. H., and R. F. Wilson. 1960. Triassic strata of the salt anticline region, Utah and Colorado. In K. G. Smith, ed., *Geology of the Paradox Basin Fold and Fault Belt*. Four Corners Geological Society, Third Field Conference Guidebook, pp. 98–106.

Thaden, R. E., A. F. Trites, and T. L. Finnell. 1964. Geology and ore deposits of the White Canyon area, San Juan and Garfield counties, Utah. *U.S. Geological Survey Bulletin* 1125, 166 pp.

Chapter 6

Alvarez, W., E. Staley, D. O'Connor, and M. A. Chan. 1998. A synsedimentary deformation in the Jurassic of southeastern Utah—A case of impact shaking? *Geology*, v. 26, pp. 579–582.

Aubrey, W. M. 1996. Stratigraphic architecture and deformational history of Early Cretaceous foreland basin, eastern Utah and southwestern Colorado. In A. C. Huffman, Jr., W. R. Lund, and L. H. Godwin, eds., *Geology and Resources of the Paradox Basin*. Utah Geological Association, Guidebook 25, pp. 211–220.

Baker, A. A., C. H. Dane, and J. B. Reeside, Jr. 1936. Correlation of the Jurassic formations of parts of Utah, Arizona, New Mexico and Colorado. *U.S. Geological Survey Professional Paper* 183, 66 pp.

Bjerrum, C. J., and R. J. Dorsey. 1995. Tectonic controls on deposition of Middle Jurassic strata in a retroarc foreland basin, Utah-Idaho trough, western interior, United States. *Tectonics*, v. 14, pp. 962–978.

Blakey, R. C. 1994. Paleogeographic and tectonic controls on some Lower and Middle Jurassic erg deposits, Colorado Plateau. In M. V. Caputo, J. A. Peterson, and K. J. Franczyk, eds., *Mesozoic Systems of the Rocky Mountain Region, USA*. Rocky Mountain Section, SEPM (Society for Sedimentary Geology), pp. 273–298.

Bowman, S. A. G., J. T. Bowman, and R. E. Drake. 1986. Interpretation of the Morrison Formation as a time-transgressive unit. Fourth North American Paleontological Convention, Abstracts with Programs, p. 5.

Bromley, M. H. 1991. Architectural features of the Kayenta Formation (Lower Jurassic), Colorado Plateau, USA: Relationship to salt tectonics in the Paradox Basin. *Sedimentary Geology*, v. 73, pp. 77–99.

Caputo, M. V., and W. A. Pryor. 1991. Middle Jurassic tide- and wave-influenced coastal facies and paleogeography, upper San Rafael Group, east-central Utah. In T. C. Chidsey, Jr., ed., *Geology of East-Central Utah*. Utah Geological Association, Publication 19, pp. 9–25.

Chenoweth, W. L. 1975. Uranium deposits of the Canyonlands area. In J. E. Fassett, ed., *Canyonlands Country*. Four Corners Geological Society Guidebook, pp. 253–260.

———. 1996. The uranium industry in the Paradox Basin. In A. C. Huffman, Jr., W. R. Lund, and L. H. Godwin, eds., *Geology and Resources of the Paradox Basin*. Utah Geological Association, Guidebook 25, pp. 95–108.

Clemmensen, L. B., H. Olsen, and R. C. Blakey. 1989. Erg-margin deposits in the Lower Jurassic Moenave Formation and Wingate Sandstone, southern Utah. *Geological Society of America Bulletin* 101, pp. 759–773.

Cohenour, R. E. 1967. History of uranium and development of Colorado Plateau ores, with notes on uranium production in Utah. In L. F. Hintze, ed., *Uranium Districts of Southeastern Utah*. Utah Geological Society, Guidebook to the Geology of Utah, no. 21, pp. 12–22.

Craig, L. C., and D. R. Shawe. 1975. Jurassic rocks of east-central Utah. In J. E. Fassett, ed., *Canyonlands Country*. Four Corners Geological Society Guidebook, 8th Field Conference, pp. 157–165.

Dane, C. H. 1935. *Geology of the Salt Valley Anticline and Adjacent areas, Grand County, Utah*. U.S. Geological Survey Bulletin 863, 184 pp.

Dickinson, W. R., and G. E. Gehrels. 2002. Provenance of Colorado Plateau Permian and Jurassic eolianites as inferred from U-Pb detrital zircon ages. Geological Society of America Annual Meeting, Abstracts with Programs, v. 34. no. 6, p. 208.

———. 2003. U-Pb ages of detrital zircons from Permian and Jurassic eolian sandstones of the Colorado Plateau, USA: Paleogeographic implications. *Sedimentary Geology*, v. 163, pp. 29–66.

Doelling, H. H. 2000. Geology of Arches National Park, Grand County, Utah. In D. A. Sprinkel, T. C. Chidsey, Jr., and P. B. Anderson, eds., *Geology of Utah's Parks and Monuments*. Utah Geological Association, Publication 28, pp. 11–36.

Dutton, C. E. 1885. Mount Taylor and the Zuni Plateau. *U.S. Geological Survey Sixth Annual Report*, pp. 105–188.

Gilland, J. K. 1979. Paleoenvironment of a carbonate lens in the lower Navajo Sandstone near Moab, Utah. *Utah Geology*, v. 6, pp. 29–38.

Gilluly, J., and J. B. Reeside, Jr. 1928. Sedimentary rocks of the San Rafael Swell and some adjacent areas in eastern Utah. *U.S. Geological Survey Professional Paper* 150-D, pp. 61–110.

Goldman, M. I., and A. C. Spencer. 1941. Correlation of Cross' La Plata Sandstone, southwestern Colorado. *American Association of Petroleum Geologists Bulletin*, v. 25, pp. 1745–1766.

Gregory, H.E. 1917. Geology of the Navajo country. *U.S. Geological Survey Professional Paper* 93. 161 pp.

———. 1938. The San Juan country. *U.S. Geological Survey Professional Paper* 188. 123 p.

———. 1950. Geology and geography of the Zion National Park region, Utah and Arizona. U.S. Geological Survey Professional Paper 220, 200 pp.

Harshbarger, J. W., C. A. Repenning, and J. H. Irwin. 1957. Stratigraphy of the uppermost Triassic and the Jurassic rocks of the Navajo country. *U.S. Geological Survey Professional Paper* 291, 74 pp.

Kocurek, G., and R. H. Dott. 1983. Jurassic paleogeography and paleoclimate of the central and southern Rocky Mountain region. In M. W. Reynolds and E. D. Dolly, eds., *Mesozoic Paleogeography of the West-central United States*. Rocky Mountain Paleogeography Symposium 2, Society of Economic Paleontologists and Mineralogists, pp. 101–118.

Kowallis, B. J., E. H. Christiansen, and A. L. Deino. 1991. Age of the Brushy Basin Member of the Morrison Formation, Colorado Plateau, western USA. *Cretaceous Research*, v. 12, pp. 483–493.

Kowallis, B. J., and J. S. Heaton. 1987. Fission–track dating of bentonites and bentonitic mudstones from the Morrison Formation in central Utah. *Geology*, v. 15, pp. 1138–1142.

Lucas, S. G., and O. J. Anderson. 1997. The Jurassic San Rafael Group, Four Corners region. In O. J. Anderson, B. S. Kues, and S. G. Lucas, eds., *Mesozoic Geology and Paleontology of the Four Corners Region*. New Mexico Geological Society Guidebook, 48th Field Conference, pp. 115–132.

Luttrell, P. R. 1993a. Basinwide sedimentation and continuum of paleoflow in an ancient river system. Kayenta Formation (Lower Jurassic), central portion Colorado Plateau. *Sedimentary Geology*, v. 85, pp. 411–434.

———. 1993b. Jurassic depositional history of the Colorado Plateau. In M. Morales, ed., Aspects of Mesozoic geology and paleontology of the Colorado Plateau. *Museum of Northern Arizona Bulletin* 59, pp. 99–110.

McKnight, E. T. 1940, Geology of the area between Green and Colorado Rivers, Grand and San Juan Counties, Utah. *U.S. Geological Survey Bulletin* 908, 147 pp.

Molenaar, C. M. 1981. Mesozoic stratigraphy of the Paradox basin—An overview. In D. L. Weigand, ed., *Geology of the Paradox Basin*. Rocky Mountain Association of Geologists, 1981 Field Conference Guidebook, pp. 119–127.

Nation, M. J. 1990. Analysis of eolian architecture and depositional systems in the Jurassic Wingate Sandstone, central Colorado Plateau. M.S. thesis, Northern Arizona University, Flagstaff, 222 pp.

O'Sullivan, R. B. 1980. Stratigraphic sections of Middle Jurassic San Rafael Group from Wilson Arch to Bluff in southeastern Utah. U.S. Geological Survey Oil and Gas Investigations Chart OC-12.

———. 1981. The Middle Jurassic San Rafael Group and related rocks in east-central Utah. In R. C. Epis and J. F. Callender, eds., *Western Slope Colorado*. New Mexico Geological Society Guidebook, 32nd Field Conference, pp. 89–95.

Peterson, F. 1986. Jurassic paleotectonics in the west-central part of the Colorado Plateau, Utah and Arizona. In J. A. Peterson, ed., *Paleotectonics and Sedimentation in the Rocky Mountain Region, United States*. American Association of Petroleum Geologists Memoir 41, pp. 563–596.

———. 1988. Stratigraphy and nomenclature of Middle and Upper Jurassic rocks, western Colorado Plateau, Utah and Arizona. *U.S. Geological Survey Bulletin* 1633-B, pp. 17–56.

———. 1994. Sand dunes, sabkhas, streams, and shallow seas: Jurassic paleogeography in the southern part of the Western Interior basin. In M. V. Caputo, J. A. Peterson, and K. J. Franczyk, eds., *Mesozoic Systems of the Rocky Mountain Region, USA*. Rocky Mountain Section, SEPM (Society for Sedimentary Geology), pp. 233–272.

Peterson, F., and G. N. Pipiringos. 1979. Stratigraphic relations of the Navajo Sandstone to Middle Jurassic Formations, southern Utah and northern Arizona. *U.S. Geological Survey Professional Paper* 1035-B, 43 pp.

Peterson, F., and C. E. Turner-Peterson. 1987. The Morrison Formation of the Colorado Plateau: Recent advances in sedimentology, stratigraphy, and paleotectonics. *Hunteria*, v. 2, pp. 1–18.

Pipiringos, G. N., and R. B. O'Sullivan. 1978. Principal unconformities in Triassic and Jurassic rocks, Western Interior United States—A preliminary survey. *U.S. Geological Survey Professional Paper* 1035-A, 29 pp.

Riggs, N. R., and R. C. Blakey. 1993. Early and Middle Jurassic paleogeography and volcanology of Arizona and adjacent areas. In G. C. Dunne and K. A. McDougall, eds., *Mesozoic Paleogeography of the Western United States II*. Pacific Section, Society of Economic Paleontologists and Mineralogists, pp. 347–375.

Stokes, W. L. 1967. A survey of southeastern Utah uranium districts. In L. F. Hintze, ed., *Uranium Districts of Southeastern Utah*. Utah Geological Society, Guidebook to the Geology of Utah, no. 21, pp. 12–22.

Tschudy, R. H., B. D. Tschudy, and L. C. Craig. 1984. Palynological evaluation of Cedar Mountain and Burro Canyon Formations, Colorado Plateau. *U.S. Geological Survey Professional Paper* 1281, 24 pp.

Turner, C. E., and N. S. Fishman. 1991. Jurassic Lake T'oo'dichi': A large alkaline, saline lake, Morrison Formation, eastern Colorado Plateau. *Geological Society of America Bulletin*, v. 103, pp. 538–558.

Turner, C. E., and F. Peterson. 2004. Reconstruction of the Upper Jurassic Morrison Formation extinct ecosystem—a synthesis. In C. E. Turner, F. Peterson, and S. P. Dunagan, eds., Reconstruction of the extinct ecosystem of the Upper Jurassic Morrison Formation. *Sedimentary Geology*, v. 167, pp. 309–355.

Wright, J. C., D. R. Shawe, and S. W. Lohman. 1962. Definition of members of Jurassic Entrada Sandstone in east-central Utah and west-central Colorado. *Bulletin of the American Association of Petroleum Geologists*, v. 46, pp. 2057–2070.

Chapter 7

Alvarez, L. A., W. Alvarez, F. Asaro, and H. V. Michel. 1980. Extraterrestrial cause for the Cretaceous-Tertiary extinction. Science, v. 208, pp. 1095–1108.

Anderson, P. B., T. C. Chidsey, Jr., and T. A. Ryer. 1997. Fluvial-deltaic sedimentation and stratigraphy of the Ferron Sandstone. In P. K. Link and B. J. Kowallis, eds., *Geological Society of America Field Trip Guidebook, 1997 Annual Meeting*. Brigham Young University Geology Studies, v. 42, part II, pp. 135–154.

Aubrey, W. M. 1996. Stratigraphic architecture and deformational history of Early Cretaceous foreland basin, eastern Utah and southwestern Colorado. In A. C. Huffman, Jr., W. R. Lund, and L. H. Godwin, eds., *Geology and Resources of the Paradox Basin*. Utah Geological Association, Guidebook 25, pp. 211–220.

Chan, M. A., and B. J. Pfaff. 1991. Fluvial sedimentology of the Upper Cretaceous Castlegate Sandstone, Book Cliffs, Utah. In T. C. Chidsey, Jr., ed., *Geology of East-central Utah*. Utah Geological Association, Publication 19, pp. 95–109.

Cifelli, R. L., J. I. Kirkland, A. Weil, A. R. Deinos, and B. J. Kowallis. 1997. High-precision 40Ar/39Ar geochronology and the advent of North America's Late Cretaceous terrestrial fauna. *Proceedings of the National Academy of Science*, USA, v. 97, pp. 11163–11167.

Cole, R. D. 1987. Cretaceous rocks of the Dinosaur Triangle. In W. R. Averitt, ed., *Paleontology and Geology of the Dinosaur Triangle*. Guidebook for 1987 Field Trip, Grand Junction, Colorado, Museum of Western Colorado, pp. 21–35.

Craig, L. C. 1981. Lower Cretaceous rocks, southwestern Colorado and southeastern Utah. In D. L. Weigand, ed., *Geology of the Paradox Basin*. Rocky Mountain Association of Geologists, 1981 Field Conference Guidebook, pp. 119–127.

Cross, C. W., and C. W. Purington. 1899. Description of the Telluride quadrangle [Colorado]. *U.S. Geological Survey Geologic Atlas of the United States*, Telluride folio, no. 57. 18 pp.

DeCourten, F. 1998. *Dinosaurs of Utah*. Salt Lake City, University of Utah Press, 300 pp.

Difley, R. L., B. B. Britt, B. W. Greenhalgh, and E. H. Christiansen. 2004. First discovery of the Cretaceous-Tertiary (K-T) boundary clay in Utah. Geological Society of America, Abstracts with Programs, v. 36. no. 4, p. 7.

Doelling, H. H. 2000. Geology of Arches National Park, Grand County, Utah. In D. A. Sprinkel, T. C. Chidsey, Jr., and P. B. Anderson, eds., *Geology of Utah's Parks and Monuments*. Utah Geological Association, Publication 28, pp. 11–36.

Fouch, T. D., T. F. Lawton, D. J. Nichols, W. B. Cashion, and W. A. Cobban. 1983. Patterns and timing of synorogenic sedimentation in Upper Cretaceous rocks of central and northeast Utah. In M. W. Reynolds and E. D. Dolly, eds., *Mesozoic Paleogeography of the West-central United States*. Rocky Mountain Section, Society of Economic Paleontologists and Mineralogists, pp. 305–336.

Garrison, J. R., Jr., T. C. V. van den Bergh, C. E. Barker, and D. E. Tabet. 1997. Depositional sequence stratigraphy and architecture of the Cretaceous Ferron Sandstone: Implications for coal and coalbed methane resources—a field excursion. In P. K. Link and B. J. Kowallis, eds., *Geological Society of America Field Trip Guidebook*, 1997 Annual Meeting. Brigham Young University Geology Studies, v. 42, part II, pp. 155–202.

Goldstrand, P. M. 1992. Evolution of Late Cretaceous and early Tertiary basins of southwest Utah based on clastic petrology. *Journal of Sedimentary Petrology*, v. 62, pp. 495–507.

Hildebrand, A. R., G. T. Penfield, D. A. Kring, M. Pilkington, A. Z. Camargo, S. B. Jacobsen, and W. V. Boynton. 1991. A possible Cretaceous-Tertiary boundary impact crater on the Yucatan Peninsula, Mexico. *Geology*, v. 19, pp. 867–871.

Johnson, J. L. 1978. Stratigraphy of the coal-bearing Blackhawk Formation on North Horn Mountain, Wasatch Plateau, Utah. *Utah Geology*, v. 5, pp. 57–77.

Kauffman, E. G. 1977. Geological and biological overview. Western Interior Cretaceous basin. In E. G. Kauffman, ed., Cretaceous facies, faunas, and paleoenvironments across the Western Interior basin. *Mountain Geologist*, v. 14. nos. 3 and 4, pp. 75–99.

Kirkland, J. I., B. B. Britt, D. L. Burge, K. Carpenter, R. L. Cifelli, D. L. DeCourten, J. G. Eaton, S. Hasiotis, and T. F. Lawton. 1997. Lower to Middle Cretaceous dinosaur faunas of the central Colorado Plateau: A key to understanding 35 million years of tectonics, evolution, and biogeography. In P. K. Link and B. J. Kowallis, eds., Geological Society of America Field Trip Guidebook, 1997 Annual Meeting. *Brigham Young University Geology Studies*, v. 42, part II, pp. 69–103.

Kirkland, J. I., D. Burge, and R. Gaston. 1993. A large dromaeosaur (Theropoda) from the Lower Cretaceous of Utah. *Hunteria*, v. 2, pp. 1–16.

Kirkland, J. I., R. L. Cifelli, B. B. Britt, D. L. Burge, F. L. DeCourten, J. G. Eaton, and J. M. Parrish. 1999. Distribution of vertebrate faunas in the Cedar Mountain Formation, east-central Utah. In D. D. Gillette, ed., *Vertebrate Paleontology in Utah*. Utah Geological Survey, Miscellaneous Publication 99-1, pp. 201–217.

Kring, D. A. 2000. Impact events and their effect on the origin, evolution, and distribution of life. *GSA Today*, v. 10, no. 8, pp. 1–7.

Lawton, T. F. 1983. Late Cretaceous fluvial systems and the age of foreland uplifts in central Utah. In J. D. Lowell, ed., *Rocky Mountain Foreland Basins and Uplifts*. Rocky Mountain Association of Geologists, pp. 181–199.

———. 1986. Fluvial systems of the Upper Cretaceous Mesaverde Group and Paleocene North Horn Formation, central Utah: A record of transition from thin-skinned to thick-skinned deformation in the foreland region. In J. A. Peterson, ed., Paleotectonics and sedimentation in the Rocky Mountain region, United States. *American Association of Petroleum Geologists Memoir* 41, pp. 423–442.

Lawton, T. F., S. L. Pollock, and R. A. J. Robinson. 2003. Integrating sandstone petrology and nonmarine sequence stratigraphy: Application to the Late Cretaceous fluvial systems of southwestern Utah, U.S.A. *Journal of Sedimentary Research*, v. 73, pp. 389–406.

Marley, W. E., R. M. Flores, and V. V. Cavaroc. 1979. Coal accumulation in Upper Cretaceous marginal deltaic environments of the Blackhawk Formation and Star Point Sandstone, Emery, Utah. *Utah Geology*, v. 6, pp. 25–40.

Matheny, J. P., and M. D. Picard. 1985. Sedimentology and depositional environments of the Emery Sandstone Member of the Mancos Shale, Emery and Sevier Counties, *Utah. Mountain Geologist*, v. 22, pp. 94–109.

McGookey, D. P., J. D. Haun, L. A. Hale, H. G. Goodell, D. G. McCubbin, R. J. Weimer, and G. R. Wulf. 1972. Cretaceous system. In W. W. Mallory, ed., *Geologic Atlas of the Rocky Mountain Region*. Rocky Mountain Association of Geologists, pp. 190–228.

Morgan, J., and M. Warner. 1999. Chicxulub: The third dimension of a multi-ring impact basin. *Geology*, v. 27, pp. 407–410.

Newman, K. F., and M. A. Chan. 1991. Depositional facies and sequences in the Upper Cretaceous Panther Tongue Member of the Star Point Formation, Wasatch Plateau. In T. C. Chidsey, Jr., ed., *Geology of East-central Utah*. Utah Geological Association, Publication 19, pp. 65–76.

O'Byrne, C. J., and S. Flint, S. 1995. Sequence, parasequence, and intraparasequence architecture of the Grassy Member, Blackhawk Formation, Book Cliffs, Utah, U.S.A. In J. C. Van Wagoner and G. T. Bertram, eds., Sequence stratigraphy of foreland basin deposits: Outcrop and subsurface examples from the Cretaceous of western North America. American *Association of Petroleum Geologists Memoir* 64, pp. 225–256.

Parker, L. R. 1976. The paleoecology of the fluvial coal-forming swamps and associated floodplain environments in the Blackhawk Formation (Upper Cretaceous) of central Utah. *Brigham Young University Studies*, v. 22, pp. 99–116.

Peterson, F., R. T. Ryder, and B. E. Law. 1980. Stratigraphy, sedimentology and regional relationships of the Cretaceous System in the Henry Mountains region. In M. D. Picard, ed., *Henry Mountains Symposium*. Utah Geological Association, Publication 8, pp. 151–170.

Ryer, T. A. 1981. Deltaic coas of the Ferron Sandstone Member of the Mancos Shale—a predictive model for Cretaceous coal-bearing strata of the western interior. *American Association of Petroleum Geologists Bulletin*, v. 65, pp. 2323–2340.

Ryer, T. A., and M. McPhillips. 1983. Early Late Cretaceous paleogeography of east-central Utah. In M. W. Reynolds and E. D. Dolly, eds., *Mesozoic Paleogeography of West-central United States*. Rocky Mountain Section, Society of Economic Paleontologists and Mineralogists, pp. 253–272.

Schwans, P. 1988. Depositional response of Pigeon Creek Formation, Utah, to initial fold-thrust belt deformation in a differentially subsiding basin. In C. J. Schmidt and W. J. Perry, Jr., eds., Interaction of the Rocky Mountain foreland and the Cordilleran thrust belt. *Geological Society of America Memoir* 171, pp. 531–556.

Smit, J. 1999. The global stratigraphy of the Cretaceous-Tertiary boundary impact ejecta. *Annual Review of Earth and Planetary Sciences*, v. 27, pp. 75–113.

Stokes, W. L. 1944. Morrison and related deposits in and adjacent to the Colorado Plateau. *Geological Society of America Bulletin*, v. 55, pp. 951–992.

———. 1952. Lower Cretaceous in Colorado Plateau. American Association of Petroleum Geologists Bulletin, v. 36, pp. 1766–1776.

Stokes, W. L., and D. A. Phoenix. 1948. Geology of the Egnar-Gypsum Valley area, San Miguel and Montrose Counties, Colorado. U.S. *Geological Survey Oil and Gas Investigation*, Preliminary Map 93.

Swisher, C. C., and eleven others. 1992. Coeval 40Ar/39Ar ages of 65.0 million years ago from Chicxulub crater melt rock and Cretaceous-Tertiary boundary tektites. *Science*, v. 257, pp. 954–958.

Tschudy, R. H., B. D. Tschudy, and L. C. Craig. 1984. Palynological evaluation of Cedar Mountain and Burro Canyon Formations, Colorado Plateau. *U.S. Geological Survey Professional Paper* 1281, 24 pp.

van de Graaf, F. R. 1972. Fluvial-deltaic facies of the Castlegate Sandstone (Cretaceous), east-central Utah. *Journal of Sedimentary Petrology*, v. 42, pp. 558–571.

Willis, A. 2000. Tectonic control of nested sequence architecture in the Sego Sandstone, Neslen Formation and upper Castlegate Sandstone (Upper Cretaceous), Sevier foreland basin, Utah, USA. *Sedimentary Geology*, v. 136, pp. 277–317.

Yoshida, S. 2000. Sequence and facies architecture of the upper Blackhawk Formation and lower Castlegate Sandstone (Upper Cretaceous), Book Cliffs, Utah, USA. *Sedimentary Geology*, v. 136, pp. 239–276.

Chapter 8

Anderson, D. W., and M. D. Picard. 1972. Stratigraphy of the Duchesne River Formation (Eocene-Oligocene?), northern Uinta Basin, northeastern Utah. *Utah Geological and Mineralogical Survey Bulletin* 97, 29 pp.

Armstrong, R. L. 1969. K-Ar dating of laccolithic centers of the Colorado Plateau and vicinity. *Geological Society of America Bulletin*, v. 80, pp. 2081–2086.

Berry, E. W. 1925. Flora and ecology of so-called Bridger beds of Wind River Basin, Wyoming. *Pan-Am Geologist*, v. 44, pp. 357–368.

Best, M. G., and E. H. Christiansen. 1991. Limited extension during peak Tertiary volcanism, Great Basin of Nevada and Utah. *Journal of Geophysical Research*, v. 96B, pp. 13509–13528.

Bradley, W. H. 1929. The varves and climate of the Green River epoch. *U.S. Geological Survey Professional Paper* 158-E, pp. 87–110.

———. 1931. Origin and microfossils of the oil shale of the Green River Formation of Colorado and Utah. *U.S. Geological Survey Professional Paper* 168, 58 pp.

———. 1964. Geology of the Green River Formation and associated Eocene rocks in southwestern Wyoming and adjacent parts of Colorado and Utah. *U.S. Geological Survey Professional Paper* 496-A, 86 pp.

Brown, R. W. 1929. Additions to the flora of the Green River formation. *U.S. Geological Survey Professional Paper* 154, pp. 279–293.

Bump, A. P. 2003. Reactivation, trishear modeling, and folded basement in Laramide uplifts: Implications for the origins of intra-continental faults. *GSA Today*, v. 13, pp. 4–10.

Cashion, W. B. 1967. Geology and fuel resources of the Green River Formation southeastern Uinta basin, Utah and Colorado. *U.S. Geological Survey Professional Paper* 548, 48 pp.

———. 1974. Revision of nomenclature of the upper part of the Green River Formation, Piceance Creek basin, Colorado, and eastern Uinta basin, Utah. *U.S. Geological Survey Bulletin* 1394-G, 9 pp.

Coven, B., K. Panter, and A. Stork. 1999. Ar/Ar age of West Elk volcano, Gunnison and Delta Counties, Colorado. Geological Society of America, Abstracts with Program, v. 31. no. 7, p. 478.

Dane, C. H. 1954. Stratigraphic and facies relationships of upper part of Green River Formation and lower part of Uinta Formation in Duchesne, Uintah, and Wasatch counties, Utah. *Bulletin of the American Association of Petroleum Geologists*, v. 38, pp. 405–425.

Davis, G. H. 1978. Monocline fold pattern of the Colorado Plateau. In V. Matthews III, ed., Laramide folding associated with basement block faulting in the western United States. *Geological Society of America Memoir* 151, pp. 215–233.

Dayvault, R. D., L. A. Codington, D. Kohls, W. D. Hawes, and P. M. Ott. 1995. Fossil insects and spiders from three locations in the Green River Formation of the Piceance basin, Colorado. In W. R. Averett, ed., *The Green River Formation in Piceance Creek and Eastern Uinta Basins*. Grand Junction Geological Society Field Trip Guide, pp. 97–115.

Dickinson, R. G., E. B. Leopold, and R. F. Marvin. 1968. Late Cretaceous uplift and volcanism on the north flank of the San Juan Mountains, Colorado. *Colorado School of Mines Quarterly*, v. 63, pp. 125–148.

Dickinson, W. R., and W. S. Snyder. 1978. Plate tectonics of the Laramide orogeny. In V. Matthews III, ed., Laramide folding associated with basement block faulting in western United States. *Geological Society of America Memoir* 151, pp. 355–366.

Dickinson, W. R., and six others. 1988. Paleogeographic and paleotectonic setting of Laramide sedimentary basins in the central Rocky Mountain region. *Geological Society of America Bulletin*, v. 100, pp. 1023–1039.

Dyni, J. R. 1974. Stratigraphy and nahcolite resources of the saline facies of the Green River Formation in northwest Colorado. In D. K. Murray, ed., *Guidebook to the Energy Resources of the Piceance Creek Basin, Colorado*. Rocky Mountain Association of Geologists, 25th Field Conference Guidebook, pp. 111–121.

Faulds, J. E., M. A. Wallace, L. A. Gonzalez, and M. T. Heisler. 2001. Depositional environment and paleogeographic implications of late Miocene Hualapai Limestone, northwestern Arizona and southern Nevada. In Young and Spamer, eds., *The Colorado River*. Grand Canyon Association Monograph 12, pp. 81–87.

Fillmore, R. 2000. *Geology of the Parks, Monuments, and Wildlands of Southern Utah*. Salt Lake City, University of Utah Press, 268 pp.

Fisher, D. J., C. E. Erdmann, and J. B. Reeside, Jr. 1960. Cretaceous and Tertiary formations of the Book Cliffs, Carbon, Emery, and Grand Counties, Utah and Garfield and Mesa Counties, Colorado. *U.S. Geological Survey Professional Paper* 332, 80 pp.

Fouch, T. D. 1976. Revision of the lower part of the Tertiary System in the central and western Uinta basin, Utah. *U.S. Geological Survey Bulletin* 1405-C, 7 pp.

Fouch, T. D., T. F. Lawton, D. J. Nichols, W. B. Cashion, and W. A. Cobban. 1983. Patterns and timing of synorogenic sedimentation in Upper Cretaceous rocks of central and northeast Utah. In M. W. Reynolds and E. D. Dolly, eds., *Mesozoic Paleogeography of the West-central United States*. Rocky Mountain Section, Society of Economic Paleontologists and Mineralogists, pp. 305–336.

Franczyk, K. J., and J. K. Pitman. 1987. Basal Tertiary conglomerate sequence, southeastern Uinta basin, Utah: A preliminary report. In J. A. Campbell, ed., *Geology of Cataract Canyon and Vicinity*. Four Corners Geological Society Guidebook, 10th Field Conference, pp. 119–126.

Franczyk, K. J., J. K. Pitman, and D. J. Nichols. 1990. Sedimentology, mineralogy, palynology, and depositional history of some uppermost Cretaceous and lowermost Tertiary rocks along the Utah and Roan Cliffs east of the Green River. *U.S. Geological Survey Bulletin* 1787-N, 27 pp.

Gilbert, G. K. 1877. *Report on the Geology of the Henry Mountains*. Department of Interior, U.S. Geographical and Geological Survey of the Rocky Mountain Region, 160 pp.

Hunt, C. B. 1953. Geology and geography of the Henry Mountains region, Utah. *U.S. Geological Survey Professional Paper* 228, 234 pp.

———. 1958. Structural and igneous geology of the La Sal Mountains, Utah. *U.S. Geological Survey Professional Paper* 294-I, pp. 305–364.

Huntoon, P. W. 1993. Influence of inherited Precambrian basement structure on the localization and form of Laramide monoclines, Grand Canyon, Arizona. In C. J. Schmidt, R. B. Chase, and E. A. Erslev, eds., Laramide basement deformation in the Rocky Mountain foreland of the western United States. *Geological Society of America Special Paper* 280, pp. 243–256.

Isby, S. J., and M. D. Picard. 1983. Currant Creek Formation: Record of tectonism in Sevier-Laramide orogenic belt, northcentral Utah. *Contributions to Geology* (University of Wyoming), v. 22, pp. 91–108.

———. 1985. Depositional setting of Upper Cretaceous–Lower Tertiary Currant Creek Formation, north-central Utah. In M. D. Picard, ed., *Geology and Energy Resources, Uinta Basin of Utah*. Salt Lake City, Utah Geological Association, pp. 39–48.

Jackson, M. 1998. Processes of laccolithic emplacement in the southern Henry Mountains, southeastern Utah. In J. D. Friedman and A. C. Huffman, Jr., eds., Laccolith complexes of southeastern Utah: Time of emplacement and tectonic setting— workshop proceedings. *U.S. Geological Survey Bulletin* 2158, pp. 51–59.

Jackson, M. D., and D. D. Pollard. 1988. The laccolith-stock controversy: New results from the southern Henry Mountains, Utah. *Geological Society of America Bulletin*, v. 100, pp. 117–139.

Johnson, K. R., and C. Plumb. 1995. Common plant fossils from the Green River Formation at Douglas Pass, Colorado and Bonanza, Utah. In W. R. Averett, ed., *The Green River Formation in Piceance Creek and Eastern Uinta Basins*. Grand Junction Geological Society, Field Trip Guide, pp. 121–130.

Johnson, R. C. 1985. Early Cenozoic history of the Uinta and Piceance Creek basins, Utah and Colorado, with special reference to the development of Eocene Lake Uinta. In R. M. Flores and S. S. Kaplan, eds., *Cenozoic Paleogeography of West-central United States*. Denver, Colorado, Rocky Mountain Section SEPM, pp. 247–276.

Jordan, T. E., and R. W. Allmendinger. 1986. The Sierra Pampeanas of Argentina: A modern analogue of Rocky Mountain foreland deformation. *American Journal of Science*, v. 286, pp. 737–764.

Jordan, T. E., and R. N. Alonso. 1987. Cenozoic stratigraphy and basin tectonics of the Andes Mountains, 20°–28° south latitude. *American Association of Petroleum Geologists Bulletin*, v. 71, pp. 49–64.
Jordan, T. E., B. L. Isacks, R. W. Allmendinger, J. A. Brewer, V. A. Ramos, and C. J. Ando. 1983. Andean tectonics related to geometry of subducted Nazca plate. *Geological Society of America Bulletin*, v. 94, pp. 341–361.
Kay, J. L. 1957. The Eocene vertebrates of the Uinta basin, Utah. In O. G. Seal, ed., *Guidebook to the Geology of the Uinta Basin*. Intermountain Association of Petroleum Geologists, 8th Annual Field Conference, pp. 110–114.
Kelley, V. C. 1955a. Regional tectonics of the Colorado Plateau and relationship to the origin and distribution of uranium. *University of New Mexico Publications in Geology*, no. 5, 120 pp.
———. 1955b. Monoclines of the Colorado Plateau. *Bulletin of the Geological Society of America*, v. 66, pp. 789–804.
Knowlton, F. H. 1923. Revision of the flora of the Green River formation. *U.S. Geological Survey Professional Paper* 131, pp. 133–197.
Lawton, T. F. 1985. Style and timing of frontal structures, thrust belt, central Utah. *American Association of Petroleum Geologists*, v. 69, p. 1145–1159.
———. 1986. Fluvial systems of the Upper Cretaceous Mesaverde Group and Paleocene North Horn Formation, central Utah: A record of transition from thin-skinned to thick-skinned deformation in the foreland region. In J. A. Peterson, ed., Paleotectonics and sedimentation in the Rocky Mountain region, United States. *American Association of Petroleum Geologists Memoir* 41, pp. 423–442.
Lipman, P. W. 1983. Tectonic setting of the Mid to Late Tertiary in the Rocky Mountain region. The genesis of Rocky Mountain ore deposits: Changes with time and tectonics. *Denver Region of Exploration Geologists Society*, pp. 125–132.
Livaccarri, R. F., and F. V. Perry. 1993. Isotopic evidence for preservation of Cordilleran lithospheric mantle during the Sevier-Laramide orogeny, western United States. *Geology*, v. 21, pp. 719–722.
Longwell, C. R. 1946. How old is the Colorado River? *American Journal of Science*, v. 244, pp. 817–835.
Lucchitta, I. 1972. Early history of the Colorado River in the basin and range province. *Geological Society of America Bulletin*, v. 83, pp. 1933–1947.
Marsh, O. C. 1871. On the geology of the eastern Uinta Mountains. *American Journal of Science*, 3rd series, v. 1, pp. 191–198.
Marshak, S., K. Karlstrom, and J. M. Timmons. 2000. Inversion of Proterozoic extensional faults: An explanation for the pattern of Laramide and Ancestral Rockies intracratonic deformation, United States. *Geology*, v. 28, pp. 735–738.
Mauger, R. L. 1977. K-Ar ages of biotites from tuffs in Eocene rocks of the Green River, Washakie, and Uinta basins, Utah, Wyoming, and Colorado. *Contributions to Geology* (University of Wyoming), v. 15, pp. 17–41.
Maxson, J., and B. Tikoff. 1996. Hit-and-run collision model for the Laramide orogeny, western United States. *Geology*, v. 24, pp. 968–972.
McDowell, F. W., J. A. Wilson, and J. Clark. 1973. K-Ar dates for biotite from two paleontologically significant localities: Duchesne River Formation, Utah and Chadron Formation, South Dakota. *Isochron/West*, no. 7, pp. 11–12.

McQuarrie, N., and C. G. Chase. 2000. Raising the Colorado Plateau. *Geology*, v. 28, pp. 91–94.

Molenaar, C. M. 1987. Tectonic map of southeastern Utah and adjacent areas. In J. A. Campbell, ed., *Geology of Cataract Canyon and Vicinity*. Four Corners Geological Society, 10th Field Conference Guidebook, p. 9.

Morgan, P. 2003. Colorado Plateau and Southern Rocky Mountains uplift and erosion. In R. G. Raynolds and R. M. Flores, eds., *Cenozoic Systems of the Rocky Mountain Region*. Denver, Colorado, Rocky Mountain Section, SEPM, pp. 1–31.

Morris, T. H., D. R. Richmond, and J. E. Marino. 1991. The Paleocene/Eocene Colton Formation: A fluvial-dominated lacustrine deltaic system, Roan Cliffs, Utah. In T. C. Chidsey, ed., *Geology of East-central Utah*. Utah Geological Association Field Symposium, Guidebook 19, pp. 129–139.

Morris, T. H., D. R. Richmond, J. E. Marino, A. Garner, M. B. Wegner, B. Thomas, and D. Tingey. 2003. The Paleocene/Eocene Colton Formation–Green River transition: Sedimentology and reservoir characterization. In R. G. Raynolds and R. M. Flores, eds., *Cenozoic Systems of the Rocky Mountain Region*. Denver, Colorado, Rocky Mountain Section, SEPM, pp. 213–225.

Murphy, R. T., A. Stork, K. Panter, W. McIntosh, and R. Esser. 2000. 40Ar/39Ar geochronology and petrogenesis of a Middle Tertiary volcano-laccolith complex. Geological Society of America, Abstracts with Program, v. 32. no. 7, pp. 495.

Nelson, S. T. 1998. Reevaluation of the central Colorado Plateau laccoliths in the light of new age determinations. In J. D. Friedman and A. C. Huffman, Jr., eds., Laccolith complexes of southeastern Utah: Time of emplacement and Tectonic setting – Workshop proceedings. *U.S. Geological Survey Bulletin* 2158, pp. 37–39.

Nelson, S. T., and J. P. Davidson. 1998. The petrogenesis of Colorado Plateau laccoliths and their relationships to regional magmatism. In J. D. Friedman and A. C. Huffman, Jr., eds., Laccolith complexes of southeastern Utah: Time of emplacement and tectonic setting—workshop proceedings. *U.S. Geological Survey Bulletin* 2158, pp. 85–100.

Nelson, S. T., J. P. Davidson, and K. R. Sullivan. 1992. New age determinations of central Colorado Plateau laccoliths, Utah—Recognizing disturbed K-Ar systematics and re-evaluating tectonomagmatic relationships. *Geological Society of America Bulletin*, v. 104, pp. 1547–1560.

Osborn, H. F. 1895. Fossil mammals of the Uinta Basin. *American Museum of Natural History Bulletin*, v. 7, pp. 72–105.

Pederson, J. L., R. D. Mackley, and J. L. Eddleman. 2002. Colorado Plateau uplift and erosion evaluated using GIS. *GSA Today*, v. 12, no. 8, pp. 4–10.

Perry, M. L. 1995. Preliminary description of a new fossil scorpion from the middle Eocene, Green River Formation, Rio Blanco County, Colorado. In W. R. Averett, ed., *The Green River Formation in Piceance Creek and Eastern Uinta Basins*. Grand Junction Geological Society, Field Trip Guide, pp. 131–133.

Ranney, W. 2005. *Carving Grand Canyon: Evidence, Theories, and Mystery*. Grand Canyon Association, 160 pp.

Remy, R. R. 1992. Stratigraphy of the Eocene part of the Green River Formation in the south-central part of the Uinta basin, Utah. *U.S. Geological Survey Bulletin* 1787-BB, 79 pp.

Ritzma, H. R. 1974. Asphalt Ridge structure, stratigraphy and oil-impregnated sands. *Energy Resources of the Uinta Basin*. Utah Geological Association Publication 4, p. 60.

Ross, M. L. 1998. Geology of the intrusive centers of the La Sal Mountains, Utah—Influence of preexisting structural features on emplacement and morphology. In J. D. Friedman and A. C. Huffman, Jr., eds., Laccolith complexes of southeastern Utah: Time of emplacement and tectonic setting—workshop proceedings. *U.S. Geological Survey Bulletin* 2158, pp. 61–83.

Rowley, P. D., C. G. Cunningham, T. A. Steven, H. H. Mehnert, and C. W. Naeser. 1998. Cenozoic igneous and tectonic setting of the Marysvale volcanic field and its relation to other igneous centers in Utah and Nevada. In J. D. Friedman and A. C. Huffman, Jr., eds., Laccolith complexes of southeastern Utah: Time of emplacement and tectonic setting—workshop proceedings. *U.S. Geological Survey Bulletin* 2158, pp. 167–201.

Ryder, R. T., T. D. Fouch, and J. T. Elison. 1976. Early Tertiary sedimentation in the western Uinta basin, Utah. *Geological Society of America Bulletin*, v. 87, pp. 496–512.

Sears, J. D. 1956. Geology of Comb Ridge and vicinity north of San Juan River, San Juan County, Utah. *U.S. Geological Survey Bulletin* 1021-E, pp. 167–207.

Spencer, J. E. 1996. Uplift of the Colorado Plateau due to lithosphere attenuation during Laramide low-angle subduction. *Journal of Geophysical Research*, v. 101, pp. 13595–13609.

Spencer, J. E., L. Peters, W. C. McIntosh, and P. J. Patchett. 2001. 40Ar/39Ar geochronology of the Hualapai Limestone and Bouse Formation and implications for the age of the lower Colorado River. In Young and Spamer, eds., *The Colorado River*. Grand Canyon Association Monograph 12, pp. 89–99.

Stanley, K. O., and J. W. Collinson. 1979. Depositional history of Paleocene-lower Eocene Flagstaff Limestone and coeval rocks, central Utah. *American Association of Petroleum Geologists Bulletin*, v. 63, pp. 311–323.

Sullivan, K. R. 1998. Isotopic ages of igneous intrusions in southeastern Utah. In J. D. Friedman and A. C. Huffman, Jr., eds., Laccolith complexes of southeastern Utah: Time of emplacement and tectonic setting—workshop proceedings. *U.S. Geological Survey Bulletin* 2158, pp. 33–35.

Surdam, R. C., and K. O. Stanley. 1980. Effects of changes in drainage-basin boundaries in sedimentation in Eocene Lakes Gosiute and Uinta of Wyoming, Utah, and Colorado. *Geology*, v. 8, pp. 135–139.

Thompson, G. A., and M. L. Zoback. 1979. Regional geophysics of the Colorado Plateau. *Tectonophysics*, v. 61, pp. 149–181.

Tikoff, B., and J. Maxson. 2001. Lithospheric buckling of the Laramide foreland during Late Cretaceous and Paleogene, western United States. *Rocky Mountain Geology*, v. 36, pp. 13–35.

Witkind, I. J. 1964. Geology of the Abajo Mountains area, San Juan County, Utah. *U.S. Geological Survey Professional Paper* 453. 110 pp.

———. 1975. The Abajo Mountains: An example of the laccolithic groups on the Colorado Plateau. In J. E. Fassett, ed., *Canyonlands Country*. Four Corners Geological Society Guidebook, 8th Field Conference, pp. 245–252.

Wolfe, J. A., C. E. Forest, and P. Molnar. 1998. Paleobotanical evidence of Eocene and Oligocene paleolatitudes in midlatitude western North America. *Geological Society of America Bulletin*, v. 110, pp. 664–678.

Young, R. A. 2001. The Laramide-Paleogene history of the western Grand Canyon region: Setting the stage. In Young and Spamer, eds., *The Colorado River*. Grand Canyon Association Monograph 12, pp. 7–15.

Young, R. A., and E. W. McKee. 1978. Early and middle Cenozoic drainage and erosion in west-central Arizona. *Geological Society of America Bulletin*, v. 89, pp. 1745–1750.

Young, R. A., and E. E. Spamer. 2001. *The Colorado River: Origin and Evolution*. Grand Canyon Association Monograph 12, 280 pp.

Zawiskie, J., D. Chapman, and R. Alley. 1982. Depositional history of the Paleocene-Eocene Colton Formation, north-central Utah. In D. L. Nielson, ed., Overthrust belt of Utah. *Utah Geological Association Publication* 10, pp. 273–284.

Chapter 9

Agenbroad, L. D., and J. I. Mead. 1987. Late Pleistocene alluvium and megafauna dung deposits of the central Colorado Plateau. In G. H. Davis and E. M. VandenDolder, eds., *Geologic Diversity of Arizona and Its Margins: Excursions to Choice Areas*. Arizona Bureau of Geology and Mineral Technology, Special Paper 5, pp. 68–84.

———. 1989. Quaternary geochronology and distribution of Mammuthus on the Colorado Plateau. *Geology*, v. 17, pp. 861–864.

Betancourt, J. L. 1984. Late Quaternary plant zonation and climate in southeastern Utah. *Great Basin Naturalist*, v. 44, pp. 1–35.

———. 1990. Late Quaternary biogeography of the Colorado Plateau. In J. L. Betancourt, D. R. VanDevender, and P. S. Martin, eds., *Packrat Middens: The Last 40,000 Years of Biotic Change*. Tucson, University of Arizona Press, pp. 259–292.

Cater, F. W. 1970. Geology of the salt anticline region in southwestern Colorado. *U.S. Geological Survey Professional Paper* 637. 80 pp.

Colman, S. M. 1983. Influence of the Onion Creek salt diaper on the late Cenozoic history of Fisher Valley, southeastern Utah. *Geology*, v. 11, pp. 240–243.

Colman, S. M., A. F. Choquette, and F. F. Hawkins. 1988. Physical, soil, and paleomagnetic stratigraphy of the Upper Cenozoic sediments in Fisher Valley, southeastern Utah. *U.S. Geological Survey Bulletin* 1686, 33 pp.

Colman, S. M., A. F. Choquette, J. N. Rosholt, G. H. Miller, and D. J. Huntley. 1986. Dating the upper Cenozoic sediments in Fisher Valley, southeastern Utah. *Geological Society of America Bulletin*, v. 97, pp. 1422–1431.

Dane, C. H. 1935. Geology of the Salt Valley anticline and adjacent areas, Grand County, Utah. *U.S. Geological Survey Bulletin* 863, 184 pp.

Davis, O. K., J. I. Mead, P. S. Martin, and L. D. Agenbroad. 1985. Riparian plants were a major component of the diet of mammoths of southern Utah. *Current Research in the Pleistocene*, v. 2.

Davis, S. W., M. E. Davis, I. Lucchitta, T. C. Hanks, R. C. Finkel, and M. Caffee. 2004. Erosional history of the Colorado River through Glen and Grand canyons. In Young and Spamer, eds., *The Colorado River*. Grand Canyon Association Monograph 12, pp. 135–139.

Hanks, T. C., and six others. 2004. The Colorado River and the age of Glen Canyon. In Young and Spamer, eds., *The Colorado River*. Grand Canyon Association Monograph 12, pp. 129–133.

Izett, G. A. 1981. Volcanic ash beds—Recorders of upper Cenozoic silicic pyroclastic volcanism in western United States. *Journal of Geophysical Research*, v. 86. no. B11, pp. 10200–10222.

Kropf, M., J. I. Mead, and R. S. Anderson. 2007. "Dung, diet, and paleoenvironment of the extinct shrup-ox (*Euceratherium collinum*) on the Colorado Plateau, USA". *Quaternary Research*, v. 67, pp. 143–151.

Lucchitta, I., G. H. Curtis, M. E. Davis, S. W. Davis, and B. Turrin. 2000. Cyclic aggradation and downcutting, fluvial response to volcanic activity, and calibration of soil-carbonate stages in western Grand Canyon. *Quaternary Research*, v. 53, pp. 23–33.

———. 2001. Rates of downcutting of the Colorado River in the Grand Canyon region. In Young and Spamer, eds., *The Colorado River*. Grand Canyon Association Monograph 12, pp. 155–157.

Martin, P. S., R. S. Thompson, and A. Long. 1985. Shasta ground sloth extinction: A test for the blitzkrieg model. In J. I. Mead and D. J. Meltzer, eds., *Environments and Extinctions: Man in Late Glacial North America*. University of Maine, Center for the Study of Early Man, pp. 5–14.

Mead, J. I., L. D. Agenbroad, O. K. Davis, and P. S. Martin. 1986. Dung of *Mammuthus* in the arid Southwest, North America. *Quaternary Research*, v. 25, pp. 121–127.

Meltzer, D. J. 2003. Peopling of North America. In A. Gillespie, S. C. Porter, and B. Atwater, eds., *The Quaternary Period in the United States*. Elsevier Science, pp. 539–563.

Oviatt, C. G. 1988. Evidence for Quaternary deformation in the Salt Valley anticline, southeastern Utah. *Utah Geological and Mineral Survey Bulletin* 122, pp. 63–76.

Pederson, J., K. Karlstrom, W. Sharp, and W. McIntosh. 2002. Differential incision of the Grand Canyon related to Quaternary faulting—Constraints from U-series and Ar/Ar dating. *Geology*, v. 30, pp. 739–742.

Richmond, G. M. 1962. Quaternary stratigraphy of the La Sal Mountains, Utah. *U.S. Geological Survey Professional Paper* 324. 135 pp.

Shroder, J. F., and R. E. Sewell. 1985. Mass movement in the La Sal Mountains, Utah. In *Contributions to Quaternary Geology of the Colorado Plateau*. Utah Geological and Mineral Survey, Special Studies 64, pp. 48–85.

Wolkowinsky, A. J., and D. E. Granger. 2004. Early Pleistocene incision of the San Juan River, Utah, dated with 26Al and 10Be. *Geology*, v. 32, pp. 749–752.

Woodward-Clyde Consultants. 1984. Geologic characterization report for the Paradox Basin study region, Utah study areas. Volume VI—Salt Valley. ONWI-290, 176 pp.

Road Logs

Bjørnstad, E. 1988. *Desert Rock*. Denver, Colorado, Chockstone Press, 453 pp.

Chenoweth, W. L., and G. J. Daub. 1983. Road log from Moab, Utah to Dewey Bridge, via Potash and Fisher Towers. In W. R. Averett, *Northern Paradox Basin: Uncompahgre Uplift*. Grand Junction Geological Society, pp. 121–132.

Doelling, H. H. 1985. Geologic map of Arches National Park and vicinity, Grand County, Utah. Utah Geological Survey, Map 74, scale: 1:50,000.

———. 2000. Geology of Arches National Park, Grand County, Utah. In D. A. Sprinkel, T. C. Chidsey, Jr., and P. B. Anderson, eds., *Geology of Utah's Parks and Monuments*. Utah Geological Association Publication 28, pp. 11–36.

———. 2001. Geological map of the Moab and eastern part of the San Rafael Desert 30' x 60' quadrangles, Grand and Emery Counties, Utah, and Mesa County, Colorado. Utah Geological Survey, Map 180, scale: 1:100,000.

———. 2004. Geologic map of the La Sal 30' x 60' quadrangle, San Juan, Wayne, and Garfield Counties, Utah, and Montrose and San Miguel Counties, Colorado. Utah Geological Survey, Map 205, scale: 1:100,000.

Doelling, H. H., and M. L. Ross. 1998. Geologic map of the Big Bend Quadrangle, Grand County, Utah. Utah Geological Survey, Map 171, scale 1:24,000.

Firmage, R. A. 1996. *A History of Grand County*. Salt Lake City, Utah State Historical Society, 438 pp.

Haynes, D. D., J. D. Vogel, and D. G. Wyant. 1972. Geology, structure, and uranium deposits of the Cortez quadrangle, Colorado and Utah. U.S. Geological Survey Miscellaneous Investigation Series, Map I-629, scale: 1:250,000.

Huntoon, P. W., G. H. Billingsley, Jr., and W. J. Breed. 1982. Geologic map of Canyonlands National Park and vicinity, Utah. Canyonlands Natural History Association, scale: 1:62,500.

Lohman, S. W. 1975. History and geography of Canyonlands National Park. In J. E. Fassett, ed., *Canyonlands Country*. Four Corners Geological Society, 8th Field Conference, pp. 35–50.

Miller, D. E. 1959. *Hole in the Rock*. Salt Lake City: University of Utah Press, 226 pp.

Nelson, S. T., J. P. Davidson, and K. R. Sullivan. 1992. New age determinations of central Colorado Plateau laccoliths, Utah: Recognizing disturbed K-Ar systematics and re-evaluating tectonomagmatic relationships. *Geological Society of America Bulletin*, v. 104, pp. 1547–1560.

Powell, J. W. 1865. *The Exploration of the Colorado River and Its Canyons*. New York: Dover, 400 pp.

Richmond, G. M. 1962. Quaternary stratigraphy of the La Sal Mountains, Utah. *U.S. Geological Survey Professional Paper* 324, 135 pp.

Sears, J. D. Geology of Comb Ridge and vicinity north of San Juan River, San Juan County, Utah. *U.S. Geological Survey Bulletin* 1021-E, pp. 167–207.

Semken, S. C., and W. L. McIntosh. 1997. 40Ar/39Ar age determinations for the Carrizo Mountains laccolith, Navajo Nation, Arizona. In O. J. Anderson, B. S. Kues, and S. G. Lucas, eds., *Mesozoic Geology and Paleontology of the Four Corners Region*. New Mexico Geological Society Guidebook, 48th Field Conference, pp. 75–80.

Smith, G. M., and C. S. Goodknight. 2005. Quaternary salt dissolution in the Moab-Spanish Valley, Utah, Pleistocene and Holocene evidence. Geological Society of America, Abstracts with Programs, v. 37. no. 6, p. 36.

Van Cott, J. W. 1990. *Utah Place Names*. Salt Lake City, University of Utah Press, 453 pp.

Williams, R. L. 1964. Geology, structure, and uranium deposits of the Moab quadrangle, Colorado and Utah. U.S. Geological Survey Miscellaneous Investigation Series, Map I-360, scale: 1:250,000.

Index

Note: Entries printed in *italics* refer to illustrations; entries printed in **boldface** refer to the road logs.